Honda VFR800FI
Service and Repair Manual

by Mark Coombs & Phil Mather

Models covered

Honda VFR800FI-W. 781cc. 1997 to 1998
Honda VFR800FI-X. 781cc. 1999
Honda VFR800FI-Y. 781 cc. 2000
Honda VFR800FI-1. 781 cc. 2001

(3703 - 272 - 7AB1)

© Haynes Publishing 2002

A book in the Haynes Service and Repair Manual Series

All rights reserved. No part of this book may be reproduced or transmitted in any form or by any means, electronic or mechanical, including photocopying, recording or by any information storage or retrieval system, without permission in writing from the copyright holder.

ISBN 1 85960 940 6

British Library Cataloguing in Publication Data
A catalogue record for this book is available from the British Library.

Library of Congress Control Number 2002111140

ABCDE
FGHIJ
KLMNO
PQ

Printed in the USA

Haynes Publishing
Sparkford, Yeovil, Somerset BA22 7JJ, England

Haynes North America, Inc
861 Lawrence Drive, Newbury Park, California 91320, USA

Editions Haynes
4, Rue de l'Abreuvoir
92415 COURBEVOIE CEDEX, France

Haynes Publishing Nordiska AB
Box 1504, 751 45 UPPSALA, Sweden

Contents

LIVING WITH YOUR HONDA VFR

Introduction
The Birth of a Dream	Page	0•4
Acknowledgements	Page	0•8
About this manual	Page	0•8
Safety first!	Page	0•9
Model development	Page	0•10
Performance data	Page	0•10
Bike spec	Page	0•11
Identification numbers	Page	0•12
Buying spare parts	Page	0•12

Daily (pre-ride) checks
Engine/transmission oil level check	Page	0•14
Coolant level check	Page	0•14
Brake and clutch fluid level checks	Page	0•15
Tyre checks	Page	0•16
Suspension, steering and final drive checks	Page	0•16
Legal and safety checks	Page	0•16

MAINTENANCE

Routine maintenance and servicing
Specifications	Page	1•2
Recommended lubricants and fluids	Page	1•2
Maintenance schedule	Page	1•3
Component locations	Page	1•4
Maintenance procedures	Page	1•6

Contents

REPAIRS AND OVERHAUL

Engine, transmission and associated systems

Engine, clutch and transmission	Page	2•1
Cooling system	Page	3•1
Fuel and exhaust systems	Page	4•1
Ignition system	Page	5•1

Chassis and bodywork components

Frame, suspension and final drive	Page	6•1
Brakes	Page	7•1
Wheels	Page	7•18
Tyres	Page	7•24
Fairing and bodywork	Page	8•1

Electrical system

Page 9•1

Wiring diagrams

Page 9•20

REFERENCE

Tools and Workshop Tips	Page	REF•2
Security	Page	REF•20
Lubricants and fluids	Page	REF•23
Conversion Factors	Page	REF•26
MOT test checks	Page	REF•27
Storage	Page	REF•32
Fault finding	Page	REF•35
Fault finding equipment	Page	REF•44
Technical terms explained	Page	REF•48

Index

Page REF•52

Introduction

The Birth of a Dream

by Julian Ryder

There is no better example of the Japanese post-war industrial miracle than Honda. Like other companies which have become household names, it started with one man's vision. In this case the man was the 40-year old Soichiro Honda who had sold his piston-ring manufacturing business to Toyota in 1945 and was happily spending the proceeds on prolonged parties for his friends. However, the difficulties of getting around in the chaos of post-war Japan irked Honda, so when he came across a job lot of generator engines he realised that here was a way of getting people mobile again at low cost.

A 12 by 18-foot shack in Hamamatsu became his first bike factory, fitting the generator motors into pushbikes. Before long he'd used up all 500 generator motors and started manufacturing his own engine, known as the 'chimney', either because of the elongated cylinder head or the smoky exhaust or perhaps both. The chimney made all of half a horsepower from its 50 cc engine but it was a major success and became the Honda A-type.

Less than two years after he'd set up in Hamamatsu, Soichiro Honda founded the Honda Motor Company in September 1948. By then, the A-type had been developed into the 90 cc B-type engine, which Mr Honda decided deserved its own chassis not a bicycle frame. Honda was about to become Japan's first post-war manufacturer of complete motorcycles. In August 1949 the first prototype was ready. With an output of three horsepower, the 98 cc D-type was still a simple two-stroke but it had a two-speed transmission and most importantly a pressed steel frame with telescopic forks and hard tail rear end. The frame was almost triangular in profile with the top rail going in a straight line from the massively braced steering head to the rear axle. Legend has it that after the D-type's first tests the entire workforce went for a drink to celebrate and try and think of a name for the bike. One man broke one of those silences you get when people are thinking, exclaiming 'This is like a dream!' 'That's it!' shouted Honda, and so the Honda Dream was christened.

Honda C70 and C90 OHV-engined models

'This is like a dream!' 'That's it' shouted Honda

Mr Honda was a brilliant, intuitive engineer and designer but he did not bother himself with the marketing side of his business. With hindsight, it is possible to see that employing Takeo Fujisawa who would both sort out the home market and plan the eventual expansion into overseas markets was a masterstroke. He arrived in October 1949 and in 1950 was made Sales Director. Another vital new name was Kiyoshi Kawashima, who along with Honda himself, designed the company's first four-stroke after Kawashima had told them that the four-stroke opposition to Honda's two-strokes sounded nicer and therefore sold better. The result of that statement was the overhead-valve 148 cc E-type which first ran in July 1951 just two months after the first drawings were made. Kawashima was made a director of the Honda Company at 34 years old.

The E-type was a massive success, over 32,000 were made in 1953 alone, a feat of mass-production that was astounding by the standards of the day given the relative complexity of the machine. But Honda's lifelong pursuit of technical innovation sometimes distracted him from commercial reality. Fujisawa pointed out that they were in danger of ignoring their core business, the motorised bicycles that still formed Japan's main means of transport. In May 1952 the F-type Cub appeared, another two-stroke despite the top men's reservations. You could buy a complete machine or just the motor to attach to your own bicycle. The result was certainly distinctive, a white fuel tank with a circular profile went just below and behind the saddle on the left of the bike, and the motor with its horizontal cylinder and bright red cover just below the rear axle on the same side of the bike. This was the machine that turned Honda into the biggest bike maker in Japan with 70% of the market for bolt-on bicycle motors, the F-type was also the first Honda to be exported. Next came the machine that would turn Honda into the biggest motorcycle manufacturer in the world.

The C100 Super Cub was a typically audacious piece of Honda engineering and marketing. For the first time, but not the last, Honda invented a completely new type of motorcycle, although the term 'scooterette' was coined to describe the new bike which had many of the characteristics of a scooter but the large wheels, and therefore stability, of a motorcycle. The first one was sold in August 1958, fifteen years later over nine-million of them were on the roads of the world. If ever a machine can be said to have brought mobility to the masses it is the Super Cub. If you add in the electric starter that was added for the C102 model of 1961, the design of the Super Cub has remained substantially unchanged ever since, testament to how right Honda got it first time. The Super Cub made Honda the world's biggest manufacturer after just two years of production.

The CB250N Super Dream became a favorite with UK learner riders of the late seventies and early eighties

Honda's export drive started in earnest in 1957 when Britain and Holland got their first bikes, America got just two bikes the next year. By 1962 Honda had half the American market with 65,000 sales. But Soichiro Honda had already travelled abroad to Europe and the USA, making a special

The GL1000 introduced in 1975, was the first in Honda's line of GoldWings

0•6 Introduction

Carl Fogarty in action at the Suzuka 8 Hour on the RC45

An early CB750 Four

point of going to the Isle of Man TT, then the most important race in the GP calendar. He realised that no matter how advanced his products were, only racing success would convince overseas markets for whom 'Made in Japan' still meant cheap and nasty. It took five years from Soichiro Honda's first visit to the Island before his bikes were ready for the TT. In 1959 the factory entered five riders in the 125 class. They did not have a massive impact on the event being benevolently regarded as a curiosity, but sixth, seventh and eighth were good enough for the team prize. The bikes were off the pace but they were well engineered and very reliable.

The TT was the only time the West saw the Hondas in '59, but they came back for more the following year with the first of a generation of bikes which shaped the future of motorcycling – the double-overhead-cam four-cylinder 250. It was fast and reliable – it revved to 14,000 rpm – but didn't handle anywhere near as well as the opposition. However, Honda had now signed up non-Japanese riders to lead their challenge. The first win didn't come until 1962 (Aussie Tom Phillis in the Spanish 125 GP) and was followed up with a world-shaking performance at the TT. Twenty-one year old Mike Hailwood won both 125 and 250 cc TTs and Hondas filled the top five positions in both races. Soichiro Honda's master plan was starting to come to fruition, Hailwood and Honda won the 1961 250 cc World Championship. Next year Honda won three titles. The other Japanese factories fought back and inspired Honda to produce some of the most fascinating racers ever seen: the awesome six-cylinder 250, the five-cylinder 125, and the 500 four with which the immortal Hailwood battled Agostini and the MV Agusta.

When Honda pulled out of racing in '67 they had won sixteen rider's titles, eighteen manufacturer's titles, and 137 GPs, including 18 TTs, and introduced the concept of the modern works team to motorcycle racing. Sales success followed racing victory as Soichiro Honda had predicted, but only because the products advanced as rapidly as the racing machinery. The Hondas that came to Britain in the early '60s were incredibly sophisticated. They had overhead cams where the British bikes had pushrods, they had electric starters when the Brits relied on the kickstart, they had 12V electrics when even the biggest British bike used a 6V system. There seemed no end to the technical wizardry. It wasn't that the technology itself was so amazing but just like that first E-type, it was the fact that Honda could mass-produce it more reliably than the lower-tech competition that was so astonishing.

When in 1968 the first four-cylinder CB750 road bike arrived the world of motorcycling changed for ever, they even had to invent a new word for it, 'Superbike'. Honda raced again with the CB750 at Daytona and won the

Introduction 0•7

World Endurance title with a prototype DOHC version that became the CB900 roadster. There was the six-cylinder CBX, the CX500T – the world's first turbocharged production bike, they invented the full-dress tourer with the GoldWing, and came back to GPs with the revolutionary oval-pistoned NR500 four-stroke, a much-misunderstood bike that was more a rolling experimental laboratory than a racer. Just to show their versatility Honda also came up with the weird CX500 shaft-drive V-twin, a rugged workhorse that powered a new industry, the courier companies that oiled the wheels of commerce in London and other big cities.

It was true, though, that Mr Honda was not keen on two-strokes – early motocross engines had to be explained away to him as lawnmower motors! However, in 1982 Honda raced the NS500, an agile three-cylinder lightweight against the big four-cylinder opposition in 500 GPs. The bike won in its first year and in '83 took the world title for Freddie Spencer. In four-stroke racing the V4 layout took over from the straight four, dominating TT, F1 and Endurance championships with the RVF750, the nearest thing ever built to a Formula 1 car on wheels. And when Superbike arrived Honda were ready with the RC30. On the roads the VFR V4 became an instant classic while the CBR600 invented another new class of bike on its way to becoming a best-seller. The V4 road bikes had problems to start with but the VFR750 sold world-wide over its lifetime while the VFR400 became a massive commercial success and cult bike in Japan. The original RC30 won the first two World Superbike Championships is 1988 and '89, but Honda had to wait until 1997 to win it again with the RC45, the last of the V4 roadsters. In Grands Prix, the NSR500 V4 two-stroke superseded the NS triple and became the benchmark racing machine of the '90s. Mick Doohan secured his place in history by winning five World Championships in consecutive years on it.

In yet another example of Honda inventing a new class of motorcycle, they came up with the astounding CBR900RR FireBlade, a bike with the punch of a 1000 cc motor in a package the size and weight of a 750. It became a cult bike as well as a best seller, and with judicious redesigns continues to give much more recent designs a run for their money.

When it became apparent that the high-tech V4 motor of the RC45 was too expensive to produce, Honda looked to a V-twin engine to power its flagship for the first time. Typically, the VTR1000 FireStorm was a much more rideable machine than its opposition and once accepted by the market formed the basis of the next generation of Superbike racer, the VTR-SP-1.

One of Mr Honda's mottos was that technology would solve the customers' problems, and no company has embraced

The CX500 – Honda's first V-Twin and a favorite choice of dispatch riders

cutting-edge technology more firmly than Honda. In fact Honda often developed new technology, especially in the fields of materials science and metallurgy. The embodiment of that was the NR750, a bike that was misunderstood nearly as much as the original NR500 racer. This limited-edition technological tour-de-force embodied many of Soichiro Honda's ideals. It used the latest techniques and materials in every component, from the oval-piston, 32-valve V4 motor to the titanium coating on the windscreen, it was – as Mr Honda would have wanted – the best it could possibly be. A fitting memorial to the man who has shaped the motorcycle industry and motorcycles as we know them today.

The VFR400R was a cult bike in Japan and a popular grey import in the UK

Introduction

The Honda VFR800FI

Honda had a serious problem: the VFR750 had been around since 1986 and had been the undisputed king of the all-rounders for over ten years. But some serious competition, notably from Triumph and Ducati, was making the old V4 look a little, well, dated. It had been condescended to as an old man's bike – the safe option – for years despite constant residence at the top of the best sellers charts. Was it possible to answer both those criticisms and if so, how?

Honda did find an answer, and they implemented it in as restrained a fashion as the bike's image. Although the result was still visually understated and obviously a VFR, it was in effect a brand new bike. Don't be fooled by the hike in capacity to 800 cc, this wasn't a simple matter of invoking the old American saying that 'You can't beat cubes', this was a new motorcycle. In reality, it would be fairer to think of the VFR800 as a detuned and civilised RC45 rather than a development of the old VFR line. In short, it's the RC45 motor stroked out by 2 mm to up capacity from 749 to 781 cc and with the crank rephased to 180 degrees. There wasn't anything wrong with the old motor but its race-ready short-stroke architecture was designed for John Kocinski to win the 1997 World Superbike Championship not for road use, so the longer stroke slowed down engine speeds and made it useable for ordinary mortals. Both power and torque were improved, but the latter by more which shows you where Honda's priorities lay: in maintaining the all-round rideability of the bike. The increased torque makes itself felt around the 6000rpm mark and the motor then surfs all the way to its 11,750 rpm red line on a seamless wave of torque.

Of course the original fuel injection system had to be remapped to cope and there are other changes both subtle and obvious, but they are enough to make the VFR as exciting a proposition as it was in 1986. Compare the 800 to a 750 motor and as well as the fuel injection, there are massive differences such as the cam drive moving from the centre of the motor to the side, steeper valve angle, enlarged airbox – as I said it's really an RC45 motor.

And even though the bike's bodywork looks the same as the old-model 750, it hides a completely new chassis derived from the VTR1000 FireStorm's. That features a 'pivotless' frame with the single-sided swinging arm pivoting in the crankcase castings which therefore means the engine casings function as a stressed member of the chassis. Incidentally, this modification signalled the end of the RC45 as we know it, for the dies back at the factory used for casting crankcases had to be heavily modified to produce the new VFR.

The resiting of the cam drive narrows the motor considerably as one main bearing can be done away with so that the already thin – compared to an in-line four – V4 motor is slimmed down even more. Add in a new-generation chassis and handling is upgraded to cope with the new power and torque. Handling is still best described as impeccably neutral; just what you want on a bike like this. To get things back under control, Honda fitted the not universally popular DCBS linked brake system which applies both front and rear brakes whether the lever is squeezed or the pedal depressed. However, this excited less criticism on the VFR than on previous applications, even being praised for dialling out fork dive during heavy braking efforts.

British customers also got a lightly-tweaked limited-edition VFR800 to celebrate the fiftieth anniversary of the Honda Motor Company.

This complete redesign immediately re-established the new VFR as the king of the all-rounders. There are very few bikes on which you can scratch with the best of them at a track day yet, with just a few adjustments to suspension settings, undertake a two-week touring holiday complete with pillion passenger. What's more, both of you would be more comfortable than on any motorcycle this side of a GoldWing. That incredible versatility had always been the hallmark of the VFR range and the VFR800FI carried on that tradition, gaining a full eco-friendly emission control package and integrated immobiliser system during its production life to fully enhance what was already an extremely rider-friendly package. History will recall that the VFR750 was admired for many things except, perhaps, the ability to make its rider smile – riding the VFR800FI, on the other hand, you know you're on one of the best all-round bikes ever built and the smile on your face is as wide as a mile!

The VFR800 50th anniversary model

Acknowledgements

Our thanks are due to Bransons Motorcycles of Yeovil who supplied the machine featured in the illustrations throughout this manual. We would also like to thank NGK Spark Plugs (UK) Ltd for supplying the colour spark plug condition photos and the Avon Rubber Company for supplying information on tyre fitting.

Thanks are due to Honda (UK) Ltd for supplying colour transparencies. The introduction 'The Birth of a Dream' was written by Julian Ryder.

About this manual

The aim of this manual is to help you get the best value from your motorcycle. It can do so in several ways. It can help you decide what work must be done, even if you choose to have it done by a dealer; it provides information and procedures for routine maintenance and servicing; and it offers diagnostic and repair procedures to follow when trouble occurs.

We hope you use the manual to tackle the work yourself. For many simpler jobs, doing it yourself may be quicker than arranging an appointment to get the motorcycle into a dealer and making the trips to leave it and pick it up. More importantly, a lot of money can be saved by avoiding the expense the shop must pass on to you to cover its labour and overhead costs. An added benefit is the sense of satisfaction and accomplishment that you feel after doing the job yourself.

References to the left or right side of the motorcycle assume you are sitting on the seat, facing forward.

We take great pride in the accuracy of information given in this manual, but motorcycle manufacturers make alterations and design changes during the production run of a particular motorcycle of which they do not inform us. No liability can be accepted by the authors or publishers for loss, damage or injury caused by any errors in, or omissions from, the information given.

Safety first! 0•9

Professional mechanics are trained in safe working procedures. However enthusiastic you may be about getting on with the job at hand, take the time to ensure that your safety is not put at risk. A moment's lack of attention can result in an accident, as can failure to observe simple precautions.

There will always be new ways of having accidents, and the following is not a comprehensive list of all dangers; it is intended rather to make you aware of the risks and to encourage a safe approach to all work you carry out on your bike.

Asbestos

● Certain friction, insulating, sealing and other products - such as brake pads, clutch linings, gaskets, etc. - contain asbestos. Extreme care must be taken to avoid inhalation of dust from such products since it is hazardous to health. If in doubt, assume that they do contain asbestos.

Fire

● Remember at all times that petrol is highly flammable. Never smoke or have any kind of naked flame around, when working on the vehicle. But the risk does not end there - a spark caused by an electrical short-circuit, by two metal surfaces contacting each other, by careless use of tools, or even by static electricity built up in your body under certain conditions, can ignite petrol vapour, which in a confined space is highly explosive. Never use petrol as a cleaning solvent. Use an approved safety solvent.

● Always disconnect the battery earth terminal before working on any part of the fuel or electrical system, and never risk spilling fuel on to a hot engine or exhaust.

● It is recommended that a fire extinguisher of a type suitable for fuel and electrical fires is kept handy in the garage or workplace at all times. Never try to extinguish a fuel or electrical fire with water.

Fumes

● Certain fumes are highly toxic and can quickly cause unconsciousness and even death if inhaled to any extent. Petrol vapour comes into this category, as do the vapours from certain solvents such as trichloro-ethylene. Any draining or pouring of such volatile fluids should be done in a well ventilated area.

● When using cleaning fluids and solvents, read the instructions carefully. Never use materials from unmarked containers - they may give off poisonous vapours.

● Never run the engine of a motor vehicle in an enclosed space such as a garage. Exhaust fumes contain carbon monoxide which is extremely poisonous; if you need to run the engine, always do so in the open air or at least have the rear of the vehicle outside the workplace.

The battery

● Never cause a spark, or allow a naked light near the vehicle's battery. It will normally be giving off a certain amount of hydrogen gas, which is highly explosive.

● Always disconnect the battery ground (earth) terminal before working on the fuel or electrical systems (except where noted).

● If possible, loosen the filler plugs or cover when charging the battery from an external source. Do not charge at an excessive rate or the battery may burst.

● Take care when topping up, cleaning or carrying the battery. The acid electrolyte, evenwhen diluted, is very corrosive and should not be allowed to contact the eyes or skin. Always wear rubber gloves and goggles or a face shield. If you ever need to prepare electrolyte yourself, always add the acid slowly to the water; never add the water to the acid.

Electricity

● When using an electric power tool, inspection light etc., always ensure that the appliance is correctly connected to its plug and that, where necessary, it is properly grounded (earthed). Do not use such appliances in damp conditions and, again, beware of creating a spark or applying excessive heat in the vicinity of fuel or fuel vapour. Also ensure that the appliances meet national safety standards.

● A severe electric shock can result from touching certain parts of the electrical system, such as the spark plug wires (HT leads), when the engine is running or being cranked, particularly if components are damp or the insulation is defective. Where an electronic ignition system is used, the secondary (HT) voltage is much higher and could prove fatal.

Remember...

✗ **Don't** start the engine without first ascertaining that the transmission is in neutral.

✗ **Don't** suddenly remove the pressure cap from a hot cooling system - cover it with a cloth and release the pressure gradually first, or you may get scalded by escaping coolant.

✗ **Don't** attempt to drain oil until you are sure it has cooled sufficiently to avoid scalding you.

✗ **Don't** grasp any part of the engine or exhaust system without first ascertaining that it is cool enough not to burn you.

✗ **Don't** allow brake fluid or antifreeze to contact the machine's paintwork or plastic components.

✗ **Don't** siphon toxic liquids such as fuel, hydraulic fluid or antifreeze by mouth, or allow them to remain on your skin.

✗ **Don't** inhale dust - it may be injurious to health (see Asbestos heading).

✗ **Don't** allow any spilled oil or grease to remain on the floor - wipe it up right away, before someone slips on it.

✗ **Don't** use ill-fitting spanners or other tools which may slip and cause injury.

✗ **Don't** lift a heavy component which may be beyond your capability - get assistance.

✗ **Don't** rush to finish a job or take unverified short cuts.

✗ **Don't** allow children or animals in or around an unattended vehicle.

✗ **Don't** inflate a tyre above the recommended pressure. Apart from overstressing the carcass, in extreme cases the tyre may blow off forcibly.

✓ **Do** ensure that the machine is supported securely at all times. This is especially important when the machine is blocked up to aid wheel or fork removal.

✓ **Do** take care when attempting to loosen a stubborn nut or bolt. It is generally better to pull on a spanner, rather than push, so that if you slip, you fall away from the machine rather than onto it.

✓ **Do** wear eye protection when using power tools such as drill, sander, bench grinder etc.

✓ **Do** use a barrier cream on your hands prior to undertaking dirty jobs - it will protect your skin from infection as well as making the dirt easier to remove afterwards; but make sure your hands aren't left slippery. Note that long-term contact with used engine oil can be a health hazard.

✓ **Do** keep loose clothing (cuffs, ties etc. and long hair) well out of the way of moving mechanical parts.

✓ **Do** remove rings, wristwatch etc., before working on the vehicle - especially the electrical system.

✓ **Do** keep your work area tidy - it is only too easy to fall over articles left lying around.

✓ **Do** exercise caution when compressing springs for removal or installation. Ensure that the tension is applied and released in a controlled manner, using suitable tools which preclude the possibility of the spring escaping violently.

✓ **Do** ensure that any lifting tackle used has a safe working load rating adequate for the job.

✓ **Do** get someone to check periodically that all is well, when working alone on the vehicle.

✓ **Do** carry out work in a logical sequence and check that everything is correctly assembled and tightened afterwards.

✓ **Do** remember that your vehicle's safety affects that of yourself and others. If in doubt on any point, get professional advice.

● If in spite of following these precautions, you are unfortunate enough to injure yourself, seek medical attention as soon as possible.

Model development and Performance data

Model development

VFR800FI–W (1998 model year)

The VFR800 was introduced in 1998. The engine, a fuel injected V4, was much revised from the earlier VFR750, with gear driven DOHC and the rear suspension pivot located in the back of the crankcases. The final drive chain was adjusted by an eccentric bearing holder in the swingarm, and the front and rear brakes were linked by a Combined Braking System (CBS). Selected pistons in the front and rear brake calipers were operated by the handlebar brake lever or brake pedal to provide balanced braking.

German and Swiss models were fitted with a catalytic converter and California models had an EVAP system to prevent fuel vapour from the tank being released into the atmosphere. 1998 machines were available in red, black or silver finish.

VFR800FI–X (1999 model year)

A limited edition 50th Anniversary model was available in 1999 only. This had marginally different power characteristics to the standard model for that year and 1960's RC-style paint and graphics. Standard machines were available in red, black or yellow finish and a pillion seat cover was available in place of the pillion grab handles.

VFR800FI–Y (2000 model year)

A number of modifications were introduced with the VFR800FI-Y model. The catalytic converter was fitted to all models. Anti-judder springs were no longer fitted in the clutch, and the clutch plates and clutch master cylinder were modified. The rear brake disc was redesigned to dispense with the disc spacer, the cable operated choke was replaced by an automatic unit and an immobiliser was fitted to the ignition system. Machines were available in red, blue or green finish.

VFR800FI–1 (2001 model year)

A silver colour option was re-introduced to the range in 2001 and the green finish was discontinued.

Performance data

Maximum power
1998 models .... 101 bhp (75.3 kW) @ 10,500 rpm
50th Anniversary model .... 108.6 bhp (81.0 kW) @ 10,500 rpm
1999/2000/2001 models .... 108.5 bhp (80.9 kW) @ 10,500 rpm

Maximum torque
1998 models .... 56.4 lbf ft (76.5 Nm) @ 8549 rpm
50th Anniversary model .... 60.5 lbf ft (82 Nm) @ 8500 rpm
1999 model .... n/a
2000/2001 .... 61.5 lbf ft (83 Nm) @ 8750 rpm

Power-to-weight ratio (approximate)
1998 models .... 0.48 bhp per kg (0.36 kW per kg)
50th Anniversary model ..... 0.52 bhp per kg (0.38 kW per kg)
1999/2000/2001 models ..... 0.42 bhp per kg (0.38 kW per kg)

Top speed .... 155.6 mph (250.4 km/h)

Acceleration
Time taken to cover a 1/4 mile from a standing start ... 11.4 seconds
Terminal speed after 1/4 mile .... 129 mph (207.6 km/h)

Average fuel consumption
Miles per Imp gal .... 40 mpg
Miles per litre .... 8.8 mpl
Litres per 100 km .... 7.1 l/100 km

Fuel tank capacity .... 21 litres (4.6 Imp gal, 5.5 US gal)

Fuel tank range .... 185 miles (298 km)

Performance data sourced from Motor Cycle News road test features. See the MCN website for up-to-date biking news.

MCN www.motorcyclenews.com

Bike spec 0•11

Bike spec

Weights and dimensions
Wheelbase . . . . .1410 mm
Overall length
   UK models . . . . .2095 mm
   US models . . . . .2100 mm
Overall height . . . . .1190 mm
Overall width . . . . .735 mm
Seat height . . . . .805 mm
Ground clearance . . . . .130 mm
Dry weight . . . . .209 kg
Wet weight (with fuel and oil) . . . . .234 kg

Engine
Type . . . . .Liquid cooled, 90° V4
Capacity . . . . .781.7 cc
Bore . . . . .72 mm
Stroke . . . . .48 mm
Compression ratio 11.6:1
Camshafts . . . . .DOHC, chain driven
Valves . . . . .4 valves per cylinder
Fuel system . . . . .PGM-F1 fuel injection
Ignition system . . . . .Computer controlled,
                           digital transistorized with electronic advance
Clutch . . . . .Wet multi-plate, hydraulically operated
Transmission . . . . .6 speed constant mesh
Final drive
   Chain . . . . .DID 50 VA7 (108 links)
   Sprockets . . . . .17 tooth front/43 tooth rear

Chassis
Type . . . . .Twin spar, box section aluminium alloy
Rake . . . . .25.5°
Trail . . . . .95 mm
Front suspension
   Type . . . . .41 mm oil-damped cartridge type hydraulic forks
   Travel
      Standard model . . . . .120 mm
      50th Anniversary model . . . . .109 mm
   Adjustments . . . . .Spring pre-load
Rear suspension
   Type . . . . .Single sided aluminium alloy arm with
                                    Pro-Link rising rate linkage
   Travel . . . . .120 mm
   Adjustments . .Spring pre-load, compression and rebound damping
Wheels . . . . .17 inch 6-spoke U-section alloys
Tyre sizes
   Front 120/70 ZR17 58W
   Rear 180/55 ZR17 73W
Brakes (DCBS)
   Front . . . . .2 x 296 mm discs with Nissin three-piston calipers
   Rear . . . . .256 mm disc with Nissin three-piston caliper

0•12 Identification numbers

The frame serial number is stamped into the right side of the steering head. The engine number is stamped onto the top of the crankcase, directly above the swingarm pivot, and is visible from the left side of the machine. The vehicle identification number (VIN) plate is riveted to the right side of the frame. All of these numbers should be recorded and kept in a safe place so they can be given to law enforcement officials in the event of a theft. The throttle body identification number is stamped onto the front of the assembly.

The frame serial number, engine serial number and vehicle identification number (VIN) should also be kept in a handy place (such as with your driver's licence) so they are always available when purchasing or ordering parts for your machine.

When ordering painted components, be sure to quote the colour code from the colour code label, which is located underneath the seat where it is stuck on the top of the rear mudguard. The model code (eg VFR800FI-**X**) can also be found on the colour code label and can be determined from the initial engine and frame serial numbers in the accompanying table.

UK models	Year	Engine number	Frame number
VFR800FI-W	1998	RC46E-2000001 on	JH2RC46A*WM000001 on
VFR800FI-X	1999	RC46E-2200001 on	JH2RC46A*XM100001 on
VFR800FI-X (50th Anniversary model)	1999	RC46E-2250001 on	JH2RC46A*XM150001 on
VFR800FI-Y	2000	RC46E-2300001 on	JH2RC46D*YM200001 on
VFR800FI-1	2001	RC46E-2350001 on	JH2RC46D*1M300001 on

US models (49-state)	Year	Engine number	Frame number
VFR800FI	1998	RC46E-2000001 on	RC460*WM000001 on
VFR800FI	1999	RC46E-2200001 on	RC460*XM100001 on
VFR800FI	2000	RC46E-2300001 on	JH2RC460*YM200001 on
VFR800FI	2001	RC46E-2350001 on	JH2RC460*1M300001 on

US models (California)	Year	Engine number	Frame number
VFR800FI	1998	RC46E-2000001 on	RC461*WM000001 on
VFR800FI	1999	RC46E-2200001 on	RC461*XM100001 on
VFR800FI	2000	RC46E-2300001 on	JH2RC461*YM200001 on
VFR800FI	2001	RC46E-2350001 on	JH2RC461*1M300001 on

Buying spare parts

Once you have found all the identification numbers, record them for reference when buying parts. Since the manufacturers change specifications, parts and vendors (companies that manufacture various components on the machine), providing the ID numbers is the only way to be reasonably sure that you are buying the correct parts.

Whenever possible, take the worn part to the dealer so direct comparison with the new component can be made. Along the trail from the manufacturer to the parts shelf, there are numerous places that the part can end up with the wrong number or be listed incorrectly.

The two places to purchase new parts for your motorcycle – the accessory store and the franchised dealer – differ in the type of parts they carry. While dealers can obtain every part for your motorcycle, the accessory dealer is usually limited to normal high wear items such as shock absorbers, tune-up parts, various engine gaskets, cables, chains, brake parts, etc. Rarely will an accessory outlet have major suspension components, camshafts, transmission gears, or cases.

Used parts can be obtained for considerably less than new ones, but you can't always be sure of what you're getting. Once again, take your worn part to the breaker for direct comparison.

Whether buying new, used or rebuilt parts, the best course is to deal directly with someone who specialises in parts for your particular make.

Identification numbers 0•13

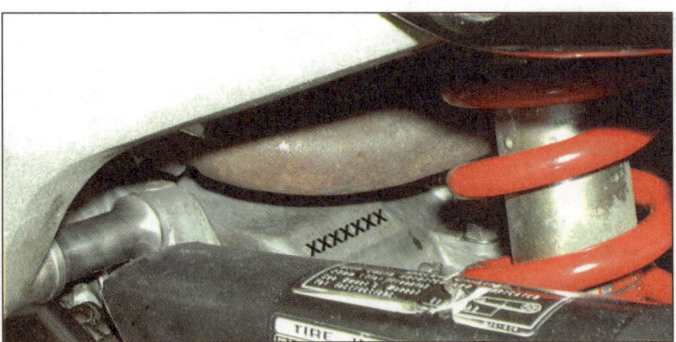

The engine number is stamped on the top of the crankcase on the left side of the engine

The frame number is stamped on the right side of the steering head

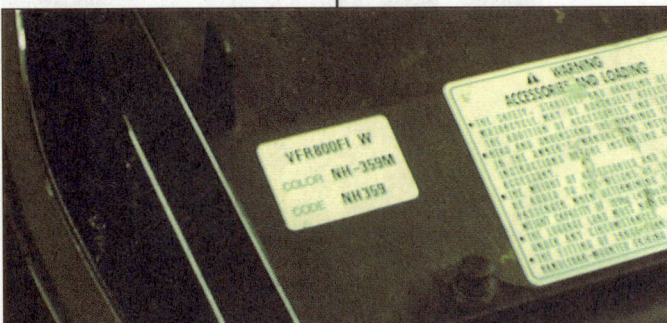

The colour code label is under the seat on the rear mudguard

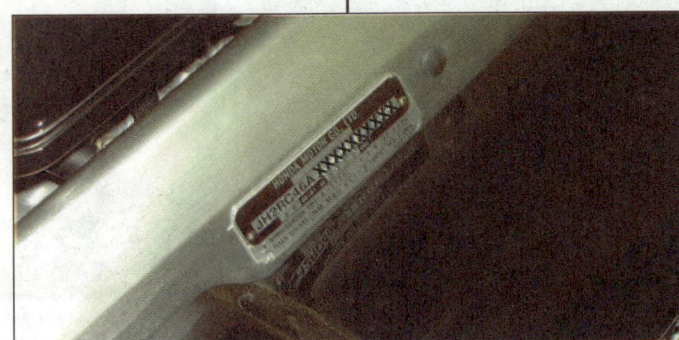

The vehicle identification number (VIN) plate is riveted to the right side of the frame

0•14 Daily (pre-ride) checks

Engine/transmission oil level check

Before you start:
✔ Start the engine and allow it to reach normal operating temperature.
Caution: Do not run the engine in an enclosed space such as a garage or workshop.
✔ Stop the engine and support the motorcycle on its centrestand. Allow it to stand undisturbed for a few minutes to allow the oil level to stabilise. Make sure the motorcycle is on level ground.

Bike care:
● If you have to add oil frequently, you should check whether you have any oil leaks. If there is no sign of oil leakage from the joints and gaskets the engine could be burning oil (see *Fault Finding*).

The correct oil
● Modern, high-revving engines place great demands on their oil. It is very important that the correct oil for your bike is used.
● Always top up with a good quality oil of the specified type and viscosity and do not overfill the engine.

Oil type	API grade SF or SG
Oil viscosity	SAE 10W40

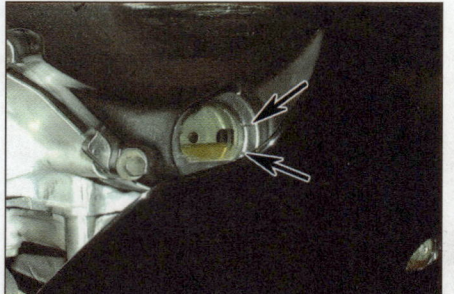

1 Check the oil level in the inspection window at the bottom of the right crankcase cover. The level should lie between the upper and lower level marks (arrows).

2 The oil filler cap is located at the top of the right crankcase cover. Unscrew the cap to add oil.

3 Add the specified oil to bring the oil level to the upper level mark on the inspection window but do not overfill. Once the level is correct, securely refit the filler cap.

Coolant level check

> **Warning: DO NOT remove the filler neck pressure cap to add coolant. Topping up is done via the coolant reservoir tank filler. DO NOT leave open containers of coolant about, as it is poisonous.**

Before you start:
✔ Make sure you have a supply of coolant available (a mixture of 50% distilled water and 50% corrosion inhibited ethylene glycol anti-freeze is needed).

✔ Always check the coolant level when the engine is at normal working temperature. Start the engine allow it to reach normal temperature, then stop the engine.
Caution: Do not run the engine in an enclosed space such as a garage or workshop
✔ Support the motorcycle on its centrestand whilst checking the level. Make sure the motorcycle is on level ground.

Bike care:
● Use only the specified coolant mixture. It is important that anti-freeze is used in the system all year round, and not just in the winter. Do not top the system up using only water, as the system will become too diluted.
● Do not overfill the reservoir tank. If the coolant is significantly above the UPPER level line at any time, the surplus should be siphoned or drained off to prevent the possibility of it being expelled out of the overflow hose.
● If the coolant level falls steadily, check the system for leaks (see Chapter 1). If no leaks are found and the level continues to fall, it is recommended that the machine is taken to a Honda dealer for a pressure test.

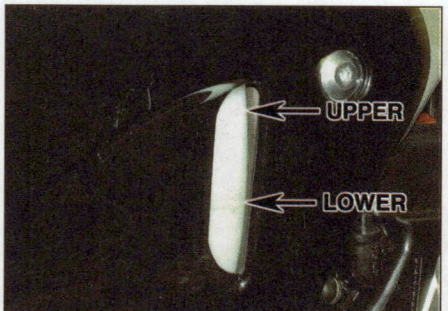

1 The coolant reservoir is located behind the fairing left lower panel. The coolant UPPER and LOWER level lines (arrowed) are visible via the fairing duct.

2 If the coolant level is not between the UPPER and LOWER markings, remove the left lower fairing panel (see Chapter 8) to gain access to the reservoir.

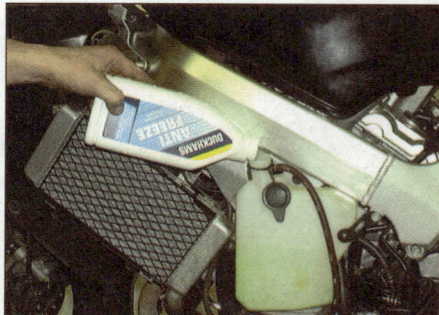

3 Top the coolant level up with the recommended coolant mixture then securely fit the cap to the reservoir before installing the fairing panel.

Daily (pre-ride) checks

Brake and clutch fluid level checks

> **Warning:** Hydraulic fluid can harm your eyes and damage painted surfaces, so use extreme caution when handling and pouring it and cover surrounding surfaces with rag. Do not use fluid that has been standing open for some time, as it absorbs moisture from the air which can cause a dangerous loss of braking effectiveness.

Before you start:
✔ Support the motorcycle on its centrestand whilst checking the levels. Make sure the motorcycle is on level ground.
✔ Make sure you have the correct hydraulic fluid. DOT 4 is recommended.
✔ Wrap a rag around the reservoir being worked on to ensure that any spillage does not come into contact with painted surfaces.
✔ Access to the front brake/clutch reservoir cover screws is restricted by the windshield. Use a short or angled screwdriver to access the screws.

Bike care:
● The fluid in the front and rear brake master cylinder reservoirs will drop slightly as the brake pads wear down.
● If any fluid reservoir requires repeated topping-up this is an indication of an hydraulic leak somewhere in the system, which should be investigated immediately.
● Check for signs of fluid leakage from the hydraulic hoses and components – if found, rectify immediately.
● Check the operation of both brakes before taking the machine on the road; if there is evidence of air in the system (spongy feel to lever or pedal), the system must be bled as described in Chapter 7.

Front brake and clutch fluid

1 The front brake and the clutch fluid levels are checked in the same way. With the reservoir as level as possible, check that the fluid level is above the LOWER level line on the inspection window.

2 If the level is below the LOWER level line, remove the two reservoir cover screws and remove the cover, the diaphragm plate and the diaphragm.

3 Top up with new DOT 4 hydraulic fluid, until the level is just below the upper level mark; this mark is in the form of a line cast on the inside of the reservoir front face (arrowed). Do not overfill the reservoir, and take care to avoid spills (see **Warning** above).

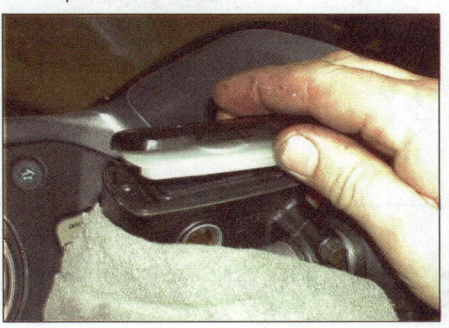

4 When the fluid level is correct, clean and dry the diaphragm, fold it into its compressed state and install it. Ensure that the diaphragm is correctly seated before installing the plate and cover. Tighten the cover screws securely and wipe off any spilt fluid.

Rear brake fluid

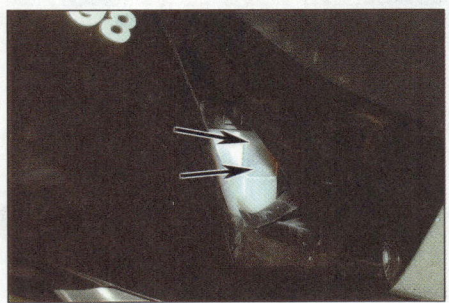

5 The rear brake fluid level is visible through the cutout on the right side of the seat cowling – the fluid level must be between the UPPER and LOWER level lines (arrowed).

6 If the level is below the LOWER level line, remove the seat cowling (see Chapter 8). Unscrew the reservoir cap and remove the diaphragm plate and diaphragm.

7 Top up with new DOT 4 hydraulic fluid, until the level is just below the UPPER level line. Do not overfill the reservoir, and take care to avoid spills (see **Warning** above).

8 When the fluid level is correct, clean and dry the diaphragm, fold it into its compressed state and install it. Ensure that the diaphragm is correctly seated before installing the plate and cap, tightening it securely.

Daily (pre-ride) checks

Tyre checks

The correct pressures:
- The tyres must be checked when **cold**, not immediately after riding. Note that low tyre pressures may cause the tyre to slip on the rim or come off. High tyre pressures will cause abnormal tread wear and unsafe handling.
- Use an accurate pressure gauge.
- Proper air pressure will increase tyre life and provide maximum stability and ride comfort.

Tyre pressures	
Front	36 psi (2.5 Bar)
Rear	42 psi (2.9 Bar)

Tyre care:
- Check the tyres carefully for cuts, tears, embedded nails or other sharp objects and excessive wear. Operation of the motorcycle with excessively worn tyres is extremely hazardous, as traction and handling are directly affected.
- Check the condition of the tyre valve and ensure the dust cap is in place.
- Pick out any stones or nails which may have become embedded in the tyre tread. If left, they will eventually penetrate through the casing and cause a puncture.
- If tyre damage is apparent, or unexplained loss of pressure is experienced, seek the advice of a tyre fitting specialist without delay.

Tyre tread depth:
- At the time of writing UK law requires that tread depth must be at least 1 mm over 3/4 of the tread breadth all the way around the tyre, with no bald patches. Many riders, however, consider 2 mm tread depth minimum to be a safer limit. Honda recommend a minimum of 1.5 mm on the front tyre and 2 mm on the rear tyre.
- Many tyres now incorporate wear indicators in the tread. Identify the triangular pointer or 'TWI' mark on the tyre sidewall to locate the indicator bar and renew the tyre if the tread has worn down to the bar.

1 Check the tyre pressures when the tyres are **cold** and keep them properly inflated.

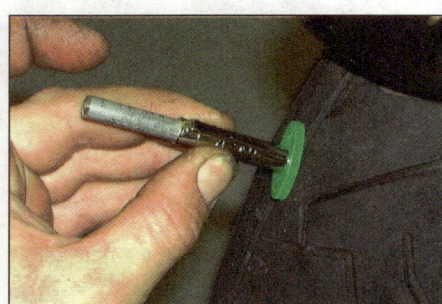

2 Measure tread depth at the centre of the tyre using a tread depth gauge.

3 Tyre tread wear indicator bar and its location marking (usually either an arrow, a triangle or the letters TWI) on the sidewall (circled).

Suspension, steering and final drive checks

Suspension and steering:
- Check that the front and rear suspension operates smoothly without binding.
- Check that the suspension is adjusted as required.
- Check that the steering moves smoothly from lock-to-lock.

Final drive:
- Check that the drive chain slack isn't excessive, and adjust if necessary (see Chapter 1).
- If the chain looks dry, lubricate it (see Chapter 1).

Legal and safety checks

Lighting and signalling:
- Take a minute to check that the headlight, tail light, brake light, instrument lights and turn signals all work correctly.
- Check that the horn sounds when the switch is operated.
- A working speedometer graduated in mph is a statutory requirement in the UK.

Safety:
- Check that the throttle grip rotates smoothly and snaps shut when released, in all steering positions. Also check for the correct amount of freeplay (see Chapter 1).
- Check that the engine shuts off when the kill switch is operated.
- Check that the sidestand and centrestand return springs hold the stands securely up when retracted.

Fuel:
- This may seem obvious, but check that you have enough fuel to complete your journey. If you notice signs of fuel leakage - rectify the cause immediately.
- Ensure you use the correct grade unleaded fuel - see Chapter 4 Specifications.

Chapter 1
Routine maintenance and servicing

Contents

Air filter – renewal	20
Battery – charging	see Chapter 9
Battery – removal, installation and inspection	see Chapter 9
Brake hoses – renewal	34
Brake master cylinder and caliper seals – renewal	33
Brake pads – wear check	2
Brake system – check	13
Brakes – fluid change	21
Clutch – fluid change	22
Coolant level check	see Daily (pre-ride) checks
Cooling system – check	10
Cooling system – draining, flushing and refilling	26
Cylinder compression – check	31
Drive chain and sprockets – check, adjustment and lubrication	1
Drive chain, slider and guide plate – check	12
Engine oil pressure – check	32
Engine/transmission oil and filter change	7
Engine/transmission oil level check	see Daily (pre-ride) checks
Evaporative emission control (EVAP) system – check (California models)	23
Exhaust system – check	11
Front forks – oil change	30
Fuel system – check	9
Headlight aim – check and adjustment	14
Idle speed – check and adjustment	3
Nuts and bolts – tightness check	17
Sidestand – check	15
Spark plugs – gap check and adjustment (European models)	5
Spark plugs – renewal (European models)	25
Spark plugs – renewal (US models)	6
Stands, pivots and cables – lubrication	4
Steering head bearings – check and adjustment	18
Steering head bearings – re-greasing	28
Suspension – check	16
Swingarm and suspension linkage bearings – re-greasing	29
Throttle and choke cable – check	8
Tyre pressure and tread depth check	see Daily (pre-ride) checks
Valve clearances – check and adjustment	24
Wheel bearings – check	27
Wheels and tyres – general check	19

Degrees of difficulty

Easy, suitable for novice with little experience 	**Fairly easy,** suitable for beginner with some experience	**Fairly difficult,** suitable for competent DIY mechanic	**Difficult,** suitable for experienced DIY mechanic 	**Very difficult,** suitable for expert DIY or professional 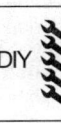

1•2 Specifications

Engine
Cylinder identification
    Number 1 .... Left rear cylinder
    Number 2 .... Left front cylinder
    Number 3 .... Right rear cylinder
    Number 4 .... Right front cylinder
Engine idle speed
    California models .... 1300 ± 100 rpm
    All other models .... 1200 ± 100 rpm
Spark plugs
    Type .... NGK CR9EH-9 or Nippondenso U27FER-9
    Electrode gap .... 0.8 to 0.9 mm
Valve clearances (COLD engine)
    Intake .... 0.16 ± 0.03 mm
    Exhaust .... 0.30 ± 0.03 mm
Cylinder compression .... 142 to 199 psi (10.0 to 14.0 bar)
Oil pressure (at 6000 rpm) .... 71 psi (5.0 bar) at 80°C (176°F)

```
        ↑
      FRONT
     ┌───┬───┐
     │ 2 │ 4 │
     ├───┼───┤
     │ 1 │ 3 │
     └───┴───┘
```

Cycle parts
Drive chain slack
    Standard .... 20 to 30 mm
    Maximum .... 50 mm
Throttle cable freeplay .... 2 to 6 mm
Tyre pressures and tyre tread depth .... see Daily (pre-ride) checks

Recommended lubricants and fluids
Engine/transmission oil type .... API grade SF or SG motor oil
Engine/transmission oil viscosity .... SAE 10W40
Engine/transmission oil capacity
    Oil change .... 2.9 litres
    Oil and filter change .... 3.1 litres
    After overhaul (dry engine, new oil filter) .... 3.8 litres
Coolant type .... 50% distilled water, 50% corrosion inhibited ethylene glycol anti-freeze
Coolant capacity
    Radiator and engine .... 2.75 litres
    Reservoir .... 0.45 litre
Front fork oil .... see Chapter 6
Brake and clutch fluid .... DOT 4
Drive chain .... SAE 80 or 90 gear oil or aerosol chain lubricant for O-ring chains
Steering head bearings .... Lithium-based multi-purpose grease
Wheel bearings (unsealed) .... Lithium-based multi-purpose grease
Swingarm pivot bearings .... Lithium-based multi-purpose grease
Suspension linkage bearings .... Lithium-based multi-purpose grease
Oil/dust seal lips .... Lithium-based multi-purpose grease
Brake pedal pivots .... Lithium-based multi-purpose grease
Brake lever pivot and piston tip .... Silicone grease
Clutch lever pivot and piston tip .... Silicone grease
Cables .... Cable lubricant or 10W40 motor oil
Sidestand/centrestand pivots .... Lithium-based multi-purpose grease or dry film lubricant
Throttle grip .... Multi-purpose grease or dry film lubricant
Shock absorber spring adjuster cam .... Molybdenum grease

Torque settings
Handlebar clamp bolt .... 26 Nm
Oil filter .... 10 Nm
Oil pan drain plug .... 29 Nm
Rear wheel bearing holder clamp bolt .... 74 Nm
Right crankcase cover inspection plug .... 18 Nm
Spark plugs .... 12 Nm
Steering head bearing adjuster nut .... 25 Nm
Steering stem nut .... 103 Nm
Top yoke fork clamp bolt .... 23 Nm

Maintenance schedule 1•3

Note: *The daily (pre-ride) checks outlined in the owner's manual covers those items which should be inspected on a daily basis. Always perform the pre-ride inspection at every maintenance interval (in addition to the procedures listed). The intervals listed below are the intervals recommended by the manufacturer for each particular operation during the model years covered in this manual. Your owner's manual may have slightly different intervals for some items.*

Daily (pre-ride)
- [ ] See *'Daily (pre-ride) checks'* at the beginning of this manual

After the initial 600 miles (1000 km)*
- [ ] Check, adjust and lubricate the drive chain (Section 1)
- [ ] Change the engine/transmission oil and filter (Section 7)
- [ ] Check the operation of the braking system and brake light switch (Section 13)
- [ ] Check the tightness of all nuts, bolts and fasteners (Section 17)
- [ ] Check and adjust the steering head bearings (Section 18)
- [ ] Check and adjust the idle speed (Section 19)

*Note: This check is usually performed by a dealer after the first 600 miles (1000 km) from new. Thereafter, maintenance should be carried out according to the following intervals.

Every 500 miles (800 km)
- [ ] Check, adjust and lubricate the drive chain (Section 1)

Every 4000 miles (6000 km) or 6 months (whichever comes sooner)
- [ ] Check the brake pads (Section 2)
- [ ] Check and adjust the idle speed (Section 3)
- [ ] Lubricate the clutch/brake lever/brake pedal/sidestand/centrestand pivots and the throttle/choke cables (Section 4)

Every 8000 miles (12,000 km) or 12 months (whichever comes sooner)
- [ ] Check the spark plugs – UK models (Section 5)
- [ ] Renew the spark plugs – US models (Section 6)
- [ ] Change the engine/transmission oil and filter (Section 7)
- [ ] Check the operation of the throttle and choke cables (Section 8)
- [ ] Check the fuel system and hoses (Section 9)
- [ ] Check the cooling system (Section 10)
- [ ] Check the exhaust system (Section 11)
- [ ] Check the drive chain, slider and guide plate (Section 12)
- [ ] Check the operation of the braking system and brake light switch (Section 13)
- [ ] Check and adjust the headlight aim (Section 14)
- [ ] Check the sidestand (Section 15)
- [ ] Check the suspension (Section 16)
- [ ] Check the tightness of all nuts, bolts and fasteners (Section 17)
- [ ] Check and adjust the steering head bearings (Section 18)
- [ ] Check the condition of the wheels and tyres (Section 19)

Every 12,000 miles (18,000 km)
- [ ] Renew the air filter element (Section 20)
- [ ] Change the brake fluid (Section 21)*
- [ ] Change the clutch fluid (Section 22)*
- [ ] Check the evaporative emission control (EVAP) system hoses – California models (Section 23)

*The brake and clutch fluid should be changed at the specified mileage or every two years, whichever comes sooner

Every 16,000 miles (24,000 km)
- [ ] Check and adjust the valve clearances (Section 24)
- [ ] Renew the spark plugs – UK models (Section 25)

Every 24,000 miles (36,000 km)
- [ ] Change the coolant (Section 26)*

*The coolant should be changed at the specified mileage or every two years, whichever comes sooner

Non-scheduled maintenance
- [ ] Check the wheel bearings (Section 27)
- [ ] Re-grease the steering head bearings (Section 28)
- [ ] Re-grease the swingarm and suspension linkage bearings (Section 29)
- [ ] Change the front fork oil (Section 30)
- [ ] Check the cylinder compression (Section 31)
- [ ] Check the engine oil pressure (Section 32)
- [ ] Renew the brake master cylinder and caliper seals (Section 33)
- [ ] Renew the brake hoses (Section 34)

1•4 Maintenance - component location

Component locations on the right side

1 Rear brake fluid reservoir
2 Fuel filter
3 Idle speed adjuster screw plug
4 Front brake fluid reservoir
5 Radiator pressure cap
6 Engine/transmission oil level inspection window
7 Engine/transmission oil filler cap
8 Battery

Maintenance - component location 1•5

Component locations on the left side

1 Clutch fluid reservoir
2 Steering head bearings
3 Air filter
4 Coolant reservoir
5 Drive chain slider

6 Drive chain guide plate
7 Clutch slave cylinder
8 Coolant drain plug
9 Engine/transmission oil drain plug
10 Engine/transmission oil filter

1•6 Maintenance procedures

Introduction

1 This Chapter is designed to help the home mechanic maintain his/her motorcycle for safety, economy, long life and peak performance.

2 Deciding where to start or plug into the routine maintenance schedule depends on several factors. If the warranty period on your motorcycle has just expired, and if it has been maintained according to the warranty standards, you may want to pick up routine maintenance as it coincides with the next mileage or calendar interval. If you have owned the machine for some time but have never performed any maintenance on it, then you may want to start at the nearest interval and include some additional procedures to ensure that nothing important is overlooked. If you have just had a major engine overhaul, then you may want to start the maintenance routine from the beginning. If you have a used machine and have no knowledge of its history or maintenance record, you may desire to combine all the checks into one large service initially and then settle into the maintenance schedule prescribed.

3 Before beginning any maintenance or repair, the machine should be cleaned thoroughly, especially around the oil filter, spark plugs, valve covers, seat cowling, etc. Cleaning will help ensure that dirt does not contaminate the engine and will allow you to detect wear and damage that could otherwise easily go unnoticed.

4 Certain maintenance information is sometimes printed on decals attached to the motorcycle. If the information on the decals differs from that included here, use the information on the decal.

Every 500 miles (800 km)

1 Drive chain and sprockets – check, adjustment and lubrication

Check

1 A neglected drive chain won't last long and can quickly damage the sprockets. Routine chain adjustment and lubrication isn't difficult and will ensure maximum chain and sprocket life.

2 To check the chain, place the bike on its centrestand and shift the transmission into neutral. Make sure the ignition switch is OFF.

3 Rotate the rear wheel whilst pushing up on the bottom run of the chain until the tightest spot on the chain is positioned at the centre of its bottom run (the chain will rarely wear evenly). Measure the chain slack midway between the two sprockets and compare your measurement to that listed in this Chapter's Specifications (see illustration). Rotate the rear wheel and repeat the measurement procedure at several other points on the chain. At the tightest point the chain slack should be within the recommended limits, at all other points it should be less than the specified maximum. If adjustment is necessary, proceed as described below.

Caution: If the machine is ridden with excessive slack in the drive chain, the chain could contact the frame and swingarm, causing severe damage. Never overtighten the chain as this will place excess strain on the transmission output shaft bearing which could lead to it failing.

4 If it is not possible to set the chain slack to recommended limits at its tightest spot whilst keeping the slack within the specified maximum limits at all other points, the chain and sprockets are worn out and must be renewed as a set (see Chapter 6).

Adjustment

5 With the motorcycle on its centrestand, rotate the rear wheel until the chain is positioned with the tightest point at the centre of its bottom run.

6 Slacken the bearing holder clamp bolt on the swingarm.

7 Using a suitable C-spanner, such as the one supplied in the bike's tool kit, rotate the bearing holder clockwise or anti-clockwise (as applicable) until the correct chain tension is obtained (see illustration). If it is not possible to adjust the chain correctly it is excessively worn and the chain and both sprockets should be replaced as a set (see Chapter 6). The chain wear decal on the chainguard will also indicate the need for chain replacement when the tip of the sprocket teeth align with the 'replace chain' zone of the decal.

8 When the chain has the correct amount of slack, tighten the bearing holder clamp bolt to the specified torque.

Lubrication

9 If required, wash the chain in paraffin (kerosene), then wipe it off and allow it to dry, using compressed air if available. If the chain is excessively dirty it should be removed from the machine and allowed to soak in the paraffin (see Chapter 6).

Caution: Don't use petrol, solvent or other cleaning fluids which might damage the internal sealing properties of the chain. Don't use high-pressure water. The entire process shouldn't take longer than ten minutes – if it does, the O-rings in the chain rollers could be damaged.

10 For routine lubrication, the best time to lubricate the chain is after the motorcycle has been ridden. When the chain is warm, the lubricant will penetrate the joints between the side plates better than when cold. **Note:** *Honda specifies SAE 80 to SAE 90 gear oil or an aerosol chain lube which is suitable for O-ring chains.*

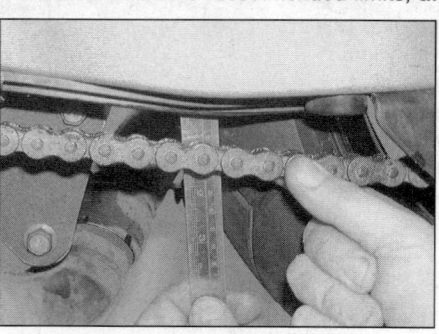

1.3 Measuring drive chain slack

1.7 Slacken the clamp bolt (arrowed) and adjust the chain slack by rotating the bearing holder using the C-spanner in the bike's tool kit

 *Apply the oil to the top of the lower chain run, so centrifugal force will work the oil into the chain when the bike is moving. After applying the lubricant, let it soak in a few minutes before wiping off any excess.*

Maintenance procedures 1•7

Every 4000 miles (6000 km) or 6 months

2 Brake pads – wear check

1 A quick check of the brake pads can be made without removing them from the caliper. The amount of pad wear can be judged by looking at the pads from the front of the caliper (both front and rear). A cut-out in the friction material indicates the wear limit **(see illustration)**.
2 If either pad has worn down to, or beyond the cut-out/grooves in the friction material, both pads must be renewed as a set. If the pads are dirty or if you are in doubt as to the amount of friction material remaining, remove them for inspection (see Chapter 7). **Note:** *Some after-market pads may use different indicators to those on the original equipment as shown.*
3 Refer to Chapter 7 for details of pad renewal.

3 Idle speed – check and adjustment

1 The idle speed should be checked and adjusted with the engine at its normal operating temperature.
2 Place the motorcycle on its centrestand, and make sure the transmission is in neutral.
3 Start the engine and turn the handlebars back-and-forth and see if the idle speed changes as this is done. If it does, the throttle cables are not adjusted or routed correctly. This is a dangerous condition, that can cause loss of control of the bike, which must be corrected before proceeding (see Section 8).
4 With the engine at its normal operating temperature, start it then snap the throttle open and shut a few times. Compare the engine idle speed to the specified idle speed given in the Specifications.
5 If adjustment is necessary, prise out the rubber plug from the right side of the frame **(see illustration)**; the idle speed adjuster is located on the right side of the throttle body assembly and is accessed through the hole. With the engine idling, use a long flat-bladed screwdriver to rotate the adjuster screw until the idle speed listed in this Chapter's Specifications is obtained **(see illustration)**. Snap the throttle open and shut a few times, then recheck the idle speed. If necessary, repeat the adjustment procedure. Once adjustment is complete, switch off the engine and fit the rubber plug to the frame.

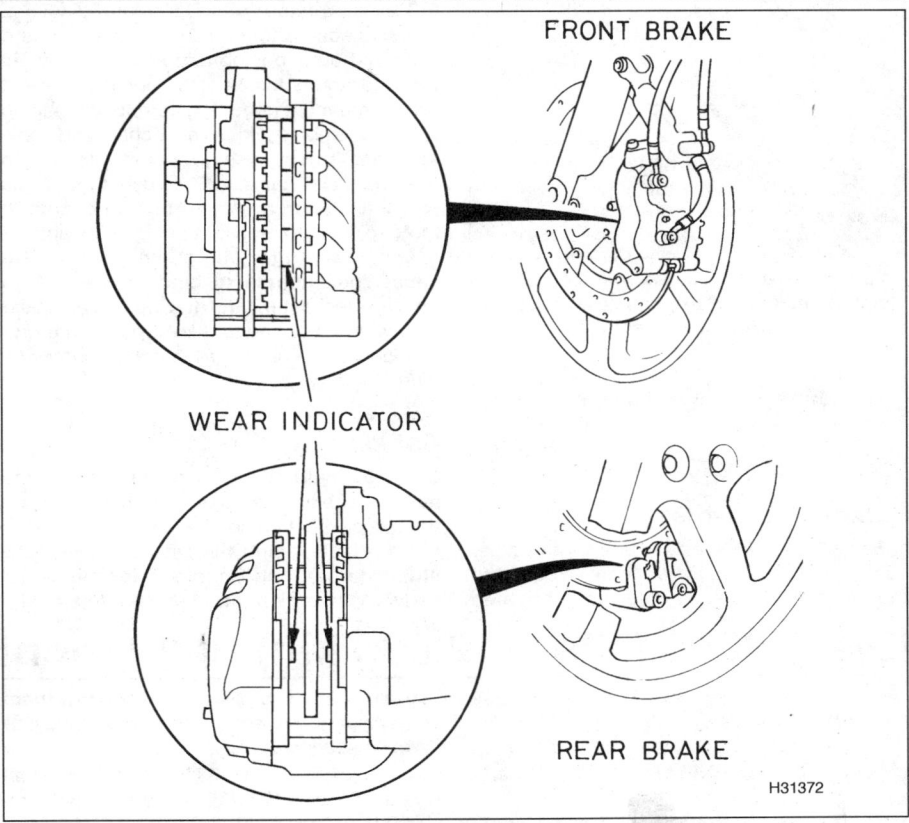

2.1 Front and rear brake pad wear indicator details

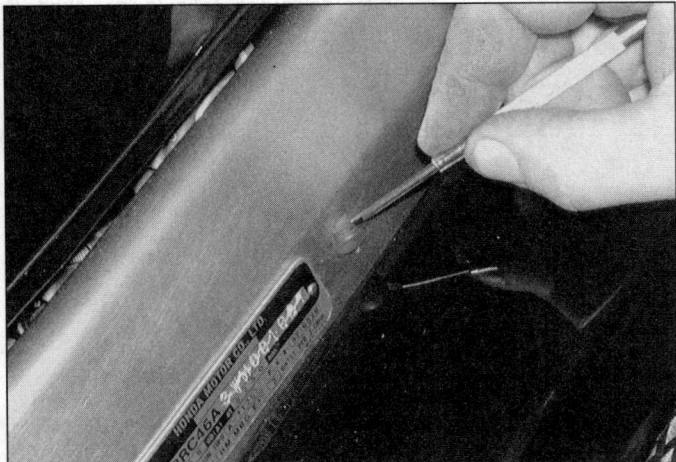

3.5a Prise out the rubber plug from the frame . . .

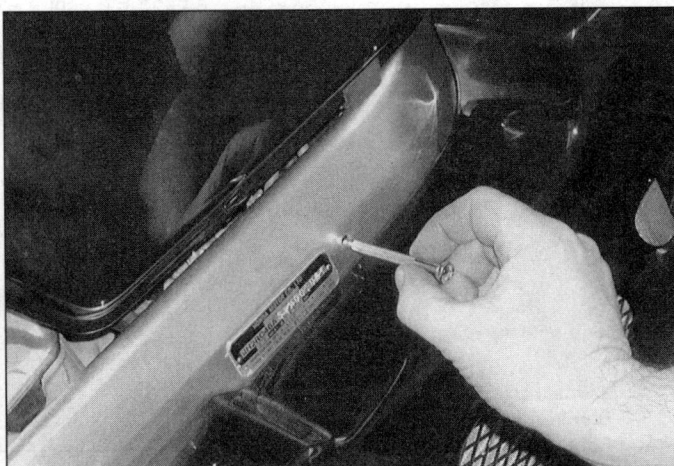

3.5b . . . and adjust the idle speed with a flat-bladed screwdriver

1•8 Every 4000 miles or 6 months

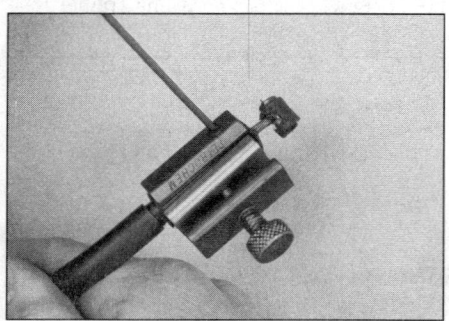

4.3a Lubricating a cable with a cable oil clamp (make sure the tool seats around the cable correctly)

4 Stands, pivots and cables – lubrication

Stands and pivots

1 Since the controls, cables and various other components of a motorcycle are exposed to the elements, they should be lubricated periodically to ensure safe and trouble-free operation.

2 The footrests, clutch and brake levers, brake pedal, sidestand and centrestand pivots should be lubricated frequently. In order for the lubricant to be applied where it will do the most good, the component should be disassembled. However, if chain and cable lubricant is being used, it can be applied to the pivot joint gaps and will usually work its way into the areas where friction occurs. If motor oil or light grease is being used, apply it sparingly as it may attract dirt (which could cause the controls to bind or wear at an accelerated rate). **Note:** *One of the best lubricants for the control lever pivots is a dry-film lubricant (available from many sources by different names).*

Cables

3 To lubricate the cables, disconnect the relevant cable at its upper end, then lubricate the cable with a cable oiler clamp, or if one is not available, using the set-up shown **(see illustrations)**. See Chapter 4 for the choke and throttle cable removal procedures.

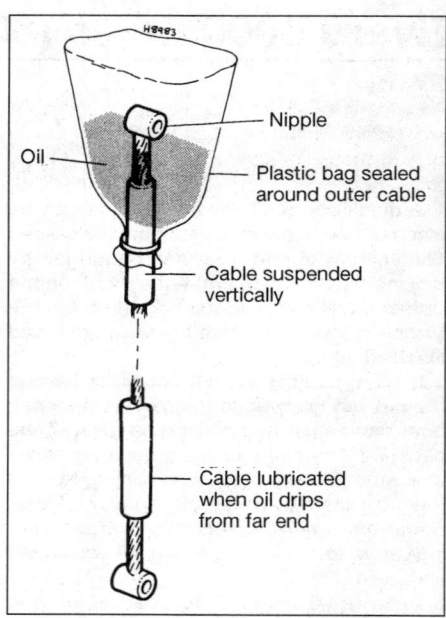

4.3b Lubricating a cable with a home-made funnel arrangement

Every 8000 miles (12,000 km) or 12 months

5 Spark plugs – gap check and adjustment (European models)

1 To gain access to the front cylinder spark plugs, remove the fairing lower panel and inner panel as described in Chapter 8. Unscrew the oil cooler mounting bolts then pivot the oil cooler forwards to allow access to the spark plug caps **(see illustration)**. Take care not to place any strain on the cooler pipes or lose the collars from the cooler mounting rubbers.

2 To gain access to the rear cylinder spark plugs, raise and support the fuel tank as described in Chapter 4.

3 Clean the area around the plug caps to prevent any dirt falling into the spark plug channels. Check that the cylinder location is marked on each plug lead (mark them accordingly if not) and pull the spark plug cap off each spark plug.

4 Clean the area around the base of the plugs to prevent any dirt falling into the engine. Using either the plug removing tool supplied in the bike's toolkit or a deep socket type wrench, unscrew the plugs from the cylinder head. Lay each plug out in relation to its cylinder; if any plug shows up a problem it will then be easy to identify the troublesome cylinder.

5 Inspect the electrodes for wear. Both the centre and side electrodes should have square edges and the side electrodes should be of uniform thickness. Look for excessive deposits and evidence of a cracked or chipped insulator around the centre electrode. Compare your spark plugs to the colour spark plug reading chart at the end of this manual. Check the threads, the washer and the ceramic insulator body for cracks and other damage.

 **HAYNES HINT** *Stripped plug threads in the cylinder head can be repaired with a thread insert – see 'Tools and Workshop Tips' in the Reference section.*

6 If the electrodes are not excessively worn, and if the deposits can be easily removed with a wire brush, the plugs can be re-gapped and re-used (if no cracks or chips are visible in the insulator). If in doubt concerning the condition of the plugs, renew them, as the expense is minimal.

7 Cleaning spark plugs by sandblasting is not recommended.

8 Before installing the plugs, make sure they are the correct type and heat range and check the gap between the electrodes **(see illustrations)**. Compare the gap to that specified and adjust as necessary. If the gap must be adjusted, bend the side electrodes

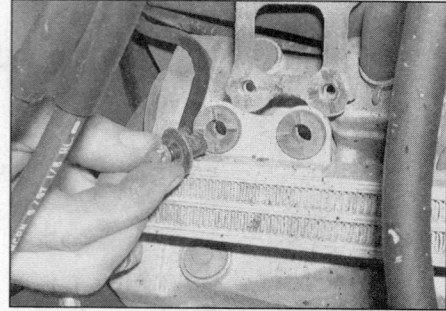

5.1 Slacken and remove the oil cooler mounting bolts and collars and pivot the cooler forwards to gain access to the front cylinder spark plugs

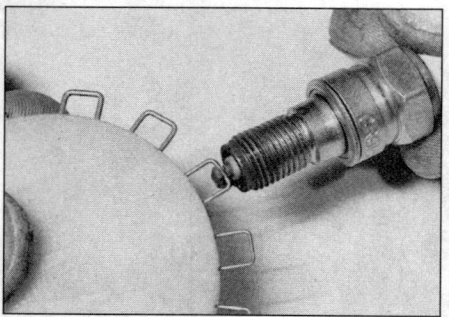

5.8a Using a wire gauge to measure a spark plug electrode gap

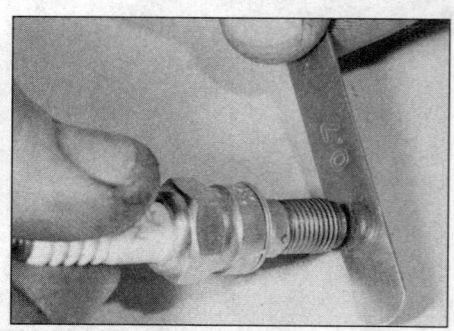

5.8b Using a feeler gauge to measure a spark plug electrode gap

Every 8000 miles or 12 months

only and be very careful not to chip or crack the insulator nose **(see illustration)**. Make sure the washer is in place before installing each plug.

9 Since the cylinder head is made of aluminium, which is soft and easily damaged, thread the plugs into the heads turning the tool by hand. Once the plugs are finger-tight, the job can be finished with a spanner on the tool supplied or a socket drive. Tighten the plugs an additional 1/4 to 1/2 turn or as directed on the plug manufacturer's packaging. If a torque wrench can be fitted to the socket/wrench, tighten the plugs to the specified torque.

> **HAYNES HINT**
> *As the plugs are recessed in the heads, slip a short length of hose over the end of the plug to use as a tool to thread it into place. The hose will grip the plug well enough to turn it, but will start to slip if the plug begins to cross-thread in the hole – this will prevent damaged threads.*

10 Reconnect each the spark plug cap to its correct cylinder making sure all caps are pushed fully onto the plugs.

11 Lower the fuel tank back into position (see Chapter 4). Align the oil cooler with its mounting bracket then, ensuring the mounting collars are in position, install the cooler mounting bolts and tighten securely. Fit the fairing panels as described in Chapter 8.

6 Spark plugs – renewal (US models)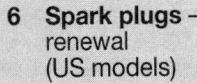

1 Renew the spark plugs as described in Section 5.

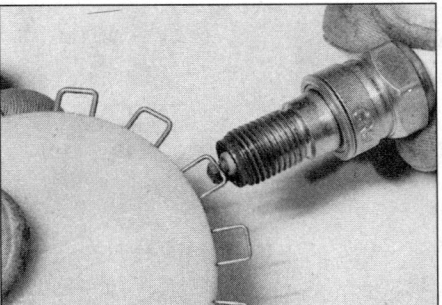

5.8c Adjust the spark plug electrode gap by bending the side electrode

7 Engine/transmission – oil and filter change

⚠ **Warning: Be careful when draining the oil, as the exhaust pipes, the engine, and the oil itself can cause severe burns.**

Note: *Oil filter removal and installation is far easier with a filter wrench, either the Honda service tool (No. 07HAA-PJ70100) or a pattern alternative (available from most good motorcycle accessory dealers).*

1 Consistent routine oil and filter changes are the single most important maintenance procedure you can perform on a motorcycle. The oil not only lubricates the internal parts of the engine, transmission and clutch, but it also acts as a coolant, a cleaner, a sealant, and a protectant. Because of these demands, the oil takes a terrific amount of abuse and should be changed often with new oil of the recommended grade and type. Saving a little money on the difference in cost between a good oil and a cheap oil won't pay off if the engine is damaged.

2 Before changing the oil, warm up the engine so the oil will drain easily. Remove the fairing lower panels and inner panel (see Chapter 8).

3 Put the motorcycle on its centrestand, and position a clean drain tray below the engine. Unscrew the oil filler cap from the right crankcase cover to vent the crankcase and to act as a reminder that there is no oil in the engine.

4 Next, unscrew the oil drain plug from the oil pan on the bottom of the engine and allow the oil to flow into the drain tray **(see illustration)**. Check the condition of the sealing washer on the drain plug and obtain a new one if it is damaged or worn.

> **HAYNES HINT** *To help determine whether any abnormal or excessive engine wear is occurring, place a strainer between the engine and the drain tray so that any debris in the oil is filtered out and can be examined. If there are flakes or chips of metal in the oil, then something is drastically wrong internally and the engine will have to be disassembled for inspection and repair. If there are pieces of fibre-like material in the oil, the clutch is experiencing excessive wear and should be checked.*

5 When the oil has completely drained, fit the plug to the oil pan, using a new sealing washer if necessary, and tighten it to the specified torque. Avoid overtightening, as damage to the oil pan will result.

6 Now place the drain tray below the oil filter. Unscrew the oil filter using a filter wrench or a strap wrench and tip any residue oil into the drain tray **(see illustration)**. Wipe clean the oil filter mating surface on the crankcase.

7 Smear clean engine oil onto the rubber seal on the new filter, then manoeuvre it into position and screw it onto the engine and tighten it to the specified torque **(see illustrations)**. If a filter

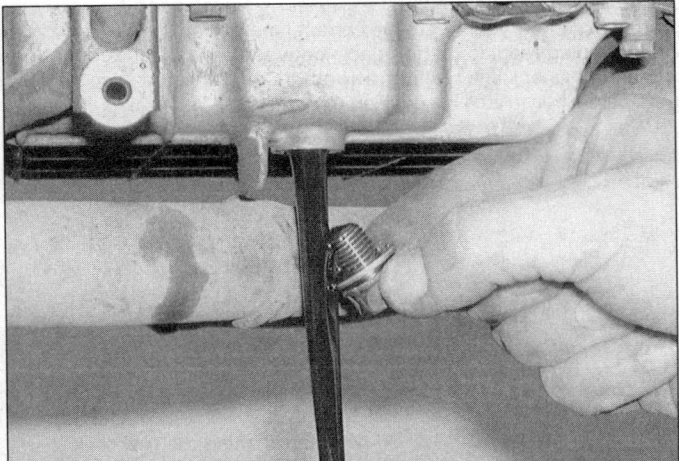

7.4 Slacken and remove the drain plug and sealing washer and allow the oil to drain into the container

7.6 Using a filter wrench to unscrew the oil filter

1•10 Every 8000 miles or 12 months

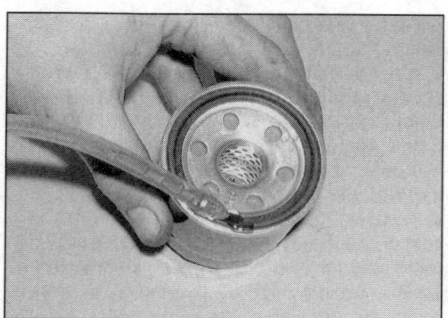

7.7a Lubricate the rubber seal with a smear of engine oil...

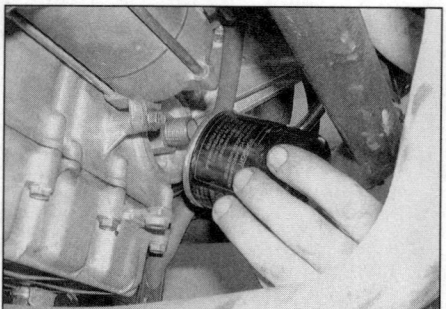

7.7b ...then screw the new filter onto the engine

7.8 Fill the engine with the recommended type and amount of oil

wrench is not being used, tighten the filter as tight as possible by hand.

8 Refill the engine to the proper level using the recommended type and amount of oil (see *Specifications*) **(see illustration)**. With the motorcycle on its centrestand, the oil level should lie between the upper and lower level lines on the inspection window in the right crankcase cover (see *Daily (pre-ride) checks*). Install the filler cap. Start the engine and let it run for two or three minutes (make sure that the oil pressure light extinguishes after a few seconds). Shut it off, wait a few minutes, then check the oil level. If necessary, add more oil to bring the level up to the upper level line on the inspection window. Check around the oil filter and drain plug for leaks before installing the fairing panels (see Chapter 8).

9 The old oil drained from the engine cannot be re-used and should be disposed of properly. Check with your local refuse disposal company, disposal facility or environmental agency to see whether they will accept the used oil for recycling. Don't pour used oil into drains or onto the ground.

Note: It is antisocial and illegal to dump oil down the drain. To find the location of your nearest oil recycling bank in the UK, call this number free. In the USA, note that any oil supplier must accept used oil for recycling.

8 Throttle and choke cable – check

Throttle cables

1 Make sure the throttle grip rotates easily from fully closed to fully open with the front wheel turned at various angles. The grip should return automatically from fully open to fully closed when released.

2 If the throttle sticks, this is probably due to a cable fault. Remove the cables (see Chapter 4) and lubricate them (see Section 4). Install the cables, making sure they are correctly routed. If this fails to improve the operation of the throttle, the cables must be renewed. Note that in very rare cases the fault could lie in the throttle body assembly rather than the cables, necessitating the removal of the throttle body and inspection of the throttle linkage (see Chapter 4).

3 With the throttle operating smoothly, check for a small amount of freeplay in the cables, measured in terms of the amount of twistgrip rotation before the throttle opens, and compare the amount to that listed in this Chapter's Specifications **(see illustration)**. If it's incorrect, adjust the cables to correct it.

4 Freeplay adjustments can be made at the throttle end of the cable. Loosen the locknut on the accelerator cable where it leaves the handlebar **(see illustration)**. Turn the adjuster until the specified amount of freeplay is obtained (see this Chapter's Specifications), then retighten the locknut.

5 If the adjuster has reached its limit of adjustment, reset it so that the freeplay is at a maximum, then raise and support the fuel tank (see Chapter 4) to gain access to the cable adjuster at the throttle body end. Slacken the adjuster locknut, then turn the adjuster out, making sure the inner nut remains captive in the bracket, thereby threading itself down the adjuster as you turn it **(see illustration)**. Turn the adjuster until the specified amount of freeplay is obtained, then securely tighten the locknut. Further adjustments can now be made at the throttle grip end. If the cable cannot be adjusted as specified, renew the cables (see Chapter 4). Once adjustment is correct, lower the fuel tank back down into position (see Chapter 4).

⚠ **Warning: Turn the handlebars all the way through their travel with the engine idling. Idle speed should not change. If it does, the cable may be routed incorrectly. Correct this condition before riding the bike.**

6 Check that the throttle twistgrip operates smoothly and snaps shut quickly when released.

Choke cable

Note: This information only applies to VFR800FI-W and X models. An automatic choke (fast idle wax unit) is fitted to VFR800FI-Y and 1 models (see Chapter 4).

7 If the choke does not operate smoothly this is probably due to a cable fault. Remove the cable (see Chapter 4) and lubricate it (see Section 4). Install the cable, routing it so it takes the smoothest route possible.

8 If this fails to improve the operation of the choke, the cable must be renewed. Note that

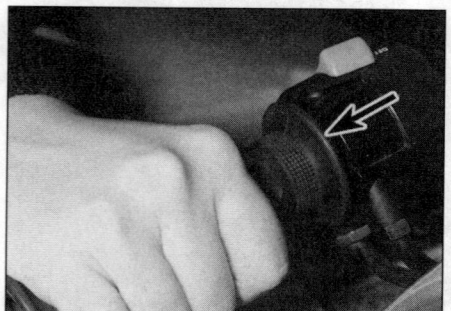

8.3 Throttle cable freeplay is measured in terms of twistgrip rotation at the grip flange (arrow)

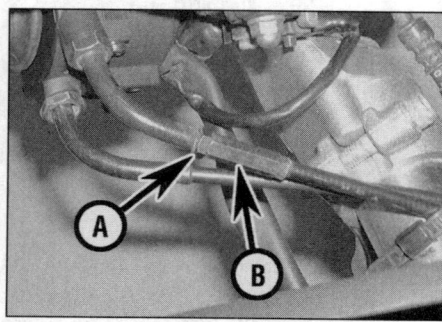

8.4 Loosen the locknut (A) and adjust throttle cable freeplay by rotating the adjuster (B)

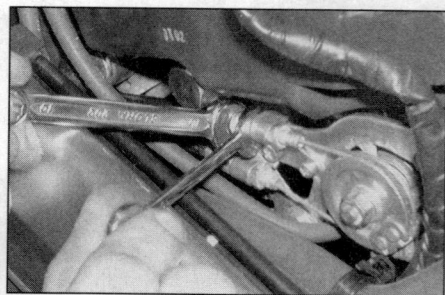

8.5 If necessary, raise the fuel tank then slacken the locknut and rotate the cable adjuster on the side of the throttle body assembly

Every 8000 miles or 12 months

in very rare cases the fault could lie in the throttle body assembly rather than the cable, necessitating the removal of the throttle body and inspection of the starter valve (see Chapter 4).

9 Make sure there is a small amount of freeplay in the cable before the starter valves move. If there isn't, check that the cable is correctly installed at both ends – remove the air filter housing to access the throttle body end of the cable (see Chapter 4) **(see illustration)**. If the cable is faulty it must be renewed, no adjustment is possible.

9 Fuel system – check

 Warning: Petrol (gasoline) is extremely flammable, so take extra precautions when you work on any part of the fuel system. Don't smoke or allow open flames or bare light bulbs near the work area, and don't work in a garage where a natural gas-type appliance is present. If you spill any fuel on your skin, rinse it off immediately with soap and water. When you perform any kind of work on the fuel system, wear safety glasses and have a fire extinguisher suitable for a Class B type fire (flammable liquids) on hand.

1 Raise and support the fuel tank (see Chapter 4) and check the fuel feed and return hoses and the tank breather/overflow hoses for signs of leakage, deterioration or damage; in particular check that there is no leakage from the fuel hose unions **(see illustrations)**. Renew any hoses which are cracked or deteriorated.

2 If there is a leak from the fuel rails or injectors, the throttle body assembly will have to be removed to allow the injectors seals to be renewed (see Chapter 4).

3 Check the pulse secondary air (PAIR) system hoses, linking the control valve assembly to the front and rear cylinder head covers, for signs of damage or deterioration and renew if necessary (see Chapter 4).

10 Cooling system – check

 Warning: The engine must be cool before beginning this procedure.

1 Remove the left and right lower fairing panels as described in Chapter 8.
2 The entire cooling system should be checked for evidence of leakage. Examine each rubber coolant hose along its entire length. Look for cracks, abrasions and other damage. Squeeze each hose at various points. They should feel firm, yet pliable, and return to their original shape when released. If

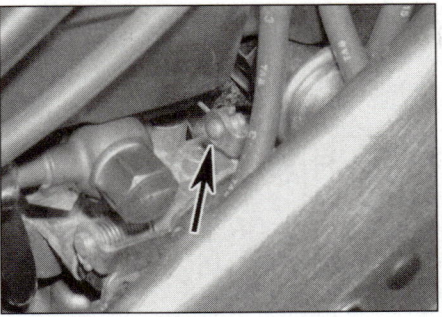

8.9 The lower end of the choke cable (arrow) is just visible with the air cleaner housing in position

they are dried out or hard, replace them with new ones.

3 Check for evidence of leaks at each cooling system joint. Tighten the hose clips carefully to prevent future leaks.
4 Check both radiators for leaks and other damage. Leaks in the radiator leave telltale scale deposits or coolant stains on the outside of the core below the leak. If leaks are noted, remove the radiator (see Chapter 3) and have it repaired by a professional.
Caution: *Do not use a liquid leak stopping compound to try to repair leaks.*
5 Check the radiator fins for mud, dirt and insects, which may impede the flow of air through the radiator. If the fins are dirty, force water or low pressure compressed air through the fins from the backside. If the fins are bent or distorted, straighten them carefully with a screwdriver.
6 Remove the pressure cap from the filler neck by turning it anti-clockwise until it reaches a stop. If you hear a hissing sound (indicating there is still pressure in the system), wait until it stops. Now press down on the cap and continue turning it until it can be removed. Check the condition of the coolant in the system. If it is rust-coloured or if accumulations of scale are visible, drain, flush and refill the system with new coolant (See Section 26). Check the cap seal for cracks and other damage. If in doubt about the pressure cap's condition, have it tested by a dealer or replace it with a new one. Install the cap by turning it clockwise until it reaches the

first stop then push down on the cap and continue turning until it can turn no further.
7 Check the antifreeze content of the coolant with an antifreeze hydrometer. Sometimes coolant looks like it's in good condition, but might be too weak to offer adequate protection. If the hydrometer indicates a weak mixture, drain, flush and refill the system (see Section 26).
8 If the coolant level is consistently low, and no evidence of leaks can be found, have the entire system pressure checked by a Honda dealer.
9 Ensure the pressure cap is correctly installed then fit the fairing panels as described in Chapter 8.

11 Exhaust system – check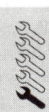

1 Periodically check all of the exhaust system joints for leaks and loose fasteners. The lower fairing panels will have to be removed to do this properly (see Chapter 8). If tightening the fasteners fails to stop any leaks, replace the gaskets with new ones (a procedure which requires disassembly of the system). Refer to Chapter 4 for further information.

12 Drive chain, slider and guide plate – check

1 Remove the fairing left side lower panel (see Chapter 8).
2 Remove the front sprocket cover from the engine as described in Chapter 6. Check the teeth on the front and rear sprockets for wear and check the entire length of the chain for damaged rollers, loose links and pins, and missing O-rings **(see illustration)**. If any damage is found, renew both sprockets and the chain. **Note:** *Never install a new chain on old sprockets, and never use the old chain if you install new sprockets – renew the chain and sprockets as a set.* Check that the front sprocket bolt and rear sprocket nuts are all tightened to the specified torque (see Chapter 6).

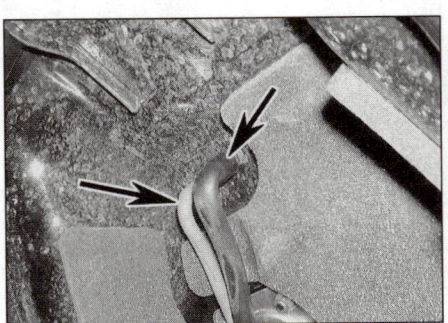

9.1a Raise and support the fuel tank and check the tank breather and overflow hoses (arrowed) . . .

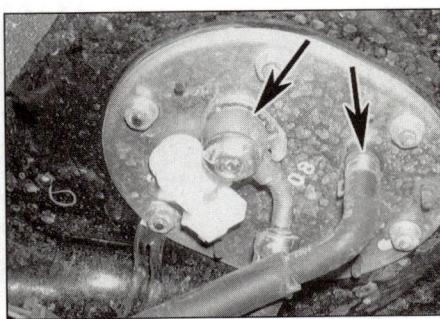

9.1b . . . and the fuel feed and return hoses (arrowed) for damage or deterioration

1•12 Every 8000 miles or 12 months

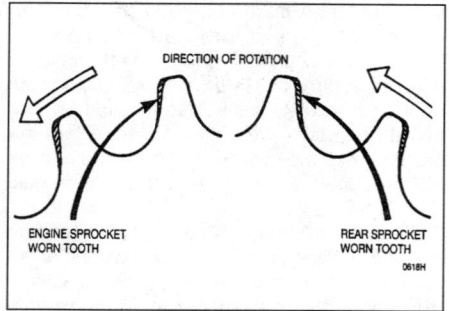

12.2 Check the sprockets in the areas indicated to see if they are worn

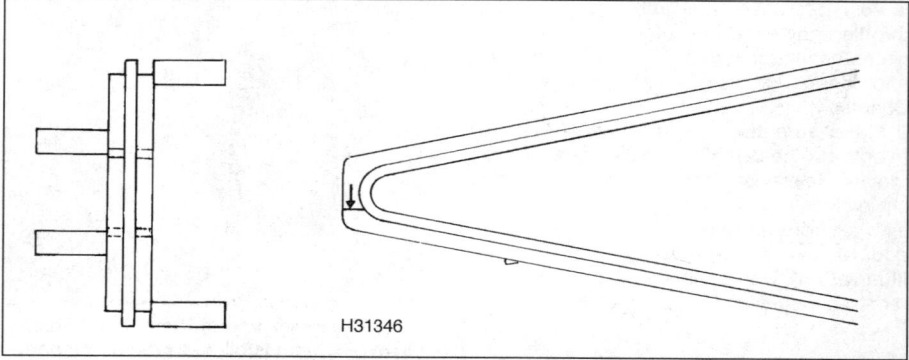

12.3 Swingarm drive chain slider wear markings

3 Clean away all traces of oil and grease from around the swingarm pivot to reveal the drive chain slider. Inspect the slider ridge for wear or damage; the slider prevents the chain contacting the swingarm. If the ridge is worn down below the wear indicator **(see illustration)**, it should be renewed. The slider is secured to the swingarm by four bolts and collars. On installation ensure the slider pins are correctly located in the swingarm then apply locking compound to the retaining bolts. Fit the collars and bolts and tighten them to the specified torque (see Chapter 6).
4 Inspect the drive chain guide plate for signs of wear or damage **(see illustration)**. The guide plate prevents the chain contacting the crankcase and must be renewed if it is damaged.
5 On completion of the check, fit the sprocket cover as described in Chapter 6.

13 Brake system – check
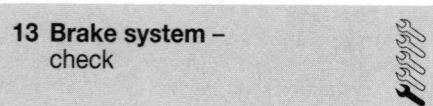

1 A routine general check of the brakes will ensure that any problems are discovered and remedied before the rider's safety is jeopardised.
2 Check the brake lever and pedal for loose connections, excessive play, bends, and other damage. Replace any damaged parts with new ones (see Chapter 7).
3 Make sure all brake fasteners are tight.

Check the brake pads for wear (see Section 2) and make sure that the fluid level in the reservoirs is correct (see 'Daily (pre-ride) checks'). Look for leaks at the hose and pipe connections and check for cracks in the hoses. If the lever or pedal is spongy, bleed the brakes as described in Chapter 7.
4 Make sure the brake light operates when the front brake lever is pulled in. The front brake light switch is not adjustable. If it fails to operate properly, replace it with a new one (see Chapter 9).
5 Make sure the brake light is activated just before the rear brake pedal takes effect. If adjustment is necessary, hold the switch and turn the adjusting ring on the switch body until the brake light is activated when required **(see illustration)**. If the switch doesn't operate the brake lights, check it as described in Chapter 9.
6 With the motorcycle on its centrestand check the operation of the dual combined brake system (DCBS) as follows.
7 Ensure the transmission is in neutral then have an assistant pivot the left front caliper bracket firmly towards the fork slider whilst you spin the rear wheel **(see illustration)**. When the caliper is forced against the fork slider the rear brake should be applied preventing rear wheel rotation. If this is not the case the secondary master cylinder circuit is not function correctly and further investigation will be required (see Chapter 7).

8 Have your assistant apply weight to the rear of the motorcycle to lift the front wheel clear of the ground. Check the front wheel spins freely then have your assistant apply the rear brake pedal. When the pedal is depressed the front brake should also be applied preventing front wheel rotation. If this is not the case the rear brake pedal hydraulic circuit is not function correctly and further investigation will be required (see Chapter 7).

14 Headlight aim – check and adjustment
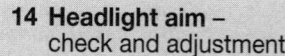

Note: *An improperly adjusted headlight may cause problems for oncoming traffic or provide poor, unsafe illumination of the road ahead. Before adjusting the headlight aim, be sure to consult with local traffic laws and regulations – refer to MOT Test Checks in the Reference section.*

1 The headlight beam can adjusted both horizontally and vertically. Before making any adjustment, check that the tyre pressures are correct and the suspension is adjusted as required. Make any adjustments to the headlight aim with the machine on level ground, with the fuel tank half full and with an assistant sitting on the seat. If the bike is usually ridden with a passenger on the back, have a second assistant to do this.

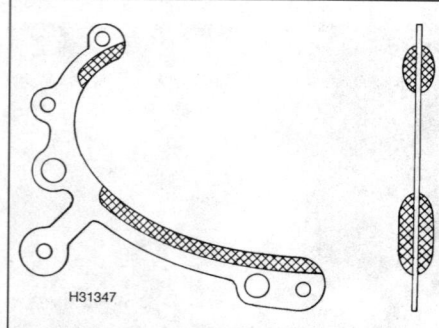

12.4 Inspect the drive chain guide plate for signs of wear. If the shaded areas show signs of damage, the plate must be replaced

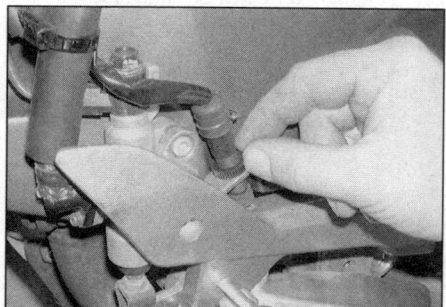

13.5 Adjust the rear brake light switch by rotating the adjusting ring whilst holding the switch stationary

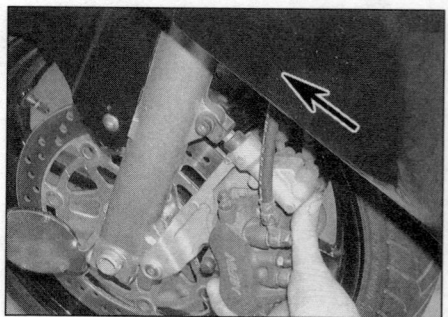

13.7 Force the left front brake caliper bracket firmly against the fork slider to check the operation of the secondary master cylinder (see text)

Every 8000 miles or 12 months

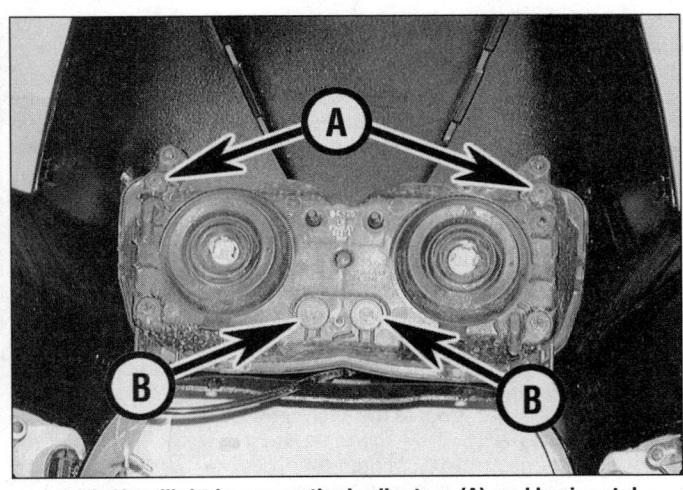

14.2 Headlight beam vertical adjusters (A) and horizontal adjusters (B) (shown with fairing removed)

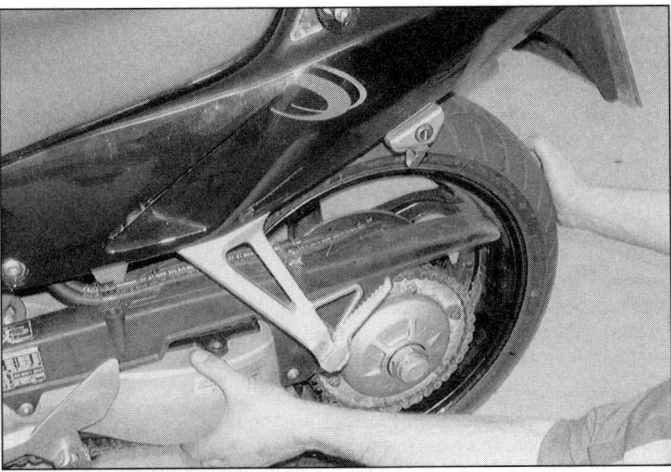

16.7a Checking for play in the swingarm bearings

2 Vertical adjustment is made by turning the adjuster screw on the top outer corner of each headlight unit **(see illustration)**. Turn it clockwise to move the beam up, and anti-clockwise to move it down. Access each adjuster screw by passing a Phillips head screwdriver through the holes provided in the fairing instrument cluster surround panel.

3 Horizontal adjustment is made by turning the adjuster screw on the bottom inner corner of each headlight unit. Turn it clockwise to move the beam to the right and anti-clockwise to move it to the left.

15 Sidestand – check

1 The sidestand return spring must be capable of retracting the stand fully and holding the stand retracted when the motorcycle is in use. If the spring is sagged or broken it must be renewed.

2 Lubricate the sidestand pivot regularly (see Section 4).

3 Check that the sidestand switch prevents the motorcycle being started if the stand is extended. Check its operation by shifting the transmission into neutral, retracting the stand and starting the engine. Pull in the clutch lever and select a gear. Extend the sidestand. The engine should stop as the sidestand is extended. If the sidestand switch does not operate as described, check its circuit (see Chapter 9).

16 Suspension – check

1 The suspension components must be maintained in top operating condition to ensure rider safety. Loose, worn or damaged suspension parts decrease the motorcycle's stability and control.

Front suspension

2 While standing alongside the motorcycle, apply the front brake and push on the handlebars to compress the forks several times. See if they move up-and-down smoothly without binding. If binding is felt, the forks should be disassembled and inspected (see Chapter 6).

3 Inspect the area around the dust seal for signs of oil leakage. If leakage is evident, the fork seals must be renewed (see Chapter 6).

4 Check the tightness of all suspension nuts and bolts to be sure none have worked loose.

Rear suspension

5 Inspect the rear shock for fluid leakage and tightness of its mountings. If leakage is found, the shock should be renewed (see Chapter 6).

6 With the aid of an assistant to support the bike, compress the rear suspension several times. It should move up and down freely without binding. If any binding is felt, the worn or faulty component must be identified and renewed. The problem could be due to either the shock absorber, the suspension linkage components or the swingarm components.

7 Support the motorcycle on its centrestand so that the rear wheel is off the ground. Grab the swingarm and attempt to rock it from side to side **(see illustration)** – there should be no

16.7b Checking for play in the rear shock absorber and suspension linkage

discernible movement at the rear. If there's a little movement or a slight clicking can be heard, inspect the tightness of all the rear suspension mounting bolts and nuts, referring to the torque settings specified at the beginning of Chapter 6, and re-check for movement. Next, grasp the top of the rear wheel and pull it upwards **(see illustration)** – there should be no discernible freeplay before the shock absorber begins to compress. Any freeplay felt in either check indicates worn bearings in the suspension linkage or swingarm, or worn shock absorber mountings. The worn components must be renewed (see Chapter 6).

8 To make an accurate assessment of the swingarm bearings, remove the rear wheel (see Chapter 7) and the bolt securing the suspension linkage assembly to the swingarm (see Chapter 6). Grasp the rear of the swingarm with one hand and place your other hand at the junction of the swingarm and the frame. Try to move the rear of the swingarm from side-to-side. Any wear (play) in the bearings should be felt as movement between the swingarm and the frame at the front. If there is any play the swingarm will be felt to move forward and backward at the front (not from side-to-side). Next, move the swingarm up and down through its full travel. It should move freely, without any binding or rough spots. If any play in the swingarm is noted or if the swingarm does not move freely, the bearings must be removed for inspection or renewal (see Chapter 6).

17 Nuts and bolts – tightness check

1 Since vibration of the machine tends to loosen fasteners, all nuts, bolts, screws, etc. should be periodically checked for proper tightness.

2 Pay particular attention to the following:

1•14 Every 8000 miles or 12 months

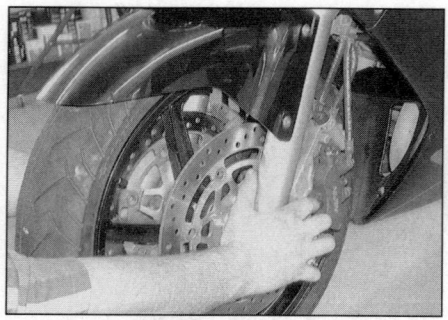

18.4 Checking for play in the steering head bearings

- Footrest and stand bolts
- Engine mounting bolts
- Shock absorber and suspension linkage bolts and swingarm pivot bolts
- Handlebar clamp bolts
- Front axle bolt and axle clamp bolts
- Front fork clamp bolts (top and bottom yoke)
- Rear wheel nuts
- Brake caliper mounting bolts
- Brake hose banjo bolts and caliper bleed valves
- Brake disc bolts
- Exhaust system bolts/nuts

3 If a torque wrench is available, use it along with the torque specifications at the beginning of this and other Chapters.

18 Steering head bearings – check and adjustment

1 This motorcycle is equipped with caged ball steering head bearings which can become dented, rough or loose during normal use of the machine. In extreme cases, worn or loose steering head bearings can cause steering wobble – a condition that is potentially dangerous.

Check

2 Position the motorcycle on its centrestand and tie down the rear of the motorcycle to raise the front wheel off the ground. Alternatively, remove the lower fairing panels (see Chapter 8) and support the bike under the engine to raise the front wheel.
3 Point the front wheel straight-ahead and slowly move the handlebars from side-to-side. Any dents or roughness in the bearing races will be felt, and if the bearings are too tight the bars will not move smoothly and freely. If the bearings are damaged or the action is rough, they should be renewed (see Chapter 6). If the bearings are too tight they should be adjusted as described below.
4 Next, grasp the fork sliders and try to move them forward and backward **(see illustration)**. Any looseness in the steering head bearings will be felt as front-to-rear movement of the forks. If play is felt in the bearings, adjust the steering head as follows.

HAYNES HINT *Freeplay in the fork due to worn fork bushes can be misinterpreted for steering head bearing play – do not confuse the two.*

Adjustment

5 Referring to Chapter 6 for further information, carefully pry off the circlip from the top of each fork tube.
6 Slacken each handlebar's clamp bolt then slide both handlebars off the fork tubes and support them to prevent straining the hydraulic hoses. Ensure the master cylinder reservoirs are kept upright to prevent possible fluid leakage.
7 Slacken the top yoke clamp bolts then remove the cap from the top of the steering stem.
8 Unscrew the steering stem nut then lift off the top yoke and place it aside, making sure no strain is placed on the ignition switch wiring.
9 Prise the lockwasher tabs out of the notches in the locknut. Slacken and remove the locknut using either a C-spanner or a suitable drift located in one of the notches.
10 Remove the lock washer and discard it; a new one must be fitted on reassembly then slacken the adjuster nut slightly.
11 If the Honda service tool (Part no. 07916-3710101 – a socket that fits the adjuster nut) is available, tighten the adjuster nut to the specified torque setting. If the service tool is not available, set the adjuster nut so that the bearings are under a very light loading, just enough to remove any freeplay.
Caution: Take great care not to apply excessive pressure because this will cause premature failure of the bearings.
12 Ensure that the bearings pivot smoothly, without any sign of tightness or freeplay then fit a new lockwasher to the adjuster nut. Bend down two opposite lockwasher tabs into the grooves of the adjuster nut.
13 Install the locknut and tighten it finger-tight only.
14 Hold the adjuster nut, to prevent it from moving, and tighten the locknut approximately 90° more until its slots align with the remaining lock washer tabs. Recheck the bearing adjustment before securing the locknut in position by bending up the lockwasher tabs into its slots.
15 Fit the top yoke to the steering stem and install the stem nut. Tighten the steering stem nut to the specified torque then install the cap in the top of the stem.
16 Tighten the top yoke clamp bolts to the specified torque.
17 Locate the handlebars on the fork tubes, ensuring that the lug on the bottom of each casting is correctly located with the cut-out on the top yoke, then tighten the handlebar clamp bolts to the specified torque.
18 Fit the circlip to each fork tube making sure they are both correctly located in their grooves.

19 Wheels and tyres – general check

Tyres

1 Check the tyre condition and tread depth thoroughly – see *Daily (pre-ride) checks*.

Wheels

2 Cast wheels are virtually maintenance free, but they should be kept clean and checked periodically for cracks and other damage. Also check the wheel runout and alignment (see Chapter 7). Never attempt to repair damaged cast wheels; they must be renewed if damaged. Check the valve rubber for signs of damage or deterioration and have it renewed if necessary. Also, make sure the valve stem cap is in place and tight.

Every 12,000 miles (18,000 km)

20 Air filter – renewal

1 Raise and support the fuel tank (see Chapter 4).
2 Disconnect the vacuum hose from the variable intake system diaphragm unit **(see illustration)**.
3 Slacken and remove the retaining screws, noting the correct fitted locations of the wiring clips, and lift off the air filter housing cover **(see illustrations)**.

20.2 Raise and support the fuel tank then disconnect the vacuum hose from the air filter housing intake system diaphragm unit

Every 12,000 miles 1•15

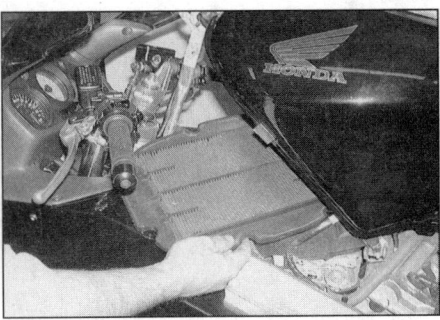

20.3a Slacken and remove the retaining screws . . .

20.3b . . . then lift off the air filter housing cover

20.4 Removing the air filter element

4 Remove the air filter element, noting which way up it is fitted, and its rubber seal from the housing and discard them **(see illustration)**.
5 Wipe clean the housing cover then install the new rubber seal in the housing groove.
6 Fit the new filter element, ensuring it is the right way up, and install the housing cover.
7 Fit the cover retaining screws, ensuring the wiring clips are correctly positioned, and tighten them securely.
8 Reconnect the vacuum hose to the intake diaphragm unit then lower the fuel tank back down into position (see Chapter 4).

21 Brakes – fluid change

Note: *To prevent damage to the paint from spilled brake fluid, always cover the fuel tank when working on the front brake master cylinder, and cover the surrounding components when working on the rear brake fluid reservoir.*

1 The procedure is similar to that for the bleeding of each hydraulic system as described in Chapter 7, except that the fluid reservoir should be emptied by siphoning, using a clean poultry baster or similar before starting, and allowance should be made for the old fluid to be expelled when bleeding a section of the circuit.
2 Working in the order described (see Chapter 7), open the first bleed valve and pump the lever/pedal gently. Be careful to keep the master cylinder reservoir topped up to above the LOWER level at all times or air may enter the system and greatly increase the length of the task. Continue pumping until new fluid can be seen emerging from the bleed valve.

> **HAYNES HiNT** *Old hydraulic fluid is usually much darker in colour than the new, making it easy to distinguish the two.*

3 When the new fluid is seen to be emerging, hold the lever/pedal and tighten the bleed valve to the specified torque (see Chapter 7). Repeat the operation on the remaining bleed valve(s) in the hydraulic circuit until all old hydraulic fluid has been replaced.
4 When the operation is complete, wash off all traces of spilt fluid then top-up the reservoir fluid level (see '*Daily (pre-ride) checks*').
5 Check the operation of the brakes before riding the motorcycle.

22 Clutch – fluid change

Note: *To prevent damage to the paint from spilled brake fluid, always cover the fuel tank when working on the clutch master cylinder.*

1 The procedure is similar to that for the bleeding of the hydraulic system as described in Chapter 2, except that the fluid reservoir should be emptied by siphoning, using a clean poultry baster or similar before starting, and allowance should be made for the old fluid to be expelled when bleeding the circuit.
2 Working as described, open the slave cylinder bleed valve and pump the lever gently. Be careful to keep the master cylinder reservoir topped up to above the LOWER level at all times or air may enter the system and greatly increase the length of the task. Continue pumping until new fluid can be seen emerging from the bleed valve.

> **HAYNES HiNT** *Old hydraulic fluid is usually much darker in colour than the new, making it easy to distinguish the two.*

3 When the new fluid is seen to be emerging, hold the lever and tighten the bleed valve to the specified torque (see Chapter 2).
4 When the operation is complete, wash off all traces of spilt fluid then top-up the reservoir fluid level (see *Daily (pre-ride) checks*).
5 Check the operation of the clutch before riding the motorcycle.

23 Evaporative emission control (EVAP) system – check (California models)

1 Periodically check the charcoal canister and its hoses for signs of damage; the lower fairing panels will have to be removed to do this properly (see Chapter 8). If any component shows signs of damage it must be renewed. Refer to Chapter 4 for further information.

Every 16,000 miles (24,000 km)

24 Valve clearances – check and adjustment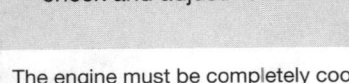

1 The engine must be completely cool for this maintenance procedure, so let the machine sit overnight before beginning.
2 Remove the front cylinder head cover and rear cylinder head cover as described in Chapter 2.
3 Unscrew the inspection plug and O-ring from the right crankcase cover **(see illustration)**.
4 Draw the valve locations on a piece of paper, referring to the Specifications for cylinder identification information.
5 Using a socket on the crankshaft bolt, rotate the crankshaft clockwise until the 'T' mark of number 1 cylinder is aligned with the index mark (cut-out or line) on the top of the crankcase cover aperture. Check the position of the index marks on the rear cylinder camshaft gears; the exhaust camshaft mark

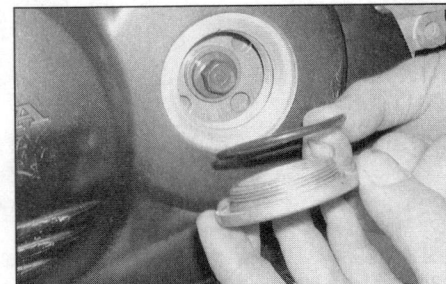

24.3 Slacken and remove the inspection plug and O-ring from the right crankcase cover

1•16 Every 16,000 miles

24.5 Valve clearances for each cylinder are checked with the piston at TDC on its compression stroke (camshaft gear and crankshaft timing marks positioned as shown - shown in firing order)

should be pointing upwards and the intake camshaft mark pointing downwards (piston number 1 is now at TDC on its compression stroke) **(see illustration)**. If the camshaft index marks are not correctly positioned, rotate the crankshaft through another 360° (one complete turn) to position them as described. **Note:** *Always turn the engine in the normal direction of rotation (clockwise), viewed from the right end of the crankshaft.*

6 With the engine in this position, all four valves of cylinder number 1 can be checked.

7 Insert a feeler gauge of the correct thickness (see Specifications) between each cam lobe and follower and check that it is a firm sliding fit. If it is not, use the feeler gauges to obtain the exact clearance. **Note:** *The intake and exhaust valve clearances are different.*

8 Record the clearance of each valve next to its relevant location on the piece of paper.

9 With the four valves of number 1 cylinder measured, rotate the crankshaft through 180° (half a turn) until the 'T' mark of number 3 cylinder is aligned with the index mark **(see illustration)**. The index marks on the rear cylinder camshaft gears should now be facing away from each other and aligned with the cylinder head upper surface (piston number 3 is now at TDC on its compression stroke) **(see illustration 24.5)**. Measure the clearances of the number 3 cylinder valves and record them on the piece of paper.

10 With the four valves of number 3 cylinder measured, rotate the crankshaft through 270° (three-quarters of a turn) until the 'T' mark of number 2 cylinder is aligned with the index mark **(see illustration)**. The index mark on the front cylinder intake camshaft gear should now be facing upwards and the mark on the exhaust camshaft gear should be facing downwards (piston number 2 is now at TDC on its compression stroke) **(see illustration 24.5)**. Measure the clearances of the number 2 cylinder valves and record them on the piece of paper.

11 With the four valves of number 2 cylinder measured, rotate the crankshaft through 180° (half a turn) until the 'T' mark of number 4 cylinder is aligned with the index mark **(see illustration)**. The index marks on the front cylinder camshaft gears should now be facing away from each other and aligned with the cylinder head upper surface (piston number 4 is now at TDC on its compression stroke) **(see illustration 24.5)**. Measure the clearances of the number 4 cylinder valves and record them on the piece of paper.

12 If any of the clearances need to be adjusted the relevant camshafts must be removed as described in Chapter 2.

13 With the camshafts removed, lift out the follower of the first valve to be adjusted out of the cylinder head and remove the shim **(see illustration)**. Note that the shim is likely to stick to the inside of the follower so take great care not to lose it as the follower is removed. **Note:** *Only remove one follower and shim at a time to avoid getting the followers and shims mixed up.*

14 The shim size should be stamped on its face, however, it is recommended that the

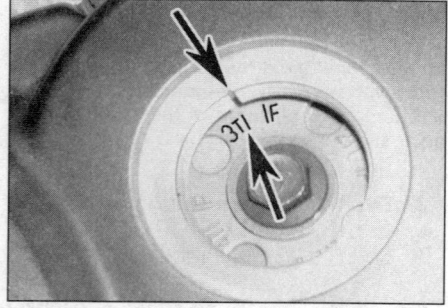

24.9 Rotate the crankshaft 180° to align number 3 cylinder 'T' mark with the index mark (arrows) then check the clearances of number 3 cylinder valves

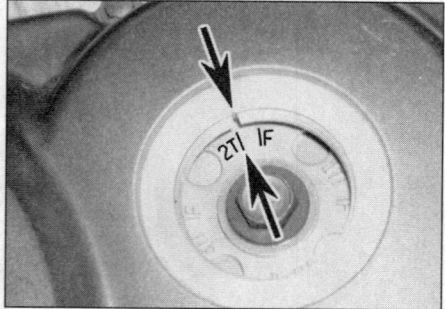

24.10 Rotate the crankshaft 270° to align number 2 cylinder 'T' mark with the index mark (arrows) then check the clearances of number 2 cylinder valves

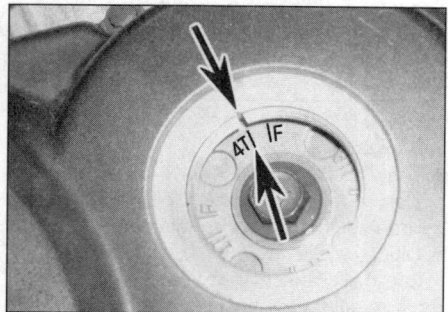

24.11 Rotate the crankshaft 180° to align number 4 cylinder 'T' mark with the index mark (arrows) then check the clearances of number 4 cylinder valves

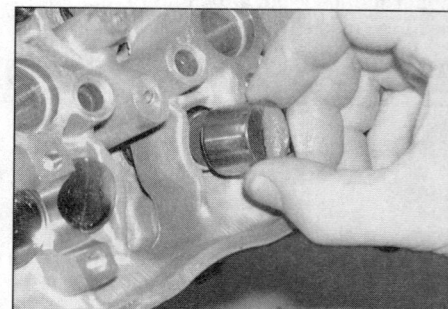

24.13 Lift out the relevant follower and recover the shim

Every 16,000 miles 1•17

shim is measured to check that it has not worn **(see illustrations)**. The size marking is in the form of a three figure number, e.g. 180 indicating that the shim is 1.800 mm thick. Shims are available in 0.025 mm increments from 1.200 to 2.900 mm. Where the number does not equal a shim thickness, it should be rounded up or down, e.g. 182 indicates that the shim is 1.825 mm thick and 188 indicates that it is 1.875 mm thick. The new shim thickness required can then be calculated as follows. **Note:** *Always aim to get the clearance as close to the specified clearance as is possible.*

15 If the valve clearance was less than specified, subtract the measured clearance from the specified clearance then deduct the result from the original shim thickness. For example:
Sample calculation –
intake valve clearance too small
 Clearance measured (A) – 0.08 mm
 Specified clearance (B) – 0.16 mm
 Difference (B – A) – 0.08 mm
 Shim thickness fitted – 2.475 mm
 Exact shim thickness required –
  2.475 – 0.08 = 2.395 mm
 Install a 2.400 mm shim (nearest size available to that required)

16 If the valve clearance was greater than specified, subtract the specified clearance from the measured clearance, and add the result to the thickness of the original shim. For example:
Sample calculation –
exhaust valve clearance too large
 Clearance measured (A) – 0.41 mm
 Specified clearance (B) – 0.30 mm
 Difference (A – B) – 0.11 mm
 Shim thickness fitted – 1.975 mm

24.14a Shim thickness is indicated by three numbers stamped on its surface; 205 indicates that the shim is 2.050 mm thick

 Ideal shim thickness required –
  1.975 + 0.11 = 2.085 mm
 Install a 2.075 mm shim (nearest size available to that required)

17 Repeat the procedure for all the other valves which require adjustment then obtain the correct thickness shims from your Honda dealer.

> **HAYNES HINT**: *If more than one valve clearance requires adjustment, before ordering all new shims, check that the shim sizes required are not fitted to any of the other valves being adjusted. Shims can be swapped between valves and this will cut down on the number of new shims required. Retain all old shims in a safe place as they may come in handy when the valve clearances are next adjusted.*

18 Install the shim in position on top of the relevant valve, making sure it is correctly

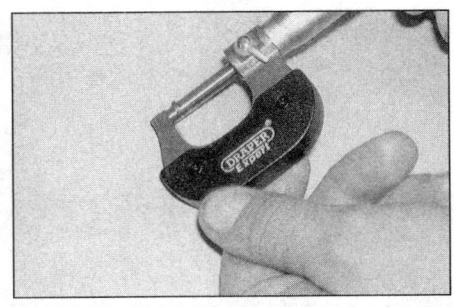

24.14b Shim thickness can also be determined by measuring it with a micrometer

seated in the valve spring retainer. Lubricate the follower with clean engine oil and insert it squarely into its relevant bore.
19 Ensure all shims and followers are correctly installed then fit the camshafts as described in Chapter 2.
20 Rotate the crankshaft a few times, to settle all disturbed components, and recheck all valve clearances as described above. If necessary, repeat the adjustment procedure.
21 When the check is complete, fit a new O-ring to the inspection plug. Lubricate the plug threads with a smear of multi-purpose grease then fit it to the crankcase cover and tighten to the specified torque.
22 Install the cylinder head covers as described in Chapter 2.

25 Spark plugs – renewal (UK models)

1 Renew the spark plugs as described in Section 5.

Every 24,000 miles (36,000 km)

26 Cooling system – draining, flushing and refilling

⚠ **Warning:** *Allow the engine to cool completely before performing this maintenance operation. Also,* don't allow antifreeze to come into contact with your skin or the painted surfaces of the motorcycle. Rinse off spills immediately with plenty of water. Antifreeze is highly toxic if ingested. Never leave antifreeze lying around in an open container or in puddles on the floor; children and pets are attracted by its sweet smell and may drink it. Check with local authorities about disposing of antifreeze. Many communities have collection centres which will see that antifreeze is disposed of safely. Antifreeze is also combustible, so don't store it near open flames.

Draining

1 Remove the lower fairing panels (see Chapter 8).
2 Remove the pressure cap from the filler neck by turning it anti-clockwise until it reaches a stop. If you hear a hissing sound (indicating there is still pressure in the system), wait until it stops. Now press down

on the cap and continue turning the cap until it can be removed.
3 Position a suitable container beneath the water pump on the left-hand side of the engine **(see illustration)**. Remove the coolant drain plug and its sealing washer and allow the coolant to completely drain from the system **(see illustration)**.

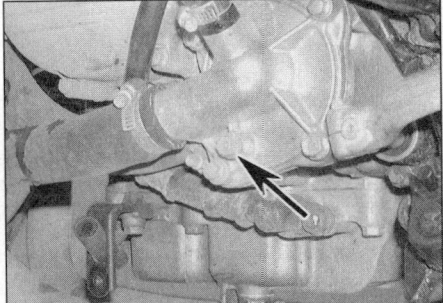

26.3a The main coolant drain plug (arrow) is on the water pump

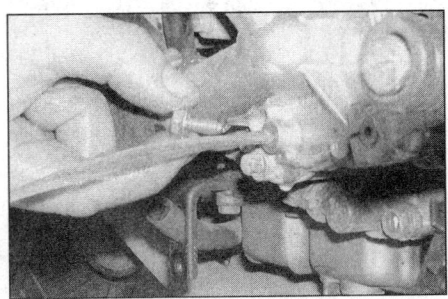

26.3b Slacken and remove the drain plug and sealing washer and allow the coolant to drain into a container

1•18 Every 24,000 miles

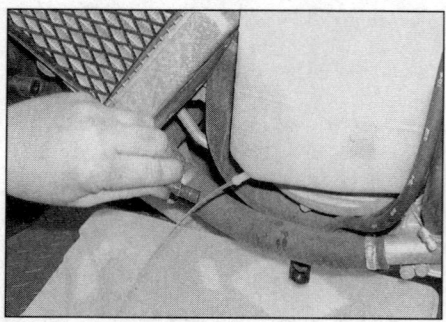

26.5 To drain the system completely, disconnect the hose from the coolant reservoir and allow its contents to drain into the container

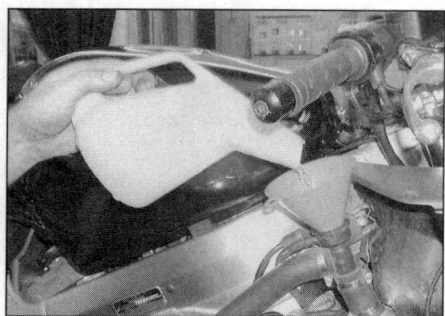

26.14 Fill the system slowly with the specified coolant mixture to prevent air being trapped in the system

4 Once the flow of coolant stops, reposition the container beneath the cylinder block coolant drain plug located beneath the front left (number 2) cylinder exhaust port. Remove the drain plug and its sealing washer and allow the coolant to completely drain from the cylinder block.

5 To completely drain the system, disconnect the hose from the base of the coolant reservoir and allow its contents to drain into the container **(see illustration)**.

Flushing

6 Flush the system with clean tap water by inserting a garden hose in the filler neck. Allow the water to run through the system until it is clear and flows cleanly out of the drain holes. If the radiators are extremely corroded, remove them (see Chapter 3) and have them cleaned at a radiator shop. Also rinse out the coolant reservoir.

7 Clean the drain holes then install the drain plugs using the old sealing washers.

8 Fill the cooling system with clean water mixed with a flushing compound. Make sure the flushing compound is compatible with aluminium components, and follow the manufacturer's instructions carefully.

9 Start the engine and allow it to reach normal operating temperature. Let it run for about ten minutes.

10 Stop the engine. Let it cool for a while, then cover the pressure cap with a heavy rag and turn it anti-clockwise to the first stop, releasing any pressure that may be present in the system. Once the hissing stops, push down on the cap and remove it completely.

11 Drain the system once again.

12 Fill the system with clean water and repeat the procedure in Paragraphs 9 to 11.

Refilling

13 Fit a new sealing washer to each drain plug then fit the drain plugs to the water pump and cylinder block and tighten securely. Securely reconnect the hose to the base of the coolant reservoir.

14 Fill the system with the proper coolant mixture (see this Chapter's Specifications) **(see illustration)**. **Note:** *Pour the coolant in slowly to minimise the amount of air entering the system.*

15 When the system is full (all the way up to the top of the filler neck), fit the pressure cap and top up the coolant reservoir to the UPPER level mark (see *Daily (pre-ride) checks*).

16 Start the engine and allow it to idle for 2 to 3 minutes. Flick the throttle twistgrip open 3 or 4 times, so that the engine speed rises to approximately 4000 – 5000 rpm, then stop the engine. This will bleed any trapped air from the cooling system.

17 Let the engine cool then remove the pressure cap as described in Paragraph 10. Check that the coolant level is still up to the radiator filler neck. If it's low, add the specified mixture until it reaches the top of the filler neck. Refit the cap.

18 Check the coolant level in the reservoir and top up if necessary.

19 Check the system for leaks.

20 Do not dispose of the old coolant by pouring it down the drain. Instead pour it into a heavy plastic container, cap it tightly and take it into an authorised disposal site or service station – see **Warning** at the beginning of this Section.

Non-scheduled maintenance

27 Wheel bearings – check

1 Wheel bearings will wear over a period of time and result in handling problems.

2 Place the motorcycle on its centrestand. Check for any play in the bearings by pushing and pulling the wheel against the hub. Also rotate the wheel and check that it rotates smoothly.

3 If any play is detected in the hub, or if the wheel does not rotate smoothly (and this is not due to brake or transmission drag), the wheel bearings must be removed and inspected for wear or damage (see Chapter 7).

28 Steering head bearings – re-greasing

1 Over a period of time the grease will harden or may be washed out of the bearings by incorrect use of jet washes.

2 Disassemble the steering head for re-greasing of the bearings. Refer to Chapter 6 for details.

29 Swingarm and suspension linkage bearings – re-greasing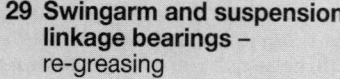

1 Over a period of time the grease will harden or dirt will penetrate the bearings due to failed dust seals.

2 Since no grease nipples are fitted, it will be necessary to remove the swingarm and suspension linkage as described in Chapter 6 to enable the bearings to be cleaned and regreased.

30 Front forks – oil change

1 Fork oil degrades over a period of time and loses its damping qualities. Since these models are not equipped with drain screws on the fork sliders the forks will have to be removed to enable the oil to be changed (see Chapter 6).

2 With both forks removed, drain all the oil from the first fork as described in Paragraphs 4 to 9 of Section 7 of Chapter 6 (it is not necessary to dismantle the fork completely) then fill it with fresh oil as described in Paragraphs 29 to 35.

3 Repeat the operation on the other fork then fit both forks to the motorcycle (see Chapter 6).

31 Cylinder compression – check

1 Among other things, poor engine performance may be caused by leaking valves, incorrect valve clearances, a leaking head gasket, or worn pistons, rings and/or cylinder walls. A cylinder compression check will help pinpoint these conditions and can also indicate the presence of excessive carbon deposits in the cylinder heads.

Non-scheduled maintenance

2 The only tools required are a compression gauge and a spark plug wrench. A compression gauge with a 10 mm threaded adapter for the spark plug hole is preferable to the type which requires hand pressure to maintain a tight seal. Depending on the outcome of the initial test, a squirt-type oil can may also be needed.

3 Ensure the valve clearances are correctly set (see Section 24) then check the compression pressures using the information given in *Fault Finding Equipment* in the Reference section.

32 Engine oil pressure – check

1 The oil pressure warning light should come on when the ignition switch is turned ON and extinguish soon after the engine is started – this serves as a check that the warning light bulb is sound. If the oil pressure light comes on whilst the engine is running, low oil pressure is indicated – stop the engine immediately and carry out an oil level check *(see Daily (pre-ride) checks)*.

2 An oil pressure check must be carried out if the warning light comes on when the engine is running yet the oil level is good (Paragraph 1). It can also provide useful information about the condition of the engine's lubrication system.

3 To check the oil pressure, a suitable gauge and adaptor will be needed. Since the oil pressure switch is located between the cylinders and cannot be accessed without removing the throttle body assembly, the only option is to obtain the special Honda service tools. Honda dealers fit a spacer (Pt. No. 07RMK-MW40100) in-between the oil filter and crankcase and attach the pressure gauge (Pt. No. 07506-3000000) to the spacer with a threaded adaptor (Pt. No. 07510-4220100).

4 Remove the oil filter (see Section 7). Lubricate the spacer O-ring with a smear of engine oil then screw it firmly onto the crankcase. Screw the adaptor and pressure gauge securely into the spacer. Ensure the rubber seal is in position then screw the oil filter securely onto the spacer.

5 Check the engine oil level *(see Daily (pre-ride) checks)* and top-up to the upper level mark.

6 Start the engine and warm it up to the normal operating temperature, ensuring there is no oil leakage from the spacer arrangement.

7 Once the engine is at operating temperature, increase the engine speed briefly to 6000 rpm whilst watching the gauge reading. The oil pressure should be similar to that given in the Specifications at the start of this Chapter.

8 If the pressure is significantly lower than the standard, either the pressure regulator is stuck open, the oil pump is faulty, the oil strainer or filter is blocked, or there is other engine damage. Begin diagnosis by checking the oil filter, strainer and regulator, then the oil pump (see Chapter 2). If those items check out okay, chances are the bearing oil clearances are excessive and the engine needs to be overhauled.

9 If the pressure is too high, either an oil passage is clogged, the regulator is stuck closed or the wrong grade of oil is being used.

10 Stop the engine, allow it to cool then remove the oil filter and spacer from the crankcase.

11 Ensure the oil filter seal is in position, lubricate it with a smear of engine oil, then refit the filter to the crankcase, tightening it to the specified torque. If a filter wrench is not being used, tighten the filter as tight as possible by hand.

12 Check the engine oil level *(see Daily (pre-ride) checks)* and top-up to the upper level mark.

13 On completion, start the engine and check for leaks.

33 Brake master cylinder and caliper seals – renewal

1 Brake seals will deteriorate over a period of time and lose their effectiveness, leading to sticking operation or fluid loss, or allowing the ingress of air and dirt. Remove the master cylinders and calipers and renew all seals as described in Chapter 7.

34 Brake hoses – renewal

1 The hoses will in time deteriorate with age and should be replaced with new ones. Refer to Chapter 7 for details noting that new banjo union sealing washers must be used.

Notes

Chapter 2
Engine, clutch and transmission

Contents

Alternator – removal and installation .... see Chapter 9
Camshaft drive gears – removal, inspection and installation 12
Camshafts and followers – removal, inspection and installation 8
Clutch – removal, inspection and installation 15
Clutch master cylinder – removal, overhaul and installation 16
Clutch slave cylinder – removal, overhaul and installation 17
Clutch system – bleeding 18
Crankcase – inspection and servicing 25
Crankcase – separation and reassembly 24
Crankshaft and main bearings – removal, inspection and
  installation 30
Cylinder compression check .... see Chapter 1
Cylinder head – removal and installation 9
Cylinder head and valves – disassembly, inspection and
  reassembly 11
Cylinder head covers – removal and installation 7
Engine – removal and installation 5
Engine disassembly and reassembly – general information 6
Gearchange mechanism – removal, inspection and installation 23
General information 1
Initial start-up after overhaul 33
Main and connecting rod bearings – general note 27
Major engine repair – general note 4
Oil and filter change .... see Chapter 1
Oil cooler and pipes – removal and installation 22
Oil pan – removal and installation 19
Oil pressure check .... see Chapter 1
Oil pressure relief valve – removal, inspection and installation 21
Oil pressure switch – check and replacement ...... see Chapter 9
Oil pump – removal, inspection and installation 20
Operations possible with the engine in the frame 2
Operations requiring engine removal 3
Piston rings – installation 29
Piston/connecting rod assemblies – removal, inspection and
  installation 28
Primary drive gear – removal, inspection and installation 14
Recommended running-in procedure 34
Selector drum and forks – removal, inspection and installation 26
Spark plug replacement .... see Chapter 1
Starter motor – removal and installation .... see Chapter 9
Starter motor clutch and idle/reduction gears – removal, inspection
  and installation 13
Transmission shafts – disassembly, inspection and reassembly ... 32
Transmission shafts – removal and installation 31
Valve clearances – checking and adjustment ...... see Chapter 1
Valves/valve seats/valves guides – servicing 10
Water pump – check, removal and installation .... see Chapter 3

Degrees of difficulty

| Easy, suitable for novice with little experience | Fairly easy, suitable for beginner with some experience | Fairly difficult, suitable for competent DIY mechanic | Difficult, suitable for experienced DIY mechanic | Very difficult, suitable for expert DIY or professional  |

Specifications

General
Capacity	781 cc
Bore	72 mm
Stroke	48 mm
Compression ratio	11.6 to 1
Cylinder identification	
Number 1	Left rear cylinder
Number 2	Left front cylinder
Number 3	Right rear cylinder
Number 4	Right front cylinder

FRONT
2 4
1 3

Cylinder head
Maximum warpage	0.10 mm
Cylinder head follower bore ID	
Standard	26.010 to 26.026 mm
Service limit	26.04 mm

Camshafts
Intake cam lobe height
  California models
    Standard .... 35.34 to 35.58 mm
    Service limit .... 35.31 mm
  All other models
    Standard .... 36.24 to 36.48 mm
    Service limit .... 36.21 mm
Exhaust cam lobe height
  California models
    Standard .... 35.18 to 35.42 mm
    Service limit .... 35.15 mm
  All other models
    Standard .... 36.08 to 36.32 mm
    Service limit .... 36.05 mm
Camshaft journal OD .... 24.959 to 24.980 mm
Camshaft bearing oil clearance
  Standard .... 0.020 to 0.062 mm
  Service limit .... 0.10 mm
Camshaft runout .... Less than 0.05 mm
Camshaft follower OD
  Standard .... 25.978 to 25.993 mm
  Service limit .... 25.97 mm

Valves, guides and springs
Intake valve stem OD
  Standard .... 4.475 to 4.490 mm
  Service limit .... 4.465 mm
Exhaust valve stem OD
  Standard .... 4.465 to 4.480 mm
  Service limit .... 4.455 mm
Valve guide ID – intake and exhaust
  Standard .... 4.500 to 4.512 mm
  Service limit .... 4.540 mm
Valve stem-to-guide clearance
  Intake .... 0.010 to 0.037 mm
  Exhaust .... 0.020 to 0.047 mm
Valve seat width .... 0.9 to 1.1 mm
Inner valve spring free length
  Standard .... 39.5 mm
  Service limit .... 37.6 mm
Outer valve spring free length
  Standard .... 42.5 mm
  Service limit .... 40.5 mm
Valve guide projection above the cylinder head .... 17.0 mm

Clutch
Friction plate thickness
  Standard .... 2.92 to 3.08 mm
  Service limit .... 2.50 mm
Plain plate maximum warpage .... 0.3 mm
Clutch spring free length
  VFR800FI-W and VFR800FI-X
    Standard .... 45.6 mm
    Service limit .... 42.6 mm
  VFR800FI-Y and VFR800FI-1
    Standard .... 46.7 mm
    Service limit .... 45.8 mm
Clutch drum centre bush ID
  Standard .... 24.995 to 25.012 mm
  Service limit .... 25.08 mm
Master cylinder bore ID
  VFR800FI-W and VFR800FI-X
    Standard .... 14.000 to 14.043 mm
    Service limit .... 14.060 mm
  VFR800FI-Y and VFR800FI-1
    Standard .... 12.043 to 12.700 mm
    Service limit .... 12.760 mm

Master cylinder piston OD
  VFR800FI-W and VFR800FI-X
    Standard .... 13.957 to 13.984 mm
    Service limit .... 13.940 mm
  VFR800FI-Y and VFR800FI-1
    Standard .... 12.657 to 12.684 mm
    Service limit .... 12.650 mm

Lubrication system

Oil pressure (at 6000 rpm) .... 71 psi (5.0 bar) at 80°C (176°F)
Oil pump inner rotor tip-to-outer rotor clearance
  Standard .... up to 0.15 mm
  Service limit .... 0.20 mm
Oil pump outer rotor-to-body clearance
  Standard .... 0.15 to 0.22 mm
  Service limit .... 0.35 mm
Oil pump rotor endfloat
  Standard .... 0.02 to 0.07 mm
  Service limit .... 0.10 mm

Starter motor clutch

Driven gear OD
  Standard .... 45.657 to 45.673 mm
  Service limit .... 45.64 mm

Cylinder block

Cylinder bore ID
  Standard .... 72.000 to 72.015 mm
  Service limit .... 72.10 mm
Maximum ovality (out-of-round) .... 0.10 mm
Maximum taper .... 0.10 mm
Cylinder-to-piston clearance .... 0.015 to 0.050 mm
Maximum gasket face warpage .... 0.10 mm

Pistons

Piston OD (measured 18 mm up from base of skirt)
  Standard .... 71.975 to 72.003 mm
  Service limit .... 71.90 mm
Piston pin bore ID
  Standard .... 17.002 to 17.008 mm
  Service limit .... 17.02 mm
Piston pin OD
  Standard .... 16.994 to 17.000 mm
  Service limit .... 16.98 mm
Piston-to-piston pin clearance .... 0.002 to 0.014 mm
  Service limit .... 0.040 mm

Piston rings

Top ring-to-groove clearance
  Standard .... 0.030 to 0.065 mm
  Service limit .... 0.11 mm
Second (middle) ring-to-groove clearance
  Standard .... 0.015 to 0.050 mm
  Service limit .... 0.10 mm
Top ring end gap
  Standard .... 0.20 to 0.30 mm
  Service limit .... 0.5 mm
Second ring end gap
  Standard .... 0.30 to 0.45 mm
  Service limit .... 0.6 mm
Oil control ring side rail end gap
  Standard .... 0.2 to 0.7 mm
  Service limit .... 0.9 mm

Selector drum and forks

Selector fork end thickness
  Standard .... 6.43 to 6.50 mm
  Service limit .... 6.40 mm
Selector fork bore ID
  Standard .... 14.000 to 14.021 mm
  Service limit .... 14.03 mm
Selector fork shaft OD
  Standard .... 13.973 to 13.984 mm
  Service limit .... 13.965 mm

Connecting rods and bearings

Connecting rod bearing side clearance .... up to 0.4 mm
Connecting rod piston pin bore ID
   Standard .... 17.016 to 17.034 mm
   Service limit .... 17.044 mm
Connecting rod crankpin bore ID
   Size group 1 .... 39.000 to 39.006 mm
   Size group 2 .... 39.006 to 39.012 mm
   Size group 3 .... 39.012 to 39.018 mm
Crankshaft crankpin OD
   Size group A .... 35.994 to 36.000 mm
   Size group B .... 35.988 to 35.994 mm
   Size group C .... 35.982 to 35.988 mm
Connecting rod bearing oil clearance
   Standard .... 0.030 to 0.052 mm
   Service limit .... 0.08 mm

Crankshaft and main bearings

Maximum crankshaft runout .... 0.03 mm
Crankcase main bearing bore ID
   Size group A .... 37.000 to 37.006 mm
   Size group B .... 37.006 to 37.012 mm
   Size group C .... 37.012 to 37.018 mm
Crankshaft journal OD
   Size group 1 .... 34.007 to 34.013 mm
   Size group 2 .... 34.001 to 34.007 mm
   Size group 3 .... 33.995 to 34.001 mm
Main bearing oil clearance
   Standard .... 0.023 to 0.041mm
   Service limit .... 0.06 mm

Transmission shafts

Ratios
   1st .... 2.846 to 1 (37/13T)
   2nd .... 2.062 to 1 (33/16T)
   3rd .... 1.631 to 1 (31/19T)
   4th .... 1.333 to 1 (28/21T)
   5th .... 1.153 to 1 (30/26T)
   6th .... 1.035 to 1 (29/28T)
Gear ID
   Input shaft 5th and 6th gears
      Standard .... 28.000 to 28.021 mm
      Service limit .... 28.04 mm
   Output shaft 1st gear
      Standard .... 26.007 to 26.028 mm
      Service limit .... 26.04 mm
   Output shaft 2nd gear
      Standard .... 31.000 to 31.016 mm
      Service limit .... 31.04 mm
   Output shaft 3rd and 4th gears
      Standard .... 31.000 to 31.025 mm
      Service limit .... 31.04 mm
Gear bushing OD
   Input shaft 5th and 6th gears
      Standard .... 27.959 to 27.980 mm
      Service limit .... 27.94 mm
   Output shaft 2nd gear
      Standard .... 30.970 to 30.995 mm
      Service limit .... 30.95 mm
   Output shaft 3rd and 4th gears
      Standard .... 30.950 to 30.975 mm
      Service limit .... 30.93 mm
Gear bushing ID
   Input shaft 5th gear
      Standard .... 24.985 to 25.006 mm
      Service limit .... 25.03 mm
   Output shaft 2nd gear
      Standard .... 28.000 to 28.021 mm
      Service limit .... 28.04 mm

Engine, clutch and transmission 2•5

Transmission shafts (continued)

Gear-to-bushing clearance
   Input shaft 5th and 6th gear .... 0.020 to 0.062 mm
   Output shaft 2nd gear .... 0.005 to 0.046 mm
   Output shaft 3rd and 4th gears .... 0.025 to 0.075 mm
Input shaft OD at 5th gear bushing point
   Standard .... 24.959 to 24.980 mm
   Service limit .... 24.950 mm
Output shaft OD at 2nd gear bushing point
   Standard .... 27.967 to 27.980 mm
   Service limit .... 27.96 mm
Shaft-to-bushing clearance
   Input shaft 5th gear .... 0.005 to 0.047 mm
   Output shaft 2nd gear .... 0.020 to 0.054 mm

Torque settings

Camshaft drive gear bolts
   Upper (6 mm) bolts .... 12 Nm
   Side (8 mm) bolt .... 22 Nm
Camshaft holder and bearing cap bolts .... 12 Nm
Clutch assembly
   Centre nut .... 127 Nm
   Spring bolts .... 12 Nm
Clutch hydraulic hose banjo fitting bolts .... 34 Nm
Clutch lever pivot bolt nut .... 6 Nm
Clutch master cylinder mounting bolts .... 12 Nm
Clutch slave cylinder bleed valve .... 9 Nm
Connecting rod bearing cap nuts .... 33 Nm
Crankcase bolts
   6 mm bolts .... 12 Nm
   7 mm bolts .... 18 Nm
   9 mm (main bearing) bolts
      Stage 1 .... 20 Nm
      Stage 2 .... Angle-tighten a further 90°
   10 mm bolts .... 39 Nm
Cylinder head bolts
   9 mm bolts .... 44 Nm
   6 mm bolts .... 12 Nm
Cylinder head cover bolts .... 10 Nm
Engine mounting bolts
   Front mounting bolt nut .... 54 Nm
   Centre and rear mounting bolts .... 44 Nm
Gearchange mechanism
   Cover retaining bolts .... 12 Nm
   Selector drum cam bolt .... 23 Nm
   Stopper arm pivot bolt .... 12 Nm
Oil pan bolts .... 12 Nm
Oil pan drain plug .... 29 Nm
Oil pump
   Mounting bolts .... 12 Nm
   Housing bolts .... 12 Nm
   Sprocket bolt .... 18 Nm
Right crankcase cover bolts .... 9 Nm
Right crankcase cover inspection plug .... 18 Nm
Timing rotor/starter clutch bolt .... 103 Nm

1 General information

The engine/transmission unit is of liquid-cooled 90° vee-four cylinder design. Its sixteen valves are operated by double overhead camshafts, gear-driven off the right end of the crankshaft; there are four camshafts (two for each pair of cylinders). The engine/transmission unit is constructed in aluminium alloy with the crankcase being divided horizontally. The crankcase incorporates a wet sump, pressure fed lubrication system, and houses a chain-driven dual rotor oil pump.

The alternator is situated on the left end of the crankshaft with the starter clutch mounted on the right end of the crankshaft. The water pump is mounted on the left side of the crankcase and is driven off the oil pump shaft.

The hydraulically-operated clutch is of the wet multi-plate type and is driven off the right end of the crankshaft by the primary drive gear. The transmission is of the six-speed constant mesh type. Final drive to the rear wheel is by chain and sprockets, the drive sprocket being mounted on the end of the output shaft.

2•6 Engine, clutch and transmission

2 Operations possible with the engine in the frame

The components and assemblies listed below can be removed without having to remove the engine/transmission assembly from the frame. If however, a number of areas require attention at the same time, removal of the engine is recommended.

Starter motor (see Chapter 9)
Alternator (see Chapter 9)
Cylinder head covers
Camshafts and followers
Front cylinder head
Front cylinder camshaft drive gears
Starter motor clutch and idle/reduction gears
Primary drive gear
Clutch assembly
Oil pan
Oil pump and pressure relief valve
Oil cooler and hoses
Gearchange mechanism components

3 Operations requiring engine removal

It is necessary to remove the engine/transmission assembly from the frame to enable the following components to be removed.

Rear cylinder head
Rear cylinder camshaft drive gears
Selector drum and forks
Piston/connecting rod assemblies and bearings
Crankshaft and bearings
Transmission shafts

4 Major engine repair – general note

1 It is not always easy to determine when or if an engine should be completely overhauled, as a number of factors must be considered.
2 High mileage is not necessarily an indication that an overhaul is needed, while low mileage, on the other hand, does not preclude the need for an overhaul. Frequency of servicing is probably the single most important consideration. An engine that has regular and frequent oil and filter changes, as well as other required maintenance, will most likely give many miles of reliable service. Conversely, a neglected engine, or one which has not been run in properly, may require an overhaul very early in its life.
3 Exhaust smoke and excessive oil consumption are both indications that piston rings and/or valve guides are in need of attention, although make sure that the fault is not due to oil leakage.
4 If the engine is making obvious knocking or rumbling noises, the connecting rod and/or main bearings are probably at fault.
5 Loss of power, rough running, excessive valve train noise and high fuel consumption rates may also point to the need for an overhaul, especially if they are all present at the same time. If a complete tune-up does not remedy the situation, major mechanical work is the only solution.
6 An engine overhaul generally involves restoring the internal parts to the specifications of a new engine. The piston rings and main and connecting rod bearings are usually renewed and the cylinder walls honed during a major overhaul. Generally the valve seats are re-ground, since they are usually in less than perfect condition at this point. The end result should be a like new engine that will give as many trouble-free miles as the original.
7 Before beginning the engine overhaul, read through the related procedures to familiarise yourself with the scope and requirements of the job. Overhauling an engine is not all that difficult, but it is time consuming. Plan on the motorcycle being tied up for a minimum of two weeks. Check on the availability of parts and make sure that any necessary special tools, equipment and supplies are obtained in advance.
8 Most work can be done with typical workshop hand tools, although a number of precision measuring tools are required for inspecting parts to determine if they must be renewed. Often a dealer will handle the inspection of parts and offer advice concerning reconditioning and renewal. As a general rule, time is the primary cost of an overhaul so it does not pay to install worn or substandard parts.
9 As a final note, to ensure maximum life and minimum trouble from a rebuilt engine, everything must be assembled with care in a spotlessly clean environment.

5 Engine – removal and installation

Caution: The engine is very heavy. Engine removal and installation should be carried out with the aid of at least one assistant; personal injury or damage could occur if the engine falls or is dropped. A mechanical or hydraulic floor jack should be used to support and lower or raise the engine if possible. Due to the design of the motorcycle (both the swingarm and centrestand are mounted onto the crankcase), note that it will also be necessary to securely support the rear of the bike prior to removing the engine (see Paragraph 10).

Removal

1 Support the bike on its centrestand. Work can be made easier by raising the machine to a suitable working height on an hydraulic ramp or a suitable platform. Make sure the motorcycle is secure and will not topple over (see *Tools and Workshop Tips* in the Reference section).
2 If the engine is dirty, particularly around its mountings, wash it thoroughly before starting any major dismantling work. This will make work much easier and rule out the possibility of caked on lumps of dirt falling into some vital component.
3 Remove the seat cowling and the fairing lower and inner panels (see Chapter 8).
4 Remove the battery (see Chapter 9).
5 Drain the engine oil (see Chapter 1). Also drain the cooling system (see Chapter 1).
6 Remove the fuel tank, throttle body assembly and complete exhaust system (see Chapter 4).
7 Remove the coolant reservoir (see Chapter 3).
8 Remove the front sprocket (see Section 16 of Chapter 6). Position the clutch slave cylinder clear of the engine unit then disconnect the speed sensor wiring connector and remove the sensor.
9 Remove the suspension linkage and swingarm (see Chapter 6).
10 It is now necessary to support the rear of the bike to enable the stand/suspension linkage bracket to be removed from the engine unit. The easiest way to do this is to support the bike on axle stands, with a plank of wood on the axle stand heads (cut the ends of the wood to shape so its sits nicely in the stands), positioned underneath the passenger footrest brackets. Have an assistant lift the rear of the bike whilst you position the stands and wood correctly underneath the footrest brackets then lower the bike carefully into position **(see illustration)**.
Caution: Ensure the bike is securely supported before proceeding.
11 Trace the cam pulse and ignition pulse generator wiring back to the connectors inside the protective cover on the right side of the frame. Disconnect both connectors so the sensor wiring is free to be removed with the engine **(see illustration)**.

5.10 Support the rear of the bike by positioning a pair of axle stands and a plank of wood underneath the passenger footrest brackets as shown

Engine, clutch and transmission 2•7

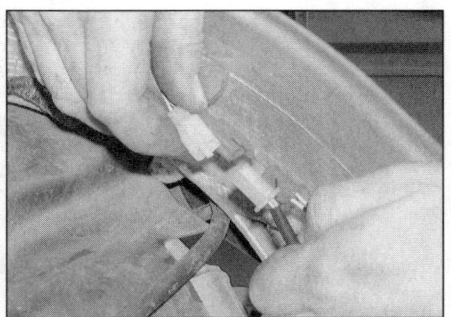

5.11 Disconnect the cam and ignition pulse generator wiring connectors . . .

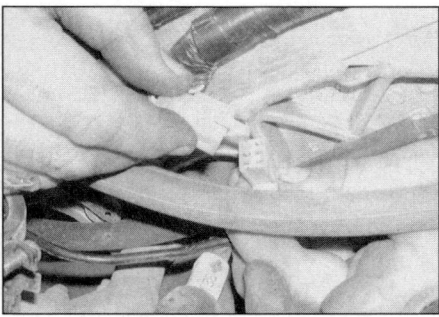

5.12 . . . and the 6-pin wiring connector located on the left side of the frame

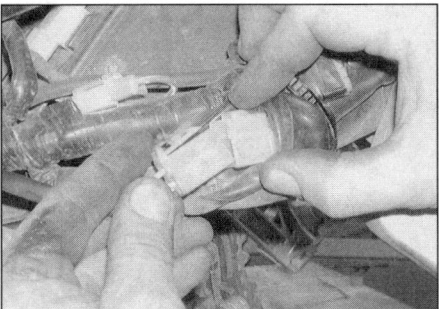

5.13 Disconnect the alternator wiring connector from the regulator/rectifier unit

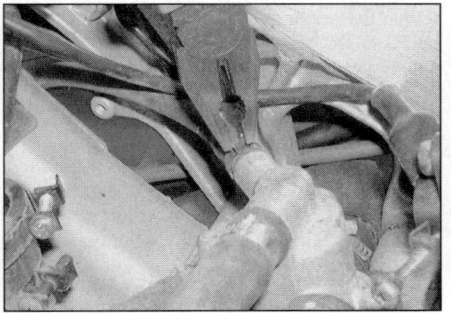

5.14a Disconnect the air bleed hose . . .

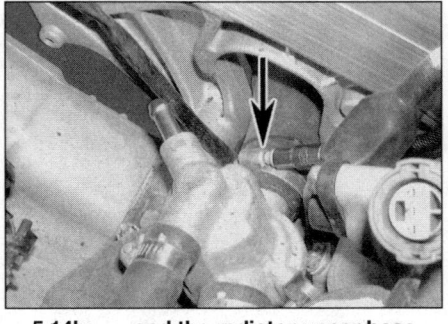

5.14b . . . and the radiator upper hose (arrow) from the thermostat housing . . .

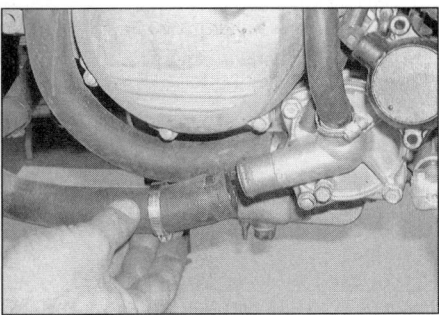

5.14c . . . and the radiator lower hose from the water pump

12 Disconnect the 6-pin (oil pressure, neutral switch and coolant temperature sensor) wiring connector located on the left side of the frame so the harness is free to be removed with the engine **(see illustration)**.

5.16 Peel back the rubber cover then unscrew the nut and detach the cable from the starter motor

13 Trace the alternator wiring back from the cover on the left side of the engine to the regulator/rectifier unit wiring cover. Disconnect the alternator wiring connector then free the wiring harness from any clips or ties, noting its routing through the frame, so that it is free to be removed with the engine **(see illustration)**.
14 Referring to Chapter 3, release the retaining clips and disconnect the air bleed hose and radiator upper hose from the thermostat housing and the radiator lower hose from the water pump cover **(see illustrations)**. Disconnect the cooling fan wiring connector (located above the left side radiator fan switch) then slacken and remove the left and right side radiator mounting bolts and collars. Unscrew the retaining bolts then remove both the left and right lower mounting brackets then manoeuvre both radiators out of position as an assembly.

15 Remove the oil cooler and pipes (see Section 22).
16 Lift the rubber cover and unscrew the nut securing the starter motor cable to the motor **(see illustration)**. Free the cable from the motor, noting its correct routing, and position it clear of the engine.
17 Referring to Chapter 4, release the retaining clips and disconnect the pulse secondary air (PAIR) hoses from the front and rear cylinder head covers **(see illustration)**. Disconnect the control valve wiring connector then remove the valve and hose assembly from the bike, noting the correct routing of each hose **(see illustrations)**.
18 Disconnect the spark plug caps and position them clear of the front and rear cylinder head covers. Free the rubber insulation cover from the rear cylinder head cover **(see illustration)**.
19 Note the correct fitted position of the

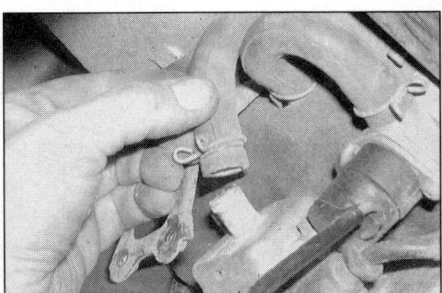

5.17a Disconnect the pulse secondary air (PAIR) hoses from the front (shown) and rear cylinder head covers . . .

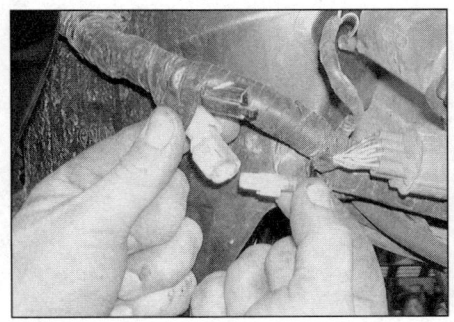

5.17b . . . then disconnect the control valve wiring connector . . .

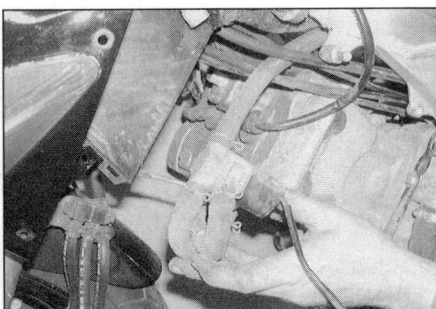

5.17c . . . and remove the valve and hose assembly from the bike

2•8 Engine, clutch and transmission

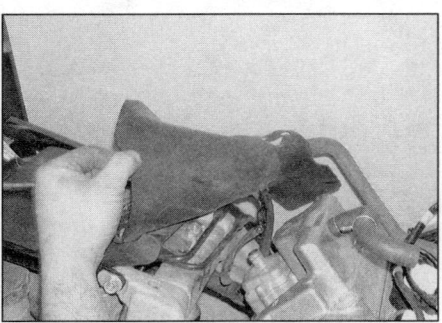

5.18 Free the rubber insulation cover from the rear cylinder head cover

5.19 Unscrew the clamp bolt and remove the gearshift lever

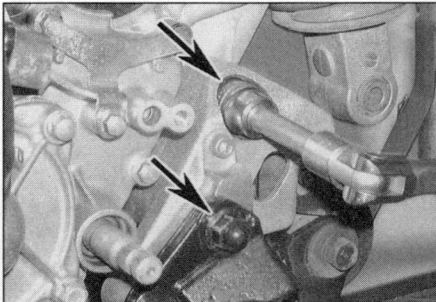

5.21a Unscrew the nuts (arrows) . . .

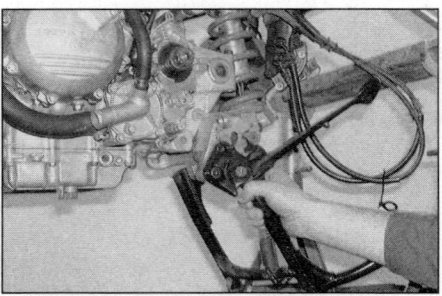

5.21b . . . then withdraw the bolts and remove the stand/suspension linkage bracket from the engine

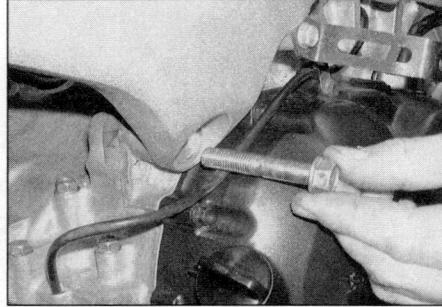

5.25 Slacken and remove the right rear engine mounting bolt . . .

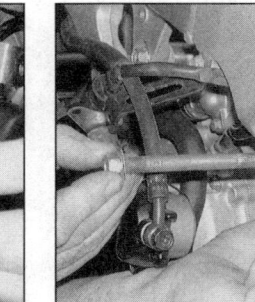

5.26 . . . then remove the left rear mounting bolt and spacer

gearchange lever on its shaft (the shaft punch mark should be aligned with the lever split) then unscrew the clamp bolt and remove the gearchange lever from the engine **(see illustration)**.

20 Remove the sidestand switch (see Chapter 9).

21 Unscrew the nuts from the stand/suspension linkage mounting bracket bolts then withdraw both bolts and remove the bracket assembly from the engine **(see illustrations)**.

22 Note the correct routing of the fuel tank breather and overflow hoses around the engine/frame then remove both hoses.

23 Unscrew the bolts and remove the fairing lower panel brackets from the oil pan to prevent these being damaged during removal.

24 At this point, position an hydraulic or mechanical jack under the engine with a block of wood between the jack head and sump.

Make sure the jack is centrally positioned so the engine will not topple when the last mounting bolt is removed. Take the weight of the engine on the jack.

25 Unscrew the engine right side rear mounting bolt **(see illustration)**.

26 Slacken and remove the engine left side rear mounting bolt and washer and recover the spacer which is fitted between the engine and frame **(see illustration)**.

27 Unscrew the engine left and right side centre mounting bolts and recover the spacer fitted between the left side of the engine and the frame **(see illustration)**.

28 Ensure the engine is securely supported then slacken and remove the nut and washer from the front mounting bolt **(see illustration)**. Withdraw the mounting bolt and recover the spacer which is fitted between the left side of the engine and frame.

29 The engine is now free to be lowered out

of the frame. Check that all relevant wiring, cables and hoses are disconnected and secured well clear, then carefully lower the engine and manoeuvre it out of the side of the frame (see **Caution** at the start of this Section) **(see illustration)**.

30 With the aid of an assistant lift the engine unit off the jack and move it carefully to the work surface.

Installation

31 With the aid of an assistant place the engine unit on top of the jack and block of wood and carefully raise the engine unit into position in the frame. Make sure no wires, cables or hoses become trapped between the engine and the frame. Prior to installing the mounting bolts, ensure the clutch slave cylinder hose is correctly routed between the front and centre mounting lugs on the left side of the frame; it is not possible to pass the

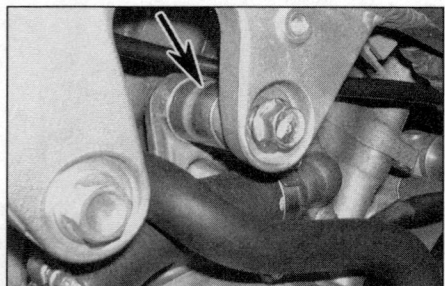

5.27 Slacken and remove the right and left centre mounting bolts and recover the spacer (arrow)

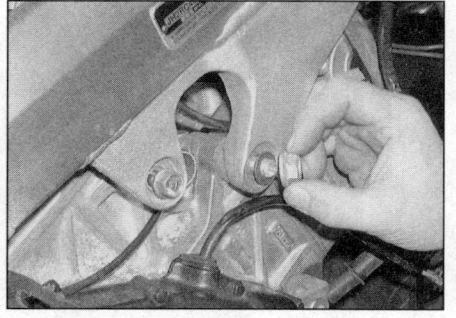

5.28 Unscrew the nut and washer and withdraw the front mounting bolt . . .

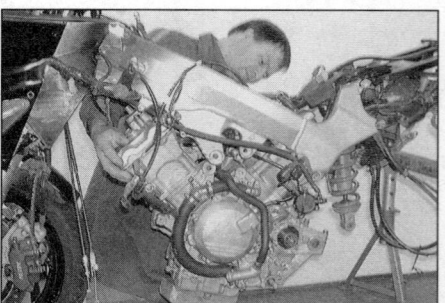

5.29 . . . then carefully lower the engine unit out from the frame

Engine, clutch and transmission 2•9

slave cylinder through once the bolts are inserted.
32 Insert the spacer in-between the engine front mounting and left side of the frame then insert the mounting bolt from the left side of the bike **(see illustration)**. Fit the washer to the bolt and screw on the nut, tightening it lightly only.
33 Insert the spacer in-between the left side of the engine centre mounting and the frame then fit the left centre mounting bolt, tightening it lightly only **(see illustration 5.27)**.
34 Insert the spacer in-between the left side of the engine rear mounting and the frame then fit the mounting bolt and washer, tightening it lightly only **(see illustration 5.26)**.
35 Fit and lightly tighten the engine right centre and rear mounting bolts.
36 With all mounting bolts in position, tighten each one to its specified torque setting in the following sequence **(see illustrations)**.
  a) *Front mounting bolt nut.*
  b) *Right centre mounting bolt.*
  c) *Right rear mounting bolt.*
  d) *Left centre mounting bolt.*
  e) *Left rear mounting bolt.*
37 The remainder of the installation procedure is the reverse of removal, noting the following points.
  a) *Make sure all wires, cables and hoses are correctly routed and connected, and secured by the relevant clips or ties.*
  b) *Tighten all nuts and bolts to the specified torque settings (where given) and renew all gaskets and O-rings disturbed on removal.*
  c) *Adjust the throttle cable freeplay (see Chapter 1).*
  d) *Adjust the drive chain (see Chapter 1).*
  e) *Refill the engine with oil and coolant (see Chapter 1).*
  f) *Prior to installing the inner and lower fairing panels start the engine and check that there are no signs of coolant/oil leakage.*

6 Engine disassembly and reassembly – general information

Disassembly

1 Before disassembling the engine, the external surfaces of the unit should be thoroughly cleaned and degreased. This will prevent contamination of the engine internals, and will also make working a lot easier and cleaner. A high flash-point solvent, such as paraffin (kerosene) can be used, or better still, a proprietary engine degreaser. Use old paintbrushes and toothbrushes to work the solvent into the various recesses of the engine casings. Take care to exclude solvent or water from the electrical components and intake and exhaust ports.

 Warning: The use of petrol (gasoline) as a cleaning agent should be avoided because of the risk of fire.

5.32 Position the spacer (arrow) between the left side of the engine and frame then insert the front mounting bolt

2 When clean and dry, arrange the unit on the workbench, leaving suitable clear area for working. Gather a selection of small containers and plastic bags so that parts can be grouped together in an easily identifiable manner. Some paper and a pen should be on hand to permit notes to be made and labels attached where necessary. A supply of clean shop towels is also required.
3 Before commencing work, read through the appropriate section so that some idea of the necessary procedure can be gained. When removing various engine components it should be noted that great force is seldom required, unless specified. In many cases, a component's reluctance to be removed is indicative of an incorrect approach or removal method. If in any doubt, re-check with the text.
4 When disassembling the engine, keep 'mated' parts together (including gears, cylinders, pistons, valves, etc. that have been in contact with each other during engine operation). These 'mated' parts must be re-used or renewed as an assembly. **Note:** *Do not interchange front and rear cylinder head components.*
5 Engine/transmission disassembly should be done in the following general order with reference to the appropriate Sections.
  Remove the camshafts
  Remove the cylinder heads
  Remove the camshaft drive gears
  Remove the starter motor (see Chapter 9)
  Remove the starter motor clutch and idler/reduction gears

5.36a Tighten the front and right side mounting bolts to the specified torque in the order shown . . .

Remove the primary drive gear
Remove the clutch
Remove the alternator (see Chapter 9)
Remove the external gearchange mechanism
Remove the oil pan
Remove the oil pump and pressure relief valve
Separate the crankcase halves
Remove the connecting rod/piston assemblies
Remove the crankshaft
Remove the transmission shafts
Remove the selector drum and forks

Reassembly

6 Reassembly is accomplished by reversing the general disassembly sequence.

7 Cylinder head covers – removal and installation

Note: *Either cylinder head cover can be removed with the engine in the frame. If the engine has been removed, ignore the paragraphs which do not apply.*

Removal

Front cylinder head cover

1 Remove the fairing left and right lower panels and the inner panel (see Chapter 8).
2 Remove the air filter housing (see Chapter 4).
3 Slacken and remove the oil cooler mounting bolts and collars then pivot the cooler forwards, taking care not to place any strain on the cooler pipes. **Note:** *There is no need to disconnect the oil pipes.*
4 Referring to Chapter 3, disconnect the cooling fan wiring connector (located above the left radiator fan switch) from the main wiring harness. Undo the retaining bolt and remove the lower mounting bracket from both left and right radiators then undo the bolt and remove the collar from each radiator upper mounting. Free both radiator upper mountings from the frame pegs and lower the radiators slightly to improve access to the front cylinder head. Support the radiators to avoid placing any strain on the coolant hoses. **Note:** *There is no need to disconnect any coolant hose.*

5.36b . . . then tighten the left centre bolt and the left rear bolt to the specified torque

2•10 Engine, clutch and transmission

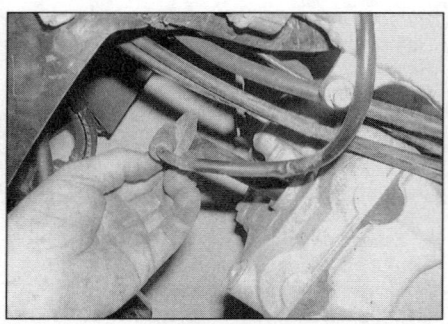

7.6 Disconnect the spark plug caps and position them clear of the cover

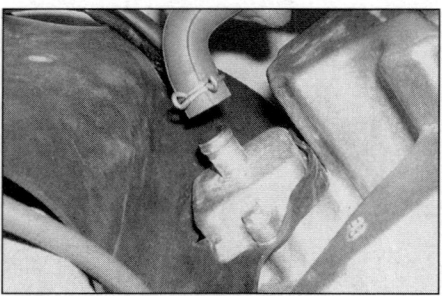

7.11 Disconnect the pulse secondary air (PAIR) hose from the rear cylinder head cover

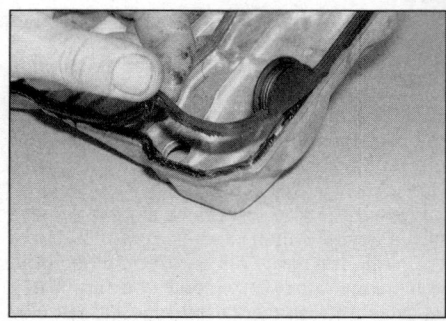

7.13a Apply sealant to the cover groove then fit the main seal . . .

5 Release the retaining clip and disconnect the pulse secondary air (PAIR) hose from cylinder head cover **(see illustration 5.17a)**.
6 Disconnect the spark plug caps from the plugs and position them clear of the cover **(see illustration)**.
7 Slacken and remove the four cylinder head cover bolts along with their sealing washers.
8 Lift the cylinder head cover away from the head and manoeuvre it out of position, complete with the cover seals (there is a large outer seal and a smaller seal around each spark plug hole). Remove the pulse secondary air (PAIR) passage collars from the top of the camshaft holder and discard the O-rings; new ones must be used on refitting.
9 Examine the cylinder head cover seals and the cover bolt sealing washers for signs of damage or deterioration and renew them as necessary. Refer to Chapter 4 for information on the pulse secondary air (PAIR) check valve assembly.

Rear cylinder head cover

10 Remove the fuel tank (see Chapter 4).
11 Release the retaining clips and disconnect the breather hose and pulse secondary air (PAIR) hose from the cylinder head cover **(see illustration)**.
12 Free the rubber insulating cover from the cylinder head cover then remove the cover as described in Paragraphs 6 to 9.

Installation

Front cylinder head cover

13 Remove all traces of sealant from the cylinder head cover groove and rubber seals.

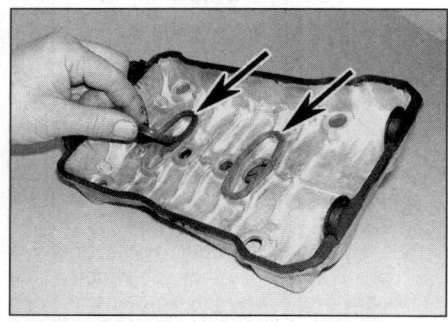

7.13b . . . and spark plug aperture seals (arrows) to the cover

Ensure the cover is clean and dry, then apply a small bead of sealant to the cover grooves (this will help hold the seals in position) and fit all the seals to the cover **(see illustrations)**.
14 Fit the pulse secondary air (PAIR) passage collars to the camshaft holder and fit a new O-ring to each collar **(see illustration)**.
15 Apply a bead of sealant to the semi-circular cutouts on each side of the cylinder head **(see illustration)**.
16 Carefully manoeuvre the cylinder head cover into position, ensuring that the seals and PAIR collars remain correctly seated, and locate it on the cylinder head.
17 Fit the sealing washers to the cylinder head cover making sure the 'UP' mark on each one is facing up **(see illustration)**.
18 Install the cover retaining bolts and tighten them to the specified torque setting, noting that the bolts fitted next to the triangular marks cast on the cover must be

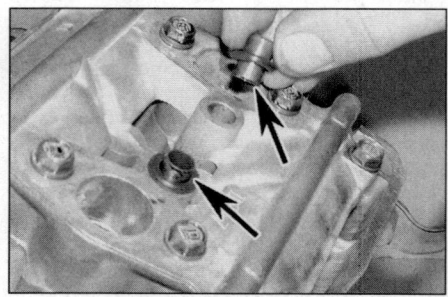

7.14 Fit the pulse secondary air (PAIR) passage collars (arrows) to the camshaft holder complete with new O-rings

tightened first **(see illustration)**.
19 Securely reconnect the spark plug caps then reconnect the hose to the PAIR valve.
20 Ensure all the mounting rubbers are in good condition then seat both radiator upper mountings on the frame pegs. Fit the collars and upper mounting bolts then install the lower mounting brackets. Securely tighten the mounting bolts then reconnect the cooling fan wiring connector.
21 Locate the oil cooler on its bracket then install the collars and mounting bolts, tightening them securely.
22 Refit the air filter housing (see Chapter 4) then install the fairing panels (see Chapter 8).

Rear cylinder head cover

23 Install the cover as described in Paragraphs 13 to 18.
24 Reconnect the spark plug caps, breather and PAIR hoses to the cover then install the fuel tank (see Chapter 4).

7.15 Apply a bead of sealant to the semi-circular cutouts (arrows) on each side of the head

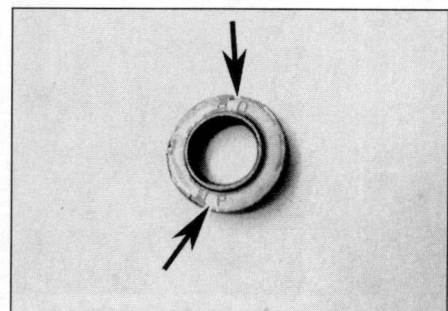

7.17 Ensure all the sealing washers are fitted with the UP marks (arrows) facing up

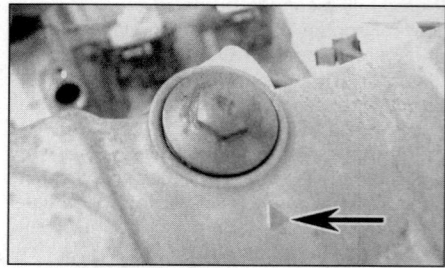

7.18 Tighten the cover bolts to the specified torque noting that the bolts fitted next to the triangular marks (arrow) must be tightened first

Engine, clutch and transmission 2•11

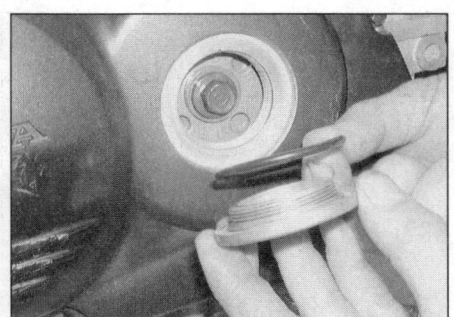

8.3 Unscrew the inspection plug and O-ring from the right crankcase cover

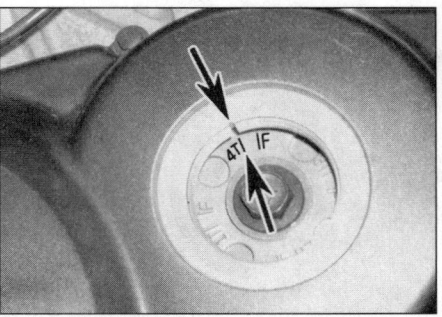

8.4a Align the timing rotor 'T' mark of number 4 cylinder with the cover index mark (arrows) . . .

8.4b . . . so the front cylinder camshaft gear index marks (arrows) are positioned as shown (number 4 piston at TDC on its compression stroke)

8 Camshafts and followers – removal, inspection and installation

Note: *This procedure can be carried out with the engine in the frame. When removing the camshafts and followers, take great care not to allow any component to fall down into the crankcase. If any component is dropped, it must be recovered before the engine is started.*

Removal

1 Remove all four spark plugs as described in Chapter 1
2 Remove the front cylinder head and rear cylinder head covers as described in Section 7 and proceed as described under the relevant sub-heading.

Front cylinder camshafts only

3 Unscrew the inspection plug and O-ring from the right crankcase cover **(see illustration)**.
4 Using a socket on the crankshaft bolt, rotate the crankshaft clockwise until the 'T' mark of number 4 cylinder is aligned with the index mark (cutout or line) on the top of the crankcase cover aperture **(see illustration)**. Check the position of the index marks on the front cylinder camshaft gears; the marks should be facing away from each other and

both be aligned with the cylinder head upper surface (piston number 4 is now at TDC on its compression stroke) **(see illustration)**. If the camshaft index marks are not correctly positioned, rotate the crankshaft through another 360° (one complete turn) to position them as described. **Note:** *Always turn the engine in the normal direction of rotation (clockwise), viewed from the right end of the engine.*
5 Note the identification markings on the camshaft holder and bearing caps. The bearing caps both have 'I' and 'E' marks cast into their upper surface; on the intake camshaft bearing cap the 'E' is crossed out leaving the 'I', and on the exhaust camshaft cap the 'I' is crossed out leaving the 'E' **(see illustration)**. The camshaft holder has an 'IN' mark cast onto its upper surface on it to show which way around it is fitted. Ensure the bearing caps and holder are clearly marked before proceeding.
6 Evenly and progressively slacken the intake and exhaust camshaft bearing cap bolts then lift both bearing caps away from the cylinder head **(see illustration)**. Recover the cap locating dowels.
7 Working in a criss-cross pattern, loosen the camshaft holder bolts by half a turn at a time to gently relieve valve spring pressure on the holder. Whilst slackening the bolts make sure that the camshaft holder lifts squarely away from the cylinder head and does not stick on

the locating dowels. Once spring pressure is relieved, remove all the bolts then lift off the holder. Discard the sealing washers which are fitted to the holder four inner retaining bolts; new ones should be used on installation.
Caution: If the bolts are carelessly loosened and the holder does not come squarely away from the head, it is likely to break. If this happens the complete cylinder head assembly must be renewed; the bearing holder is matched to the cylinder head and cannot be obtained separately.
8 Remove the holder locating dowels from the spark plug holes and pulse secondary air (PAIR) passages. Discard the O-ring fitted to each dowel; new ones must be used on installation.
9 Lift out the camshafts from the cylinder head. **Note:** *The camshafts are not interchangeable. Both camshafts are marked 'FR' indicating that they belong to the front cylinders. The intake camshaft is marked 'IN' and the exhaust camshaft 'EX'* **(see illustration)**.
10 If the followers are to be removed, obtain a container which is divided into eight compartments and label each compartment with the number of its corresponding valve in the cylinder head.
11 Using a magnet, lift each follower out of the cylinder head and store it in its corresponding compartment in the container.

8.5 Note the identification markings on the camshaft bearing caps (arrows - see text) prior to removal

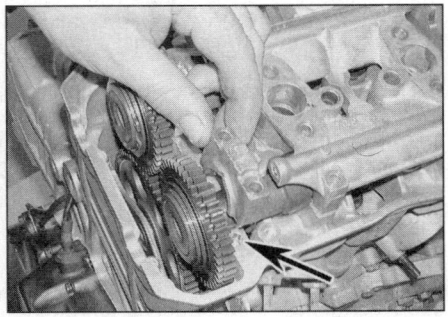

8.6 Remove the bearing caps and recover the dowels

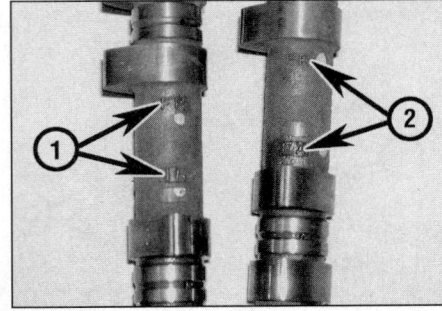

8.9 Front cylinder camshafts can be identified from their markings; the intake camshaft is marked FR and IN (1) and the exhaust camshaft FR and EX (2)

2•12 Engine, clutch and transmission

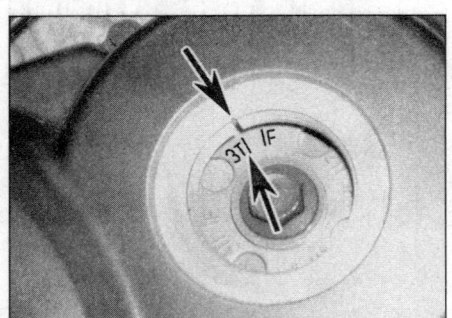

8.13a Align the timing rotor 'T' mark of number 3 cylinder with the cover index mark (arrows) . . .

8.13b . . . so the rear cylinder camshaft gear index marks (arrows) are positioned as shown (number 3 piston at TDC on its compression stroke)

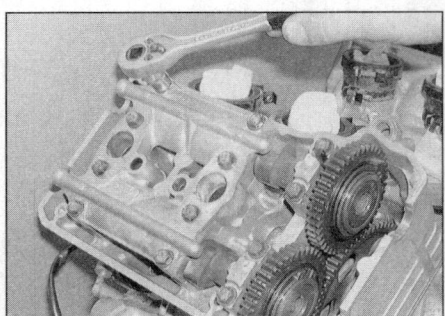

8.14 Remove the camshaft holder and bearing caps as described in text

Note that the shim is likely to stick to the inside of the follower so take great care not to lose it as the follower is removed. Remove the shims and store each one with its respective follower.

Rear cylinder camshafts only

12 Unscrew the timing hole cap and O-ring from the right crankcase cover.
13 Using a socket on the crankshaft bolt, rotate the crankshaft clockwise until the 'T' mark of number 3 cylinder is aligned with the index mark (cutout or line) on the top of the crankcase cover aperture (see illustration). Check the position of the index marks on the rear cylinder camshaft gears; the marks should be facing away from each other and both be aligned with the cylinder head upper surface (piston number 3 is now at TDC on its compression stroke) (see illustration). If the camshaft index marks are not correctly positioned, rotate the crankshaft through another 360° (one complete turn) to position them as described. **Note:** *Always turn the engine in the normal direction of rotation (clockwise), viewed from the right end of the engine.*
14 Remove the camshaft holder and bearing caps as described in Paragraphs 5 to 8 (see illustration).
15 Lift out the camshafts from the cylinder head. **Note:** *The camshafts are not interchangeable. Both camshafts are marked 'RR' indicating that they belong to the rear*

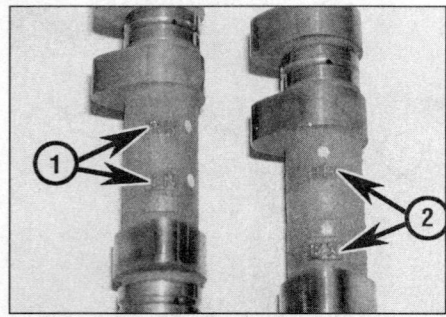

8.15 Rear cylinder camshafts can be identified from their markings; the intake camshaft is marked RR and IN (1) and the exhaust camshaft RR and EX (2)

cylinders. The intake camshaft is marked 'IN' and the exhaust camshaft 'EX'; the intake camshaft is also easily identified by the cam pulse generator triggers on the rear of the gear (see illustration).
16 Remove the followers and shims as described in Paragraphs 10 and 11.

Front and rear cylinder camshafts

Note: *Ensure that the front and rear cylinder components are kept separate are not interchanged.*

17 Remove the front camshafts and (if necessary) followers as described in Paragraphs 3 to 11.
18 Using a suitable socket, rotate the crankshaft 270° (three-quarters of a turn) clockwise until the 'T' mark of number 3 cylinder is aligned with the index mark on the top of the crankcase cover aperture. Piston number 3 is now at TDC on its compression stroke (rear camshaft gear index marks facing away from each other and aligned with cylinder head upper surface – see Paragraph 13).
19 Since both front and rear cylinder camshaft holders and bearing caps carry identical markings, great care must be taken to ensure they are not interchanged. To remove any possibility of this happening, mark the holder and caps of the front cylinder using a dab of white paint or a suitable marker.
20 Remove the rear camshaft holder and bearing caps as described in Paragraphs 5 to 8.
21 Remove the rear camshafts (see Paragraph 15).
22 If necessary, remove the rear followers as described in Paragraphs 10 and 11. Keep the followers and shims for the front and rear cylinders in two separate containers, one for the front cylinders and one for the rear.

Inspection

23 Inspect the cam bearing surfaces of the holders and camshafts. Look for score marks, deep scratches and evidence of pitting. Check the camshaft lobes for heat discoloration (blue appearance), score marks, chipped areas, flat spots and pitting.
24 Camshaft runout can be checked by

supporting each end bearing of the camshaft on V-blocks, and measuring any runout at the centre bearing using a dial gauge. If the runout exceeds the specified limit the camshaft must be renewed.
25 Measure the height of each lobe with a micrometer and compare the results to the lobe height service limit listed in this Chapter's Specifications (see illustration). If damage is noted or wear is excessive, the camshaft must be renewed.
26 The camshaft bearing oil clearance should then be checked. There are two possible ways of checking this, the first method by direct measurement (see Paragraphs 27 and 31) and the second by the use of a product known as Plastigauge (see Paragraphs 28 to 31).
27 If the first method is to be used, make sure the locating dowels are in position then fit the camshaft holder and bearing caps to the head, ensuring they are correctly positioned (see Paragraphs 43 and 44). Working in a criss-cross pattern, tighten the holder retaining bolts evenly and progressively to the specified torque then tighten the bearing cap bolts to the specified torque. Measure the diameter of each bearing journal (Honda do not give specifications for the bearing bores) and then calculate the camshaft bearing oil clearance by subtracting the camshaft bearing journal diameter from the bearing holder journal diameter. Compare the measurements obtained with the service limit given in the Specifications at the start of this Chapter.
28 If the second method is to be used, clean the camshafts and the bearing surfaces in the cylinder head, camshaft holder and bearing

8.25 Measuring a camshaft lobe height with a micrometer

Engine, clutch and transmission 2•13

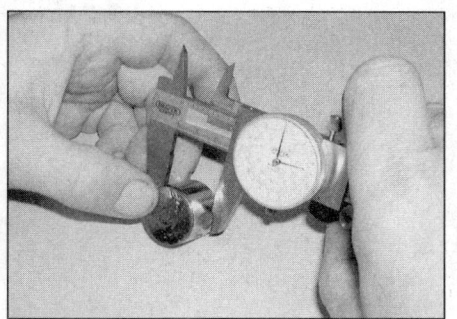

8.33 Measuring a camshaft follower diameter

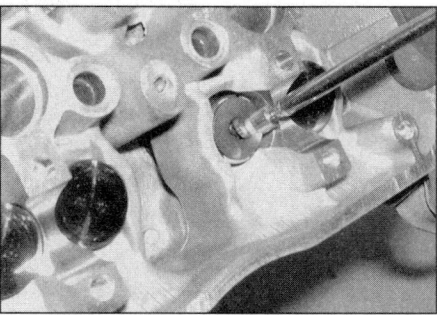

8.34 Fit each shim correctly to its relevant valve spring retainer . . .

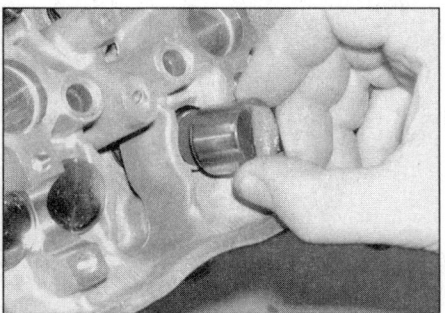

8.35 . . . then insert the follower

caps with a clean, lint-free cloth. Ensure all the shims and followers are correctly installed then make sure that the crankshaft is correctly positioned; if the front camshafts are being checked ensure the 'T' mark of number 4 cylinder is aligned with the index mark, if the rear camshafts are being checked align the 'T' mark of number 3 cylinder. Lay the camshafts in place in the cylinder head and engage them with the drive gear as described in Paragraph 41.

29 Cut strips of Plastigauge and lay one piece on each bearing journal, parallel with the camshaft axis. Make sure the locating dowels are all installed then fit the camshaft holder and bearing caps in their proper positions (see note in Paragraphs 43 and 44). Ensuring the camshafts are not rotated at all, tighten the holder and bearing cap retaining bolts to the specified torque as described in Paragraphs 46 to 48.

30 Now unscrew the bolts and carefully lift off the camshaft holder and bearing caps as described in Paragraphs 6 and 7, again making sure the camshafts are not rotated. To determine the oil clearance, compare the crushed Plastigauge (at its widest point) on each journal to the scale printed on the Plastigauge container.

31 Compare the results to this Chapter's Specifications. If the oil clearance is greater than specified, measure the diameter of the cam bearing journal with a micrometer. If the journal diameter is less than the specified limit, renew the camshaft and recheck the clearance. If the clearance is still too great, renew the cylinder head assembly.

HAYNES HiNT *Before renewing camshafts or cylinder head because of damage, check with local machine shops specialising in motorcycle engine work. In the case of the camshafts, it may be possible for cam lobes to be welded, reground and hardened, at a cost far lower than that of a new camshaft. If the bearing surfaces in the cylinder head, holder or caps are damaged, it may be possible for them to be bored out to accept bearing inserts. Due to the cost of new components it is recommended that all options be explored.*

32 Check the camshaft gear teeth for signs of damage such as chipped or missing teeth. If damage is found the camshaft must be renewed. Note: *Do not attempt to remove the circlip and dismantle the camshaft gears; no individual components are available.*

33 Check each follower for wear by measuring its outside diameter; renew any follower which exceeds the service limit given in this Chapter's Specifications (see illustration). Check for wear of the follower bores by measuring their inside diameters; if worn to or beyond the service limit, cylinder head renewal is required.

Installation

Front cylinder camshafts only

34 Fit each shim to the top of its correct valve making sure it is correctly seated in the valve spring retainer (see illustration). Note: *It is most important that the shims are returned to their original valves otherwise the valve clearances will be inaccurate.*

35 Lubricate each follower with clean engine oil then install each one into its respective position in the cylinder head, making sure its enters its bore squarely (see illustration).

36 Rotate the crankshaft clockwise until the 'T' mark of number 3 cylinder is aligned with the index mark (cutout or line) on the top of the crankcase cover aperture. Check the position of the index marks on the rear cylinder camshaft gears; the marks should be facing away from each other and aligned with the cylinder head upper surface (piston number 3 is now at TDC on its compression stroke) (see illustrations 8.13a and 8.13b). If the camshaft index marks are not correctly positioned, rotate the crankshaft through another 360° (one complete turn) to position them as described. Note: *Always turn the engine in the normal direction of rotation (clockwise), viewed from the right end of the engine.*

37 From this point, rotate the crankshaft a further 450° (one and a quarter turn) clockwise until the 'T' mark of number 4 cylinder is aligned with the index mark on the top of the crankcase cover aperture (see illustration 8.4a).

38 Apply a smear of clean engine oil to the cylinder head camshaft bearings and followers (see illustration).

39 Fit the locating dowels to the spark plug holes and PAIR passages and fit a new O-ring to each dowel (see illustration).

40 Referring to Paragraph 9, identify the intake and exhaust camshafts. Bearing in mind that the index marks on the camshaft gears will be level with the cylinder head and facing away from each other with the camshafts installed, apply paint to one of the upper pairs of teeth on each gear (see illustration). The marks can then be used to ensure the spring-loaded teeth engage correctly with the drive gear.

41 Ensure the crankshaft is still correctly positioned then lay the intake and exhaust camshafts correctly in position in the cylinder head. Engage each camshaft with the drive gear so the camshaft gear index marks are

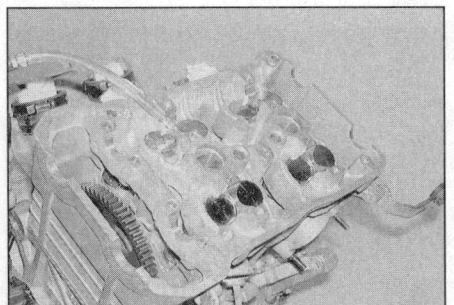

8.38 Lubricate the camshaft followers and bearings with clean engine oil

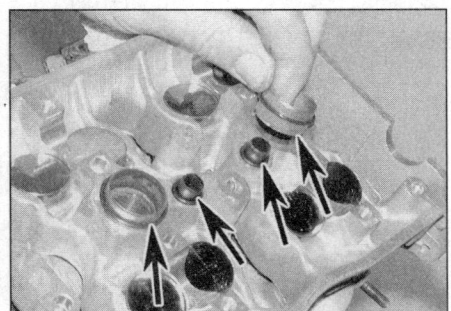

8.39 Fit the locating dowels (arrows) to the spark plug holes and PAIR passages and fit a new O-ring to each dowel

2•14 Engine, clutch and transmission

8.40 Apply paint to one of the upper pairs of teeth on each camshaft gear

facing away from each other and are both aligned with the cylinder head upper surface **(see illustration)**.

42 With both camshafts correctly positioned, apply a smear of clean engine oil to the camshaft bearing journals.

43 Ensure the locating dowels are in position then fit the camshaft holder to the cylinder head making sure its 'IN' marking is positioned on the intake side **(see illustration)**.

44 Ensure the locating dowels are in position and fit the intake camshaft (marked 'I') bearing cap and exhaust camshaft (marked 'E') bearing cap. Ensure that both caps are fitted the correct way around with the cap boss facing towards the camshaft holder **(see illustration)**.

45 Lubricate the threads and heads of the holder and bearing cap bolts with clean engine oil.

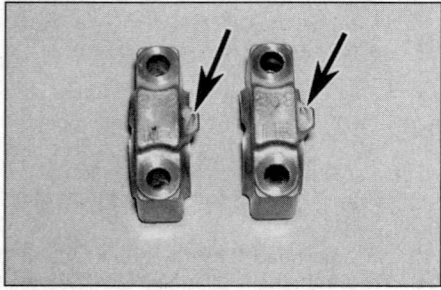

8.44 Ensure that each bearing cap is correctly installed with its boss (arrow) facing towards the camshaft holder

8.46 Remember to fit new sealing washers (arrows) to the camshaft holder four inner bolts

8.41 Lay the camshafts in the cylinder head making sure the index marks (arrows) are correctly positioned

46 Fit new sealing washers to the four inner retaining bolts then install all the camshaft holder bolts, tightening them by hand until all bolts are contacting the holder **(see illustration)**. Starting with the inner bolts, and working outwards in a criss-cross pattern, tighten the bolts by half a turn at a time to gradually draw the holder into position. Whilst tightening the bolts make sure that the camshaft holder is being pulled squarely down onto the cylinder head and is not sticking on the locating dowels.

Caution: If the bolts are carelessly tightened and the camshaft holder is not drawn squarely onto the head, it is likely to break. If this happens the complete cylinder head assembly must be renewed; the camshaft holder is matched to the cylinder head and cannot be obtained separately.

47 Once the camshaft holder is in contact with the head, go around again in the same sequence and tighten the bolts to the specified torque setting **(see illustration)**.

48 Evenly and progressively tighten the intake and exhaust camshaft bearing cap bolts to the specified torque.

49 Check the 'T' mark of number 4 cylinder is still correctly aligned with the cutout in the crankcase cover, then check that the front camshaft gear index marks are correctly aligned with the cylinder head upper surface **(see illustrations 8.4a and 8.4b)**. Also check the paint marks made on the gear teeth are all correctly aligned. If any of the marks aren't correctly positioned, unbolt the camshaft

8.47 Tighten the camshaft holder bolts to the specified torque as described in text

8.43 Ensure the camshaft holder is fitted with its IN marking (arrow) on the intake side

holder and bearing caps from the cylinder head and repeat the operations in Paragraphs 41 through 48.

50 With all the marks correctly positioned, rotate the crankshaft through a few rotations to settle all disturbed components in position.

51 Check and, if necessary, adjust the valve clearances as described in Chapter 1.

52 Lubricate all bearing surfaces with clean engine oil and fit the cylinder head covers as described in Section 7.

53 Fit a new O-ring to the inspection plug. Lubricate the plug threads with a smear of multi-purpose grease then fit it to the crankcase cover and tighten to the specified torque **(see illustration)**. Install the spark plugs as described in Chapter 1.

Rear cylinder camshafts only

54 Install the shims and followers as described in Paragraphs 34 and 35.

55 Rotate the crankshaft clockwise until the 'T' mark of number 4 cylinder is aligned with the index mark (cutout or line) on the top of the crankcase cover aperture. Check the position of the index marks on the front cylinder camshaft gears; the marks should be facing away from each other and both be aligned with the cylinder head upper surface (piston number 4 is now at TDC on its compression stroke) **(see illustrations 8.4a and 8.4b)**. If the camshaft index marks are not correctly positioned, rotate the crankshaft through another 360° (one complete turn) to position them as described. **Note:** *Always turn*

8.53 Tighten the cover inspection plug to the specified torque

Engine, clutch and transmission 2•15

the engine in the normal direction of rotation (clockwise), viewed from the right end of the engine.
56 From this point, rotate the crankshaft a further 270° (three-quarters of a turn) clockwise until the 'T' mark of number 3 cylinder is aligned with the index mark on the top of the crankcase cover aperture **(see illustration 8.13a)**.
57 Install the camshafts as described in Paragraphs 38 to 48, referring to Paragraph 15 for information on camshaft identification **(see illustrations)**.
58 Check the 'T' mark of number 3 cylinder is still correctly aligned with the cutout in the crankcase cover, then check that the rear camshaft gear index marks are correctly aligned with the cylinder head upper surface **(see illustrations 8.13a and 8.13b)**. Also check the paint marks made on the gear teeth are all correctly aligned. If any of the marks aren't correctly positioned, unbolt the camshaft holder from the cylinder head and repeat the operations in Paragraphs 41 through 48.
59 With all the marks correctly positioned, rotate the crankshaft through a few rotations to settle all disturbed components in position.
60 Check and, if necessary, adjust the valve clearances as described in Chapter 1.
61 Lubricate all bearing surfaces with clean engine oil and fit the cylinder head covers as described in Section 7.
62 Install the spark plugs as described in Chapter 1.

Front and rear cylinder camshafts

63 Install the shims and followers as described in Paragraphs 34 and 35.
64 Rotate the crankshaft clockwise until the 'T' mark of number 3 cylinder is aligned with the index mark on the top of the crankcase cover aperture **(see illustration 8.13a)**.
65 Install the rear camshafts as described in Paragraphs 38 to 48, referring to Paragraph 15 for information on camshaft identification.
66 Check the 'T' mark of number 3 cylinder is still correctly aligned with the cutout in the crankcase cover, then check that the rear camshaft gear index marks are correctly aligned with the cylinder head upper surface **(see illustrations 8.13a and 8.13b)**. Also check the paint marks made on the gear teeth are all correctly aligned. If any of the marks aren't correctly positioned, unbolt the camshaft holder from the cylinder head and repeat the operations in Paragraphs 41 through 48.
67 Once the rear camshafts are correctly installed, rotate the crankshaft a further 450° (one and a quarter turn) clockwise until the 'T' mark of number 4 cylinder is aligned with the index mark on the top of the crankcase cover aperture **(see illustration 8.4a)**.
68 Install the front camshafts as described in Paragraphs 38 to 53.

8.57a Ensure the dowels (arrows) are all correctly installed and each one is fitted with a new O-ring . . .

9 Cylinder head – removal and installation

Caution: *The engine must be completely cool before beginning this procedure or the cylinder head may become warped.*
Note: *When removing the cylinder head, take great care not to allow any component to fall down into the crankcase. If any component is dropped, it must be recovered before the engine is started.*

Front cylinder head

Note: *This procedure can be carried out with the engine in the frame.*

Removal

1 Remove the seat and fairing lower and inner panels (see Chapter 8). Release the two clips then lift the battery cover and disconnect the battery negative (-) lead.
2 Remove the throttle body assembly as described in Chapter 4.
3 Remove the exhaust system front pipe assembly as described in Chapter 4.
4 Drain the cooling system as described in Chapter 1.
5 Remove the oil cooler and pipes (see Section 22).
6 Referring to Chapter 3, release the retaining clips and disconnect the air bleed hose and radiator upper hose from the thermostat housing and the radiator lower hose from the

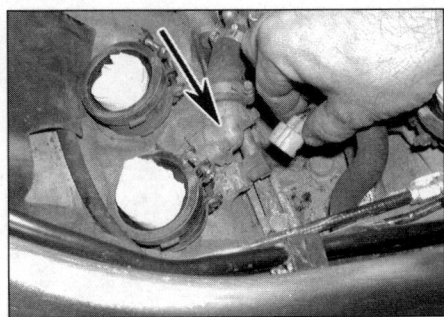

9.7 Disconnect the wiring connector from the coolant temperature sensor (coolant union arrowed)

8.57b . . . and ensure new sealing washers are fitted to the holder four inner retaining bolts (arrows)

water pump cover **(see illustrations 5.14a, 5.14b and 5.14c)**. Disconnect the cooling fan wiring connector (located above the left side radiator fan switch) then slacken and remove the left and right side radiator mounting bolts and collars. Unscrew the retaining bolts then remove both the left and right lower mounting brackets then manoeuvre both radiators out of position as an assembly.
7 Disconnect the wiring connector from the coolant temperature sensor which is fitted to the rear of the front cylinder head **(see illustration)**.
8 Unscrew the two bolts securing the coolant union to the rear of the front cylinder head. Recover the O-ring from the union and discard it; a new one must be used on installation.
9 Remove the camshafts as described in Section 8.
10 Slacken and remove the two 6 mm bolts securing the right side of the head to the crankcase.
11 Working from the outside to the inside in a criss-cross pattern, slacken the six 9 mm cylinder head bolts by half a turn at a time. Once all pressure is released from the bolts, fully unscrew them and remove along with their sealing washers. Discard the sealing washers; new ones should be used on installation.
12 Lift the head off the crankcase and remove it from the engine **(see illustration)**. If it is stuck, tap around the joint faces of the cylinder head with a soft-faced mallet to free

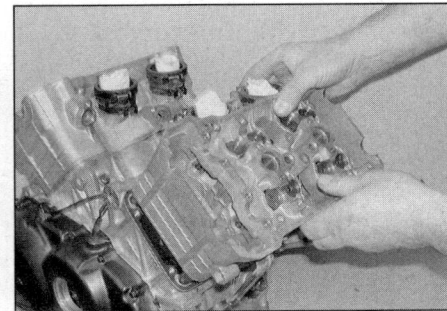

9.12 Removing the cylinder head

2•16 Engine, clutch and transmission

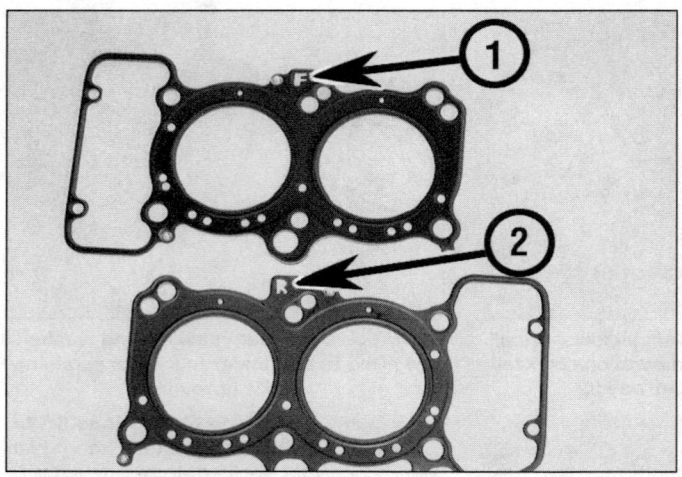

9.17a Front cylinder head gasket is marked with an 'F' (1) and the rear gasket with an 'R' (2)

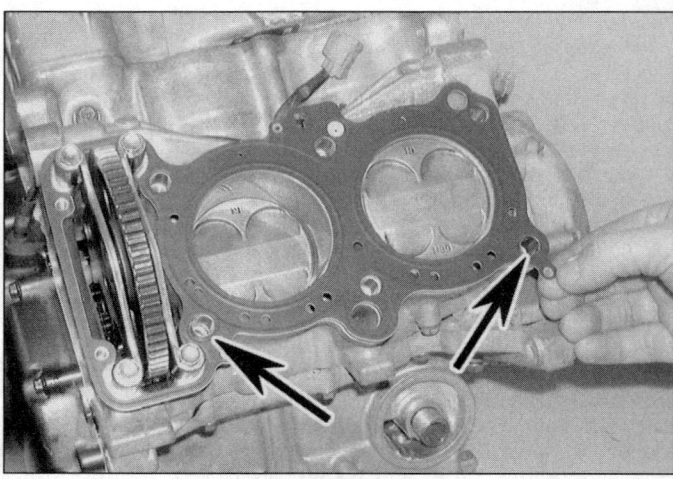

9.17b Locate the new gasket on the dowels (arrows)

the head. Don't attempt to free the head by inserting a screwdriver between the head and cylinder block – you'll damage the sealing surfaces.

13 Remove the old head gasket and discard it. If they are loose, remove the cylinder head locating dowels from the crankcase/cylinder head and store them with the head for safe-keeping.

14 Check the cylinder head gasket and the mating surfaces on the cylinder head and crankcase for signs of leakage, which could indicate warpage. Check the flatness of the head as described in Section 11.

Installation

15 Clean all traces of old gasket material from the cylinder head and crankcase. If a scraper is used, take care not to scratch or gouge the soft aluminium. Also ensure that none of the gasket material is allowed to fall into the cylinder bores, oil passages or water jacket.

16 Ensure both cylinder head and crankcase mating surfaces are clean and fit the locating dowels to the crankcase (if removed). Apply a smear of engine oil to the surface of each cylinder bore.

17 Fit the new head gasket over the locating dowels making sure its 'F' mark is the right way up. **Note:** *The front and rear cylinder head gaskets are different, the front gasket has an 'F' identification mark and the rear gasket an 'R'* **(see illustrations)**.

18 Carefully manoeuvre the cylinder head into position and lower it onto the locating dowels.

19 Lubricate the threads and undersides of the 9 mm cylinder head bolts with clean engine oil then fit the new sealing washers to the bolts **(see illustration)**. Insert the bolts into position and tighten them all by hand.

20 Fit the two 6 mm cylinder head bolts and tighten by hand.

21 Working from the inside to the outside in a criss-cross pattern, tighten the six 9 mm cylinder head bolts to approximately half the specified torque setting given in the Specifications **(see illustration)**. Then go around in the same sequence and tighten the bolts to the full specified torque setting.

22 With the 9 mm bolts tightened, tighten the two 6 mm bolts to their specified torque setting.

23 Install the camshafts as described in Section 8.

24 Fit a new O-ring to the coolant union groove then seat the union on the rear of the cylinder head and securely tighten its retaining bolts.

25 Reconnect the wiring connector to the coolant temperature sensor.

26 Manoeuvre the radiator assembly into position and locate both the left and right radiators on their mounting pegs. Engage the lower mounting brackets with the radiators and securely tighten their mounting bolts. Fit the upper mounting collars and bolts, tightening them securely, and reconnect the fan wiring connector. Reconnect all the coolant hoses and secure them in position with the retaining clips.

27 Install the oil cooler and hoses (see Section 22).

28 Fit the exhaust system and throttle body assembly as described in Chapter 4.

29 Refill the engine with coolant and oil (see Chapter 1).

30 Prior to installing the inner and lower fairing panels start the engine and check that there are no signs of coolant/oil leakage.

Rear cylinder head

Removal

31 It is not possible to remove the rear cylinder head with the engine installed in the frame. If the rear cylinder head is to be removed, first remove the engine unit (see Section 5).

32 Remove the camshafts (see Section 8).

33 Unscrew the two bolts securing the coolant union to the front of the rear cylinder head. Recover the O-ring from the union and discard it; a new one must be used on installation.

34 Remove the head as described in Paragraphs 10 to 14.

Installation

35 Clean all traces of old gasket material from the cylinder head and crankcase. If a scraper is used, take care not to scratch or gouge the soft aluminium. Also ensure that none of the gasket material is allowed to fall into the cylinder bores, oil passages or water jacket.

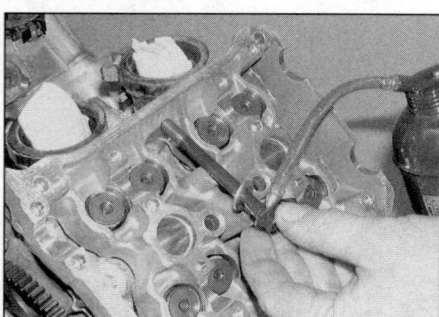

9.19 Lubricate the threads and heads of the 9 mm head bolts with oil and fit a new sealing washer to each bolt

9.21 Tighten the 9 mm cylinder head bolts to the specified torque as described in the text

Engine, clutch and transmission 2•17

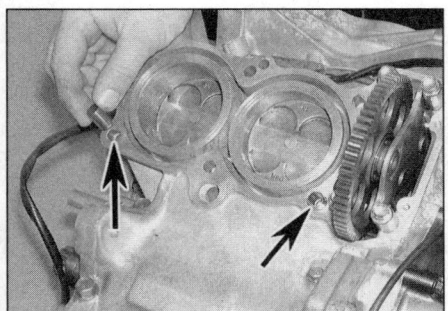

9.36 Fit the locating dowels (arrows) . . .

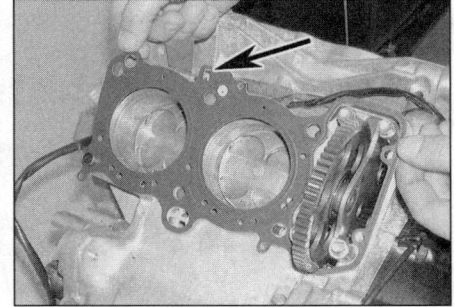

9.37 . . . then fit the rear cylinder head gasket ensuring the identification marking (arrow) is the right way up

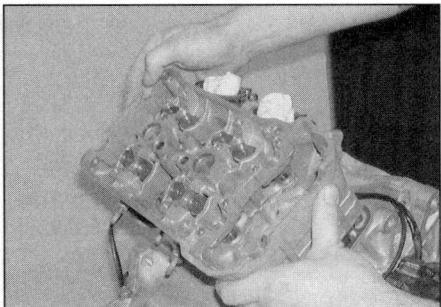

9.38 Locate the rear cylinder head on its dowels

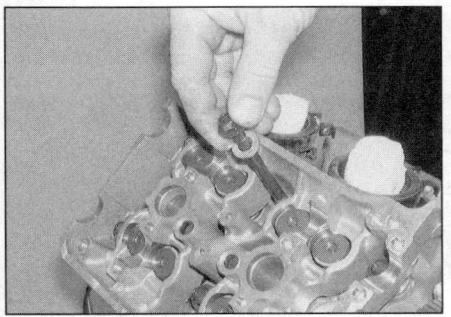

9.39a Fit the 9 mm cylinder head bolts with new sealing washers . . .

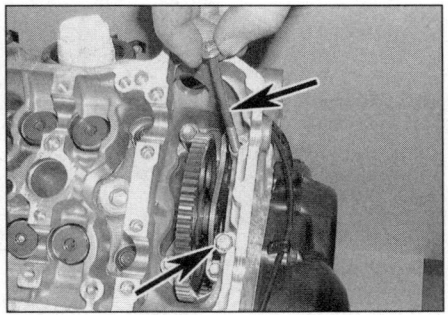

9.39b . . . then install the 6 mm bolts (arrows) . . .

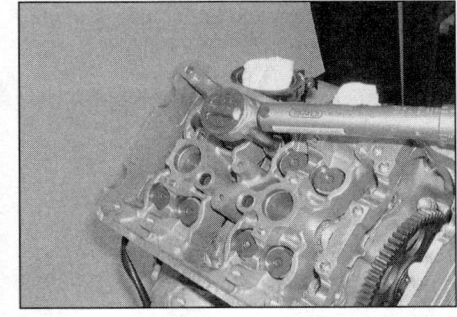

9.39c . . . and tighten them to the specified torque as described in the text

36 Ensure both cylinder head and crankcase mating surfaces are clean and fit the locating dowels to the crankcase (if removed) (see illustration). Apply a smear of engine oil to the surface of each cylinder bore.
37 Fit the new head gasket over the locating dowels making sure its 'R' mark is the right way up (see illustration). Note: *The front and rear cylinder head gaskets are different, the rear gasket has an 'R' identification mark and the front gasket an 'F'* (see illustration 9.17a).
38 Carefully lower the cylinder head onto the crankcase, aligning it with the locating dowels (see illustration).
39 Install and tighten the cylinder head bolts as described in Paragraphs 19 to 22 (see illustrations).
40 Install the camshafts and cylinder head cover (see Section 8).
41 Fit a new O-ring to the coolant union groove then seat the union on the front of the cylinder head and securely tighten its retaining bolts.
42 Install the engine (see Section 5).

10 Valves/valve seats/valves guides – servicing

1 Because of the complex nature of this job and the special tools and equipment required, most owners leave servicing of the valves, valve seats and valve guides to a professional.
2 The home mechanic can, however, remove the valves from the cylinder head, clean and check the components for wear and assess the extent of the work needed, and, unless a valve service is required, grind in the valves (see Section 11).
3 The dealer or motorcycle engineer will remove the valves and springs, renew the valves and guides, recut the valve seats, check and renew the valve springs, spring retainers and collets (as necessary), renew the valve seals and reassemble the valve components.
4 After the valve service has been performed, the head will be in like-new condition. When the head is returned, be sure to clean it again very thoroughly before installation on the engine to remove any metal particles or abrasive grit that may still be present from the valve service operations. Use compressed air, if available, to blow out all the holes and passages.

11 Cylinder head and valves – disassembly, inspection and reassembly

1 As mentioned in the previous section, valve servicing, valve seat re-cutting and valve guide renewal should be left to a dealer or motorcycle engineer. However, disassembly, cleaning and inspection of the valves and related components can be done (if the necessary special tools are available) by the home mechanic. This way no expense is incurred if the inspection reveals that overhaul is not required at this time.

2 To disassemble the valve components without the risk of damaging them, a valve spring compressor is absolutely necessary. Another useful tool is a follower bore protector which prevents the cylinder head surface being scratched by the valve spring compressor. In the absence of the special Honda protector (07HMG-MR70002), a suitable substitute can be made using a 35 mm film canister cut to the dimensions shown (see illustrations).

Disassembly

3 Before proceeding, arrange to label and store the valves along with their related components in such a way that they can be returned to their original locations without getting mixed up. A good way to do this is to use the same container as the followers and shims are stored in (see Section 8), or to

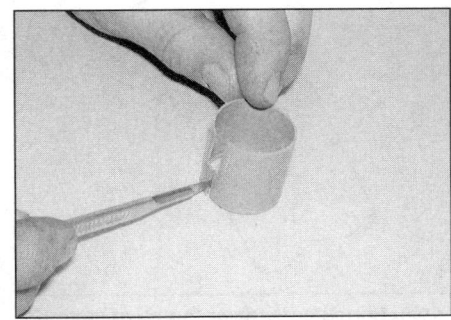

11.2a A follower bore protector can be fabricated from a 35 mm film canister

2•18 Engine, clutch and transmission

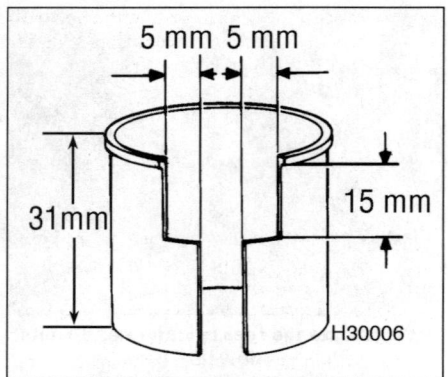

11.2b Cut a 35 mm film canister to the dimensions shown to form a home-made follower bore protector

When compressed, the outside diameter of the film canister should be 25 mm

obtain a separate container which is divided into compartments, and to label each compartment with the identity of the valve which will be stored in it (ie number of cylinder, intake or exhaust side, inner or outer valve). Alternatively, labeled plastic bags will do just as well.

4 If not already done, clean all traces of old gasket material from the cylinder head. If a scraper is used, take care not to scratch or gouge the soft aluminium **(see illustration)**

 **HAYNES HiNT** *Refer to Tools and Workshop Tips for details of gasket removal methods.*

5 Fit the follower bore protector (if available) then compress the valve spring on the first valve with a spring compressor, making sure it is correctly located onto each end of the valve assembly. Do not compress the springs any more than is absolutely necessary. Remove the collets, using either needle-nose pliers, tweezers, a magnet or a screwdriver with a dab of grease on it. Carefully release the valve

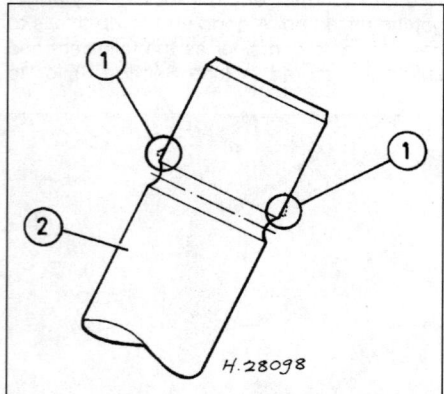

11.5 Check the area of the valve stem (2) around the collet groove (1) for burrs and remove any that you find

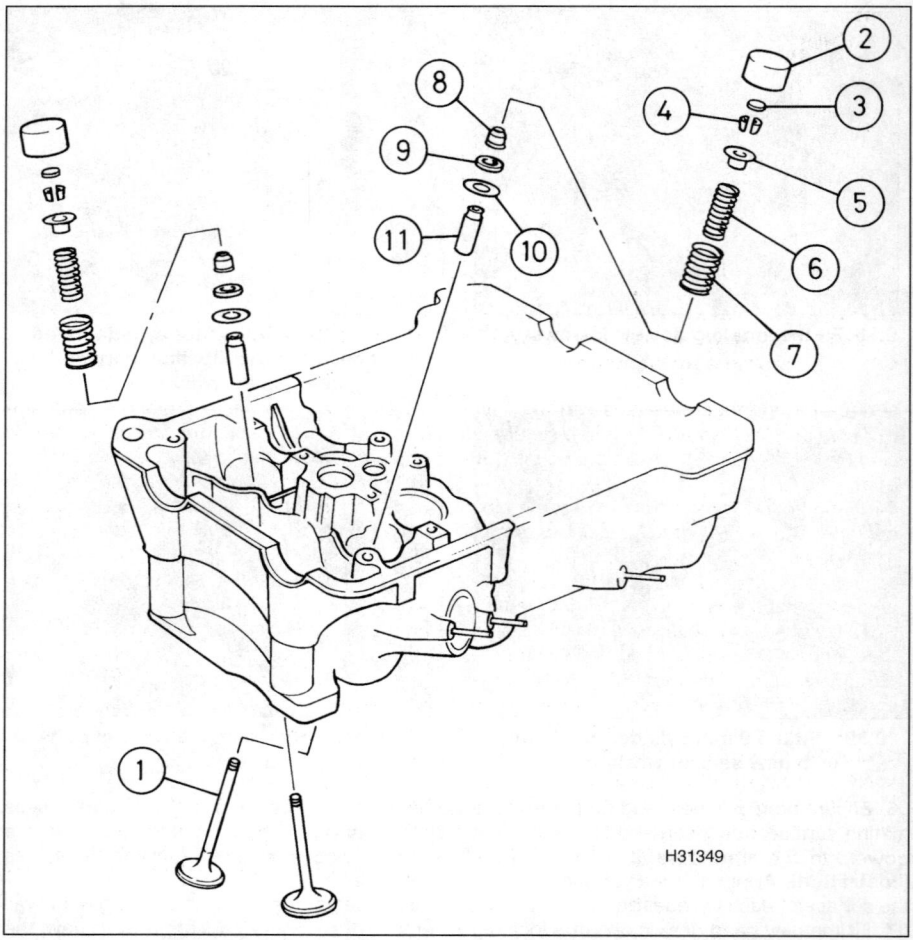

11.4 Exploded view of cylinder head components (rear head shown)

1 Valve	4 Collets	7 Outer spring	9 Inner spring seat
2 Follower	5 Spring retainer	8 Valve stem oil	10 Outer spring seat
3 Shim	6 Inner spring	seal	11 Valve guide

spring compressor and remove the spring retainer, noting which way up it fits, the springs, the spring seats, and the valve from the head. If the valve binds in the guide (won't pull through), push it back into the head and deburr the area around the collet groove with a very fine file or whetstone **(see illustration)**.

6 Repeat the procedure for the remaining valves. Remember to keep the parts for each valve together and separate from the other valves so they can be reinstalled in the same location.

7 Once the valves have been removed and labeled, pull the valve stem seals off the top of the valve guides with pliers and discard them (the old seals should never be reused).

8 Next, clean the cylinder head with solvent and dry it thoroughly. Compressed air will speed the drying process and ensure that all holes and recessed areas are clean.

9 Clean all of the valve springs, collets, retainers and spring seats with solvent and dry them thoroughly. Do the parts from one valve at a time so that no mixing of parts between valves occurs.

10 Scrape off any deposits that may have formed on the valve, then use a motorised wire brush to remove deposits from the valve heads and stems. Again, make sure the valves do not get mixed up.

Inspection

11 Inspect the head very carefully for cracks and other damage. If cracks are found, a new head will be required. Check the cam bearing surfaces for wear and evidence of seizure. Check the camshafts for wear as well (see Section 8).

12 Inspect the outer surfaces of the cam followers for evidence of scoring or other damage. If a follower is in poor condition, it is probable that the bore in which it works is also damaged. Check for clearance between the followers and their bores. If the bores are seriously out-of-round or tapered, the cylinder head and the followers must be renewed.

13 Using a precision straight-edge and a feeler gauge set to the warpage limit listed in the specifications at the beginning of the Chapter, check the head gasket mating

Engine, clutch and transmission 2•19

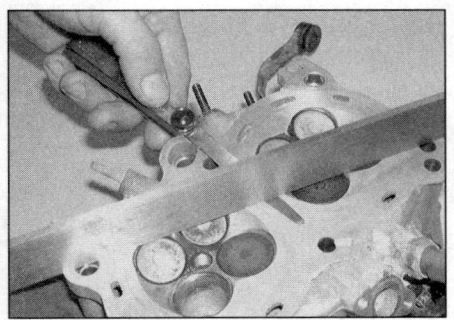

11.13 Checking cylinder head gasket surface for warpage

11.14 Measuring the width of a valve seat

surface for warpage **(see illustration)**. Refer to *Tools and Workshop Tips* in the Reference section for details of how to use the straight-edge.

14 Examine the valve seats in the combustion chamber. If they are pitted, cracked or burned, the head will require work beyond the scope of the home mechanic. Measure the valve seat width and compare it to this Chapter's Specifications **(see illustration)**. If it exceeds the service limit, or if it varies around its circumference, valve overhaul is required. If available, use Prussian blue to determine the extent of valve seat wear. Uniformly coat the seat with the Prussian blue, then install the valve and rotate it back and forth using a lapping tool. Remove the valve and check whether the ring of blue on the valve is uniform and continuous around the valve, and of the correct width as specified.

15 Measure the valve stem diameter **(see illustration)**. Clean the valve guides to remove any carbon build-up, then measure the inside diameters of the guides (at both ends and the centre of the guide) with a small-bore gauge and micrometer **(see illustrations)**. The guides are measured at the ends and at the centre to determine if they are worn in a bell-mouth pattern (more wear at the ends). Subtract the stem diameter from the valve guide diameter to obtain the valve stem-to-guide clearance. If the stem-to-guide clearance is greater than listed in this Chapter's Specifications, the guides and valves will have to be renewed. If the valve stem or guide is worn beyond its limit, or if the guide is worn unevenly, it must be renewed.

16 Carefully inspect each valve face for cracks, pits and burned spots. Check the valve stem and the collet groove area for cracks **(see illustration)**. Rotate the valve and check for any obvious indication that it is bent. Check the end of the stem for pitting and excessive wear. The presence of any of the above conditions indicates the need for valve servicing.

17 Check the end of each valve spring for wear and pitting. Measure the spring free length and compare it to that listed in the specifications **(see illustration)**. If any spring is shorter than specified it has sagged and must be renewed. Also place the spring upright on a flat surface and check it for bend by placing a ruler against it **(see illustration)**. If the bend in any spring is excessive, it must be renewed.

18 Check the spring retainers and collets for obvious wear and cracks. Any questionable parts should not be reused, as extensive damage will occur in the event of failure during engine operation.

19 If the inspection indicates that no overhaul work is required, the valve components can be reinstalled in the head.

Reassembly

20 Unless a valve service has been performed, before installing the valves in the head they should be ground in (lapped) to ensure a positive seal between the valves and seats. This procedure requires coarse and fine valve grinding compound and a valve grinding tool. If a grinding tool is not available, a piece of rubber or plastic hose can be slipped over the valve stem (after the valve has been installed in the guide) and used to turn the valve.

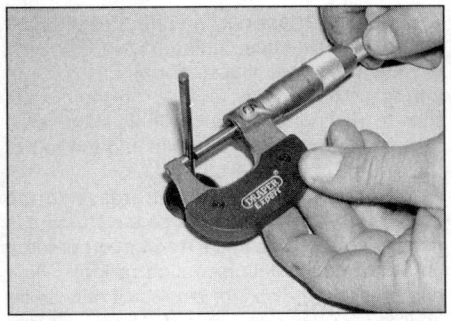

11.15a Measuring a valve stem diameter with a micrometer

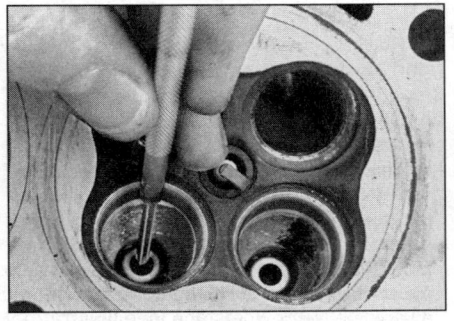

11.15b Insert a small-bore gauge into the valve guide and expand it so there is a slight drag when its pulled out . . .

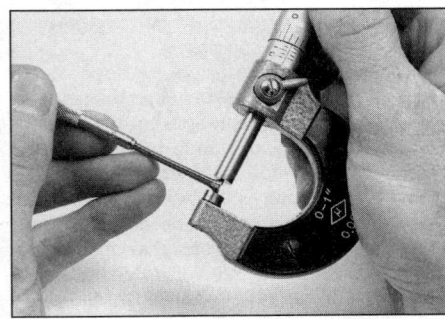

11.15c . . . then measure the gauge diameter with a micrometer to determine the valve guide diameter

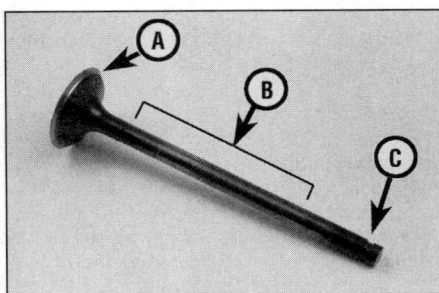

11.16 Check the valve face (A), stem (B) and collet groove (C) for signs of wear or damage

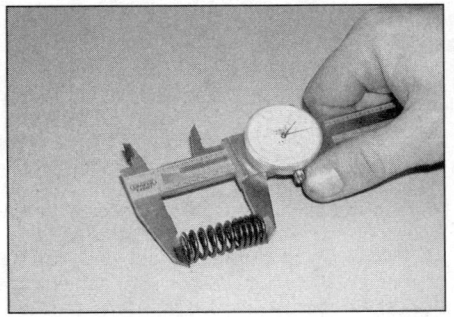

11.17a Measuring valve spring free length with a vernier caliper

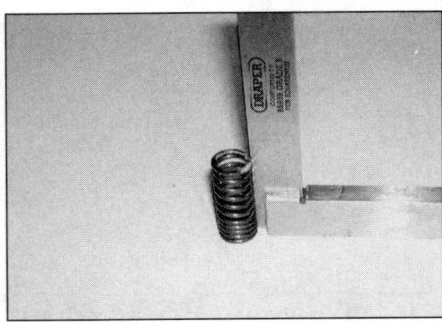

11.17b Checking a valve spring for squareness

2•20 Engine, clutch and transmission

11.21 Apply grinding compound sparingly to the valve face

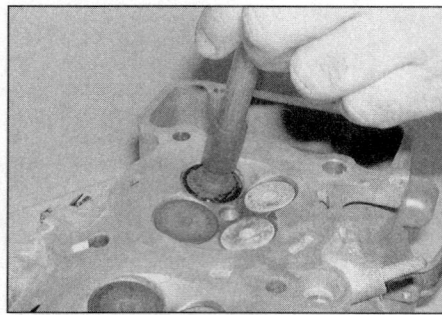

11.22a Attach a grinding tool to the valve and grind the valve in with a back-and-forth motion

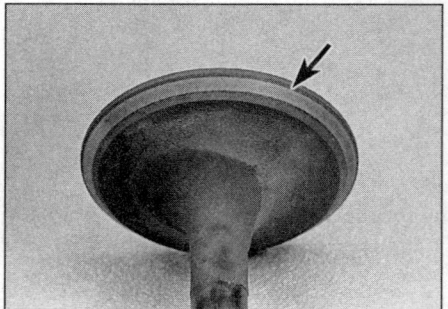

11.22b After grinding, the valve face should exhibit a uniform, unbroken contact pattern (arrow) . . .

11.22c . . . and the seat (arrow) should be the specified width with a smooth, unbroken appearance

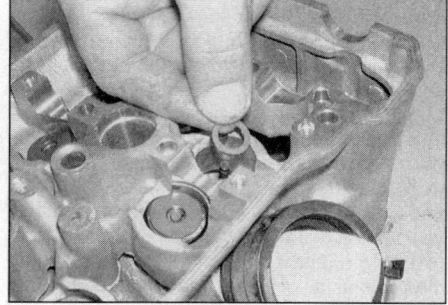

11.25a Fit the inner spring seat . . .

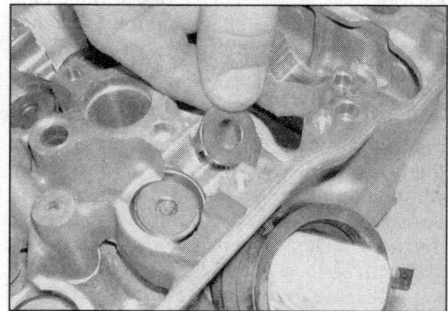

11.25b . . . and outer spring seat to the cylinder head . . .

21 Apply a small amount of coarse grinding compound to the valve face, then slip the valve into the guide (see illustration). **Note:** *Make sure each valve is installed in its correct guide and be careful not to get any grinding compound on the valve stem.*
22 Attach the grinding tool (or hose) to the valve and rotate the tool between the palms of your hands. Use a back-and-forth motion (as though rubbing your hands together) rather than a circular motion (ie so that the valve rotates alternately clockwise and anti-clockwise rather than in one direction only) (see illustration). Lift the valve off the seat and turn it at regular intervals to distribute the grinding compound properly. Continue the grinding procedure until the valve face and seat contact area is of uniform width and unbroken around the entire circumference of the valve face and seat (see illustrations).
23 Carefully remove the valve from the guide and wipe off all traces of grinding compound. Use solvent to clean the valve and wipe the seat area thoroughly with a solvent soaked cloth.
24 Repeat the procedure with fine valve grinding compound, then repeat the entire procedure for the remaining valves.
25 Lay the inner and outer spring seats for the first valve in place in the cylinder head then install a new valve stem seal onto the guide (see illustrations). Use an appropriate size deep socket to push the seal over the end of the valve guide until it is felt to clip into place. Don't twist or cock it, or it will not seal properly against the valve stem. Also, don't remove it again or it will be damaged.
26 Coat the valve stem with engine oil, then install it into its guide, rotating it slowly to avoid damaging the seal (see illustration). Check that the valve moves up and down freely in the guide.
27 Fit the follower bore protector (if available) then install the inner and outer springs, ensuring both springs are fitted with their closer-wound coils facing down into the cylinder head (see illustrations). Ensure both springs are correctly located on their seats then fit the spring retainer, with its shouldered side facing down so that it fits into the top of the springs (see illustration).
28 Apply a small amount of grease to the collets to help hold them in place. Compress the spring with the valve spring compressor and install the collets (see illustration). When compressing the spring, depress it only as far as is absolutely necessary to slip the collets into place. Make certain that the collets are securely locked in the retaining groove.

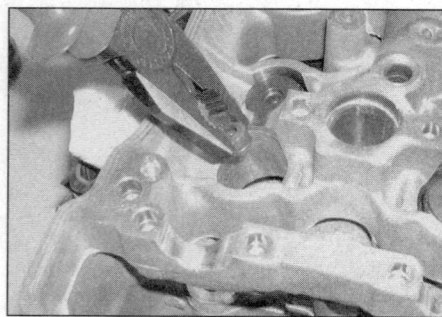

11.25c . . . then press a new seal onto the valve guide

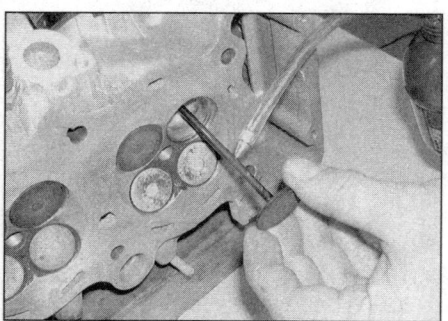

11.26 Lubricate the valve stem with oil and insert the valve into its guide

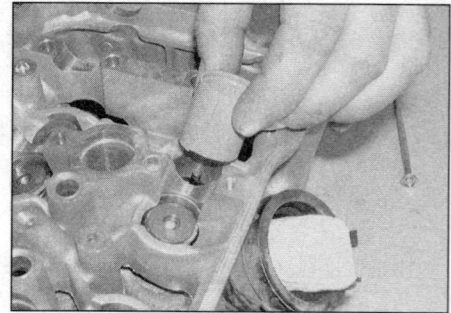

11.27a Fit the protector to the follower bore . . .

Engine, clutch and transmission 2•21

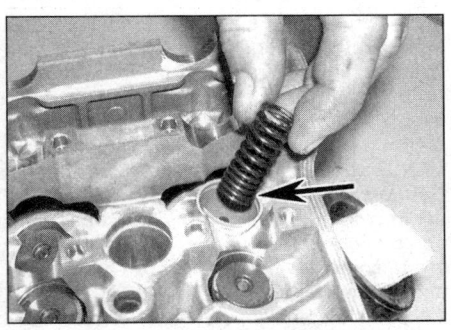

11.27b ... and install the inner spring, ensuring its closer-wound coils (arrow) are at the bottom ...

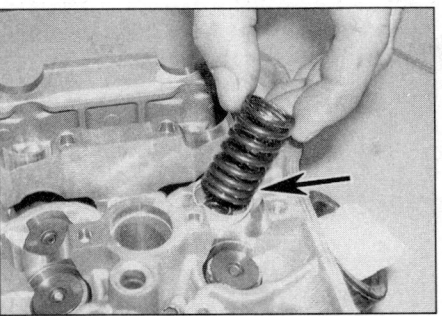

11.27c ... and the outer spring also with its closer-wound coils (arrow) at the bottom

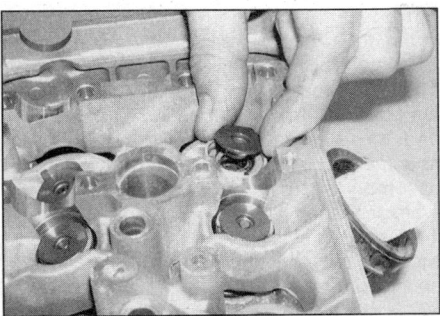

11.27d Ensure both springs are correctly seated then fit the spring retainer

29 Repeat the procedure for the remaining valves. Remember to keep the parts for each valve together and separate from the other valves so they can be reinstalled in the same location.

30 Support the cylinder head on blocks so the valves can't contact the workbench top, then very gently tap each of the valve stems with a soft-faced hammer. This will help seat the collets in their grooves.

 **Check for proper sealing of the valves by pouring a small amount of solvent into each of the valve ports. If the solvent leaks past any valve into the combustion chamber area the valve grinding operation on that valve should be repeated.**

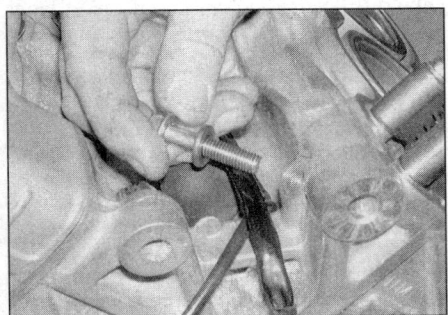

12.2 Slacken and remove the 8 mm bolt and sealing washer ...

12.3 ... then undo the four 6 mm bolts (arrows) securing the camshaft drive gear in position

12 Camshaft drive gears – removal, inspection and installation

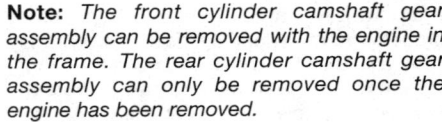

Note: *The front cylinder camshaft gear assembly can be removed with the engine in the frame. The rear cylinder camshaft gear assembly can only be removed once the engine has been removed.*

Removal

1 Remove the cylinder head(s) (see Section 9).
2 Slacken and remove the 8 mm bolt and sealing washer securing the drive gear assembly to the crankcase **(see illustration)**. Discard the sealing washer; a new one should be used on installation.
3 Evenly and progressively slacken and remove the four 6 mm bolts securing the gear assembly to the top of the crankcase **(see illustration)**.
4 Carefully lift the gear assembly out of position and recover the locating dowels from the crankcase **(see illustration)**. **Note:** *The front and rear cylinder gear assemblies are different and are not interchangeable; the front cylinder assembly has an 'F' stamped on its side and the rear cylinder assembly an 'R' **(see illustration)**. If both gear assemblies are removed at the same time, ensure the identifications markings are clearly visible.*

Inspection

5 Wash the assembly in clean solvent and dry it off.

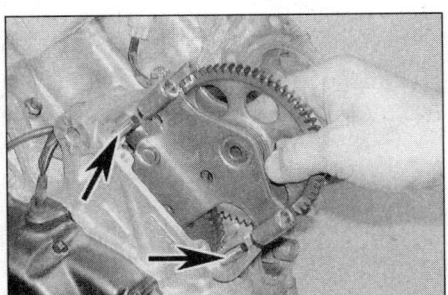

12.4a Lift out the camshaft drive gear assembly and recover the locating dowels (arrows) from the crankcase

11.28 Compress the valve springs and install the collets, using grease to hold them in position

6 Check the gear teeth for cracking and other obvious damage and make sure each gear turns smoothly.
7 If wear or damage of the gears is noted the complete assembly must be renewed. *Do not attempt to dismantle the gear assembly*. If the damage is severe, the camshaft and crankshaft gears will also need to be checked for wear and damage (see Sections 8 and 14) and renew if necessary.

Refitting

8 Insert the locating dowels into the top of the crankcase.
9 Manoeuvre the gear assembly into position and locate it on the dowels making sure it is correctly engaged with the crankshaft gear. If

12.4b The front and rear drive gear assemblies are different; the rear gear assembly is marked 'R' (1) and the front gear assembly with an 'F' (2)

2•22 Engine, clutch and transmission

both gear assemblies have been removed, make sure the assembly is being installed in its original location (see Paragraph 4).

10 Fit the four 6 mm bolts (the longer bolts are fitted to the dowel locations), then evenly and progressively tighten them to the specified torque, making sure the gear assembly is drawn squarely down onto the crankcase **(see illustration)**.

11 Fit a new sealing washer to the 8 mm bolt then install the bolt and tighten to the specified torque.

12 Fit the cylinder head (see Section 9).

13 Starter motor clutch and idle/reduction gears – removal, inspection and installation

Note: *The starter motor clutch and gears can be removed with the engine in the frame. If the engine has been removed, it will be necessary prevent crankshaft rotation to enable the ignition rotor/starter clutch bolt to be slackened (see Paragraph 9).*

Removal

1 Remove the right lower fairing panel as described in Chapter 8.
2 Drain the engine oil as described in Chapter 1.
3 Remove the air filter housing as described in Chapter 4.
4 Locate the ignition pulse generator (red) wiring connector, which is inside the protective cover on the right side of the frame, and disconnect it from the main wiring harness **(see illustration)**.

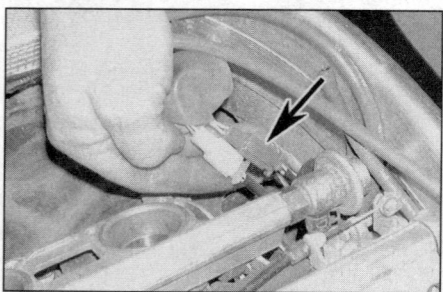

13.4 Locate the ignition pulse generator wiring connector (arrow) and disconnect it from the main wiring harness

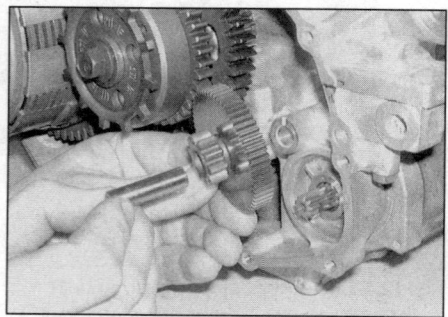

13.8b . . . then remove the shaft and reduction gear

12.10 Tighten the 6 mm bolts (arrows) evenly and progressively to the specified torque, ensuring the gear is pulled squarely into position

5 Working in a criss-cross pattern, evenly slacken the right crankcase cover retaining bolts, noting the correct fitted position of the wiring retaining clamp.
6 Lift the cover away from the engine, being prepared to catch any residual oil which may be released as the cover is removed.
7 Remove the gasket and discard it. Note the two locating dowels fitted to the crankcase; remove these for safe-keeping if they are loose.
8 Withdraw the starter idler gear shaft and remove the idler gear, then withdraw the reduction gear shaft and remove the gear **(see illustrations)**.
9 To slacken the timing rotor/starter clutch bolt the crankshaft must be prevented from turning. If the engine is still in the frame, select top gear and have an assistant apply the rear

13.8a Withdraw the shaft and remove the starter idler gear . . .

13.9 Using a peg spanner to retain the timing rotor whilst the rotor/starter clutch bolt is slackened

brake. If the engine has been removed, the timing rotor can be retained using a peg spanner which locates snugly in the timing rotor holes or alternatively remove the alternator cover from the left-hand side of the engine and hold the alternator rotor using a rotor strap (see Chapter 9) **(see illustration)**.

Caution: *Do not prevent rotation by using a socket on the alternator rotor bolt as it could become loose.*

10 Unscrew the bolt and remove the washer and the timing rotor.
11 Pull the complete starter clutch assembly off the end of the crankshaft, and remove the flanged spacer from behind it **(see illustration)**. Note how the wider grooves in the timing rotor and starter clutch fit over the wider spline on the crankshaft.
12 To dismantle the starter clutch assembly, rotate the driven gear in a clockwise direction and pull it out then remove the needle roller bearing. Using circlip pliers, extract the large circlip from the rear of the starter clutch outer then remove the one-way clutch assembly by rotating it in an anti-clockwise direction, noting which way around it is installed.

Inspection

13 Inspect the starter clutch driven gear, idler gear and reduction gear teeth and renew them as a set if any teeth are chipped or missing. Check the gear shafts and bearing surfaces for signs of wear or damage, and renew if necessary.
14 Inspect the one-way clutch rollers and driven gear contact surfaces for signs of wear and scoring. The one-way clutch rollers should be unmarked with no signs of wear such as pitting or flat spots. The degree of wear on the driven gear can be assessed by measuring the outside diameter of its boss and comparing it to the service limit given in the Specifications **(see illustration)**.
15 Inspect the needle roller bearing and driven gear contact surfaces for wear or scoring. Renew worn components as necessary.

Installation

16 Lubricate the one-way clutch assembly with clean engine oil then ease the assembly into the clutch outer, ensuring its flanged

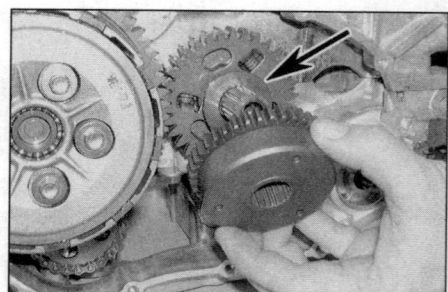

13.11 Slide the starter clutch assembly off the crankshaft and remove the flanged spacer (arrow)

Engine, clutch and transmission 2•23

13.14 Measuring starter clutch driven gear diameter

13.16a Fit the one-way clutch assembly to the clutch outer . . .

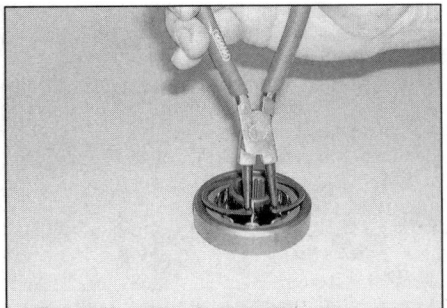

13.16b . . . and secure it in position with the circlip

13.16c Fit the driven gear with a twisting motion . . .

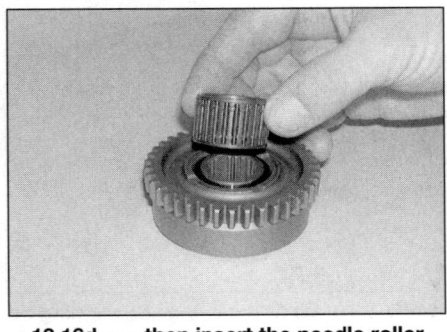

13.16d . . . then insert the needle roller bearing in between the gear and clutch outer

13.17 Fit the flanged spacer to the crankshaft ensuring its flange is facing outwards

(circlip) side is facing towards the clutch outer. Rotate the one-way clutch in an anti-clockwise direction until it is fully home then

secure it in position with the circlip (see illustrations). Fit the driven gear, whilst rotating it in a clockwise direction, then slide the needle roller bearing into position between the driven gear and clutch outer (see illustrations).
17 Fit the flanged spacer to the crankshaft with its flange facing outwards (see illustration).
18 Ensure the starter clutch assembly is correctly assembled then align its wide (master) spline with that of the crankshaft and slide it into position.
19 Ensure the ignition rotor markings are facing outwards then align its wide (master) spline with that of the crankshaft and locate it on the crankshaft end (see illustration).
20 Lubricate the threads and head of the timing rotor/starter clutch bolt with clean oil then fit the washer to the bolt. Screw the bolt

into position and tighten it to the specified torque, preventing rotation using the method employed on removal (see Paragraph 9) (see illustrations).
21 Apply a smear of clean engine oil to the idler and reduction gear shafts. Slide the shaft into the reduction gear then engage the gear with the starter motor and locate the shaft in the crankcase (see illustration 13.8b). Engage the idler gear with the reduction gear and starter clutch then insert the shaft and locate it correctly in the crankcase (see illustration 13.8a).
22 Ensure the mating surfaces are clean and dry then apply a smear of sealant (5 to 15 mm wide) to the area of the crankcase mating surface on each side of the crankcase half joints. Fit both locating dowels to the crankcase and locate the new gasket on the dowels (see illustration).

13.19 Align the wide (master) spline of the timing rotor with that of the crankshaft (arrows) and fit the rotor

13.20a Lubricate the threads and head of the timing rotor/starter clutch bolt with oil . . .

13.20b . . . then fit the bolt and tighten it to the specified torque

13.22 Locate the new crankcase cover gasket on the dowels (arrowed)

2•24 Engine, clutch and transmission

14.2a Lever the primary drive gear teeth into alignment either with a screwdriver located in the circular hole in the gear . . .

14.2b . . . or with a large, flat-bladed screwdriver positioned between the gear teeth

14.3 Slide off the camshaft drive gear, noting which way around it is fitted

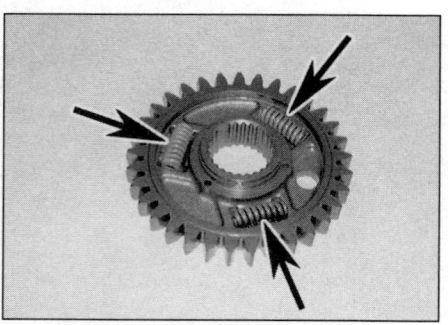

14.6a Ensure the springs (arrows) are correctly located . . .

14.6b . . . then fit the sub-gear, aligning its circular hole with that of the main gear (arrows)

14.6c Ensure the primary drive gear is correctly assembled then fit the circlip

23 Apply a smear of sealant to the ignition pulse generator wiring grommet then manoeuvre the right crankcase cover into position and locate it on the dowels.
24 Install the cover retaining bolts, making sure the wiring clamp is correctly positioned, and tighten them evenly and progressively to the specified torque setting.
25 Ensure the ignition pulse generator wiring is correctly routed then securely reconnect it to the main wiring harness. Install the air filter housing (see Chapter 4).
26 Fill the engine with the correct type and amount of oil as described in Chapter 1.
27 Install the fairing panel(s) as described in Chapter 8.

14 Primary drive gear –
removal, inspection and installation

Note: *The primary drive gear can be removed with the engine in the frame.*

Removal

1 Remove the starter clutch assembly and flanged spacer as described in Section 13.
2 In order to allow the primary drive gear to be slid off the crankshaft it is necessary to align the sub-gear teeth with the main gear. To do this, either insert a screwdriver into the circular hole in the gear or locate a large, flat-bladed screwdriver in-between the gear teeth and lever against the spring pressure to bring the gear teeth into alignment **(see illustrations)**. The primary drive gear can then be slid off the crankshaft, noting which way around it is fitted.
3 Slide the camshaft drive gear off the end of the crankshaft, noting which way around it is fitted **(see illustration)**. Note how the wider groove in the gear fits over the wider spline on the crankshaft.

Inspection

4 Inspect the primary drive gear and sub-gear teeth for signs of wear or damage. If the drive gear springs are thought to be suspect, remove the circlip and lift off the sub-gear. Inspect the springs for signs of fatigue or damage.
5 If the primary drive gear assembly is worn or damaged it must be renewed as an assembly; no components are available separately. If the damage to the gear teeth is severe be sure to check the clutch drum gear for damage.

6 If all components are in good condition, ensure the springs are correctly located in the main gear cutouts then fit the sub-gear, aligning its circular hole with that of the main gear **(see illustrations)**. Secure the sub-gear in position with the circlip making sure it is correctly located in the main gear groove **(see illustration)**.

Refitting

7 Ensure the camshaft drive gear is the correct way around (unsplined area of its internal bore innermost) then align its wide (master) spline with that of the crankshaft and slide it into position **(see illustration)**.
8 Position the primary drive gear so its sub-gear is outermost then align its wide (master) spline with that of the crankshaft and slide the gear onto the shaft **(see illustration)**. As the gear engages with the clutch drum gear, align the main gear and sub-gear teeth with a

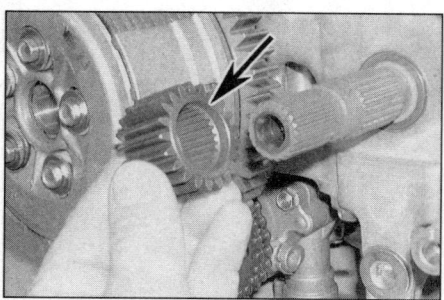

14.7 Fit the camshaft drive gear ensuring the unsplined area of its internal bore (arrow) is innermost

14.8 Fit the primary drive gear with its sub-gear outermost

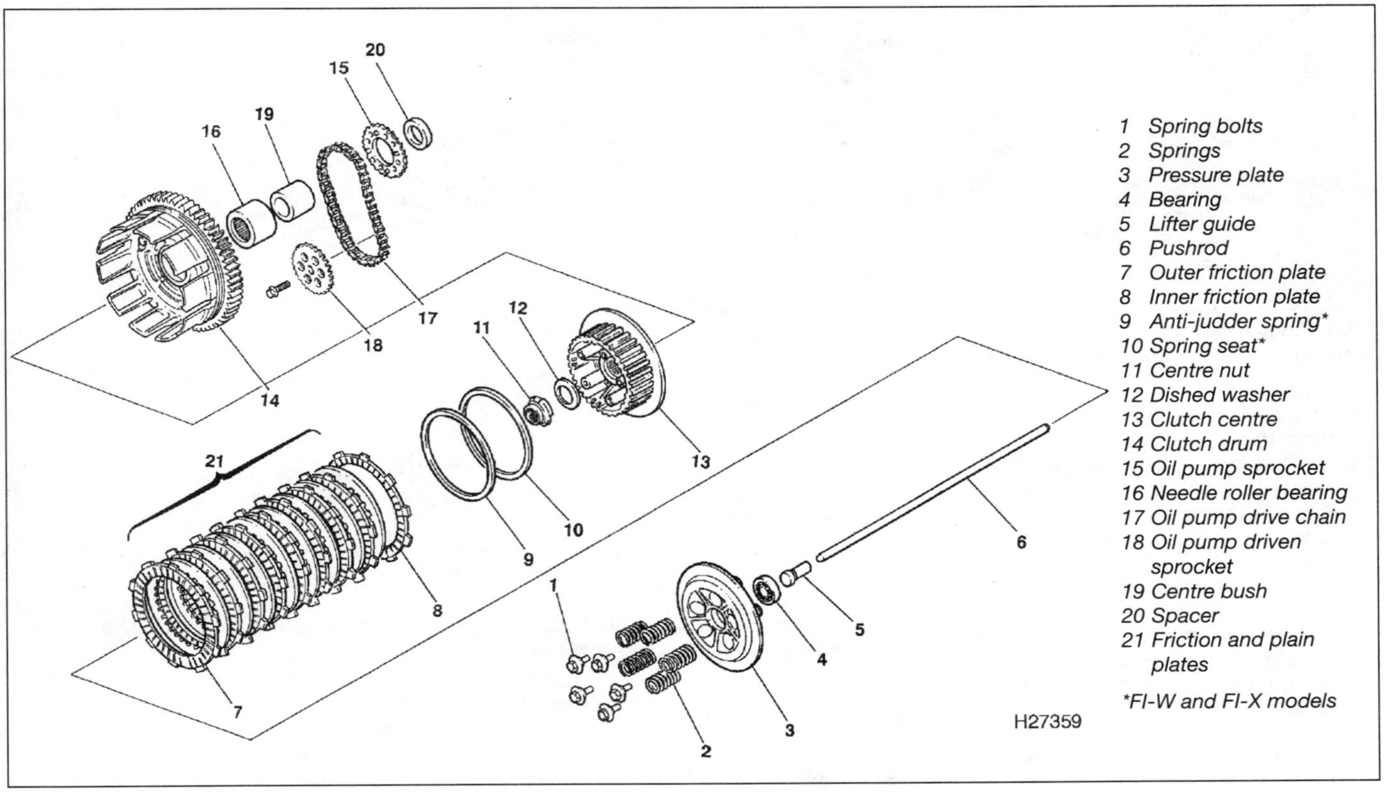

Engine, clutch and transmission 2•25

15.2 Exploded view of the clutch assembly

1. Spring bolts
2. Springs
3. Pressure plate
4. Bearing
5. Lifter guide
6. Pushrod
7. Outer friction plate
8. Inner friction plate
9. Anti-judder spring*
10. Spring seat*
11. Centre nut
12. Dished washer
13. Clutch centre
14. Clutch drum
15. Oil pump sprocket
16. Needle roller bearing
17. Oil pump drive chain
18. Oil pump driven sprocket
19. Centre bush
20. Spacer
21. Friction and plain plates

*Fl-W and Fl-X models

screwdriver (see Paragraph 2) to allow the primary drive gear to be slid fully onto the crankshaft.

9 Install the flanged spacer and starter clutch (see Section 13).

15 Clutch –
removal, inspection and installation

Note: *The clutch assembly can be removed with the engine in the frame. If the engine has been removed, some means of holding the clutch centre will be required to enable the centre nut to be slackened (see Paragraph 8).*

Removal

1 Remove the right crankcase cover as described in Paragraphs 1 to 7 of Section 13.
2 Working in a criss-cross pattern, gradually slacken the clutch spring retaining bolts until spring pressure is released. Unscrew the bolts and remove them along with the springs **(see illustration)**.
Caution: Do not operate the clutch lever from this point in the removal procedure.
3 Remove the pressure plate from the clutch and recover the lifter guide from the plate bearing or pushrod (as applicable) **(see illustrations)**.
4 Withdraw the pushrod from the centre of the input shaft.
5 Withdraw the clutch friction plates and plain

plates. **Note:** *On VFR800FI-W and X models, the inner friction plate is different to the others. Mark the plate in some way to ensure that it is fitted in its original position. On VFR800FI-Y and 1 models, the inner and outer friction plates are coloured blue, all the others are brown.*
6 On VFR800FI-W and X models, remove the anti-judder spring and spring seat from the clutch centre, noting which way around the spring is fitted.
7 Unstake the clutch centre nut using a hammer and suitable pointed-nose chisel, taking care not to damage the input shaft end **(see illustration)**.
8 To slacken the clutch centre nut the input shaft/clutch centre must be locked. This can be done in several ways.

15.3a Remove the pressure plate from the clutch . . .

15.3b . . . and recover the lifter guide from the pushrod

15.7 Unstake the centre nut from the input shaft groove with a pointed-nose chisel

2•26 Engine, clutch and transmission

15.8 Using a clutch centre holding tool to prevent rotation whilst the centre nut is slackened

a) If the engine is in the frame, lock the clutch through the transmission, by selecting top gear and applying the rear brake hard whilst the nut is slackened.
b) Retain the clutch centre with Honda service tool (07724-0050002) or a universal clutch centre holding tool (available from most good motorcycle accessory dealers). Alternatively, a clutch centre holding tool can be fabricated from some steel strap, bent at the ends and bolted together in the middle **(see illustration and Tool Tip)**.
c) If the engine is out of the frame, fit a close-fitting ring spanner over the output shaft splines then lock the clutch through the transmission by selecting top gear. Hold the spanner whilst the centre nut is slackened.

9 Remove the centre nut and discard it; a new one must be used on installation.
10 Remove the dished washer, noting which way around it is fitted.
11 Remove the clutch centre.
12 In order to allow the clutch drum needle roller bearing to be slid off the crankshaft it is necessary to align the primary drive sub-gear teeth with the main gear. To do this, locate a large, flat-bladed screwdriver in-between the gear teeth and lever against the spring pressure to bring the gear teeth into alignment. The needle roller bearing can then be manoeuvred out from the centre of the clutch drum **(see illustration)**.

15.12 Align the primary drive gear teeth with a screwdriver (arrow) then slide the needle roller bearing out from the clutch drum

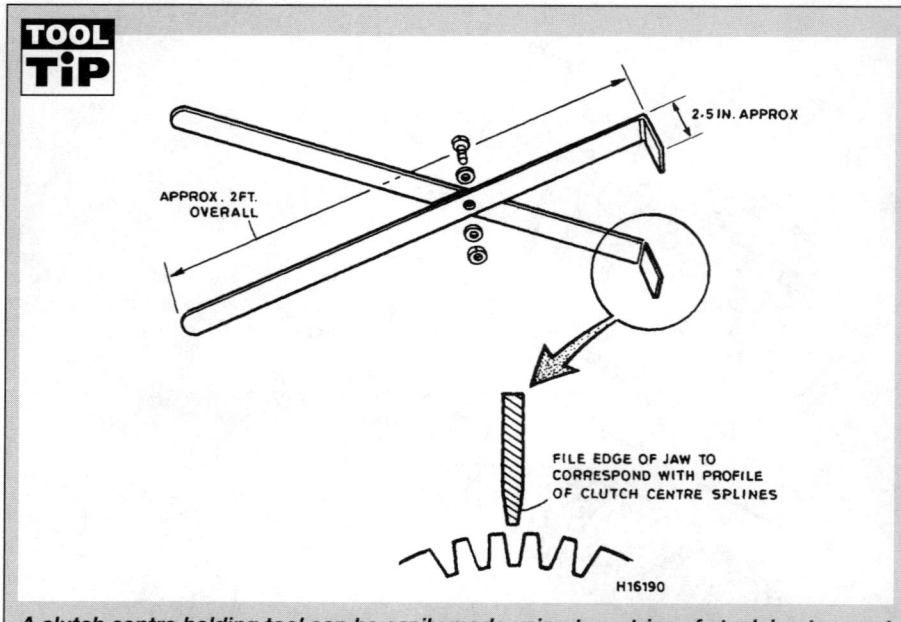

A clutch centre holding tool can be easily made using two strips of steel, bent over at the ends and bolted together in the middle

13 Disengage the clutch drum from the oil pump drive sprocket and manoeuvre it out of position.
14 Slacken and remove the oil pump sprocket retaining bolt whilst preventing the sprocket turning by inserting a screwdriver or extension bar through one of the sprocket holes.
15 Slide the drive sprocket off the input shaft and remove the drive sprocket, chain and pump sprocket as an assembly.
16 Remove the centre bush from the input shaft and slide off the spacer.

Inspection

17 After an extended period of service the clutch friction plates will wear and promote clutch slip. Measure the thickness of each friction plate using a vernier caliper **(see illustration)**. If any plate has worn to or beyond the service limit given in the Specifications, the friction plates must be renewed as a set.
18 The plain plates should not show any signs of excess heating (bluing). Check for warpage using a flat surface and feeler gauges **(see illustration)**. If any plate exceeds the maximum permissible amount of warpage, or shows signs of bluing, all plain plates must be renewed as a set.
19 Inspect the clutch assembly for burrs and indentations on the edges of the protruding tangs of the friction plates and/or slots in the edge of the clutch drum with which they engage. Similarly check for wear between the inner tongues of the plain plates and the slots in the clutch centre. Wear of this nature will cause clutch drag and slow disengagement during gearchanging, since the plates will snag when the pressure plate is lifted. With care, a small amount of wear can be corrected by dressing with a fine file, but if this is excessive the worn components must be renewed.
20 On VFR800FI-W and X models, also inspect the anti-judder spring and seat for signs of wear or distortion and renew if necessary.
21 Inspect the input shaft, centre bush and

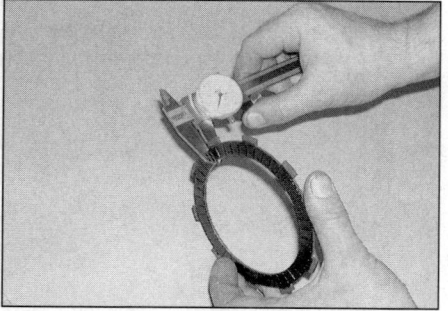

15.17 Measuring the thickness of a clutch friction plate

15.18 Checking a plain plate for warpage

Engine, clutch and transmission 2•27

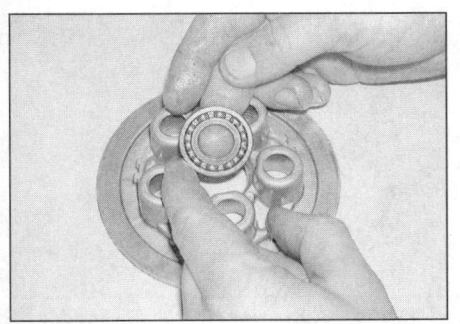

15.22 Check the pressure plate bearing for wear and renew if necessary

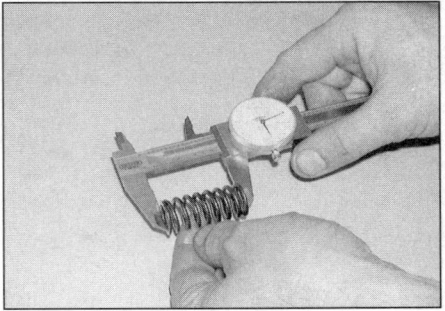

15.23 Measuring the free length of a clutch spring

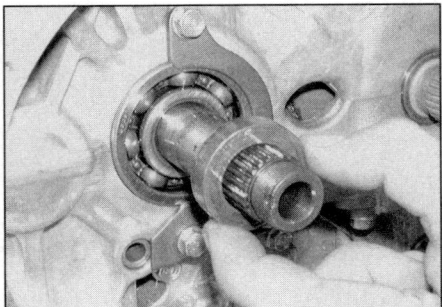

15.26 Slide the spacer onto the input shaft

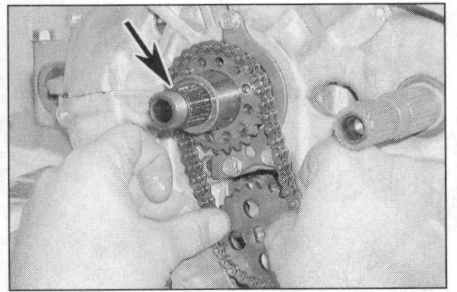

15.27 Engage the oil pump sprockets correctly with the drive chain then locate the drive sprocket on the centre bush (arrow)

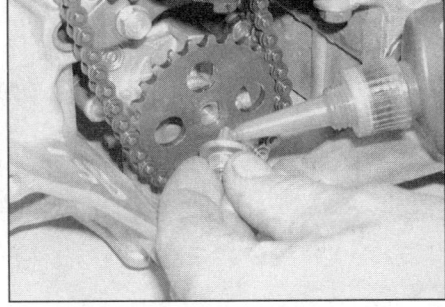

15.28a Apply thread locking compound to the pump sprocket bolt . . .

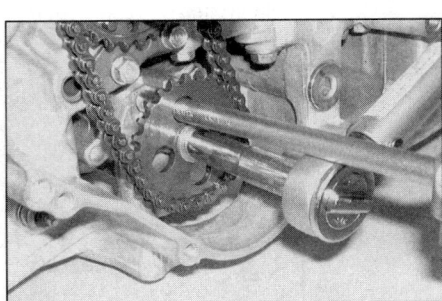

15.28b . . . the fit and tighten the bolt to the specified torque, using an extension bar to prevent rotation

clutch drum bearing surfaces, along with the needle bearing, for signs of wear and damage. If any component shows signs of wear or damage, it must be replaced.

22 Check the clutch pressure plate bearing for wear (see illustration). Ensure that the inner race of the bearing spins freely without any sign of notchiness and that there is no freeplay between the inner and outer races or the outer race and plate. If necessary, renew the bearing by driving the old bearing out of the plate and tapping the new bearing into position using a hammer and suitable tubular drift which bears only on the bearing's outer race.

23 Measure the free length of each clutch spring (see illustration). If any one has settled to less than the service limit, the clutch springs must be renewed as a set.

24 Check the pushrod for signs of damage and straightness by rolling it on a flat surface. Renew the pushrod if it's bent.

Installation

25 Remove all traces of gasket from the crankcase and cover sealing surfaces.
26 Apply a smear of clean engine oil to the spacer and centre bush and slide them onto the input shaft (see illustration).
27 Engage the oil pump and drive sprockets with the chain then slide the drive sprocket and chain assembly onto the input shaft (see illustration). Ensure the drive sprocket dogs are facing outwards then engage the oil pump sprocket with the pump driveshaft.
28 Clean the oil pump sprocket bolt and apply a few drops of a suitable locking compound to its threads (see illustration). Fit the bolt and tighten it to the specified torque, preventing rotation by inserting a screwdriver or extension bar through one of the sprocket holes (see illustration).
29 Manoeuvre the clutch drum into position and engage it with the oil pump drive sprocket dogs. To enable this it will be necessary to align the primary drive main gear and sub-gear teeth with a screwdriver (see Paragraph 12). Ensure the drum is correctly engaged with the sprocket then slide the needle roller bearing into position (see illustration 15.12). Note: Rotate the oil pump sprocket to check the drum is correctly engaged with its drive sprocket before proceeding.
30 Fit the clutch centre (see illustration).
31 Fit the dished washer with its convex face outermost (see illustration).
32 Fit the new clutch centre nut and tighten it to the specified torque setting whilst holding the clutch centre using the method employed on removal (see illustration). Secure it in

2

15.30 Ensure the drum and oil pump sprocket are correctly engaged then fit the clutch centre

15.31 Fit the dished washer with its convex face outwards

15.32a Fit a new centre nut and tighten it to the specified torque . . .

2•28 Engine, clutch and transmission

15.32b . . . then stake it firmly into the input shaft groove

15.33a On VFR800FI-W and X models fit the spring seat to the clutch centre . . .

15.33b . . . then install the anti-judder spring with its convex side facing towards the seat

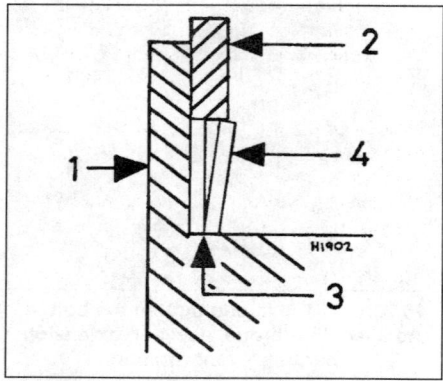

15.33c Anti-judder spring fitting details (VFR800FI-W and X models)

1 Clutch centre 3 Spring seat
2 Inner friction plate 4 Anti-judder spring

15.34a Install the inner friction plate first

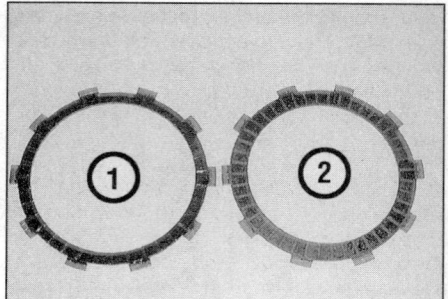

15.34b On VFR800FI-W and X models the inner friction plate (1) has a larger internal diameter than the other friction plates (2)

15.35a Once the inner friction plate is in position, alternately fit the plain plates . . .

position by staking it into the groove in the input shaft using a suitable hammer and punch (see illustration).

33 On VFR800FI-W and X models, fit the spring seat to the clutch centre followed by the anti-judder spring (see illustrations). Note: *The anti-judder spring must be fitted with its convex side facing the spring seat.*

34 Using the identification mark made on dismantling, fit the inner friction plate to the clutch centre. Note: *On VFR800FI-W and X models the inner friction plate has a slightly larger internal diameter than the other plates to allow it to fit over the anti-judder spring and its seat* (see illustrations).

35 Install a plain plate followed by one of the ordinary friction plates, then alternately install all the remaining plain and friction plates (see illustrations). Note: *If new clutch plates are being fitted, apply a coating of engine oil to their surfaces to prevent seizure.*

36 Insert the pushrod into the centre of the input shaft, ensuring it is installed the correct way around. Lubricate the end of the pushrod with a smear of grease then fit the clutch lifter guide to the pushrod (see illustrations).

37 Ensure the pressure plate bearing is correctly installed then fit the pressure plate, making sure the lifter guide engages correctly with the bearing.

38 Install the clutch springs and retaining

15.35b . . . and remaining friction plates

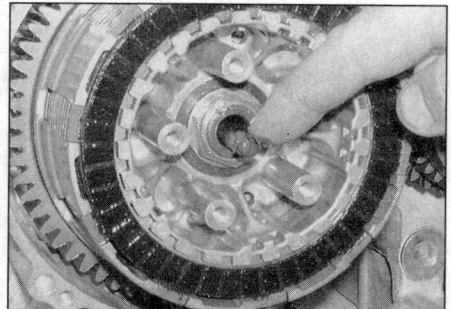

15.36a Fit the pushrod and lubricate its end with a smear of grease . . .

15.36b . . . before fitting the lifter guide to the rod

Engine, clutch and transmission 2•29

bolts. Gradually tighten the bolts evenly in a criss-cross pattern until they are all tightened to the specified torque **(see illustrations)**.
39 Install the crankcase cover as described in Paragraphs 22 to 27 of Section 13.

16 Clutch master cylinder – removal, overhaul and installation

1 If the master cylinder is leaking fluid, or if the clutch does not function and bleeding does not help (see Section 18) then master cylinder overhaul is recommended.
2 Before disassembling the master cylinder, read through the entire procedure and make sure that you have obtain all new parts required. Also, you will need some new DOT4 brake fluid, some clean rags and internal circlip pliers.

Removal

Caution: Disassembly, overhaul and reassembly of the master cylinder must be done in a spotlessly clean work area to avoid contamination and possible failure of the hydraulic system components. To prevent damage to the paint from spilled hydraulic fluid, always cover the fuel tank and fairing panels when working on the master cylinder.
3 Loosen, but do not remove, the screws holding the reservoir cover in place.
4 Disconnect the electrical connectors from the clutch switch.
5 Note the correct fitted location of the clutch hose end fitting then unscrew the banjo bolt and separate the hose from the master cylinder. Plug the hose end or wrap a plastic bag tightly around it to minimise fluid loss and prevent dirt entering the system. Discard the sealing washers as new ones must be used on installation.
6 Unscrew the mounting clamp bolts and remove the master cylinder from the handlebar.
7 Remove the reservoir cover retaining screws and lift off the cover, the diaphragm plate and the rubber diaphragm and empty the reservoir contents into a suitable container.

Overhaul

8 Unscrew the locknut from the clutch lever pivot screw then remove the screw and lever from the master cylinder **(see illustration)**.
9 Undo the screw and remove the clutch switch from the master cylinder.
10 Carefully remove the dust boot and pushrod from the end of the master cylinder bore.
11 Using circlip pliers, remove the circlip. On VFR800FI-W and X models, slide out the washer, piston assembly, primary cup and spring, noting how they fit. On VFR800FI-Y and 1 models, slide out the piston assembly and spring. Lay the parts out in the correct

15.38a Fit the pressure plate and install the clutch springs and bolts . . .

15.38b . . . tightening them evenly and progressively to the specified torque

fitted order to prevent confusion during reassembly.
12 Clean all parts with clean brake fluid or denatured alcohol. If compressed air is available, use it to dry the parts thoroughly (make sure it's filtered and unlubricated).
Caution: Do not, under any circumstances, use a petroleum-based solvent to clean master cylinder parts.
13 Check the master cylinder bore for corrosion, scratches, nicks and score marks. If the necessary measuring equipment is available, compare the dimensions of the piston and bore to those given in the Specifications Section of this Chapter. If damage or wear is evident, the master cylinder must be renewed. If the master cylinder is in poor condition, then the slave cylinder should be checked as well. Check that the fluid inlet and outlet ports in the master cylinder are clear. Inspect the reservoir cover rubber diaphragm and renew it if it is damaged or deteriorated.

14 Depending upon the original fitment, the dust boot, circlip, washer, piston assembly and primary cup are included in the new piston/seal kit **(see illustration 16.8)**. Use all of the new parts, regardless of the apparent condition of the old ones. If the seals are not already on the piston, fit them according to the layout of the old piston assembly ensuring both are installed the correct way around.
15 Where fitted, install the new primary cup to the tapered end of the new spring then insert the assembly into the master cylinder so that the primary cup is towards the piston.
16 Lubricate the new piston assembly with clean brake fluid. Ensure the assembly is the correct way round, then carefully ease it into the master cylinder. Where fitted, install the new washer, then depress the piston and install the new circlip, making sure that it locates in the master cylinder groove.
17 Fit the new rubber dust boot to the pushrod then lubricate the pushrod ends with silicone grease. Fit the pushrod and boot

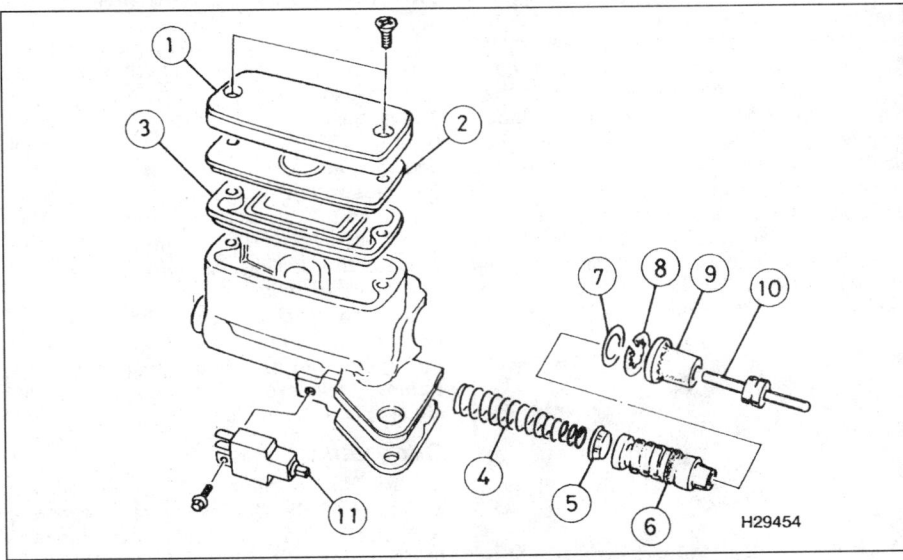

16.8 Exploded view of the clutch master cylinder

1 Reservoir cover	5 Primary cup (VFR800FI-W and X models only)	8 Circlip
2 Diaphragm plate		9 Dust boot
3 Rubber diaphragm	6 Piston assembly	10 Pushrod
4 Spring	7 Washer (VFR800FI-W and X models only)	11 Clutch switch

2•30 Engine, clutch and transmission

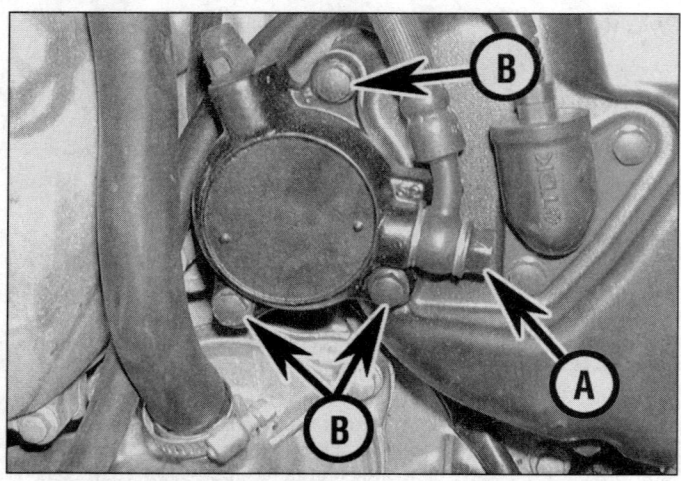

17.2 Clutch slave cylinder hydraulic hose banjo bolt (A) and cylinder retaining bolts (B)

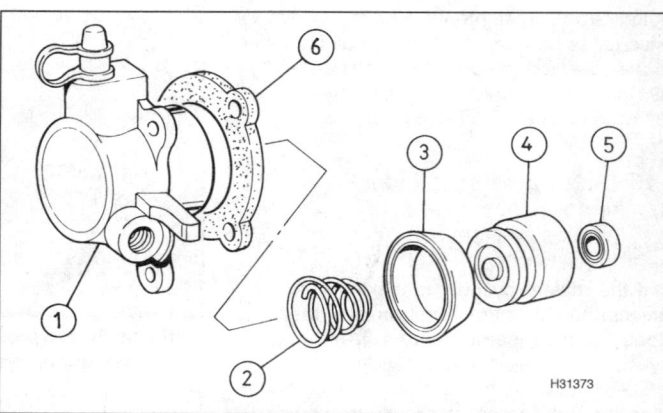

17.5 Exploded view of the clutch slave cylinder

1 Slave cylinder body
2 Spring
3 Fluid seal
4 Piston
5 Pushrod seal
6 Gasket

assembly to the master cylinder making sure the boot lip is seated correctly in the piston bore groove.
18 Fit the clutch switch to the master cylinder and securely tighten its retaining screw.
19 Fit the lever, ensuring it engages correctly with the pushrod end, then install the pivot screw. Securely tighten the pivot screw then fit the locknut and tighten to the specified torque.

Installation

20 Locate the master cylinder on the handlebar and fit the mounting clamp with its UP mark facing upwards. Align the outer corner of the master cylinder mounting with the punch mark on the top of the handlebar then tighten the mounting clamp upper bolt to the specified torque followed by the lower bolt.
21 Position a new sealing washer on each side of the clutch hose end fitting then connect the hose to the master cylinder. Ensure the hose end fitting is correctly positioned against its stop then tighten the banjo bolt to the specified torque.
22 Connect the wiring to the clutch switch.
23 Fill the fluid reservoir with new DOT4 hydraulic fluid as described in *Daily (pre-ride) checks*. Refer to Section 18 of this Chapter and bleed the air from the hydraulic system.

17.15a Ensure the locating dowels (arrows) are in position . . .

24 Ensure the fluid level is correct then fit the reservoir rubber diaphragm, diaphragm plate and cover and securely tighten its retaining screws.
25 Thoroughly check the operation of the clutch before riding the motorcycle.

17 Clutch slave cylinder – removal, overhaul and installation

Caution: Disassembly, overhaul and reassembly of the slave cylinder must be done in a spotlessly clean work area to avoid contamination and possible failure of the hydraulic system components.

Removal

1 Remove the left lower fairing panel as described in Chapter 8.
2 Note the correct fitted location of the clutch hose end fitting then unscrew the banjo bolt and separate the hose from the slave cylinder **(see illustration)**. Plug the hose end or wrap a plastic bag tightly around it to minimise fluid loss and prevent dirt entering the system. Discard the sealing washers as new ones must be used on installation.
3 Unscrew the clutch slave cylinder bolts and remove the cylinder from the sprocket cover, taking care not to lose the locating dowels. Remove the gasket and discard it; a new one should be used on refitting.

Overhaul

4 Clean the exterior of the cylinder with denatured alcohol or brake system cleaner.
5 Withdraw the piston from cylinder **(see illustration)**.

Haynes Hint: If the piston cannot be withdrawn by hand, it can be pushed out by applying compressed air to the clutch hose union hole. Only low pressure should be required, such as is generated by a foot pump. Wrap the slave cylinder in a wad of rag to prevent the piston being forcibly expelled.

6 Recover the spring from the cylinder, noting which way around it is fitted.
7 Carefully remove the pushrod seal and fluid seal from the piston, noting which way around each one is fitted.
8 Clean all parts with clean brake fluid or denatured alcohol. If compressed air is available, use it to dry the parts thoroughly (make sure it's filtered and unlubricated).

Caution: Do not, under any circumstances, use a petroleum-based solvent to clean master cylinder parts.

9 Inspect the cylinder bore and piston for signs of corrosion, nicks and burrs and loss of plating. If surface defects are present, the cylinder assembly must be renewed. If the cylinder is in bad shape the master cylinder should also be checked. No specifications are given to check piston and bore wear.
10 Lubricate the new piston fluid seal with clean fluid. Fit the seal to the piston groove making sure it is fitted the correct way around with its grooved face (wider end) facing away from the piston.
11 Fit the new pushrod seal to the piston, making sure its grooved face is facing towards the piston.
12 Fit the spring to the cylinder with its tapered end facing the piston.
13 Lubricate the piston with clean fluid and install it in the cylinder bore. Make sure the spring remains correctly positioned and take great care to ensure the fluid seal is not damaged as it enters the bore. Using your thumbs, push the piston all the way in, making sure it enters the bore squarely.

Installation

14 Remove all traces of gasket from the cylinder and cover sealing surfaces.

Engine, clutch and transmission 2•31

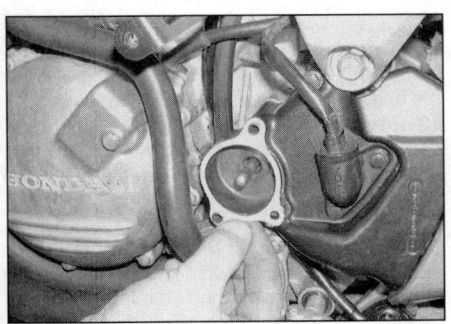

17.15b ... then fit a new gasket to the sprocket cover

15 Ensure the locating dowels are correctly fitted and fit a new gasket to the sprocket cover **(see illustrations)**.
16 Apply a smear of grease to the pushrod end then fit the clutch slave cylinder, aligning it with the pushrod and locating dowels. Install the retaining bolts and tighten them securely.
17 Position a new sealing washer on each side of the clutch hose end fitting then connect the hose to the slave cylinder. Ensure the hose end fitting is correctly positioned against its stop then tighten the banjo bolt to the specified torque.
18 Fill the master cylinder with new DOT4 hydraulic fluid (see *Daily pre-ride checks*) and bleed the hydraulic system as described in Section 18.
19 Check for leaks and thoroughly test the operation of the clutch before installing the fairing panel.

18 Clutch system – bleeding

Caution: *To prevent damage to the paint from spilled hydraulic fluid, always cover the fuel tank and fairing panels when working on the master cylinder.*

1 Bleeding the clutch is simply the process of removing all the air bubbles from the clutch master cylinder, the hose and the slave cylinder. Bleeding is necessary whenever a hydraulic connection is loosened, when a component or hose is renewed, or when the master cylinder or slave cylinder is overhauled.

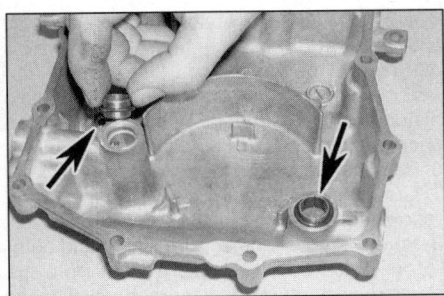

19.10 Fit a new O-ring to each locating dowel (arrows) and fit the dowels to the oil pan/crankcase

Leaks in the system may also allow air to enter, but leaking fluid will reveal their presence and warn you of the need for repair.
2 To bleed the clutch, you will need some new DOT4 hydraulic fluid, a length of clear vinyl or plastic tubing, a small container partially filled with clean hydraulic fluid, some rags and a spanner to fit the slave cylinder bleed valve.
3 Remove the left side lower fairing panel (see Chapter 8) to gain access to the slave cylinder bleed valve.
4 Undo the retaining screws and remove the master cylinder reservoir cover, diaphragm plate and diaphragm and slowly pump the clutch lever a few times, until no air bubbles can be seen floating up from the holes in the bottom of the reservoir. Doing this bleeds the air from the master cylinder end of the line. Loosely refit the reservoir cover.
5 Pull the dust cap off the slave cylinder bleed valve. Attach one end of the clear vinyl or plastic tubing to the bleed valve and submerge the other end in the brake fluid in the container.
6 Remove the reservoir cover and check the fluid level. Do not allow the fluid level to drop below the lower mark during the bleeding process.
7 Carefully pump the clutch lever three or four times and hold it in while opening the bleed valve. When the valve is opened, fluid will flow out of the cylinder into the clear tubing.
8 Retighten the bleed valve, then release the clutch lever gradually. Repeat the process until no air bubbles are visible in the fluid leaving the valve and the clutch action feels smooth and progressive.
9 Disconnect the bleeding equipment, then tighten the bleed valve to the specified torque and install the dust cap.
10 Top the fluid level up to the upper level mark (see *Daily (pre-ride) checks*) then install the diaphragm, diaphragm plate and reservoir cover assembly and securely tighten the retaining screws. Wipe up any spilled fluid and check the system for leaks before installing the fairing panel (see Chapter 8).

 *Failure to bleed satisfactorily after a reasonable repetition of the bleeding procedure may be due to worn master cylinder seals.*

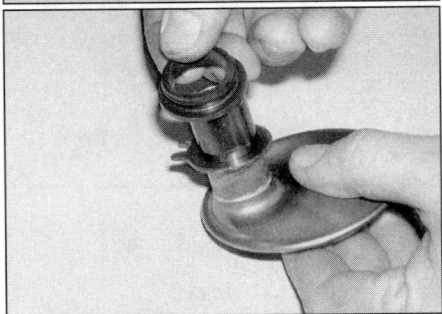

19.11a Fit the new seal with its flange facing the strainer ...

19 Oil pan – removal and installation

Note: *The oil pan can be removed with the engine in the frame.*

Removal

1 Remove the exhaust system front pipe assembly as described in Chapter 4.
2 Drain the engine oil as described in Chapter 1.
3 Unscrew the bolts securing the oil cooler pipe end fittings to the left and right of the oil pan and disconnect both pipes. Remove the O-ring from each end fitting and discard them; new ones must be used on installation.
4 Working in a criss-cross pattern, gradually loosen the oil pan retaining bolts.
5 Remove all the bolts and lower the oil pan away from the crankcase. If the engine is in the frame, note that as the oil pan is removed, the oil pump pick-up strainer and oil pressure relief valve may fall out of the bottom of the crankcase.
6 Recover the oil pan locating dowels and O-rings; discard the O-rings, new ones must be used on installation.
7 If they were not released when the oil pan was removed, remove the oil pump pick-up strainer and the oil pressure relief valve from the base of the engine. Remove the seal from the pick-up and the O-ring from the pressure relief valve and discard the; a new seal and O-ring will be needed for installation.
8 Clean the pick-up strainer mesh in solvent. Check the strainer for clogging or splitting and renew it if necessary.

Installation

9 Remove all traces of sealant from the oil pan and crankcase mating surfaces.
10 Fit a new O-ring to each end of the locating dowels and install them in their original locations in the crankcase/oil pan (as applicable) **(see illustration)**.
11 Fit a new seal to the oil pump pick-up strainer making sure its flange is facing the strainer **(see illustration)**. Fit the strainer to the oil pump, ensuring the locating peg on the pump is correctly positioned in the strainer lug cutout **(see illustration)**.

19.11b ... then fit the strainer to the oil pump ensuring the pump peg (arrow) is correctly located in the strainer lug cutout

2•32 Engine, clutch and transmission

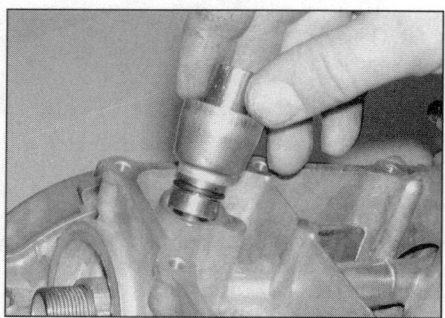

19.12 Fit a new O-ring to the oil pressure relief valve and insert the valve into the crankcase

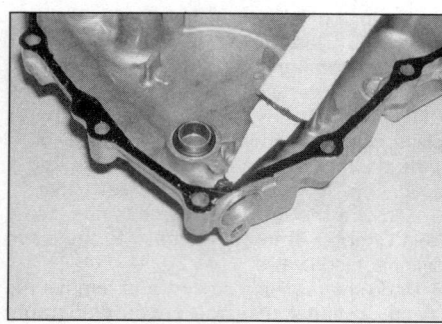

19.13 Apply a thin coat of sealant to the oil pan mating surface

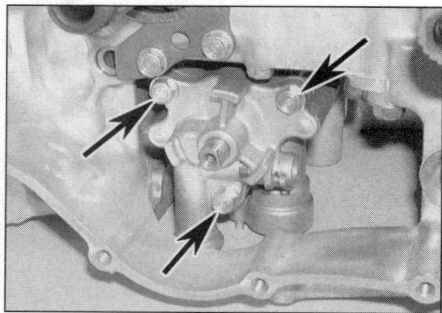

20.3a Unscrew the mounting bolts (arrows) then remove the oil pump . . .

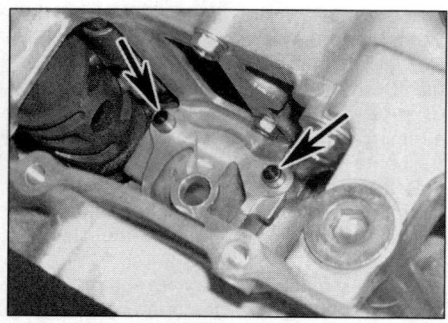

20.3b . . . and recover the locating dowels (arrows)

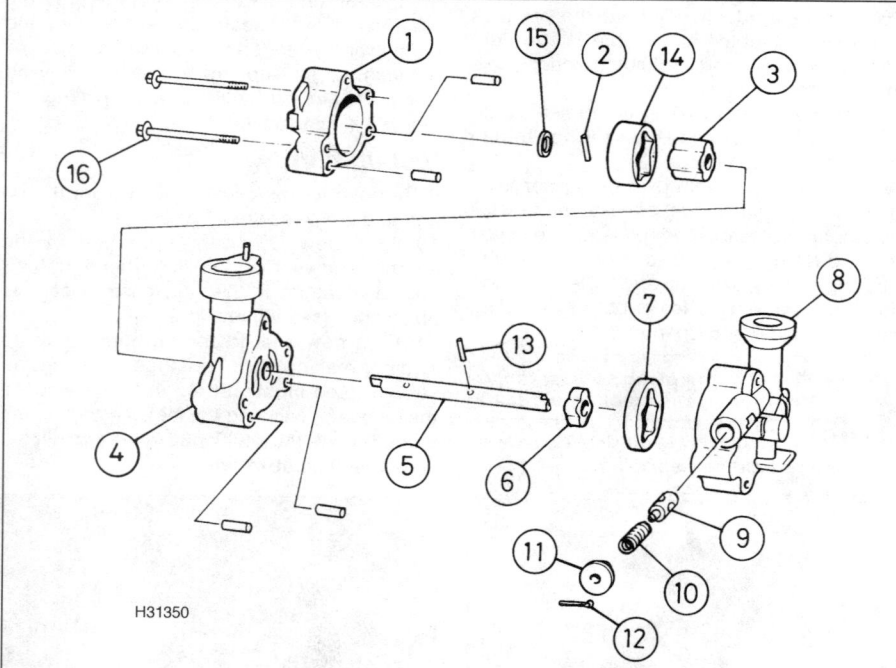

20.4 Exploded view of the oil pump

1 Main pump housing
2 Drive pin
3 Main pump inner rotor
4 Main pump housing
5 Driveshaft
6 Oil cooler pump inner rotor
7 Oil cooler pump outer rotor
8 Oil cooler pump housing
9 Oil cooler relief valve piston
10 Spring
11 Spring retainer
12 Split pin
13 Drive pin
14 Main pump outer rotor
15 Thrust washer
16 Bolts

12 Fit a new O-ring to the oil pressure relief valve groove. Apply a smear of oil to the O-ring and ease the relief valve into position in the base of the crankcase (see illustration).
13 Ensure the oil pan and crankcase mating surfaces are clean and dry then apply a thin coat of sealant (Honda recommend the use of Three Bond 1207B) to the mating surface of the oil pan (see illustration).
14 Ensure the locating dowels, relief valve and strainer are all correctly positioned then fit the oil pan. Insert all the retaining bolts and tighten them by hand.
15 Working in a criss-cross pattern, tighten all the oil pan retaining bolts to the specified torque setting.
16 Fit a new O-ring to each oil cooler pipe end fitting then locate the end fittings in the oil pan and securely tighten their retaining bolts.
17 Install the exhaust front pipe assembly as described in Chapter 4.
18 Fill the engine with the correct type and quantity of oil as described in 'Daily (pre-ride) checks'. Start the engine and check that there are no leaks.
19 If all is well, fit the fairing panels as described in Chapter 8.

20 Oil pump – removal, inspection and installation

Note: *The oil pump can be removed with the engine in the frame.*

Removal

1 Remove the oil pan and associated components as described in Section 19.
2 Remove the clutch assembly and the oil pump drive chain and sprockets as described in Section 15.
3 Unscrew the three mounting bolts and remove the pump assembly from the crankcase (see illustration). Recover the locating dowels and store them with the pump for safe-keeping.

Inspection

Note: *The pump has two sets of rotors. The thinner set circulates oil around the oil cooler and the thicker set functions as the main oil pump supply to the engine.*

4 Wash the oil pump in solvent, then dry it off (see illustration).
5 Remove the split pin and withdraw the spring seat, spring and oil cooler relief valve piston from the oil cooler pump cover, noting each component's correct fitted direction.
6 Remove the two pump cover bolts.
7 Lift off the oil cooler pump housing and recover the locating dowels.
8 Remove the oil cooler pump inner and outer rotors then slide the drive pin out from the pump driveshaft. **Note:** *Prior to removal, make a note which way around the pump outer rotor is fitted (there is a punch mark on one side of the rotor).*

Engine, clutch and transmission 2•33

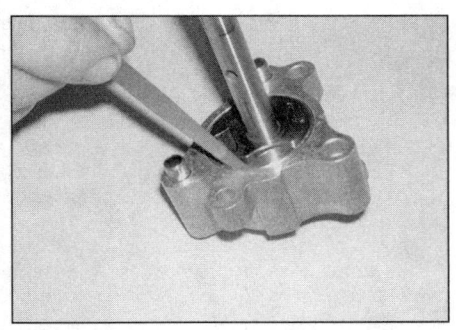

20.12 Using a feeler gauge to measure outer rotor-to-pump housing clearance

20.13 Measuring inner rotor tip-to-outer rotor clearance

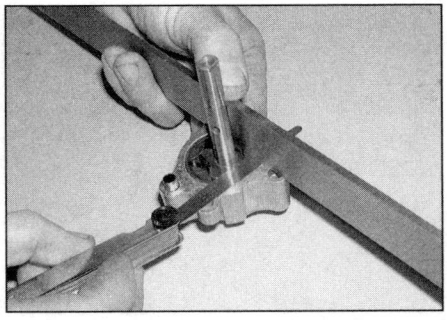

20.14 Checking oil pump rotor endfloat with a feeler gauge and straight edge

9 Remove the main pump housing from the opposite end of the pump spacer. Recover the housing locating dowels and remove the thrustwasher from the driveshaft end.

10 Remove the main pump outer rotor then slide out the driveshaft and inner rotor assembly. Slide the inner rotor off the driveshaft and recover the drive pin. **Note:** *Prior to removal, make a note which way around the pump outer rotor is fitted (there is a punch mark on one side of the rotor).*

11 Check the pump housings and rotors for scoring and wear. If any damage or uneven or excessive wear is evident, renew the pump (individual parts aren't available). If you are rebuilding the engine, it's a good idea to install a new oil pump.

12 Install the rotors in their housings and insert the driveshaft. Measure the clearance between the outer rotor and housing with a feeler gauge and compare it to the value listed in this Chapter's Specifications **(see illustration)**. If the clearance is excessive, renew the pump.

13 Measure the clearance between the inner rotor tip and the outer rotor **(see illustration)**. Again, renew the pump if the clearance is excessive.

14 Lay a straight-edge across the rotors and pump housing and measure the rotor endfloat (gap between the rotors and pump housing) with a feeler gauge **(see illustration)**. If the gap is outside the limits listed in this Chapter's Specifications, renew the pump.

15 Inspect the pump drive chain and sprockets for wear and damage and renew if necessary. If replacement is necessary note that both sprockets and the chain should be renewed as a set.

16 If the pump is good make sure all components are clean and lubricate them with clean engine oil. Reassemble the pump as follows.

17 Fit the main pump drive pin to the driveshaft then slide on the main pump inner rotor and engage it with the drive pin **(see illustrations)**.

18 Fit the thrustwasher to the driveshaft end and insert the driveshaft assembly into the main pump housing **(see illustrations)**.

19 Fit the main pump outer rotor to its housing, ensuring its punch mark is facing the same way as was noted prior to removal, and engage it with the inner rotor **(see illustration)**. **Note:** *If the outer rotor is fitted the wrong way around, the rate of pump wear will be accelerated.*

20 Ensure the locating dowels are in position then slide the pump spacer onto the driveshaft and locate it on the main pump housing **(see illustration)**.

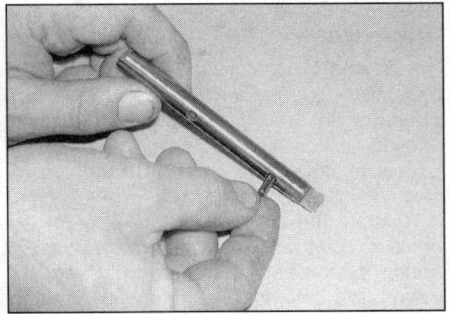

20.17a Fit the main pump drive pin to the driveshaft . . .

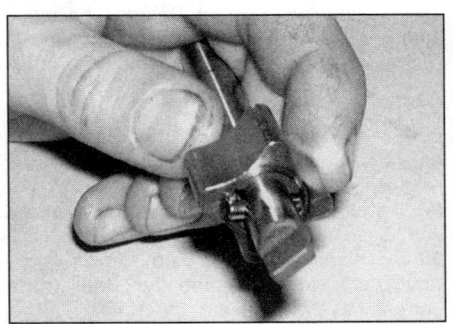

20.17b . . . then slide on the inner rotor and locate the drive pin in the rotor slot

20.18a Fit the thrust washer to the driveshaft end . . .

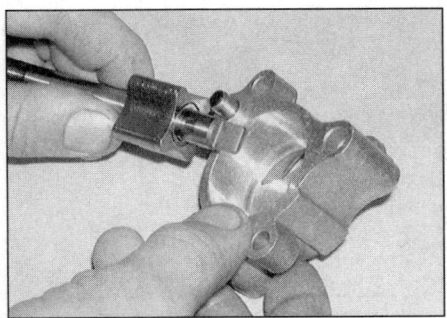

20.18b . . . then locate the driveshaft assembly in the main pump housing

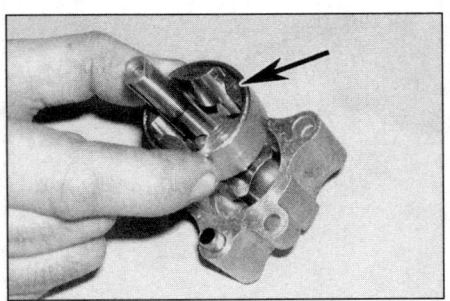

20.19 Fit the main pump outer rotor ensuring the punch mark (arrow) is the correct way around

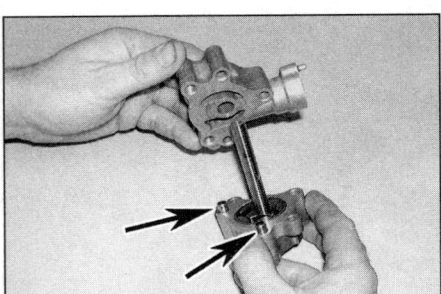

20.20 Ensure the locating dowels (arrows) are in position then slide the pump spacer into position

2•34 Engine, clutch and transmission

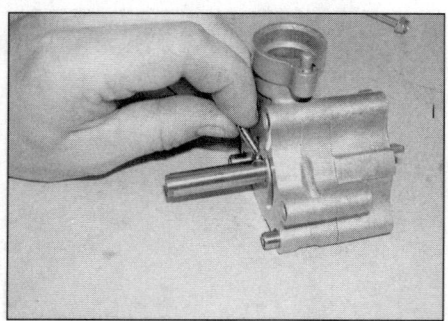

20.21a Insert the oil cooler pump drive pin in to the driveshaft . . .

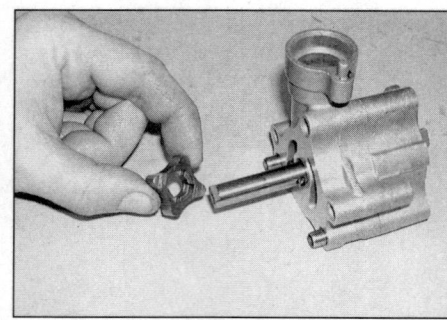

20.21b . . . then slide on the inner rotor and engage it with the pin

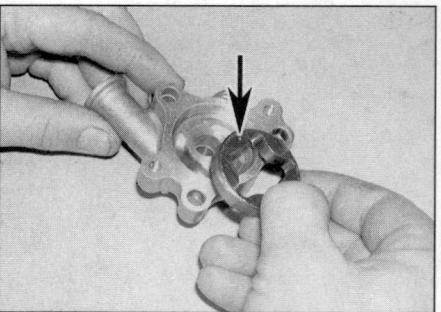

20.22 Fit the oil cooler pump outer rotor to the housing making sure its punch mark (arrow) is the correct way around

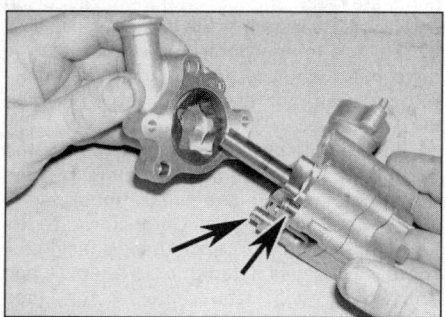

20.23 Ensure the locating dowels (arrows) are in position then fit the housing and outer rotor to the pump spacer

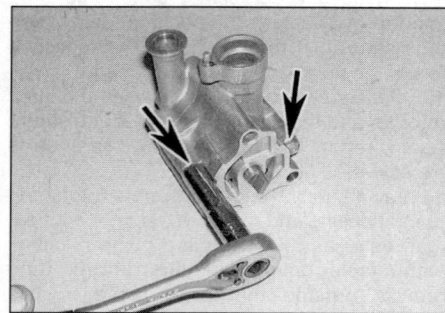

20.24 Ensure the pump is correctly assembled then fit the pump bolts (arrows) and tighten to the specified torque

21 Insert the second drive pin into the driveshaft then slide on the oil cooler pump inner rotor, making sure the drive pin is correctly located in the rotor slot **(see illustrations)**.
22 Fit the oil cooler pump outer rotor to its housing, ensuring its punch mark is facing the same way as was noted prior to removal **(see illustration)**. **Note:** *If the outer rotor is fitted the wrong way around, the rate of pump wear will be accelerated.*
23 Ensure the locating dowels are in position then fit the oil cooler pump housing and outer rotor assembly to the pump spacer **(see illustration)**.
24 With the oil pump correctly assembled, fit the pump bolts and tighten them to the specified torque **(see illustration)**. Check the driveshaft rotates freely before proceeding.
25 Insert the oil cooler relief valve piston, spring and spring seat making sure the piston and spring seat are fitted the correct way around **(see illustrations)**. Depress the spring seat and secure the relief valve components in position with new split pin **(see illustration)**.

Installation

26 Lubricate the pump rotors with clean engine oil and check that the pump driveshaft rotates freely.
27 Install the pump locating dowels in the crankcase.
28 Position the water pump and oil pump driveshaft slots so that they will mesh correctly as the pump is installed.
29 Manoeuvre the pump into position, engaging it with the water pump shaft, and locate it on the dowels.
30 Fit the pump mounting bolts and tighten them to the specified torque setting.
31 Fit the oil pump drive chain and sprockets and the clutch assembly as described in Section 15.
32 Install the oil pan and associated components as described in Section 19.

21 Oil pressure relief valve – removal, inspection and installation

Note: *The pressure relief valve can be removed with the engine in the frame.*

Removal

1 Remove the oil pan and pressure relief valve as described in Section 19.

Inspection

2 Push the piston into the relief valve body and check for free movement. If the valve operation is sticky it must be renewed (individual parts are not available). If necessary, the valve can be dismantled for cleaning/inspection as follows.
3 Remove the circlip from the base of the pressure relief valve and withdraw the washer, spring, piston and piston stop pin from the valve body. Inspect all components for signs of wear or damage **(see illustration)**.

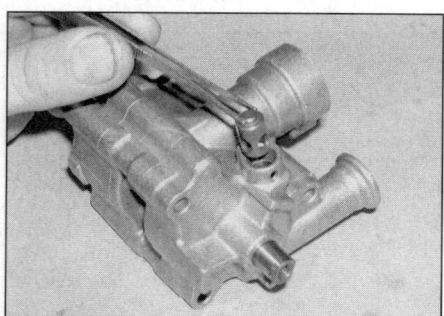

20.25a Insert the oil cooler relief valve piston . . .

20.25b . . . and fit the valve spring and spring retainer

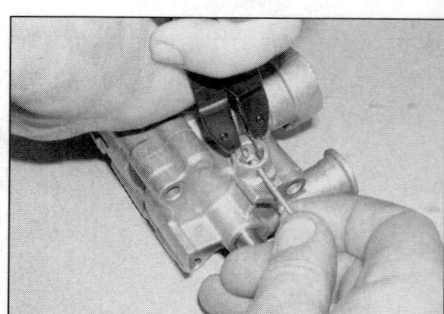

20.25c Depress the spring and secure the relief valve components in position with a new split pin

Engine, clutch and transmission 2•35

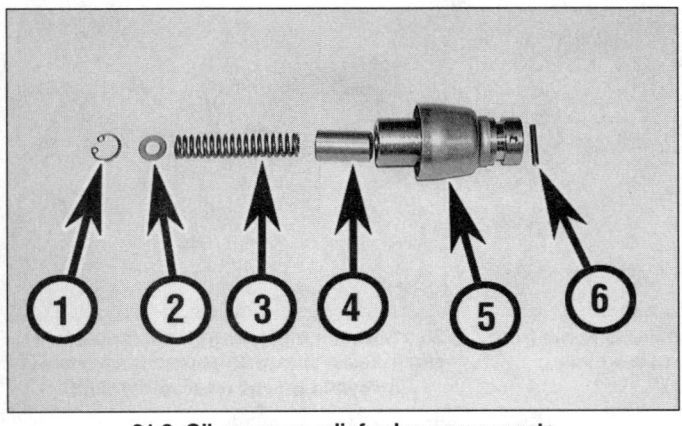

21.3 Oil pressure relief valve components

1 Circlip   3 Spring   5 Valve body
2 Washer   4 Piston   6 Piston stop pin

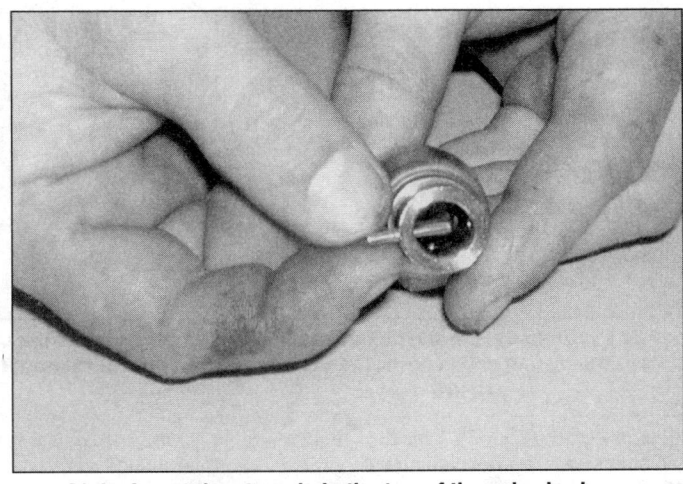

21.4a Insert the stop pin in the top of the valve body . . .

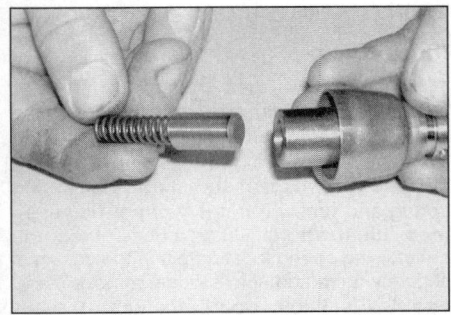

21.4b . . . then insert the piston and spring from the bottom

4 If the assembly is in good condition, insert the piston stop pin into the top of the valve body then engage the piston with the spring and insert the two into the base of the valve **(see illustrations)**. Ensure the piston is against the stop pin then fit the washer and secure all components in position with the circlip **(see illustration)**. Ensure the circlip is correctly located in its groove and check that the piston is free to move before installing the valve.

Installation

5 Install the pressure relief valve and oil pan as described in Section 19.

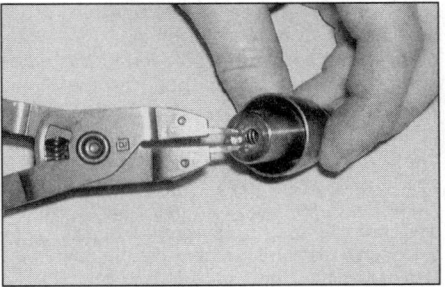

21.4c Fit the washer then compress the spring and secure all components in position with the circlip

22 Oil cooler and pipes – removal and installation

Note: *The oil cooler and pipes can be removed with the engine in the frame.*

Removal

Oil cooler

1 Remove the lower fairing panels and the inner panel as described in Chapter 8.
2 Drain the engine oil as described in Chapter 1.
3 Unscrew the oil cooler mounting bolts and recover the collars from the mounting rubbers. Pivot the oil cooler assembly forwards to allow improved access to the pipe end fitting bolts **(see illustration)**.
4 Undo the two bolts securing each pipe end fitting to the oil cooler **(see illustration)**. Free each end fitting from the cooler then remove the cooler from the bike. Recover the O-rings and discard them; new ones must be used on installation.
5 Check the oil cooler rubber mountings for signs of damage or deterioration and renew them if necessary.

Oil cooler pipes

6 Detach the pipe upper end from the oil cooler as described in Paragraphs 1 to 4. If both pipes are being removed, remove the cooler.
7 Undo the bolt securing the lower end fitting to the oil pan. Recover the O-ring from the end fitting then remove the pipe from the bike **(see illustrations)**.

Installation

Oil cooler

8 Ensure the oil cooler mounting rubbers are correctly fitted then fit a new O-ring to each oil cooler pipe end fitting.
9 Manoeuvre the cooler into position and

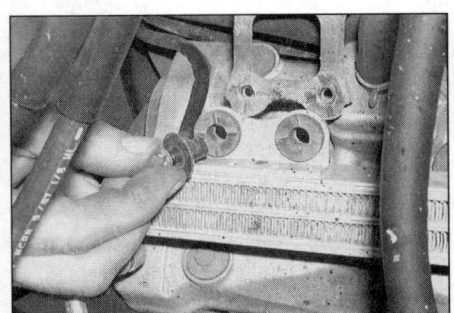

22.3 Slacken and remove the oil cooler mounting bolts and collars and free the cooler from its bracket

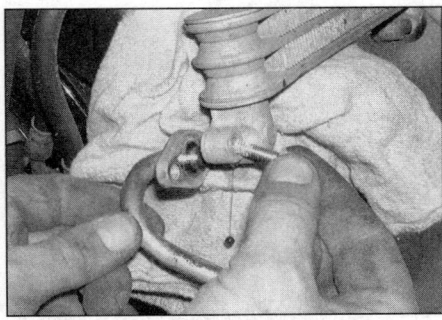

22.4 Unscrew the bolts then detach the oil pipes from the cooler and recover the O-ring

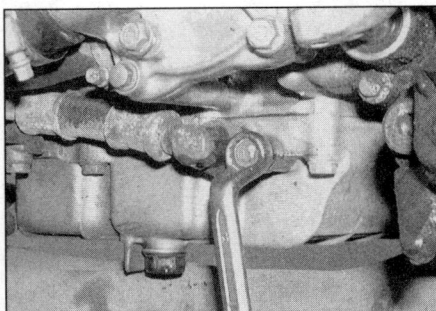

22.7a Unscrew the retaining bolt . . .

2

2•36 Engine, clutch and transmission

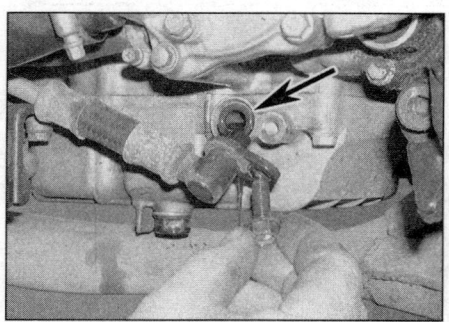

22.7b ... then detach the oil cooler pipe from the oil pan and recover the O-ring (arrow)

23.10 Remove the cam locating pin from the shift drum for safe-keeping

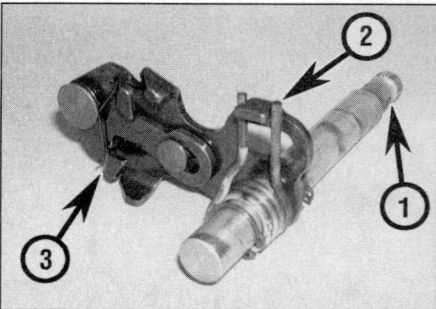

23.11 Check the gearshift shaft splines (1), centralising spring (2) and shift mechanism (3) for signs of wear or damage

engage it with the pipe end fittings, tightening the pipe end fitting bolts securely.
10 Align the cooler with its mounting bracket. Fit the collars to the rubber mountings then install the mounting bolts, tightening them securely.
11 Fill the engine with the correct type and quantity of oil as described in 'Daily (pre-ride) checks'. Start the engine and check that there are no leaks. If all is well, fit the fairing panels as described in Chapter 8.

Oil cooler pipes

12 Manoeuvre the pipe into position, ensuring it is correctly routed, and fit a new O-ring to its lower end fitting. Seat the end fitting correctly in the oil pan and securely tighten its retaining bolt.
13 Reconnect the pipe to the oil cooler as described in Paragraphs 8 to 11.

23 Gearchange mechanism – removal, inspection and installation

Note: *The gearchange mechanism components can be removed with the engine in the frame.*

Removal

1 Remove the front sprocket as described in Chapter 6.
2 Drain the engine oil as described in Chapter 1.

3 Note the correct fitted position of the gearchange lever on its shaft (the shaft punch mark should be aligned with the lever split). Ensure the transmission is in neutral then unscrew the clamp bolt and remove the gearchange lever from the engine.
4 Unscrew the two bolts securing the water pump to the crankcase then free the pump assembly from the engine and position it clear of the gearchange mechanism cover. Remove the O-ring from the pump and discard it; a new one must be used on installation. **Note:** *There is no need to drain the cooling system or disconnect the hoses from the pump.*
5 Unscrew the bolt and remove the cover bolt retaining plate, then slacken and remove the six gearchange mechanism cover retaining bolts.
6 Remove all traces of dirt from the gearshift shaft then slide the cover off the shaft and remove it from the engine. Retrieve the outer washer from the gearchange shaft and store it with the cover for safe-keeping. If the cover locating dowels are loose, remove them and store them with the cover.
7 Remove the gasket and discard it; a new one must be used on refitting.
8 Remove the gearchange shaft assembly from the engine and recover the inner washer from its shaft.
9 Unscrew the pivot bolt and remove the selector drum stopper arm, washer and spring, noting each component's correct fitted location.

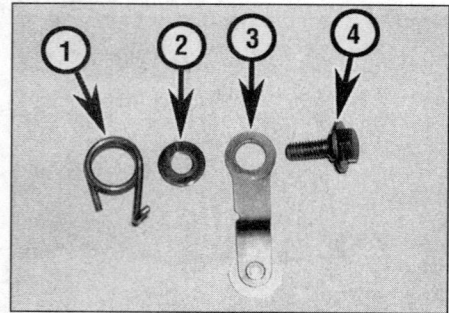

23.12 Stopper arm components
1 Spring   3 Stopper arm
2 Washer   4 Pivot bolt

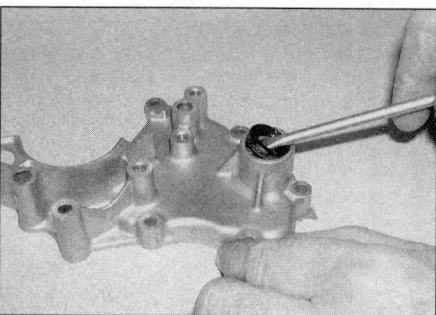

23.13 Renew the gearshift shaft oil seal if it is damaged

10 If necessary, unscrew the retaining bolt and remove the selector cam from the drum. If the cam locating pin is a loose-fit in the drum, remove it and store it with the cam for safe-keeping **(see illustration).**

Inspection

11 Check the gearchange shaft for straightness and damage to the splines. If the shaft is bent you can attempt to straighten it, but if the splines are damaged the shaft must be renewed. Inspect the shaft centralising spring and renew it if it is worn or damaged **(see illustration).** Also check that the centralising spring locating pin in the casing is securely tightened. If it is loose, remove it and apply a suitable non-permanent thread-locking compound before installing and tightening it securely.
12 Inspect the stopper arm and selector cam components, paying particular attention to the arm spring, and renew any defective item **(see illustration).**
13 Check the condition of the gearchange shaft oil seal and needle roller bearing in the cover. To renew the oil seal, lever the old seal out of position with a flat-bladed screwdriver then press the new one into position, ensuring its sealing lip is facing inwards **(see illustration).** If the needle roller bearing requires renewal, remove the oil seal and note the correct fitted location of the bearing before pressing/drifting it out of position. The new bearing should be pressed or drawn into its bore rather than driven into position to prevent possible damage. In the absence of a press, a suitable drawbolt arrangement can be made up as described in *Tools and Workshop Tips* in the Reference section. Ensure the bearing is correctly positioned in its bore then press the new oil seal into position.

Installation

14 Ensure the locating pin is in position then fit the selector cam to the drum, ensuring the pin locates correctly in the cam cutout **(see illustration).** Apply a few drops of suitable locking compound to the retaining bolt then fit the bolt and tighten it to the specified torque **(see illustrations).**

Engine, clutch and transmission 2•37

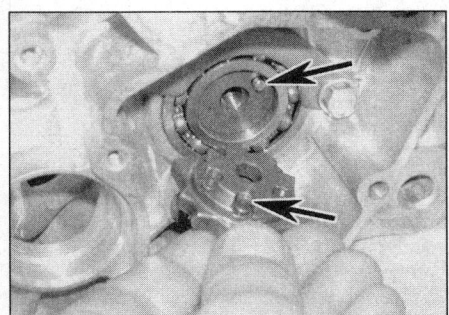

23.14a Fit the cam to the gearshift drum ensuring that the pin locates correctly in the cam slot (arrows)

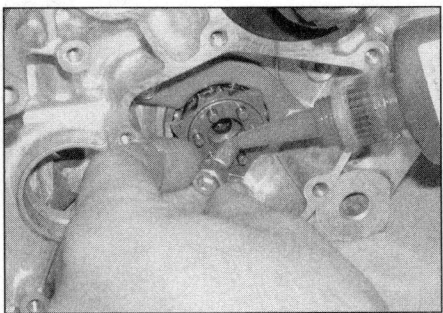

23.14b Apply locking compound to the shift cam bolt . . .

23.14c . . . then fit the bolt and tighten to the specified torque

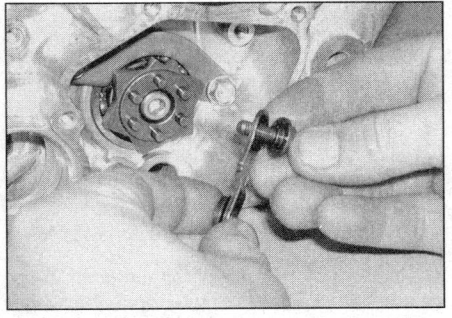

23.15a Fit the stopper arm to the pivot bolt . . .

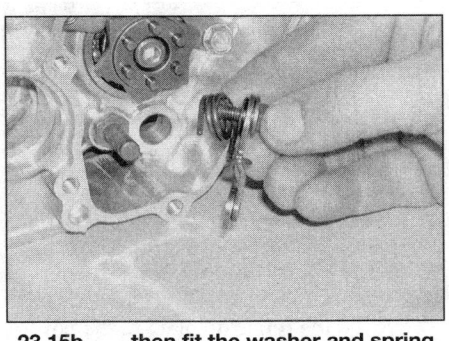

23.15b . . . then fit the washer and spring

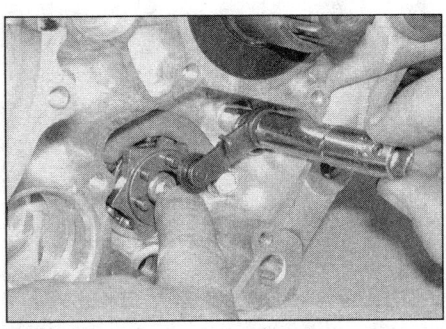

23.15c Screw the pivot bolt in a few turns . . .

15 Locate the stopper arm on the pivot bolt, ensuring it is the correct way around, then fit the washer and spring **(see illustrations)**. Manoeuvre the assembly into position and screw in the pivot bolt a few turns. Locate the stopper arm correctly on the selector cam and pivot bolt collar then tighten the pivot bolt to the specified torque **(see illustrations)**. Ensure the stopper arm pivots smoothly and is securely held against the selector cam by its spring before proceeding.

16 Fit the washer to the inner end of the gearchange shaft and locate the shaft in its crankcase bore **(see illustration)**. Align the shaft centralising spring with its locating pin then push the shaft fully into position **(see illustration)**.

17 Ensure the mating surfaces are clean and dry then fit the cover locating dowels and a new gasket to the crankcase **(see illustration)**.

18 Fit the outer washer to the gearchange shaft then carefully slide the cover along the gearchange shaft and locate it on the dowels **(see illustrations)**.

23.15d . . . then locate the stopper arm correctly on the cam (arrow) before tightening the pivot bolt to the specified torque

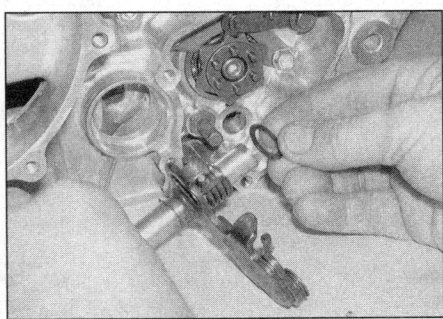

23.16a Fit the inner washer . . .

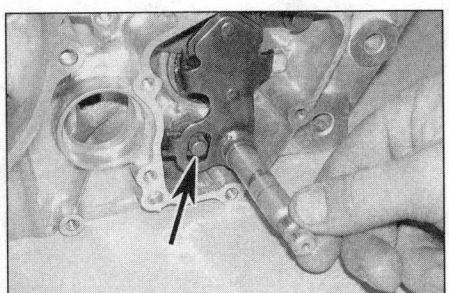

23.16b . . . then install the gearshift shaft ensuring the centralising spring engages correctly with the pin (arrow)

23.17 Ensure the dowels (arrows) are in position then fit a new gasket

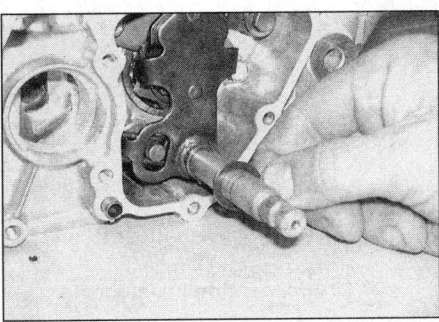

23.18a Slide the outer washer onto the gearshift shaft . . .

2•38 Engine, clutch and transmission

23.18b ... then fit the cover

23.19a Apply sealant to the threads of the two upper bolts (arrows) ...

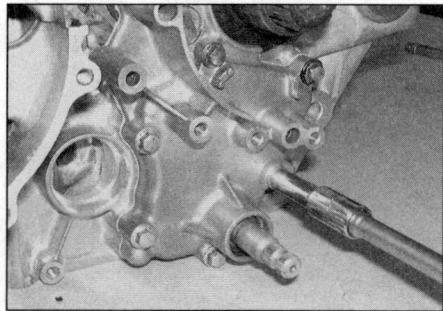

23.19b ... then evenly tighten all the cover bolts to the specified torque

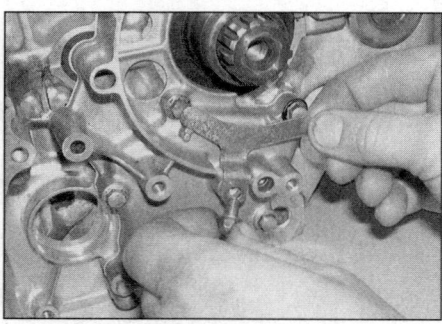

23.20 Fit the retaining plate to the cover and securely tighten its bolt

24 Crankcase – separation and reassembly

Separation

1 To examine and repair or renew the crankshaft, pistons and connecting rods, bearings and transmission components, the crankcase must be split into two parts.
2 To enable the crankcases to be split the engine must be removed from the frame (see Section 5) and the following components first removed with reference to the relevant Sections.
   a) Camshafts and followers.
   b) Cylinder heads.
   c) Camshaft drive gears.
   d) Clutch.
   e) Oil pan.
   f) Oil pump.
   g) Gearchange mechanism components.
   h) Alternator rotor (Chapter 9).
   j) Starter motor (Chapter 9).
3 With all the relevant components removed proceed as follows.
4 Disconnect the wiring connector from the neutral light switch then peel back the rubber cover and undo the screw securing the wiring connector to the oil pressure switch. Remove the engine wiring harness noting the correct routing of the neutral switch wire.
5 Undo the three retaining bolts and remove the input shaft bearing retaining plate from the right side of the crankcase **(see illustration)**.
6 With the crankcase the right way up, slacken and remove the six 6 mm bolts and the five 10 mm bolts from the top of the upper crankcase half **(see illustration)**. *Note: As each bolt is removed, store it in its relative position in a cardboard template of the crankcase halves. This will ensure that each bolt is returned to its original location on reassembly.* Discard the sealing washer fitted to the forward-most 6 mm

19 Apply a smear of sealant to the threads of the gearchange mechanism cover's two upper (shorter) retaining bolts, then fit all the bolts and tighten them evenly and progressively to the specified torque **(see illustrations)**.
20 Fit the retaining plate to the mechanism cover and securely tighten its retaining bolt **(see illustration)**.
21 Align the gearchange lever split with the punch mark on the shaft and engage the lever with the gearchange shaft splines. Fit the clamp bolt to the lever, tightening it securely, and check the operation of the gearchange

mechanism before proceeding further.
22 Referring to Chapter 3, fit a new O-ring to the water pump body then install the pump into the crankcase, aligning the slot in the impeller shaft with the tab on the oil pump shaft. Ensure the pump is correctly located then securely tighten its mounting bolts.
23 Install the front sprocket as described in Chapter 6.
24 Fill the engine with the correct type and quantity of oil as described in '*Daily (pre-ride) checks*'. Start the engine and check that there are no leaks before installing the fairing panel.

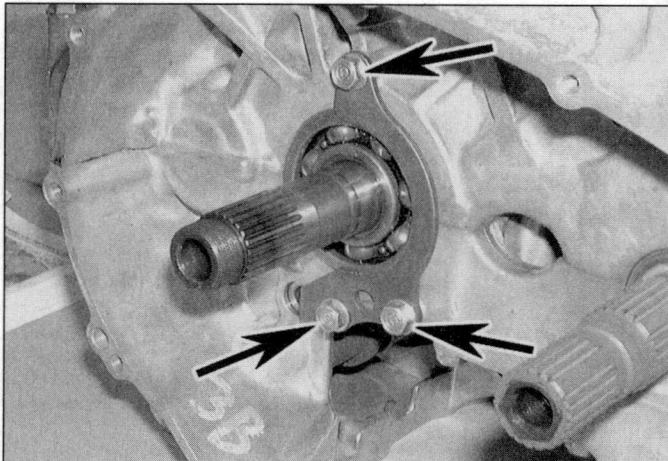

24.5 Undo the bolts (arrows) and remove the input shaft bearing retaining plate

24.6a Upper crankcase bolt locations

1 6 mm bolts   2 10 mm bolts

Engine, clutch and transmission 2•39

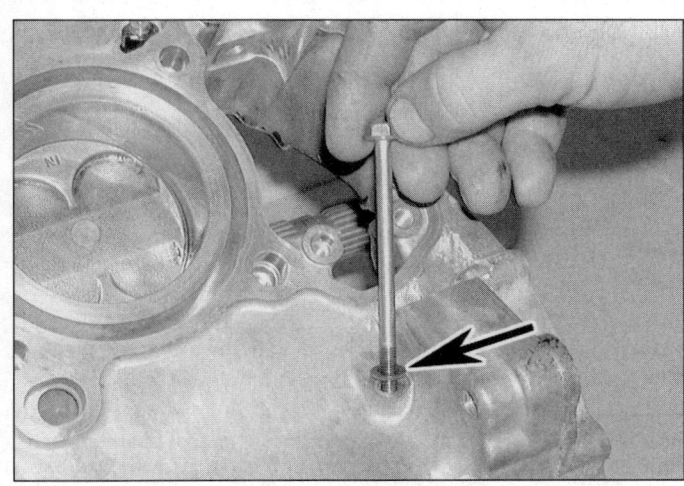

24.6b Note the front 6 mm upper bolt has a sealing washer fitted to it

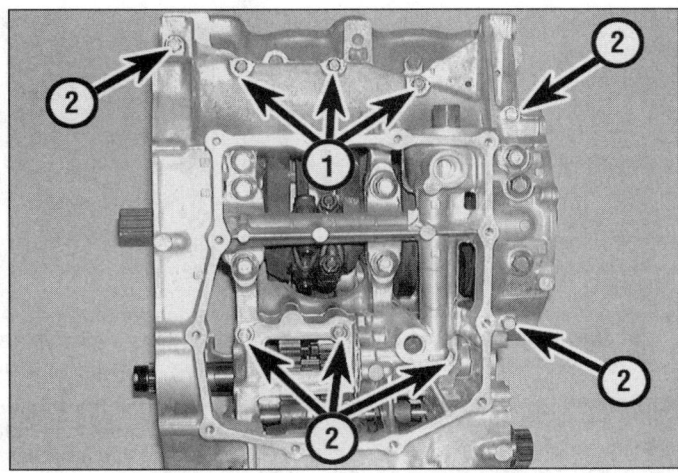

24.7 Lower crankcase bolt locations

1 7 mm bolts   2 6 mm bolts

bolt; a new one must be used on reassembly **(see illustration)**.

7 Turn the crankcase upside down, and unscrew the six 6 mm lower crankcase bolts and the three 7 mm bolts from the crankcase (see Note in Paragraph 6) **(see illustration)**.

8 Working in the reverse of the tightening sequence (numbers cast on lower crankcase – see Paragraph 24), gradually slacken the eight 9 mm (main bearing) bolts. Once all the bolts are loose, unscrew and remove them **(see illustration)**. Discard the 9 mm bolts; new ones must be used on reassembly (the bolts are angle-tightened and should never be reused).

9 Carefully lift the lower crankcase half off, leaving the crankshaft and transmission shafts in the upper half of the crankcase **(see illustration)**. As the lower half is lifted away take care not to dislodge or lose any main bearing inserts. **Note:** *If it won't come away easily, make sure all fasteners have been removed. Don't pry against the crankcase mating surfaces or they will leak; initial separation can be achieved by tapping gently with a soft-faced mallet.*

10 Remove the swingarm pivot collars from the rear of the crankcase.

11 Remove the three locating dowels from the upper crankcase half.

12 Remove the two oil jets and the two oil orifices from the upper crankcase half, noting which way around each orifice is fitted. Also remove and discard the clutch pushrod oil seal from the left of the crankcase; a new seal should be fitted on reassembly.

Reassembly

13 Check that the transmission shafts and crankshaft are correctly installed in the upper crankcase half as described in Sections 30 and 31.

14 Remove all traces of sealant from the crankcase mating surfaces, being careful not to let any fall into the case as this is done.

15 Check that all components are installed and that they can rotate smoothly and easily.

16 Lubricate the transmission shafts and crankshaft with clean engine oil then use a rag soaked in high flash-point solvent to wipe over the gasket surfaces of both halves to remove all traces of oil **(see illustration)**.

17 Make sure the oil jet and oil orifice holes are clear. Fit both oil jets to the upper crankcase half so their ends are flush with the crankcase mating surface then fit the oil orifices making sure their larger diameter internal bores are facing towards the lower crankcase half **(see illustrations)**.

24.8 Lower crankcase 9 mm (main bearing) bolt locations (arrows)

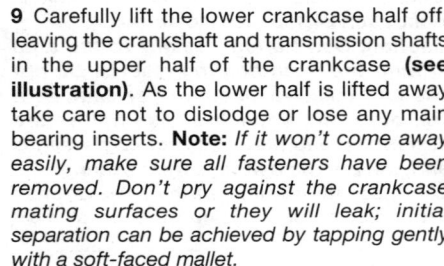

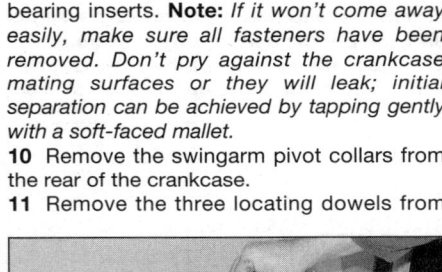

24.9 Lift the lower crankcase half off leaving all components in the upper half

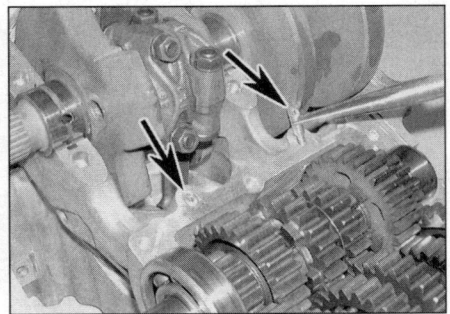

24.16 Lubricate all components with clean engine oil before assembling the crankcase halves

24.17a Insert the oil jets (arrows) . . .

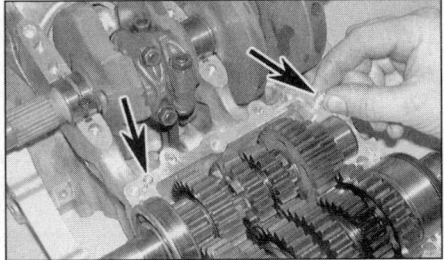

24.17b . . . and oil orifices (arrows) into the crankcase upper half, ensuring their larger diameter internal bores are facing towards the lower crankcase half

2•40 Engine, clutch and transmission

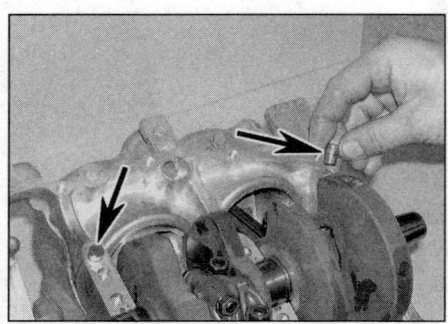

24.18a Install the two front locating dowels (arrows) . . .

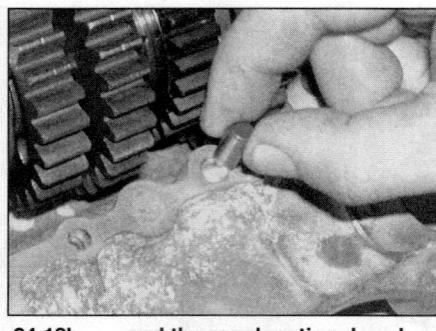

24.18b . . . and the rear locating dowel . . .

24.18c . . . and fit the swingarm pivot collars to the rear of the crankcase

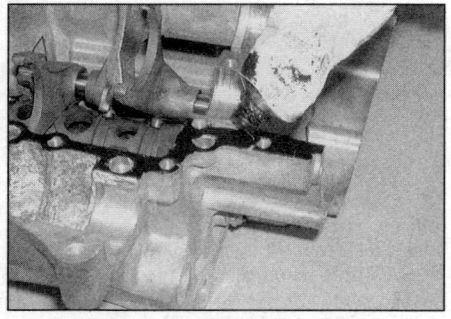

24.19a Apply a thin coat of sealant to the mating surfaces of the lower crankcase half

18 Install the three locating dowels in the upper crankcase half and fit the swingarm pivot collars (see illustrations).
19 Ensure the crankcase half mating surfaces are clean and dry then apply a thin smear of suitable sealant to the areas of the lower crankcase half mating surface shown in illustration 24.19b (see illustrations).
Caution: Don't apply sealant to the area around the main bearing inserts. Do not use an excessive amount of sealant, as it will ooze out when the case halves are assembled and may obstruct oil passages and prevent the bearings from seating.
20 Check the position of the selector drum cam, selector forks and transmission shafts – make sure they're in the neutral position.
21 Make sure that the main bearing inserts are in position and carefully install the lower crankcase half on the upper half. The selector forks must engage with their respective slots in the transmission gears as the halves are joined.
22 Check that the lower crankcase half is correctly seated and that all shafts are free to rotate. Note: If the casings are not correctly seated, remove the lower crankcase half and investigate the problem. Do not attempt to pull them together using the crankcase bolts as the casing will crack and be ruined.
23 Install the eight new 9 mm (main bearing) bolts (the two longer bolts are installed in locations 3 and 5) and tighten them all by hand (see illustration). Note: Each bolt is supplied pre-coated with an oil additive to ensure it is correctly tightened. Do not remove this additive from the bolt.
24 Note the number cast next to each 9 mm (main bearing) bolt on the lower crankcase; these numbers are the tightening sequence (see illustration). Working in the correct sequence, tighten each of the 9 mm bolts to its specified stage 1 torque setting.
25 Once all bolts have been tightened to the stage 1 torque, again working in the specified sequence, tighten each bolt through its specified stage 2 angle, using a socket and extension bar. It is recommended that an angle-measuring gauge is used during this stage of the tightening, to ensure accuracy (see illustration).

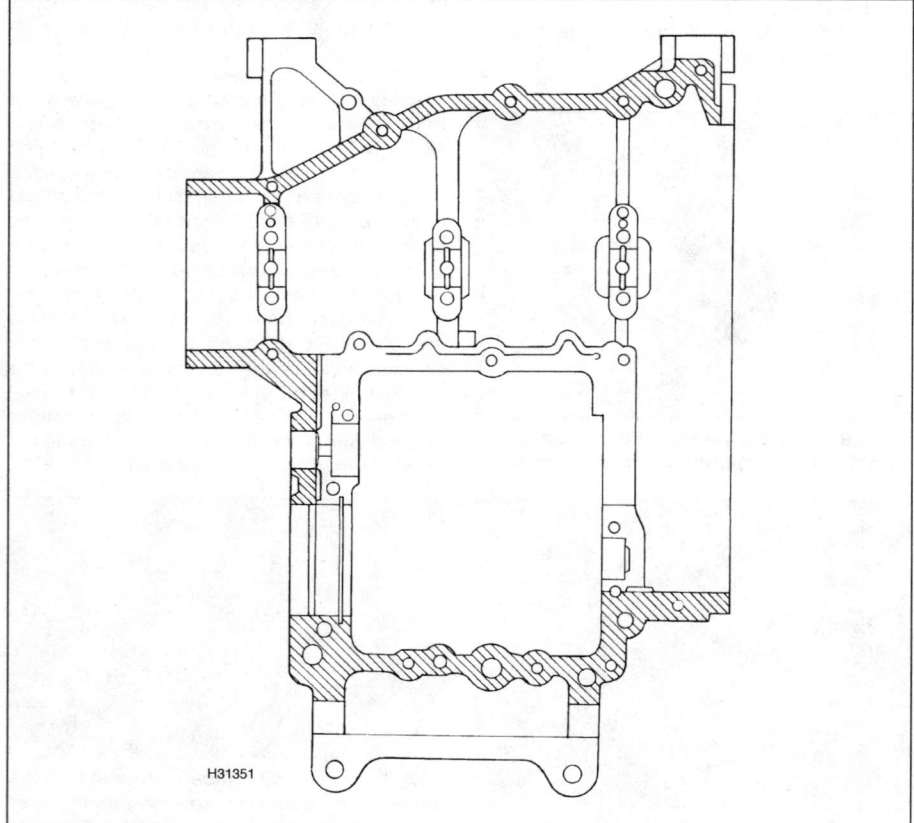

24.19b Apply sealant only to the shaded areas of the lower crankcase half

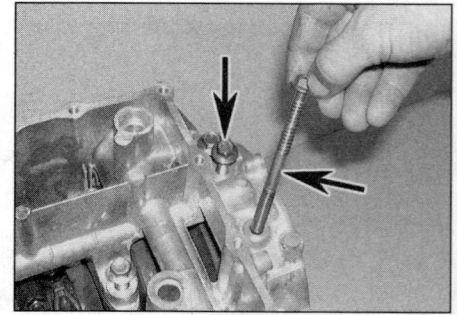

24.23 Fit the new 9 mm (main bearing) bolts ensuring the longer bolts (arrows) are installed in locations 3 and 5

Engine, clutch and transmission 2•41

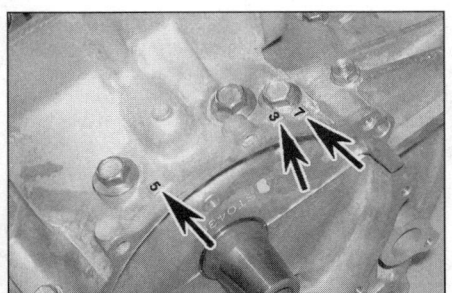

24.24 Tighten the bolts to the specified stage 1 torque setting in the numerical sequence marked on the lower crankcase (arrows)

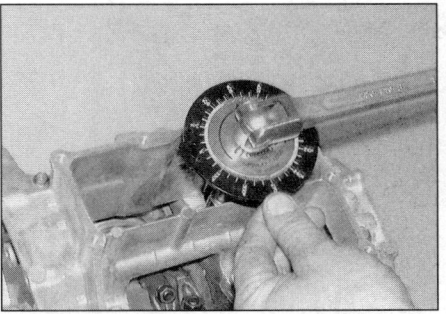

24.25 Use an angle-measuring gauge to ensure accuracy when tighten the bolts through the specified stage 2 angle

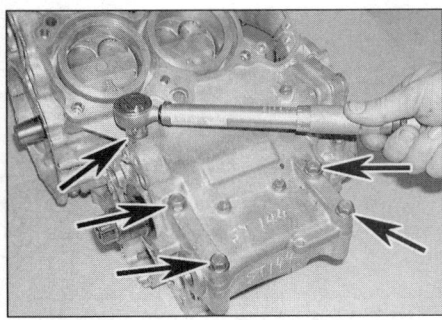

24.28 Ensure the swingarm pivot collars are correctly positioned then tighten the 10 mm bolts (arrows) to the specified torque

26 Fit the six 6 mm bolts and the three 7 mm bolts to the lower crankcase (see illustration 24.7). Make sure the swingarm pivot collars are fully in position then evenly and progressively tighten the 6 mm and 7 mm bolts to their specified torque settings, working in a criss-cross pattern.
27 Turn the crankcase over so that it is upright.
28 Recheck that the swingarm pivot collars are fully in position then fit the five 10 mm upper crankcase bolts; evenly and progressively tighten the bolts to the specified torque setting (see illustration).
29 Fit a new sealing washer to the forward-most 6 mm bolt (whose location is indicated by the triangular mark cast on the upper crankcase surface) then fit six 6 mm upper crankcase bolts in their original locations (see illustration). Tighten all bolts by hand, then tighten them to the specified torque setting (see illustration).
30 Clean the threads of the input shaft bearing retaining plate bolts and apply locking compound to their threads (see illustration). Install the retaining plate then fit the retaining bolts and tighten securely (see illustration).
31 With all crankcase fasteners tightened, check that the crankshaft and transmission shafts rotate smoothly and easily. If there are any signs of undue stiffness or of any other problem, the fault must be rectified before proceeding further.
32 Route the neutral switch wire through the hole in the rear of the crankcase and reconnect it to the switch. Reconnect the wire to the oil pressure switch, tightening its retaining screw securely, and seat the rubber cover over the switch.
33 Fit a new clutch pushrod oil seal to the left side of the crankcase, ensuring its sealing lip is facing inwards (see illustration).
34 Install all other removed assemblies in the reverse of the sequence given in Paragraph 2.

25 Crankcase – inspection and servicing

1 After the crankcases have been separated and the crankshaft and transmission components have been removed, the crankcases should be cleaned thoroughly with new solvent and dried with compressed air.

Cylinder bores

Note: *Don't attempt to separate the liners from the cylinder block.*
2 Check the cylinder walls carefully for scratches and score marks.
3 Using the appropriate precision measuring tools, check each cylinder's diameter. Measure near the top, centre and bottom of the cylinder bore, parallel to the crankshaft axis. Next, measure each cylinder's diameter at the same three locations across the crankshaft axis. Compare the results to this Chapter's Specifications. If the cylinder bores are tapered, out-of-round, worn beyond the specified limits, or badly scuffed or scored, have them rebored and honed by a dealer. If a rebore is done, oversize pistons and rings will be required as well. Honda produce two sizes of oversize pistons (see Section 28).

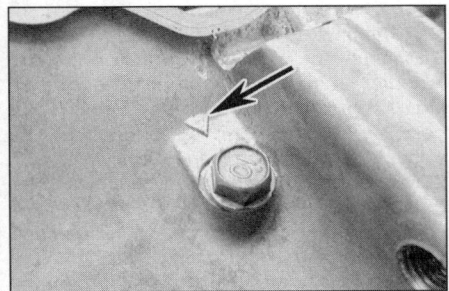

24.29a Ensure a new sealing washer is fitted to the 6 mm bolt next to the triangular mark (arrow) . . .

24.29b . . . then tighten all 6 mm bolts (arrows) to the specified torque

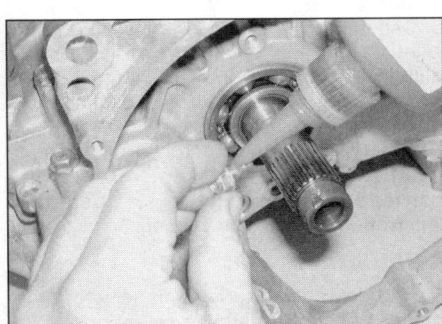

24.30a Apply thread locking compound to the retaining bolt threads . . .

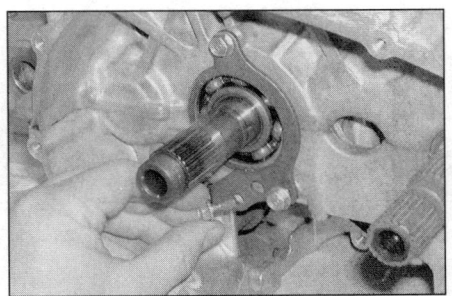

24.30b . . . then refit the bearing retaining plate to the crankcase and securely tighten the bolts

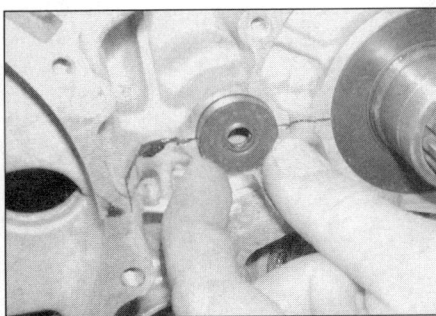

24.33 Fit a new clutch pushrod oil seal ensuring is sealing lip is facing inwards

2•42 Engine, clutch and transmission

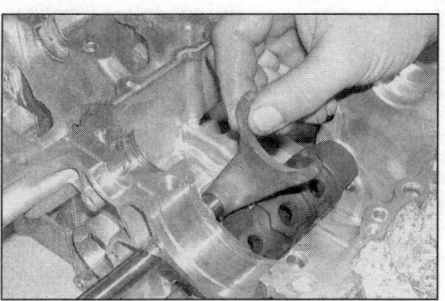

26.3 Slide out the shift fork shaft and remove the forks from the lower crankcase half

4 As an alternative, if the precision measuring tools are not available, a dealer will make the measurements and offer advice concerning servicing of the cylinders.
5 If they are in reasonably good condition and not worn to the outside of the limits, and if the piston-to-cylinder clearances can be maintained properly (see Section 28), then the cylinders do not have to be rebored; honing is all that is necessary.
6 To perform the honing operation you will need the proper size flexible hone with fine stones, or a 'bottle brush' type hone, plenty of light oil or honing oil, some shop towels and an electric drill motor. Hold the upper crankcase half in a vice (cushioned with soft jaws or wood blocks) when performing the honing operation. Mount the hone in the drill motor, compress the stones and slip the hone into the top of the cylinder. Lubricate the cylinder thoroughly, turn on the drill and move the hone up and down in the cylinder at a pace which will produce a fine crosshatch pattern on the cylinder wall with the crosshatch lines intersecting at approximately a 60° angle. Be sure to use plenty of lubricant and do not take off any more material than is absolutely necessary to produce the desired effect. Do not withdraw the hone from the cylinder while it is running. Instead, shut off the drill and continue moving the hone up and down in the cylinder until it comes to a complete stop, then compress the stones and withdraw the hone. Wipe the oil out of the cylinder and repeat the procedure on the other cylinders. Remember, do not remove too much material from the cylinder wall. If

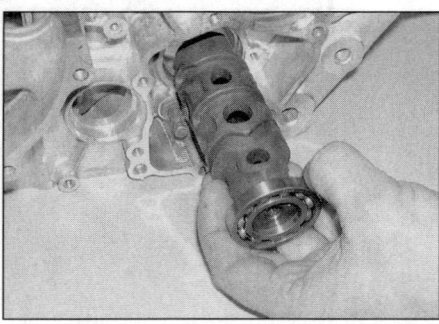

26.4 Removing the shift drum and bearing

you do not have the tools, or do not desire to perform the honing operation, a dealer will generally do it for a reasonable fee.
7 Next, the cylinders must be thoroughly washed with warm soapy water to remove all traces of the abrasive grit produced during the honing operation. Be sure to run a brush through the bolt holes and flush them with running water. After rinsing, dry the cylinders thoroughly and apply a coat of light, rust-preventative oil to all machined surfaces.

Crankcase castings

8 Remove any oil passage plugs that haven't already been removed. All oil passages should be blown out with compressed air.
9 All traces of old gasket sealant should be removed from the mating surfaces. Minor damage to the surfaces can be cleaned up with a fine sharpening stone or grindstone.
Caution: Be very careful not to nick or gouge the crankcase mating surfaces or leaks will result. Check both crankcase halves very carefully for cracks and other damage.
10 Small cracks or holes in aluminium castings may be repaired with an epoxy resin adhesive as a temporary measure. Permanent repairs can only be effected by argon-arc welding, and only a specialist in this process is in a position to advise on the economy or practical aspect of such a repair. If any damage is found that can't be repaired, renew the crankcase halves as a set.
11 Damaged threads can be economically reclaimed by using a diamond section wire insert, of the Helicoil type, which is easily

fitted after drilling and re-tapping the affected thread. Sheared studs or screws can usually be removed with screw extractors, which consist of a tapered, left thread screw of very hard steel. These are inserted into a pre-drilled hole in the stud, and usually succeed in dislodging the most stubborn stud or screw.

HAYNES HiNT *Refer to Tools and Workshop Tips in the Reference section for details of thread repair methods and using screw extractors.*

26 Selector drum and forks – removal, inspection and installation

Removal

1 Separate the crankcase halves as described in Section 24.
2 Undo the retaining bolts and remove the selector drum bearing retaining plate from the left side of the lower crankcase half.
3 Withdraw the selector fork shaft slowly and remove the selector forks from the crankcase as they are released from the end of the shaft. When all three selector forks have been removed, fully withdraw the shaft from the crankcase **(see illustration)**.
4 Remove the selector drum and bearing from the crankcase **(see illustration)**.

Inspection

5 The selector forks and shaft should be closely inspected to ensure that they are not badly damaged or worn.
6 Measure the width of both fork ends and the internal diameter of the shaft bore **(see illustration)**. If either fork end or the shaft bore has worn beyond its service limit the selector fork(s) must be renewed.
7 The selector fork shaft can be checked for trueness by rolling it along a flat surface. A bent shaft will cause difficulty in selecting gears and make the gearchange action heavy. Measure the diameter of the shaft at the points where it is in contact with the selector forks. If the shaft is bent or has worn beyond its service limit at any point it must be renewed.
8 Inspect the selector drum grooves and selector fork guide pins for signs of wear or damage. If either component shows signs of wear or damage the selector fork(s) and drum must be renewed.
9 Check that the selector drum bearing rotates freely and has no sign of freeplay between its inner and outer race **(see illustration)**. Renew the bearing if necessary.

Installation

10 Lubricate the bearing and selector drum grooves with clean engine oil and insert the shift drum and bearing into position in the crankcase.

26.6 Measuring a shift fork end with a micrometer

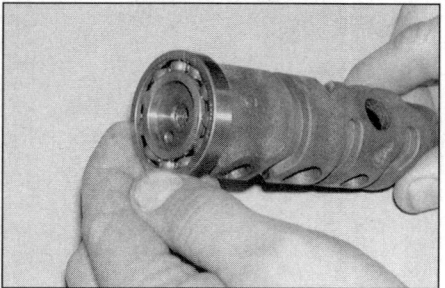

26.9 Check the shift drum bearing for signs of wear or roughness and renew if necessary

Engine, clutch and transmission 2•43

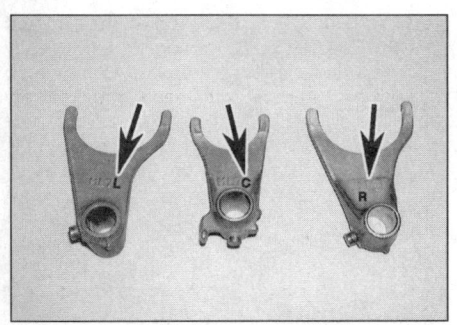

26.12 The shift forks can be identified from the marks (arrows); the left fork is marked 'L', the centre fork 'C' and the right fork 'R'

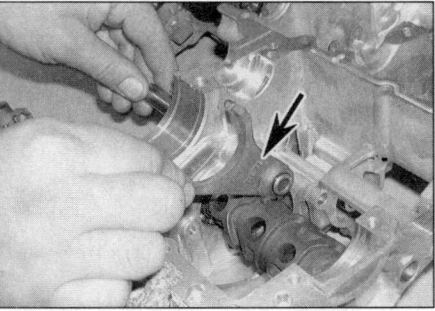

26.13a Locate the left fork in its shift drum groove and slide the shaft into position (identification marking arrowed) . . .

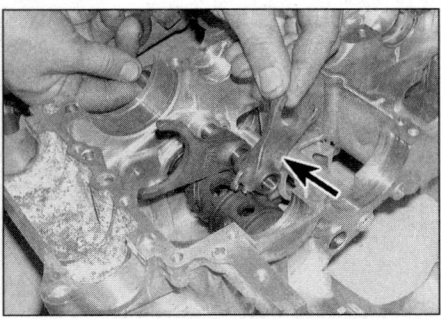

26.13b . . . then fit the centre fork (identification marking arrowed) . . .

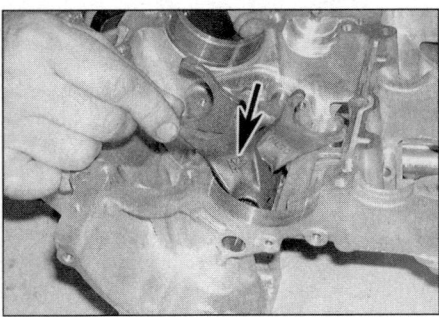

26.13c . . . and right fork and fully insert the shift fork shaft. Ensure all the identification markings (arrows) are facing towards the clutch side

11 Apply a smear of engine oil to the selector fork shaft and slide the shaft partially into the crankcase.
12 The selector forks can be identified by the letter cast on each one; L denotes the left fork, C the centre, and R the right **(see illustration)**. **Note:** *All selector forks must be installed in the crankcase so that the letter on each one faces towards the right side of the casing (clutch).*
13 Locate the left fork with its groove in the selector drum and slide in the selector fork shaft until it engages the fork. Repeat the process for the centre and right fork and push the shaft fully home, making sure each fork is positioned as described in Paragraph 12 **(see illustrations)**.
14 Clean the threads of the selector drum bearing retaining plate bolts and apply locking compound to their threads. Install the retaining plate then fit the retaining bolts and tighten securely **(see illustrations)**.
15 Join the crankcase halves as described in Section 24.

27 Main and connecting rod bearings – general note

1 Even though main and connecting rod bearings are generally renewed during the engine overhaul, the old bearings should be retained for close examination as they may reveal valuable information about the condition of the engine.
2 Bearing failure occurs mainly because of lack of lubrication, the presence of dirt or other foreign particles, overloading the engine and/or corrosion. Regardless of the cause of bearing failure, it must be corrected before the engine is reassembled to prevent it from happening again.
3 When examining the bearings, remove the main bearings from the case halves and the rod bearings from the connecting rods and caps and lay them out on a clean surface in the same general position as their location on the crankshaft journals. This will enable you to match any noted bearing problems with the corresponding crankshaft journal.

4 Dirt and other foreign particles get into the engine in a variety of ways. It may be left in the engine during assembly or it may pass through filters or breathers. It may get into the oil and from there into the bearings. Metal chips from machining operations and normal engine wear are often present. Abrasives are sometimes left in engine components after reconditioning operations such as cylinder honing, especially when parts are not thoroughly cleaned using the proper cleaning methods. Whatever the source, these foreign objects often end up imbedded in the soft bearing material and are easily recognised. Large particles will not imbed in the bearing and will score or gouge the bearing and journal. The best prevention for this cause of bearing failure is to clean all parts thoroughly and keep everything spotlessly clean during engine reassembly. Frequent and regular oil and filter changes are also recommended.
5 Lack of lubrication or lubrication breakdown has a number of interrelated causes. Excessive heat (which thins the oil), overloading (which squeezes the oil from the bearing face) and oil leakage or throw off from excessive bearing clearances, worn oil pump or high engine speeds all contribute to lubrication breakdown. Blocked oil passages will also starve a bearing and destroy it. When lack of lubrication is the cause of bearing failure, the bearing material is wiped or extruded from the steel backing of the bearing. Temperatures may increase to the point where the steel backing and the journal

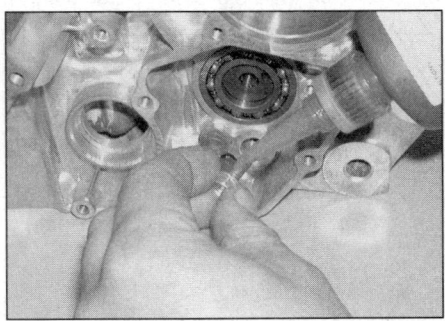

26.14a Apply thread locking compound to the bolt threads . . .

turn blue from overheating.
6 Riding habits can have a definite effect on bearing life. Full throttle low speed operation, or labouring (lugging) the engine, puts very high loads on bearings, which tend to squeeze out the oil film. These loads cause the bearings to flex, which produces fine cracks in the bearing face (fatigue failure). Eventually the bearing material will loosen in pieces and tear away from the steel backing. Short trip riding leads to corrosion of bearings, as insufficient engine heat is produced to drive off the condensed water and corrosive gases produced. These products collect in the engine oil, forming acid and sludge. As the oil is carried to the engine bearings, the acid attacks and corrodes the bearing material.

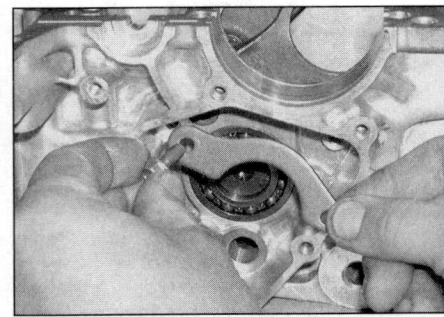

26.14b . . . then fit the shift drum bearing retaining plate and tighten the bolts securely

2•44 Engine, clutch and transmission

28.1 Checking connecting rod side clearance with a feeler gauge

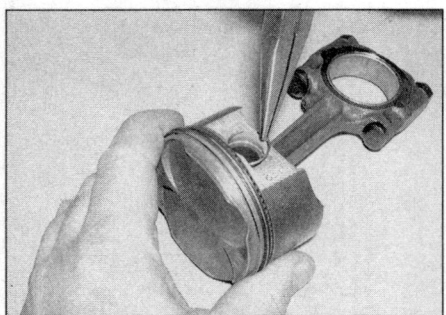

28.8 Remove the circlip from the piston then push out the piston pin

28.11 Using a piston ring removal/installation to remove a piston ring

7 Incorrect bearing installation during engine assembly will lead to bearing failure as well. Tight fitting bearings which leave insufficient bearing oil clearances result in oil starvation. Dirt or foreign particles trapped behind a bearing insert result in high spots on the bearing which lead to failure.
8 To avoid bearing problems, clean all parts thoroughly before reassembly, double check all bearing clearance measurements and lubricate the new bearings with clean engine oil during installation.

28 Piston/connecting rod assemblies – removal, inspection and installation

Removal

1 Separate the crankcase halves as described in Section 24. Before removing the piston/connecting rods from the crankshaft measure the side clearance of each pair of rods with a feeler gauge **(see illustration)**. If the clearance is greater than the service limit listed in this Chapter's Specifications, both rods will have to be renewed.
2 Using paint or a suitable marker pen, mark the relevant cylinder number on each connecting rod and bearing cap (see Specifications for cylinder identification details).
3 Unscrew the bearing cap nuts and withdraw the cap, complete with the lower bearing insert, from each of the four connecting rods. Push the connecting rods up and off their crankpins, then remove the upper bearing insert. Keep the cap, nuts and (if they are to be re-used) the bearing inserts together in their correct sequence.
4 Remove the ridge of carbon from the top of each cylinder bore. If there is a pronounced wear ridge on the top of each bore, remove it with a ridge reamer.
5 Push each piston/connecting rod assembly up and remove it from the top of the bore.
Caution: Take great care when removing the piston/connecting rod assemblies as the connecting rods can easily damage the bore surface.
6 Immediately install the relevant bearing cap, inserts and nuts on each piston/connecting rod assembly so that they are all kept together as a matched set.
7 Using a sharp scriber, scratch the number of each piston into its crown (or use a suitable marker pen or paint if the piston is clean enough).
8 Support the first piston and, using a small screwdriver or a pair of pointed-nose pliers, carefully remove the circlip from the piston groove **(see illustration)**.
9 Push the piston pin out from the opposite end to free the piston from the rod. You may have to deburr the area around the groove to enable the pin to slide out (use a triangular file for this procedure). If the pin is tight, tap it out using a suitable hammer and punch, taking care not to damage the piston. Repeat the procedure for the other pistons.

Inspection

Pistons

10 Before the inspection process can be carried out, the pistons must be cleaned and the old piston rings removed.
11 Using a piston ring removal and installation tool, carefully remove the rings from the pistons **(see illustration)**. Do not nick or gouge the pistons in the process.
12 Scrape all traces of carbon from the tops of the pistons. A hand-held wire brush or a piece of fine emery cloth can be used once most of the deposits have been scraped away. Do not, under any circumstances, use a wire brush mounted in a drill motor to remove deposits from the pistons; the piston material

28.18 Measuring piston ring-to-groove clearance with a feeler gauge

is soft and will be eroded away by the wire brush.
13 Use a piston ring groove cleaning tool to remove any carbon deposits from the ring grooves. If a tool is not available, a piece broken off an old ring will do the job. Be very careful to remove only the carbon deposits. Do not remove any metal and do not nick or gouge the sides of the ring grooves.
14 Once the deposits have been removed, clean the pistons with solvent and dry them thoroughly. Make sure the oil return holes below the oil ring grooves are clear.
15 If the pistons are not damaged or worn excessively and if the cylinders are not to be rebored, new pistons will not be necessary. Normal piston wear appears as even, vertical wear on the thrust surfaces of the piston and slight looseness of the top ring in its groove. New piston rings, on the other hand, should always be used when an engine is rebuilt.
16 Carefully inspect each piston for cracks around the skirt, at the pin bosses and at the ring lands.
17 Look for scoring and scuffing on the thrust faces of the skirt, holes in the piston crown and burned areas at the edge of the crown. If the skirt is scored or scuffed, the engine may have been suffering from overheating and/or abnormal combustion, which caused excessively high operating temperatures. The oil pump and oil cooler should be checked thoroughly. A hole in the piston crown, an extreme to be sure, is an indication that abnormal combustion (pre-ignition) was occurring. Burned areas at the edge of the piston crown are usually evidence of spark knock (detonation). If any of the above problems exist, the causes must be corrected or the damage will occur again.
18 Measure the piston ring-to-groove clearance by fitting a new piston ring in the ring groove and slipping a feeler gauge in beside it **(see illustration)**. Check the clearance at three or four locations around the groove. Be sure to use the correct ring for each groove; they are different (see Section 29). If the clearance is greater than the service limit, new pistons will have to be used when the engine is reassembled.
19 Calculate the piston-to-bore clearance by measuring the bore (see Section 25) and the

Engine, clutch and transmission 2•45

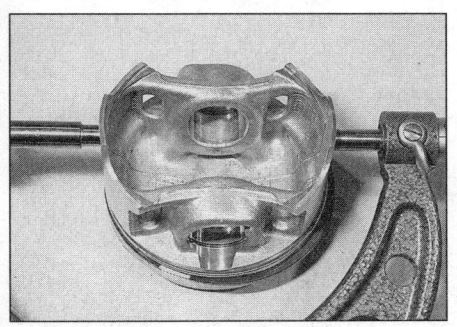

28.19 Measuring piston diameter with a micrometer

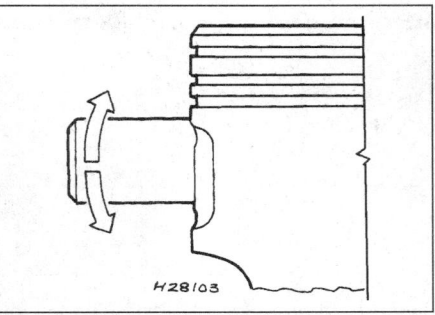

28.20 Fit the pin into the piston and move it back-and-forth to check for wear

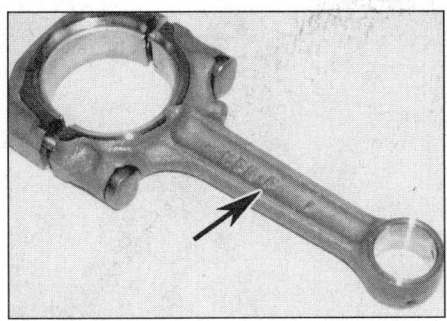

28.26 Each connecting rod has an identification marks (arrow); front cylinder rods are marked with an 'F' and rear cylinder rods with an 'R'

piston diameter. Make sure that the pistons and cylinders are correctly matched. Measure the piston across the skirt on the thrust faces at a 90° angle to the piston pin, 18 mm up from the bottom of the skirt **(see illustration)**. Subtract the piston diameter from the bore diameter to obtain the clearance. If it is greater than specified in the Specifications at the beginning of this Chapter, the cylinders will have to be rebored and new oversized pistons and rings installed.

20 Apply clean engine oil to the pin, insert it into the piston and check for freeplay by rocking the pin back-and-forth **(see illustration)**. If the pin is loose, new pistons and pins must be installed. If the necessary measuring equipment is available measure the pin diameter and piston pin bore and check the readings obtained do not exceed the limits given in this Chapter's Specifications. Renew components that are worn beyond the specified limit.

21 If the pistons are to be renewed, ensure the correct size of piston is ordered. Honda produce two oversizes of piston as well as standard pistons. The piston oversizes available are: +0.25 mm and +0.50 mm. **Note:** *Oversize pistons have their size stamped on top of the piston crown, e.g. a +0.25 mm oversize piston will be marked 0.25.*

22 Install the rings on the pistons as described in Section 29.

Connecting rods

23 Check the connecting rods for cracks and other obvious damage. Lubricate the piston pin for each rod, install it in its original rod and check for play. If it wobbles, renew the connecting rod and/or the pin. If the necessary measuring equipment is available measure the pin diameter and connecting rod bore and check the readings obtained do not exceed the limits given in this Chapter's Specifications. Renew components that are worn beyond the specified limit.

24 Refer to Section 27 and examine the connecting rod bearing inserts. If they are scored, badly scuffed or appear to have been seized, new bearings must be installed. Always renew the bearings in the connecting rods as a set. If they are badly damaged, check the corresponding crankpin. Evidence of extreme heat, such as discoloration, indicates that lubrication failure has occurred. Be sure to thoroughly check the oil pump and pressure relief valve as well as all oil holes and passages before reassembling the engine.

25 Have the rods checked for twist and bending by a dealer or automotive engineer.

26 If a connecting rod is to renewed, be sure to state whether it is a front or rear cylinder connecting rod when ordering. The front and rear connecting rods are not interchangeable as their oil holes are offset; the front cylinder connecting rods are marked 'MBG-F' on their sides, and the rear connecting rods are marked 'MBG-R' **(see illustration)**. Each connecting rod is marked with a weight code (in the form of a letter – A, B or C) on its bearing cap; all new connecting rods supplied by Honda are in weight group B which can be fitted in place of a group A or C rod.

Bearing selection

27 The connecting rod bearing running clearance is controlled in production by selecting one of five grades of bearing insert. The grades are indicated by a color-coding marked on the edge of each insert (this colour-code may no longer be visible on the original bearings) **(see illustration)**. **Note:** *The front and rear connecting rod bearing inserts are not interchangeable because the insert oil holes are offset. The front cylinder bearing inserts have one paint marking, and the rear bearing inserts have two paint markings.* In order, from the thickest to the thinnest, the insert grades are: Blue, Black, Brown, Green and Yellow. New bearing inserts are selected as follows using the crankpin and connecting rod size group markings.

28 The standard crankpin journal diameter is divided into three size groups to allow for manufacturing tolerances. The size group of each crankpin can be determined by the letters which are stamped on the crank web next to the crankpin (either A, B or C) **(see illustration)**. **Note:** *Ignore the numbers as these refer to the main bearing journals.* If the equipment is available, these marks can be checked by direct measurement **(see illustration)**.

29 The connecting rods are also divided into three size groups to allow for manufacturing tolerances. The size group is in the form of a number stamped on the rod (either 1, 2 or 3) **(see illustration)**. **Note:** *Ignore the letter as*

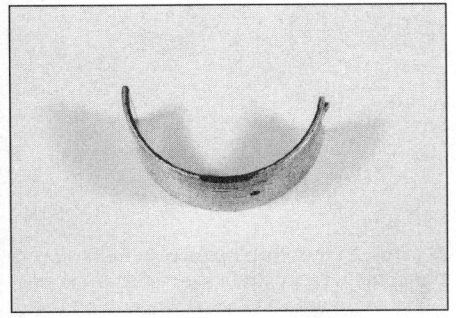

28.27 The colour code is painted on the side of the bearing insert

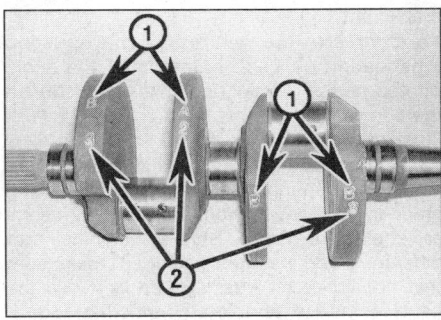

28.28a The crankpin size groups can be determined using the letters (1) on the crankshaft webs. The numbers (2) identify the main bearing journal size groups

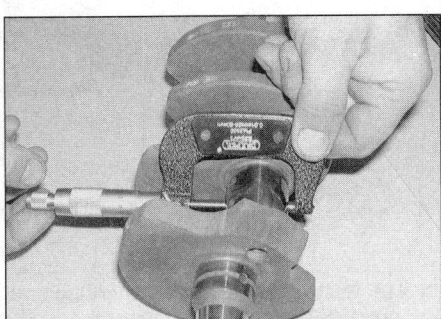

28.28b Using a micrometer to measure a crankpin journal diameter

2•46 Engine, clutch and transmission

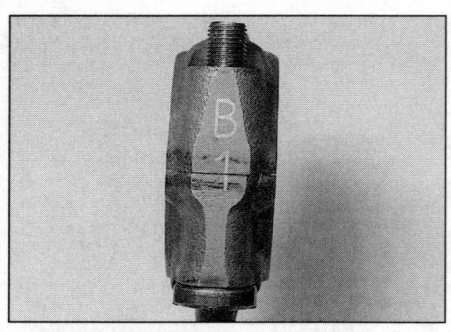

28.29 Connecting rod size group number and weight group letter

Connecting rod size group	Crankpin size group	Bearing insert required
1	A	Yellow
1	B	Green
1	C	Brown
2	A	Green
2	B	Brown
2	C	Black
3	A	Brown
3	B	Black
3	C	Blue

this indicates the weight group of the connecting rod. If the equipment is available, these marks can be checked by direct measurement.

30 Match the relevant connecting rod code with its crankshaft code and select a new set of bearing inserts using the table at the top of the page.

Oil clearance check

31 Whether new bearing inserts are being fitted or the original ones are being re-used, the connecting rod bearing oil clearance should be checked prior to reassembly.

32 Clean the backs of the bearing inserts and the bearing locations in both the connecting rod and bearing cap.

33 Press the bearing inserts into their locations, ensuring that the tab on each insert engages the notch in the connecting rod/bearing cap. Make sure the bearings are fitted in the correct locations and take care not to touch any insert's bearing surface with your fingers.

34 There are two possible ways of checking the oil clearance. The first method is by direct measurement (see Paragraphs 35 and 38) and the second by the use of a product known as Plastigauge (see Paragraphs 36 to 38).

35 If the first method is to be used, fit the bearing cap to the connecting rod, with the bearing inserts in place. Make sure the cap is fitted the correct way around so the connecting rod and bearing cap weight/size markings are correctly aligned. Tighten the cap retaining nuts to the specified torque and measure the internal diameter of each assembled pair of bearing inserts. If the diameter of each corresponding crankpin journal is measured and then subtracted from the bearing internal diameter, the result will be the connecting rod bearing oil clearance.

36 If the second method is to be used, cut lengths of the appropriate size Plastigauge (they should be slightly shorter than the width of the crankpin). Place a strand of Plastigauge on each (cleaned) crankpin journal and fit the (clean) piston/connecting rod assemblies, inserts and bearing caps. Make sure the cap is fitted the correct way around so the connecting rod and bearing cap weight/size markings are correctly aligned and tighten the bearing cap nuts to the specified torque whilst ensuring that the connecting rod does not rotate. Take care not to disturb the Plastigauge. Slacken the bearing cap nuts and remove the connecting rod assemblies, again taking great care not to rotate the crankshaft.

37 Compare the width of the crushed Plastigauge on each crankpin to the scale printed on the Plastigauge envelope to obtain the connecting rod bearing oil clearance **(see illustration)**.

38 If the clearance is not within the specified limits, the bearing inserts may be the wrong grade (or excessively worn if the original inserts are being re-used). Before deciding that different grade inserts are needed, make sure that no dirt or oil was trapped between the bearing inserts and the connecting rod or bearing cap when the clearance was measured. If the clearance is excessive, even with new inserts (of the correct size), the crankpin is worn and the crankshaft should be renewed.

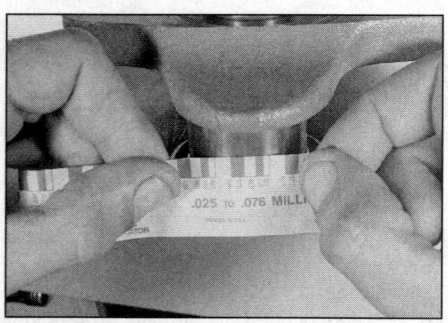

28.37 Place the Plastigauge scale next to the flattened Plastigauge to measure the bearing clearance

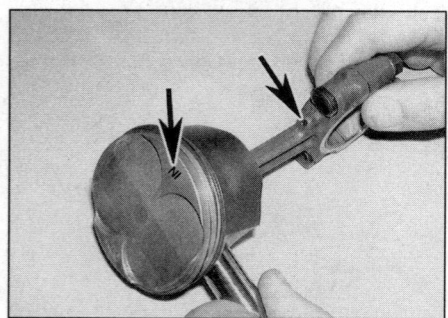

28.41a Ensure the piston 'IN' marking and connecting rod oilway (arrows) are correctly positioned then insert the piston pin

39 On completion, carefully scrape away all traces of the Plastigauge material from the crankpin and bearing inserts using a fingernail or other object which is unlikely to score the inserts.

Installation

40 Check that each piston has one new circlip fitted to it and that it is correctly seated in the piston groove with its gap away from the removal notch in the piston. Insert the piston pin from the opposite side. If it is a tight fit, the piston should be warmed first. If the original pistons/connecting rods are being installed, use the marks made on disassembly to ensure each piston is fitted to its correct connecting rod (see Paragraph 26).

41 Lubricate the piston pin and connecting rod bores with clean engine oil. On the front cylinders, fit the piston to its respective connecting rod making sure that the IN mark on the crown of the piston is on the **same** side as the connecting rod oilway **(see illustration)**. On the rear cylinders, fit the piston to its respective connecting rod making sure that the IN mark on the crown of the piston is on the **opposite** side to the connecting rod oilway **(see illustration)**.

42 Push the piston pin through both piston bosses and the connecting rod bore. If necessary the pin can be tapped carefully into position, using a hammer and suitable drift, whilst supporting the connecting rod and piston. Secure each piston pin in position with a second new circlip, making sure it is correctly seated in the piston groove with its end gap away from the removal notch in the piston.

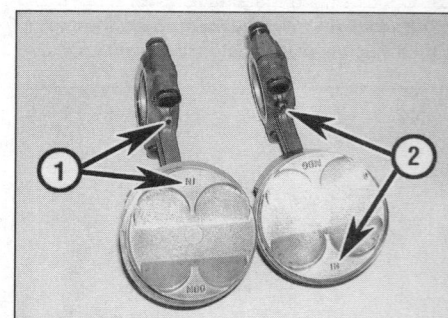

28.41b On the front piston/connecting rod assembly (1) position the 'IN' mark is on the same side as the oilway and on the rear assembly (2) position it on the opposite side

Engine, clutch and transmission 2•47

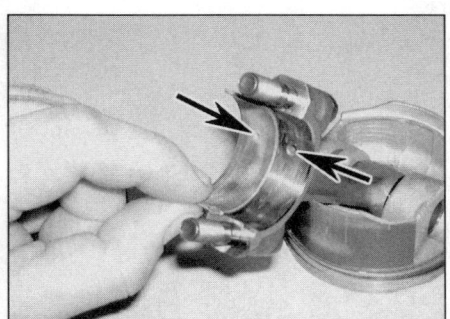

28.44 Ensure the correct bearing inserts are fitted to the connecting rods so that the oil holes (arrows) are correctly aligned

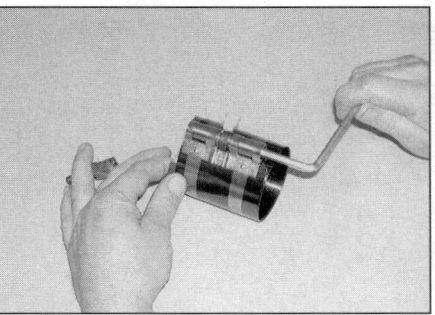

28.46 Ensure the piston ring end gaps are correctly spaced then clamp them in position with a ring compressor

28.47 Ensure the piston 'IN' marking is on the intake side then tap the assembly carefully into the bore

43 Clean the backs of the bearing inserts and the bearing recesses in both the connecting rod and bearing cap. If new inserts are being fitted, ensure that all traces of the protective grease are cleaned off using kerosene (paraffin). Wipe dry the inserts and connecting rods with a lint-free cloth.
44 Press the bearing inserts into their locations. Make sure the tab on each insert engages the notch in the connecting rod or bearing cap (see illustration). Make sure the bearings are fitted in the correct locations (see Paragraph 27 – the upper bearing oil hole must be correctly aligned with the connecting rod oil hole) and take care not to touch any insert's bearing surface with your fingers.
45 Lubricate the cylinder bores, the pistons and piston rings then lay out each piston/connecting rod assembly in its respective position.

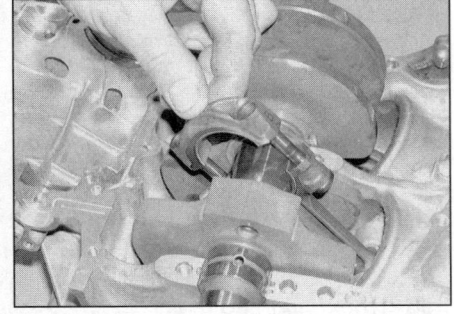

28.49a Fit the bearing cap and lower bearing insert to the connecting rod . . .

46 Starting with assembly number 1, make sure that the piston rings are still spaced as described in Section 29, then clamp them in position with a piston ring compressor (see illustration).
47 Insert the piston/connecting rod assembly into the top of cylinder No 1, ensuring that the IN marking on the piston crown is facing towards the intake side of the bore. Using a block of wood or hammer handle against the piston crown, tap the assembly gently (do not force it) into the cylinder until the piston crown is flush with the top of the cylinder (see illustration). As the piston is fitted make sure the connecting rod lower end remains correctly aligned with the crankshaft to avoid damage.

Caution: Take great care when fitting the piston/connecting rod assemblies as the connecting rods can easily damage the bore surface.

48 Ensure the bearing insert is still correctly fitted, then liberally lubricate the crankpin and bearing and pull the piston/connecting rod assembly down onto the crankpin.
49 Ensure the connecting rod and bearing cap mating surfaces are completely clean and dry then fit the bearing cap and insert to the connecting rod. Make sure the cap is fitted the correct way around so the connecting rod and bearing cap weight/size markings are correctly aligned (see illustrations).
50 Apply a smear of clean engine oil to the threads and underside of the bearing cap nuts. Fit the nuts to the connecting rod and tighten them evenly and progressively, in two

or three stages, to the specified torque (see illustrations).
51 Check that the crankshaft is free to rotate easily, then install the three remaining assemblies in the same way.

29 Piston rings – installation

1 Before installing new piston rings, their end gaps must be checked.
2 Lay out the pistons and the new ring sets so the rings will be matched with the same piston and cylinder during the end gap measurement procedure and engine assembly.
3 Insert the top ring into the top of the first cylinder and square it up with the cylinder walls by pushing it in with the top of the piston. The ring should be about 25 mm below the top edge of the cylinder. To measure the end gap, slip a feeler gauge between the ends of the ring and compare the measurement to that given in the Specifications at the beginning of this Chapter (see illustration).
4 If the gap is larger or smaller than specified, double check to make sure that you have the correct rings before proceeding.
5 Repeat the procedure for each ring that will be installed in the first cylinder and for each ring in the remaining cylinders. Remember to keep the rings, pistons and cylinders matched up.

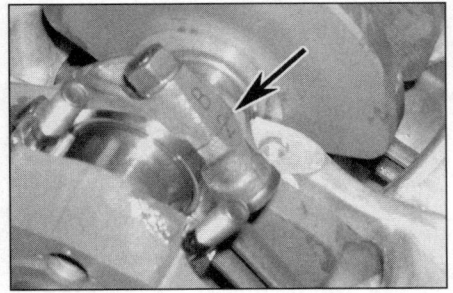

28.49b . . . ensuring that the weight/size group markings (arrow) are correctly aligned

28.50a Oil the threads and undersides of the bearing cap nuts . . .

28.50b . . . then tighten the nuts evenly and progressively to the specified torque

2•48 Engine, clutch and transmission

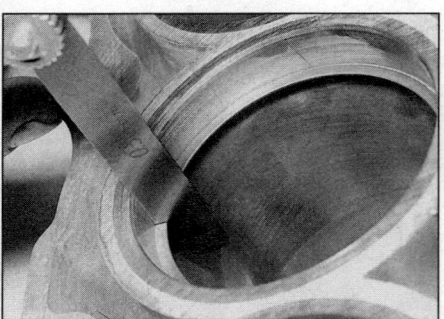

29.3 Measuring piston ring end gap

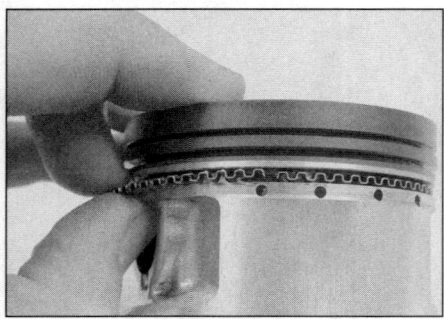

29.7a Install the oil expander ring, ensuring the ends don't overlap . . .

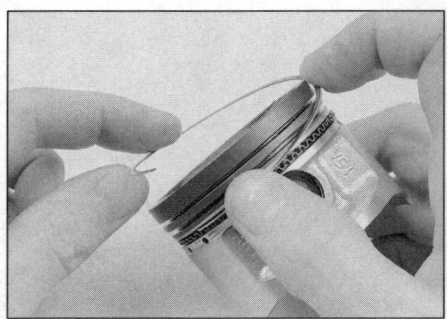

29.7b . . . then fit the side rails using only your fingers to manoeuvre them into position

6 Once the ring end gaps have been checked/corrected, the rings can be installed on the pistons.

7 The oil control ring (lowest on the piston) is installed first. It is composed of three separate components. Slip the expander into the groove, then install the upper side rail **(see illustrations)**. Do not use a piston ring installation tool on the oil ring side rails as they may be damaged. Instead, place one end of the side rail into the groove between the expander and the ring land. Hold it firmly in place and slide a finger around the piston while pushing the rail into the groove. Next, install the lower side rail in the same manner.

8 After the three oil ring components have been installed, check to make sure that both the upper and lower side rails can be turned smoothly in the ring groove.

9 Install the second (middle) ring next **(see illustration)**. **Note:** *The second ring and top ring are different and cannot be interchanged. The second ring is easily identified by its tapered outer edge.* To avoid breaking the ring, use a piston ring installation tool and make sure that the ring is fitted the correct way up with its widest point at the bottom and its identification mark (R or RN) facing up. Fit the ring into the middle groove on the piston. Do not expand the ring any more than is necessary to slide it into place.

10 Finally, install the top ring in the same manner. Make sure the identification mark (T or R) is facing up.

11 With the piston rings correctly installed, check that each ring is free to rotate easily in its groove. Check the ring-to-groove clearance of each ring using feeler gauges and check that the clearance is within the specified range then position the ring end gaps as shown **(see illustration 29.10)**.

12 Repeat the procedure for the remaining pistons and rings.

30 Crankshaft and main bearings – removal, inspection and installation

Removal

1 Separate the crankcase halves as described in Section 24.

2 Remove the piston/connecting rod assemblies as described in Section 28. **Note:** *If no work is to be carried out on the piston/connecting rod assemblies there is no need to remove them from the bores. However, the connecting rod bearing caps should be removed and the pistons pushed up to the top of the bores so that the connecting rod ends are positioned clear of the crankshaft.*

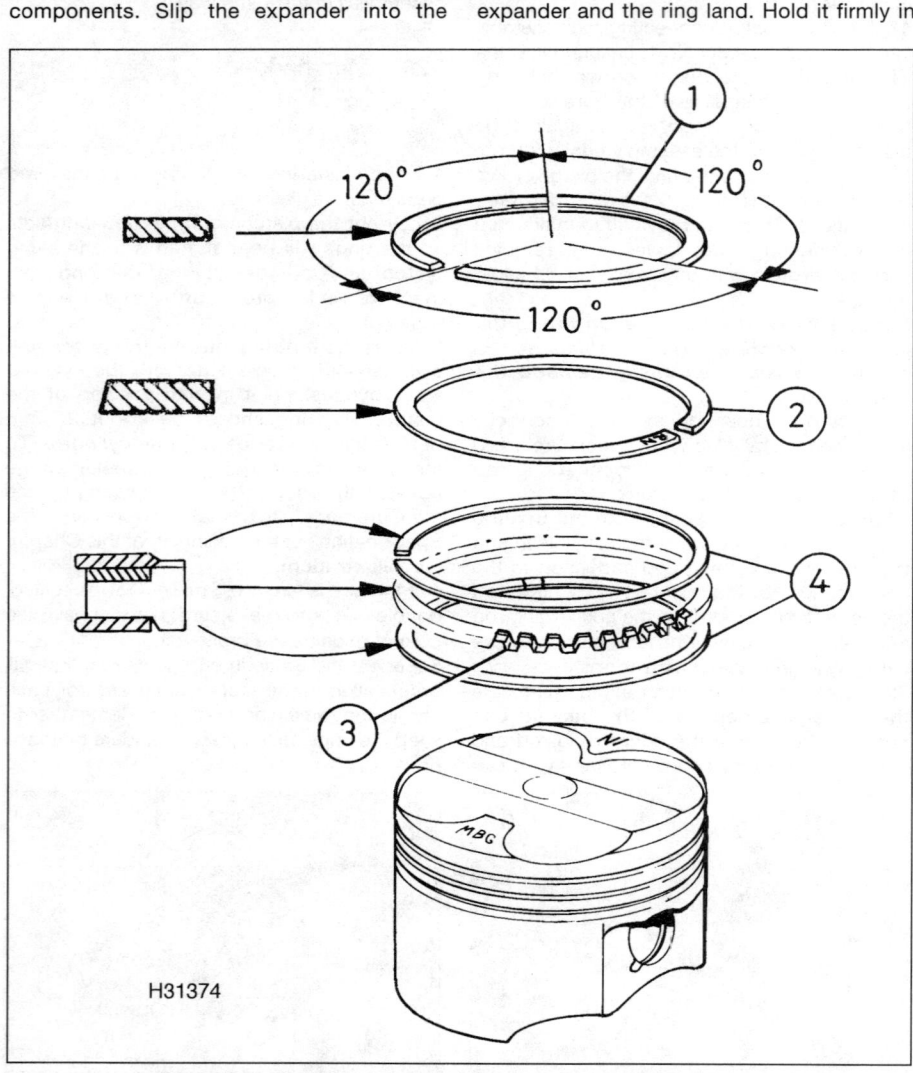

29.9 Piston ring fitting details. Space the end gaps as shown

1 Top ring
2 Second (middle) ring
3 Oil control ring expander
4 Oil control ring side rails

Engine, clutch and transmission 2•49

30.3 Lift the crankshaft out of position taking care not to dislodge the bearing inserts

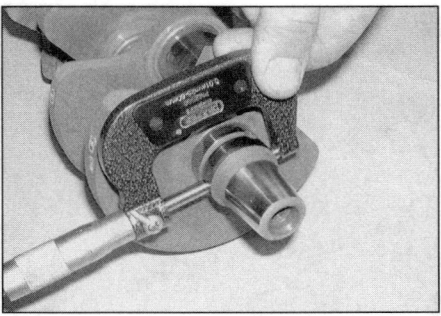

30.10 Measuring a crankshaft (main bearing) journal diameter with a micrometer

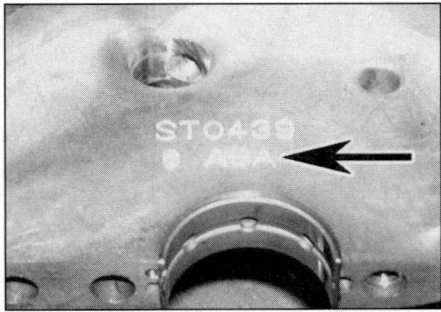

30.11 The crankcase main bearing bore size group letters (arrow) are stamped on the left end of the crankcase upper half

3 Lift the crankshaft out of the upper crankcase half, taking care not to dislodge the bearing inserts **(see illustration)**.
4 The main bearing inserts can be removed from the crankcase halves by pushing their centres to the side, then lifting them out. Keep the bearing inserts in order.

Inspection

Crankshaft and bearing inserts

5 Clean the crankshaft with solvent, using a rifle-cleaning brush to scrub out the oil passages. If available, blow the crank dry with compressed air.
6 Refer to Section 27 and examine the main bearing inserts. If they are scored, badly scuffed or appear to have been seized, new bearings must be installed. Always renew the main bearings as a set. If they are badly damaged, check the corresponding crankshaft journal. Evidence of extreme heat, such as discoloration, indicates that lubrication failure has occurred. Be sure to thoroughly check the oil pump and pressure relief valve as well as all oil holes and passages before reassembling the engine.
7 The crankshaft journals should be given a close visual examination, paying particular attention where damaged bearing inserts have been discovered. If the journals are scored or pitted in any way a new crankshaft will be required. Note that undersizes are not available, precluding the option of re-grinding the crankshaft.
8 Set the crankshaft on V-blocks and check the runout with a dial gauge touching the centre main bearing journal, comparing your findings with this Chapter's Specifications. If the runout exceeds the limit, renew the crankshaft.

Bearing selection

9 The main bearing running clearance is controlled in production by selecting one of five grades of bearing insert. The grades are indicated by a colour-coding marked on the edge of each insert (this colour-code may no longer be visible on the original bearings) **(see illustration 28.27)**. In order, from the thickest to the thinnest, the insert grades are: Blue, Black, Brown, Green and Yellow. New bearing inserts are selected as follows using the crankshaft journal and crankcase main bearing bore size group markings.
10 The standard crankshaft (main bearing) journal diameter is divided into three size groups to allow for manufacturing tolerances. The size group of each journal can be determined by the numbers (either 1, 2 or 3) which are stamped on the crankshaft webs next to each journal **(see illustration 28.28a)**. **Note:** *Ignore the letters as these refer to the crankpin journals.* If the equipment is available, these marks can be checked by direct measurement **(see illustration)**.
11 The crankcase main bearing bore diameters are divided into three size groups to allow for manufacturing tolerances. The size group of each main bearing bore is stamped on the left side of the upper crankcase half, directly above the main bearing journal **(see illustration)**. The first letter (either A, B or C) indicates the diameter of the left journal, and the last the diameter of the right journal. If the equipment is available, these marks can be checked by direct measurement.
12 Match the relevant crankcase code with its crankshaft code and select a new set of bearing inserts using the table below.

Oil clearance check

13 Whether new bearing inserts are being fitted or the original ones are being re-used, the main bearing oil clearance should be checked prior to reassembly.
14 Clean the backs of the bearing inserts and the bearing locations in both crankcase halves.
15 Press the bearing inserts into their locations, ensuring that the tab on each insert engages in the notch in the crankcase. Make sure the bearings are fitted in the correct locations and take care not to touch any insert's bearing surface with your fingers. Note that the centre bearing inserts are different to those fitted to the outer bearings (outer bearing inserts each have three oil holes).
16 There are two possible ways of checking the oil clearance. The first method is by direct measurement (see Paragraphs 17 and 23) and the second by the use of a product known as Plastigauge (see Paragraphs 18 to 23).
17 If the first method is to be used, with the main bearing inserts in position, carefully lower the lower crankcase half onto the upper half. Make sure that the selector forks (if fitted) engage with their respective slots in the transmission gears as the halves are joined. Check that the lower crankcase half is correctly seated then install the eight 9 mm (main bearing) bolts in their original locations (use the original bolts for the check, not the new ones). **Note:** *Do not tighten the crankcase bolts if the casing is not correctly seated.* Working in the sequence marked on the crankcase, tighten all the bolts to the specified stage 1 torque and then tighten them through the specified stage 2 angle (see Paragraphs 23 to 25 of Section 24). Measure the internal diameter of each assembled pair of bearing inserts. If the diameter of each corresponding crankshaft journal is measured and then subtracted from the bearing internal diameter, the result will be the connecting rod bearing oil clearance.
18 If the second method is to be used, ensure the main bearing inserts are correctly fitted and that the inserts and crankshaft are clean and dry. Lay the crankshaft in position in the upper crankcase.

Crankshaft journal size group	Crankcase bearing bore size group	Bearing insert required
1	A	Yellow
1	B	Green
1	C	Brown
2	A	Green
2	B	Brown
2	C	Black
3	A	Brown
3	B	Black
3	C	Blue

2•50 Engine, clutch and transmission

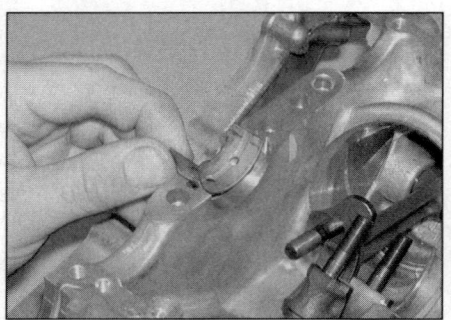

30.26 Fit the upper bearing inserts to the crankcase aligning their tabs with the crankcase notches

30.27 Lubricate all bearing inserts with clean engine oil before fitting the crankshaft

19 Cut several lengths of the appropriate size Plastigauge (they should be slightly shorter than the width of the crankshaft journal). Place a strand of Plastigauge on each (cleaned) crankshaft journal.
20 Carefully position the lower crankcase half onto the upper half. Make sure that the selector forks (if fitted) engage with their respective slots in the transmission gears as the halves are joined. Check that the lower crankcase half is correctly seated then install the eight 9 mm (main bearing) bolts in their original locations (use the original bolts for the check, not the new ones). **Note:** *Do not tighten the crankcase bolts if the casing is not correctly seated.* Working in the sequence marked on the crankcase, tighten all the bolts to the specified stage 1 torque and then tighten them through the specified stage 2 angle (see Paragraphs 23 to 25 of Section 24).

Make sure that the crankshaft is not rotated as the bolts are tightened.
21 Slacken and remove the crankcase bolts, working in the reverse of the sequence marked on the crankcase, then carefully lift off the lower crankcase half, making sure the Plastigauge is not disturbed.
22 Compare the width of the crushed Plastigauge on each crankshaft journal to the scale printed on the Plastigauge envelope to obtain the main bearing oil clearance **(see illustration 28.37)**.
23 If the clearance is not within the specified limits, the bearing inserts may be the wrong grade (or excessively worn if the original inserts are being re-used). Before deciding that different grade inserts are needed, make sure that no dirt or oil was trapped between the bearing inserts and the crankcase halves when the clearance was measured. If the clearance is excessive, even with new inserts (of the correct size), the crankshaft journal is worn and the crankshaft should be renewed.
24 On completion carefully scrape away all traces of the Plastigauge material from the crankshaft journal and bearing inserts; use a fingernail or other object which is unlikely to score the inserts.

Installation

25 Clean the backs of the bearing inserts and the bearing recesses in both crankcase halves. If new inserts are being fitted, ensure that all traces of the protective grease are cleaned off. Wipe dry the inserts and crankcase halves with a lint-free cloth.

26 Press the bearing inserts into their locations. Make sure the tab on each insert engages in the notch in the casing **(see illustration)**. Make sure the bearings are fitted in the correct locations and take care not to touch any insert's bearing surface with your fingers. Note that the centre bearing inserts are different to those fitted to the outer bearings (outer bearing inserts each have three oil holes).
27 Lubricate the bearing inserts in the upper crankcase with clean engine oil **(see illustration)**.
28 Lower the crankshaft into position in the upper crankcase making sure it is fitted the correct way around.
29 Fit the piston/connecting rod assemblies to the crankshaft as described in Section 28.
30 Reassemble the crankcase halves as described in Section 24.

31 Transmission shafts – removal and installation

Removal

1 Separate the crankcase halves as described in Section 24.
2 Lift the input shaft and output shaft out of the crankcase **(see illustration)**.
3 Recover the output shaft bearing half-ring and dowel pin from the upper crankcase half and store them with the transmission shafts for safe-keeping **(see illustrations)**.
4 Remove the oil seal from the end of the output shaft and discard it, as a new one must be used on installation.
5 If necessary, the transmission shafts can be disassembled and inspected for wear or damage as described in Section 32.

Installation

6 Install the output shaft bearing half-ring and dowel pin in the upper crankcase.
7 Slide a new oil seal on the end of the output shaft making sure it is fitted with its sealing lip facing in **(see illustration)**.
8 Lower the output shaft into position in the crankcase half. As the shaft is fitted, align the bearing groove with the half-ring, making sure

31.2 Lift out the transmission shafts from the upper crankcase half

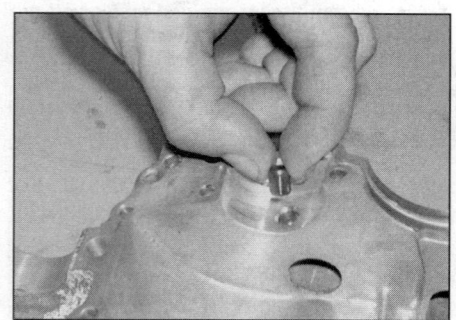

31.3a Remove the output shaft dowel pin . . .

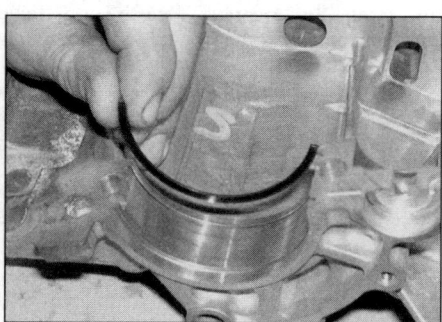

31.3b . . . and bearing half-ring from the crankcase

31.7 Fit a new oil seal to the output shaft making sure its sealing lip is facing inwards

Engine, clutch and transmission 2•51

31.8a When lowering the output shaft into position, ensure the needle bearing outer race engages correctly with the dowel pin (arrows) ...

31.8b ... the bearing groove (1) engages with the half-ring, the pin (2) locates in the crankcase cutout and the oil seal lip (3) sits in the groove

the bearing pin is correctly aligned with the casing cutout. At the same time, align the oil seal lip with the casing groove and engage the needle bearing race with the dowel pin **(see illustrations)**.
9 Lower the input shaft into position in the upper crankcase.
10 Make sure both transmission shafts are correctly seated.
Caution: If the output shaft bearing pin, half ring or dowel pin are not correctly engaged, the crankcase halves will not seat correctly.
11 Position the gears in the neutral position and check that the shafts are free to rotate freely before proceeding further.

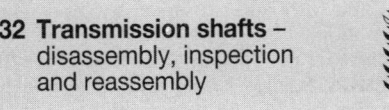

32 Transmission shafts –
disassembly, inspection
and reassembly

When disassembling the transmission shafts, place the parts on a long rod or thread a wire through them to keep them in order and facing the proper direction (see illustration).

1 Remove the shafts from the casing as described in Section 31.

Input shaft
Disassembly
2 Slide off the needle roller bearing outer race, bearing and thrust washer from the left end of the shaft **(see illustration)**.
3 Slide off the 2nd gear.
4 Disengage the lock washer from the special splined washer and slide both off the input shaft.
5 Remove the 6th gear followed by its splined bush and thrust washer.
6 Remove the circlip using a suitable pair of circlip pliers.
7 Remove the 3rd/4th gear, noting which way around it is fitted.
8 Remove the second circlip.
9 Slide off the splined thrust washer followed by the 5th gear, 5th gear bush and thrust washer.

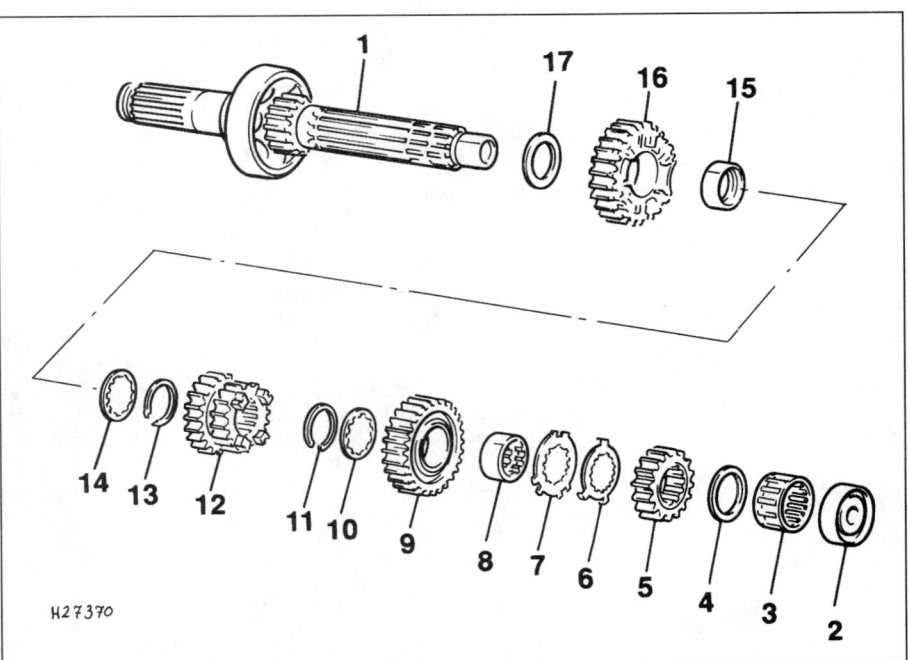

32.2 Transmission input shaft components

1 Input shaft and bearing	6 Lock washer	12 3rd/4th gear
2 Needle roller bearing outer race	7 Special splined washer	13 Circlip
3 Needle roller bearing	8 6th gear splined bushing	14 Splined thrust washer
4 Thrust washer	9 6th gear	15 5th gear bushing
5 2nd gear	10 Splined thrust washer	16 5th gear
	11 Circlip	17 Thrust washer

2•52 Engine, clutch and transmission

32.13 Measuring input shaft 5th gear bushing area diameter

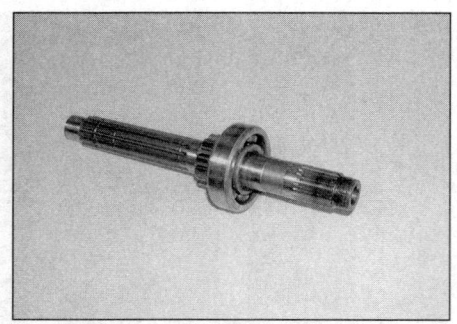

32.15 Input shaft bearing can be renewed if worn

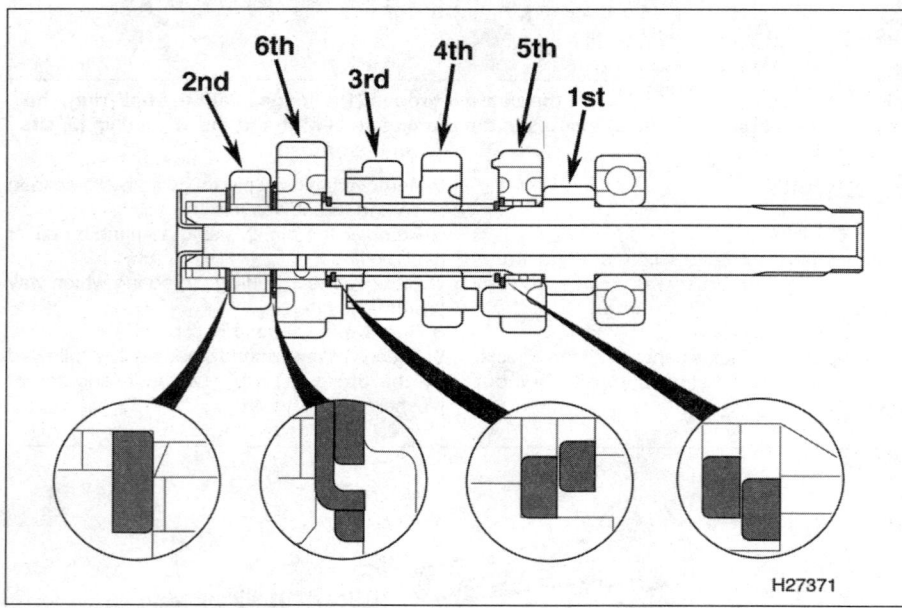

32.16 Correct fitted orientation of input shaft thrust washers and circlips

Inspection

10 Wash all of the components in clean solvent and dry them off.
11 Check the gear teeth for cracking and other obvious damage. Check the gear bushes and the inner surface of each gear for scoring or heat discoloration. If the gear or bush is damaged, renew it.
12 Inspect the dogs and the dog holes in the gears for excessive wear. Replace the paired gears as a set if necessary.
13 The shaft is unlikely to sustain damage unless the engine has seized, placing an unusually high loading on the transmission, or the machine has covered a very high mileage. Check the surface of the shaft, especially where a pinion turns on it, and renew the shaft if it has scored or picked up **(see illustration)**. Inspect the threads of the shafts and check them for trueness by setting them up in V-blocks and measuring any runout with a dial gauge. Damage of any kind can only be cured by renewal.
14 Measure the internal diameter of all gears which run on bushes and the external diameter of the bushes which they run on. If either component has worn to or beyond its service limit it must be renewed. Using the above measurements calculate the gear-to-bush clearance; if this exceeds the specified limit renew the relevant gear and bush as a pair.
15 Check that the outer race of the bearing fitted to the end of the shaft rotates freely and has no sign of freeplay between its inner and outer races **(see illustration)**. If the bearing requires renewal, a bearing puller will be required to extract the bearing from its shaft. Pull the bearing off of the shaft and fit the new bearing using a hammer and tubular drift which bears only on the inner race of the bearing.

Reassembly

16 During reassembly, lubricate all components with engine oil before assembling them. **Note:** *If the thrust washers and circlips are examined closely it will be seen that they are chamfered on one side. During reassembly it is essential that each thrust washer and circlip is fitted so its chamfer is on the correct side and the circlip ends are positioned in the shaft grooves. When fitting splined bushes/gears, always try to align one of the components oil holes with one of the shaft oilways (this may not always be possible)* **(see illustration)**.
17 Slide on the thrust washer **(see illustration)**.
18 Slide on the 5th gear bush then fit the 5th gear with its dogs facing away from the 1st gear. Fit the splined thrust washer with its chamfered edge facing towards the 5th gear **(see illustrations)**.
19 Secure the 5th gear components in position with the circlip making sure its chamfered edge is facing towards the splined thrust washer. Check the circlip is correctly located in the input shaft groove **(see illustration)**.

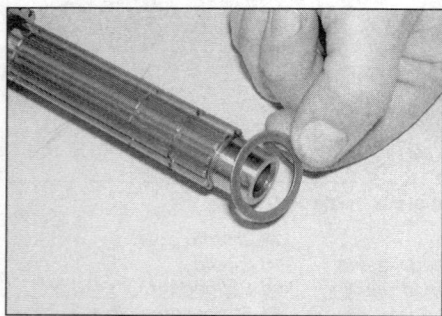

32.17 Slide the thrust washer onto the input shaft

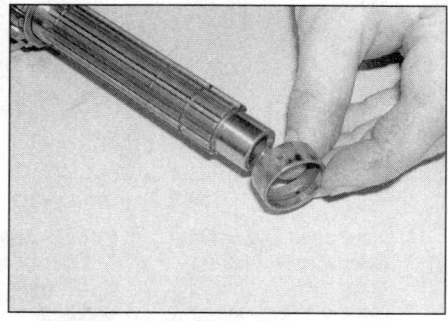

32.18a Fit the 5th gear bushing . . .

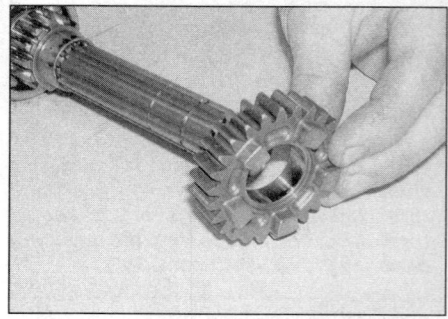

32.18b . . . then fit the 5th gear with its dogs facing away from the 1st gear

Engine, clutch and transmission 2•53

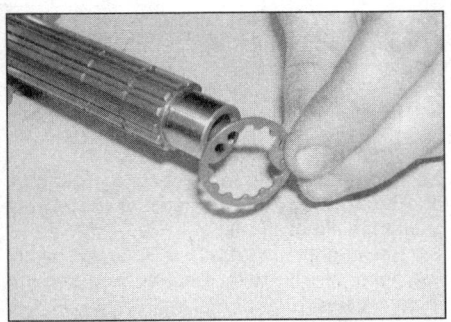

32.18c Fit the splined thrust washer . . .

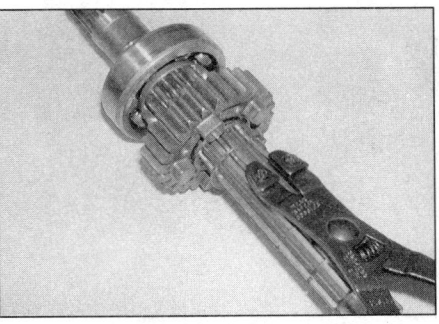

32.19 . . . and secure the 5th gear components in position with a circlip, ensuring it is correctly located in the input shaft groove

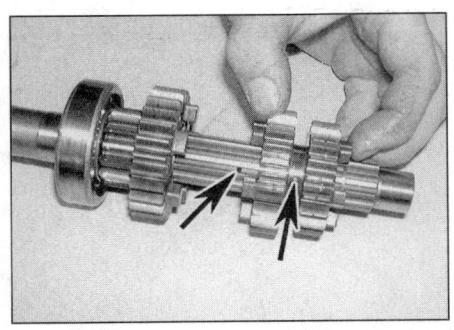

32.20 Fit the 3rd/4th gear with its larger 4th gear facing the 5th gear (oil holes arrowed)

20 Install the 3rd/4th gear with its larger 4th gear facing the 5th gear. Engage the gear on the shaft splines, aligning its oil hole with the shaft oilway (see illustration).
21 Fit the second circlip to the shaft with its

chamfered edge facing away from the 3rd/4th gear pinion. Make sure the circlip is correctly located in the shaft groove (see illustration).

22 Slide on the splined thrust washer with its chamfered edge facing away from the circlip (see illustration).
23 Align the 6th gear splined bush oil hole with the shaft oilway and slide it along the shaft. Fit the 6th gear so that its dogs are facing the 3rd/4th gear (see illustrations).
24 Fit the special splined washer to the shaft with its chamfered edge facing away from the 6th gear then slide it along the shaft until it abuts the gear. Fit the lock washer then rotate the splined washer until its cutouts align with the lock washer tabs. Engage the lock washer with the splined washer to lock it in position (see illustrations).
25 Fit the 2nd gear followed by the thrust washer, making sure the thrust washer chamfered edge faces the gear (see illustrations).

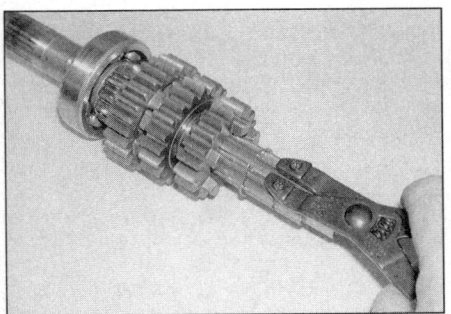

32.21 Fit a second circlip to the next groove on the shaft . . .

32.22 . . . then slide on the splined thrust washer

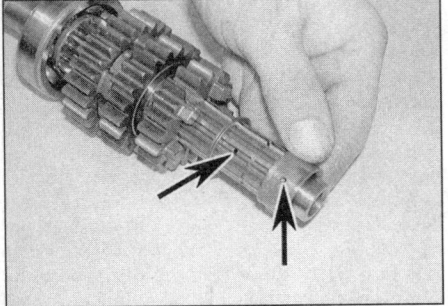

32.23a Align the oil holes (arrows) and slide on the 6th gear splined bushing . . .

32.23b . . . then fit the 6th gear with its dogs facing the 3rd/4th gear

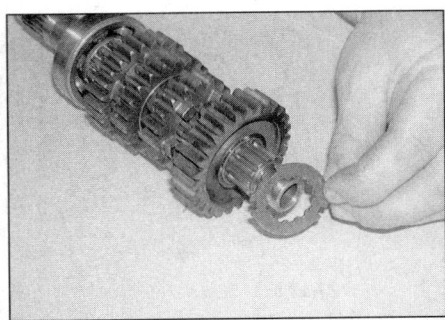

32.24a Fit the special splined washer . . .

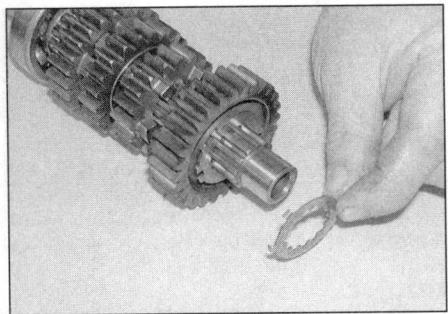

32.24b . . . then fit the lock washer . . .

32.24c . . . engaging its tabs with the washer cutouts (arrows)

32.25a Fit the 2nd gear . . .

2

2•54 Engine, clutch and transmission

32.25b ... followed by the thrust washer ...

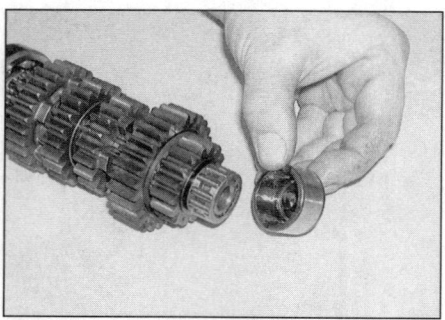

32.26 ... then install the needle roller bearing and outer race

26 Liberally oil the needle roller bearing assembly and install the bearing and outer race on the end of the shaft (see illustration).

Output shaft

Disassembly

27 Slide the needle roller bearing outer race and bearing off the right end of the output shaft (see illustration).

28 Remove the thrust washer followed by the 1st gear, needle roller bearing and second thrust washer.

29 Remove the 5th gear noting which way around it is fitted.

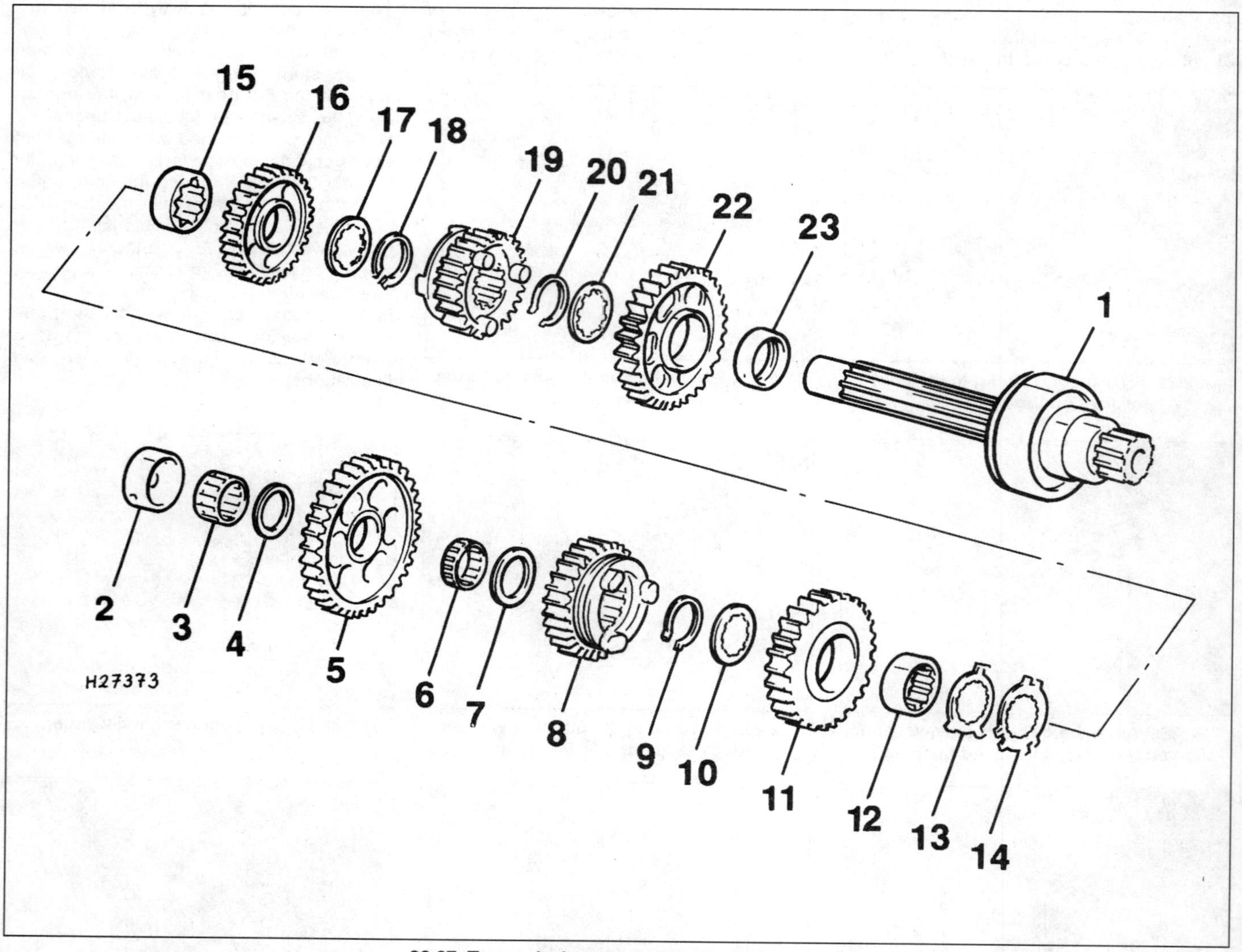

32.27 Transmission output shaft components

1 Output shaft
2 Needle roller bearing outer race
3 Needle roller bearing
4 Thrust washer
5 1st gear
6 Needle roller bearing
7 Thrust washer
8 5th gear
9 Circlip
10 Splined thrust washer
11 4th gear
12 4th gear splined bushing
13 Lock washer
14 Special splined washer
15 3rd gear splined bushing
16 3rd gear
17 Splined thrust washer
18 Circlip
19 6th gear
20 Circlip
21 Splined thrust washer
22 2nd gear
23 2nd gear bushing

Engine, clutch and transmission 2•55

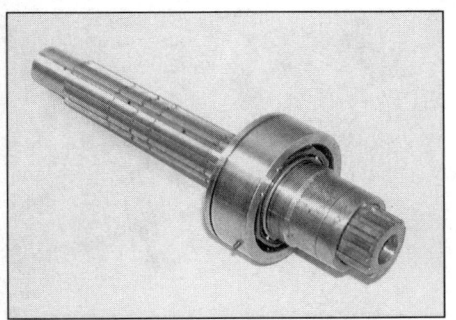

32.36 If the output shaft bearing shows signs of roughness or wear the complete shaft must be renewed

30 Remove the circlip with a suitable pair of circlip pliers.
31 Slide off the splined washer followed by the 4th gear and splined bush.
32 Disengage the lock washer from the special splined washer and slide both off the output shaft.
33 Remove the 3rd gear along with its splined bush and thrust washer.
34 Remove the second circlip and slide off the 6th gear.
35 Remove the third circlip and slide off the splined thrust washer followed by the 2nd gear, noting which way around it is fitted, and 2nd gear bush

Inspection

36 Refer to Paragraphs 10 through 15 noting that it is not possible to obtain the shaft bearing separately. If the bearing is worn, the complete shaft assembly must be renewed (see illustration).

Reassembly

37 During reassembly, lubricate all components with engine oil before assembling them. **Note:** *If the thrust washers and circlips are examined closely it will be seen that they are chamfered on one side. During reassembly it is essential that each thrust washer and circlip is fitted so its chamfer is on the correct side and the circlip ends are positioned in the shaft grooves. When fitting splined bushes/gears, always try to align one of the components oil holes with one of the shaft oilways (this may not always be possible)* (see illustration).

38 Slide on the 2nd gear bush and install the 2nd gear so that its side on which the gear hub protrudes the most is facing the output shaft bearing. Fit the splined thrust washer with its chamfered edge facing the 2nd gear then secure the 2nd gear components with the circlip. Ensure the chamfered edge of the circlip is facing the thrust washer and make sure it is correctly located in the output shaft groove (see illustrations).

39 Fit the 6th gear to the shaft so that its selector fork groove is facing away from the 2nd gear. Align the gear oil hole with the shaft oilway and slide the gear onto the shaft (see illustration).

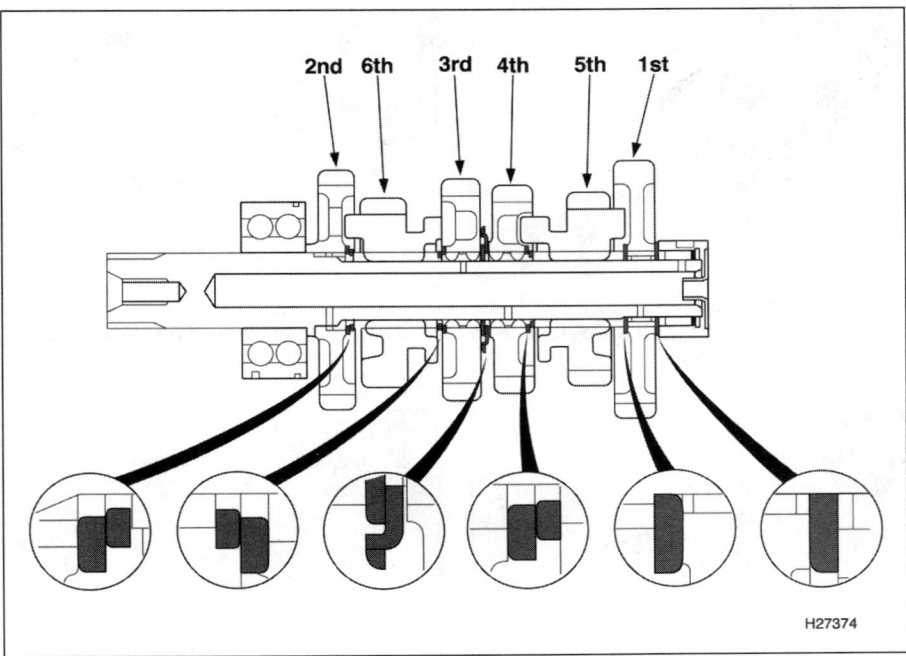

32.37 Correct fitted orientation of output shaft thrust washers and circlips

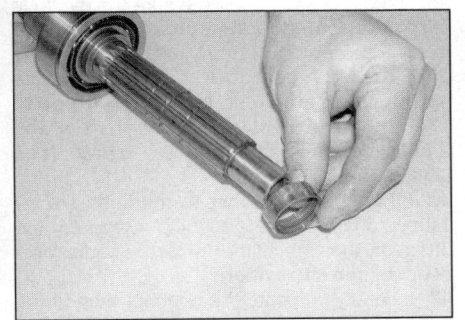

32.38a Fit the 2nd gear bushing . . .

32.38b . . . and install the 2nd gear so that its side on which the gear hub protrudes the most is facing the output shaft bearing

32.38c Slide on a splined thrust washer . . .

32.38d . . . and secure the 2nd gear components with the circlip

32.39 Slide on the 6th gear as shown making sure its oil holes are correctly aligned with the shaft oilways (arrows)

2•56 Engine, clutch and transmission

32.40 Fit another circlip to the output shaft groove . . .

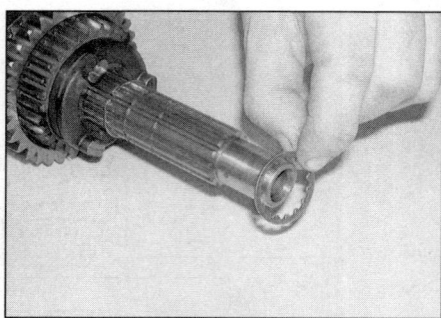

32.41 . . . then slide on a splined thrust washer

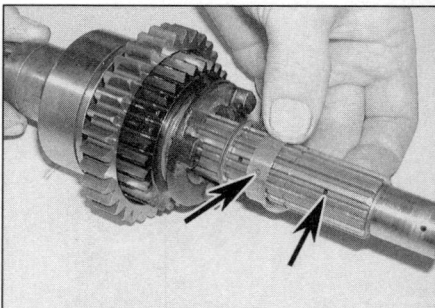

32.42a Fit the 3rd gear splined bushing, aligning its oil hole with the shaft oilway (arrows) . . .

32.42b . . . and install the 3rd gear with its dog holes facing the 6th gear

32.43a Fit the special splined washer . . .

32.43b . . . and lock washer, engaging the lock washer tabs with the splined washer cutouts

40 Fit a second new circlip with its chamfered edge facing away from the 6th gear. Ensure the circlip is correctly located in the shaft groove **(see illustration)**.

32.44a Slide on the 4th gear splined bushing, aligning its oil holes with the shaft oilways (arrows) . . .

41 Slide on the splined thrust washer with its chamfered edge facing away from the circlip **(see illustration)**.
42 Align the 3rd gear splined bush oil hole with the shaft oilway and slide it along the shaft. Fit the 3rd gear so that its dog holes are facing the 6th gear **(see illustrations)**.
43 Fit the special splined washer to the shaft with its chamfered edge facing towards the 3rd gear then slide it along the shaft until it abuts the gear. Fit the lock washer then rotate the splined washer until its cutouts align with the lock washer tabs. Engage the lock washer with the splined washer to lock it in position **(see illustrations)**.
44 Align the 4th gear splined bush oil hole with the shaft oilway and slide it along the shaft. Fit the 4th gear so that its dog holes are facing away from the 3rd gear, then slide on

the splined thrust washer with its chamfered edge facing the 4th gear. Secure the 4th gear components in position with the circlip ensuring its chamfered edge is facing the splined thrust washer. Make sure the circlip is correctly located in the output shaft groove **(see illustrations)**.
45 Fit the 5th gear to the shaft so that its selector fork groove is facing the 4th gear. Align the gear oil hole with the shaft oilway and slide the gear on the shaft **(see illustration)**.
46 Fit the thrust washer to the output shaft with its chamfered edge facing away from the 5th gear, then install the 1st gear needle roller bearing **(see illustrations)**.
47 Liberally lubricate the bearing, then fit the 1st gear so that its side on which the gear hub protrudes the most is facing away from the

32.44b . . . and fit the 4th gear with its dog holes facing away from the 3rd gear

32.44c Slide on a splined thrust washer . . .

32.44d . . . and secure it in position with another snap-ring

Engine, clutch and transmission 2•57

32.45 Fit the 5th gear as shown, aligning its oil hole with the shaft oilway (arrows)

32.46a Install the thrust washer . . .

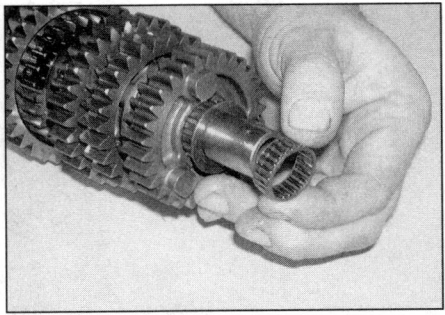

32.46b . . . followed by the 1st gear needle roller bearing

5th gear. Slide on the second thrust washer with its chamfered edge facing the 1st gear **(see illustrations)**.
48 Lubricate the needle roller bearing assembly and fit the bearing and outer race to the end of the output shaft **(see illustration)**.

33 Initial start-up after overhaul

Note: *Do not install the inner and lower fairing panels until after the engine has been run.*
1 Make sure the engine oil and coolant levels are correct (see 'Daily (pre-ride) checks').
2 Raise and support the fuel tank and disconnect the fuel pump wiring connector from the base of the tank (see Chapter 4).
3 Turn on the ignition switch and crank the engine over with the starter a few times to prime the lubrication system of the engine.
4 Reconnect the wiring connector to the fuel pump then lower the fuel tank back down into position (see Chapter 4).
5 Make sure there is fuel in the tank then start the engine and allow it idle slowly.

⚠ *Warning: If the oil pressure warning light doesn't go off immediately or soon after the engine starts (a delay of a few seconds is usual), or it comes on while the engine is running, stop the engine immediately and investigate the cause.*

6 Once the oil pressure warning light goes out, bleed the air from the cooling system (see Chapter 1) and top-up the radiator and coolant reservoir fluid levels.
7 Warm the engine up to normal operating temperature whilst checking carefully that there are no oil and coolant leaks. Check the operation of the clutch and transmission then switch off the engine.

32.47a Fit the 1st gear so that its side on which the gear hub protrudes the most is facing away from the 5th gear

8 Make sure the transmission and controls, especially the brakes, function properly before road testing the machine. Refer to Section 34 for the recommended break-in procedure.
9 Upon completion of the road test, and after the engine has cooled down completely, check the engine oil and the coolant level in both the radiator and reservoir before installing the fairing panels.

34 Recommended running-in procedure

1 Any rebuilt engine needs time to break-in, even if parts have been installed in their original locations. For this reason, treat the machine gently for the first few miles to make sure oil has circulated throughout the engine and any new parts installed have started to seat.
2 Even greater care is necessary if the engine has been rebored or a new crankshaft has been installed. In the case of a rebore, the engine will have to be broken in as if the machine were new. This means greater use of the transmission and a restraining hand on the

32.47b Slide on another thrust washer . . .

32.48 . . . and install the needle roller bearing and outer race

throttle until at least 300 miles (500 km) have been covered. There's no point in keeping to any set speed limit – the main idea is to keep from labouring the engine and to gradually increase performance until the 300 mile (500 km) mark is reached. Experience is the best guide, since it's easy to tell when an engine is running freely.
3 If a lubrication failure is suspected, stop the engine immediately and try to find the cause. If an engine is run without oil, even for a short period of time, severe damage will occur.

Notes

Chapter 3
Cooling system

Contents

Coolant hoses – removal and installation 9	Cooling system draining, flushing and refilling .... see Chapter 1
Coolant level check .... see Daily (pre-ride) checks	General information 1
Coolant reservoir – removal and installation 3	Radiator pressure cap – check 2
Coolant temperature gauge and sensor – check and replacement . 5	Radiators – removal and installation 7
Cooling fan and fan switch – check and replacement 4	Thermostat and housing – removal, check and installation 6
Cooling system checks .... see Chapter 1	Water pump – check, removal and installation 8

Degrees of difficulty

Easy, suitable for novice with little experience		Fairly easy, suitable for beginner with some experience		Fairly difficult, suitable for competent DIY mechanic		Difficult, suitable for experienced DIY mechanic		Very difficult, suitable for expert DIY or professional	

Specifications

Coolant
Mixture type and capacity .... see Chapter 1

Pressure cap
Cap valve opening pressure .... 16 to 20 psi (1.1 to 1.4 bar)

Fan switch
Switches on (fan cut-in temperature) .... 98 to 102°C
Switches off (fan cut-out temperature) .... 93 to 97°C

Coolant temperature sensor
Temperature gauge circuit resistance
  At 80°C .... 2.1 to 2.6 K-ohms
  At 120°C .... 620 to 760 ohms

Thermostat
Opening temperature .... 80 to 84°C
Valve lift .... 8 mm (min) at 95°C

Torque settings
Fan switch .... 18 Nm
Water pump cover bolts .... 13 Nm

3•2 Cooling system

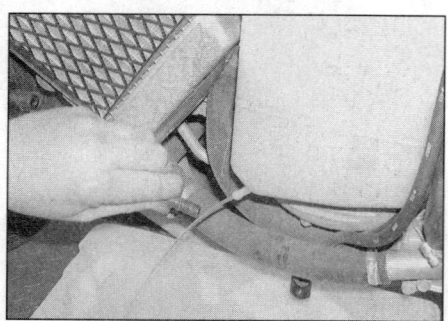

3.2 Disconnect the overflow hose from the reservoir and drain its contents into a suitable container

3.3 Slacken the bolt and pivot the mounting bracket clear of the reservoir

1 General information

The cooling system uses a water/antifreeze coolant to carry away excess energy in the form of heat. The cylinders are surrounded by a water jacket from which the heated coolant is circulated by thermo-syphonic action in conjunction with a water pump which is driven off the oil pump driveshaft. The hot coolant passes upwards to the thermostat and filler neck then through to the radiators (one on each side of the frame). The coolant then flows down each radiator core, where it is cooled by the passing air, to the water pump and back to the engine where the cycle is repeated.

A thermostat, located in-between the cylinders, is fitted in the system to prevent the coolant flowing through the radiator when the engine is cold, therefore accelerating the speed at which the engine reaches normal operating temperature. The fuel injection system coolant temperature sensor (see Chapter 4) controls the temperature gauge on the instrument panel. An electrically-operated cooling fan is fitted to aid cooling in extreme conditions. The fan is fitted to the left side radiator and is controlled by a thermostatic switch (fan switch) which is screwed into the top of the left radiator.

On VFR800FI-Y and 1 models, an automatic choke fast idle wax unit is connected to the cooling system and operates according to coolant temperature (see Chapter 4).

The complete cooling system is partially sealed and pressurised, the pressure being controlled by a valve contained in the spring-loaded pressure cap. By pressurising the coolant the boiling point is raised, preventing premature boiling in adverse conditions. The overflow pipe from the system is connected to a reservoir into which excess coolant is expelled under pressure. The discharged coolant automatically returns to the radiator when the engine cools.

⚠ **Warning:** Do not remove the pressure cap from the filler neck when the engine is hot. Scalding hot coolant and steam may be blown out under pressure, which could cause serious injury. When the engine has cooled, place a thick rag, like a towel over the pressure cap; slowly rotate the cap anti-clockwise to the first stop. This procedure allows any residual pressure to escape. When the steam has stopped escaping, press down on the cap while turning it anti-clockwise and remove it.

⚠ **Warning:** Do not allow antifreeze to come in contact with your skin or painted surfaces of the motorcycle. Rinse off any spills immediately with plenty of water. Antifreeze is highly toxic if ingested. Never leave antifreeze lying around in an open container or in puddles on the floor; children and pets are attracted by its sweet smell and may drink it. Check with the local authorities about disposing of used antifreeze. Many communities will have collection centres which will see that antifreeze is disposed of safely.

Caution: At all times use the specified type of antifreeze, and always mix it with distilled water in the correct proportion. The antifreeze contains corrosion inhibitors which are essential to avoid damage to the cooling system. A lack of these inhibitors could lead to a build-up of corrosion which would block the coolant passages, resulting in overheating and severe engine damage. Distilled water must be used as opposed to tap water to avoid a build-up of scale which would also block the passages.

2 Pressure cap – check

1 If problems such as overheating or loss of coolant occur, check the entire system as described in Chapter 1. The filler neck cap opening pressure should be checked by a dealer with the special tester required to do the job. If the cap is defective renew it.

3 Coolant reservoir – removal and installation

Removal

1 Remove the left side lower fairing panel (see Chapter 8).
2 Place a suitable container underneath the reservoir. Release the clip then detach the radiator overflow hose from the base of the reservoir and allow the coolant to drain into the container **(see illustration)**.
3 Loosen the fairing mounting bracket bolt and pivot the bracket away from the reservoir locating peg **(see illustration)**.
4 Free the overflow hose, noting its correct routing, then unscrew the mounting bolt and remove the coolant reservoir from the bike **(see illustrations)**.

Installation

5 Installation is the reverse of removal. Make sure the hoses are correctly routed and secured with their clips. On completion top-up the reservoir as described in *Daily (pre-ride) checks*.

4 Cooling fan and fan switch – check and replacement

Cooling fan motor

Check

1 If the engine is overheating and the cooling fan isn't coming on, first check the cooling fan circuit fuse (see Chapter 9) and then the fan switch as described in Paragraphs 11 to 16 below.

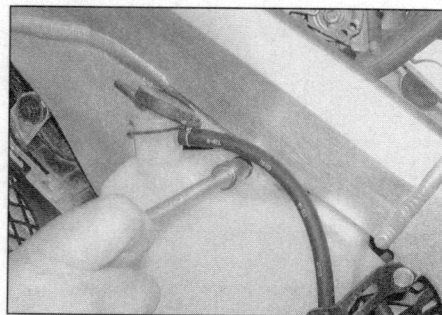

3.4a Unscrew the mounting bolt . . .

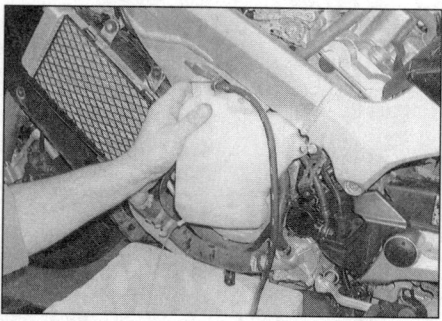

3.4b . . . and remove the coolant reservoir from the bike

Cooling system 3•3

2 If the fan does not come on (and the fan switch is good) the fault lies in either the cooling fan motor or the relevant wiring. Test all the wiring and connections as described in Chapter 9.
3 To test the cooling fan motor, remove the left side lower fairing panel (see Chapter 8) and locate the cooling fan motor (black) wiring connector, which is directly above the fan switch. Disconnect the wiring connector and the connector from the fan switch **(see illustration)**. Using a 12 volt battery and two jumper wires, connect the battery positive (+) lead to the black/blue wire of the motor connector and the battery negative (–) lead to the fan switch wiring connector. Once connected the fan should operate. If it does not, and the wiring is all good, then the fan motor is faulty.

Replacement

⚠ *Warning: The engine must be completely cool before carrying out this procedure.*

4 Remove the left radiator (see Section 7).
5 Disconnect the wiring connector from the fan switch and unclip the motor wiring from the fan shroud cover **(see illustration)**.
6 Unscrew the three bolts securing the fan shroud to the radiator, noting that one bolt also secures the earth (ground) cable, and remove the fan. The fan shroud cover and radiator grille can then be unclipped and removed from the radiator.
7 If necessary unscrew the nut and remove the fan blade from the motor. Release the motor wiring from the shroud then unscrew the retaining nuts and separate the motor and shroud.
8 Reassemble the fan motor and shroud and securely tighten the motor nuts. Ensure the motor wiring is clipped securely onto the shroud. Fit the fan blade to the motor and securely tighten its retaining nut, have applied a drop of thread-locking compound to the nut prior to installation.
9 Ensure the fan shroud cover and grille are correctly installed, then fit the fan assembly to

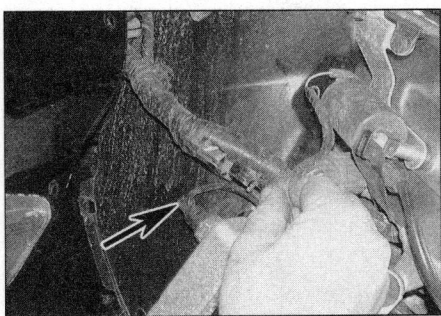

4.3 Disconnecting the cooling fan motor wiring connector (fan switch wiring connector arrowed)

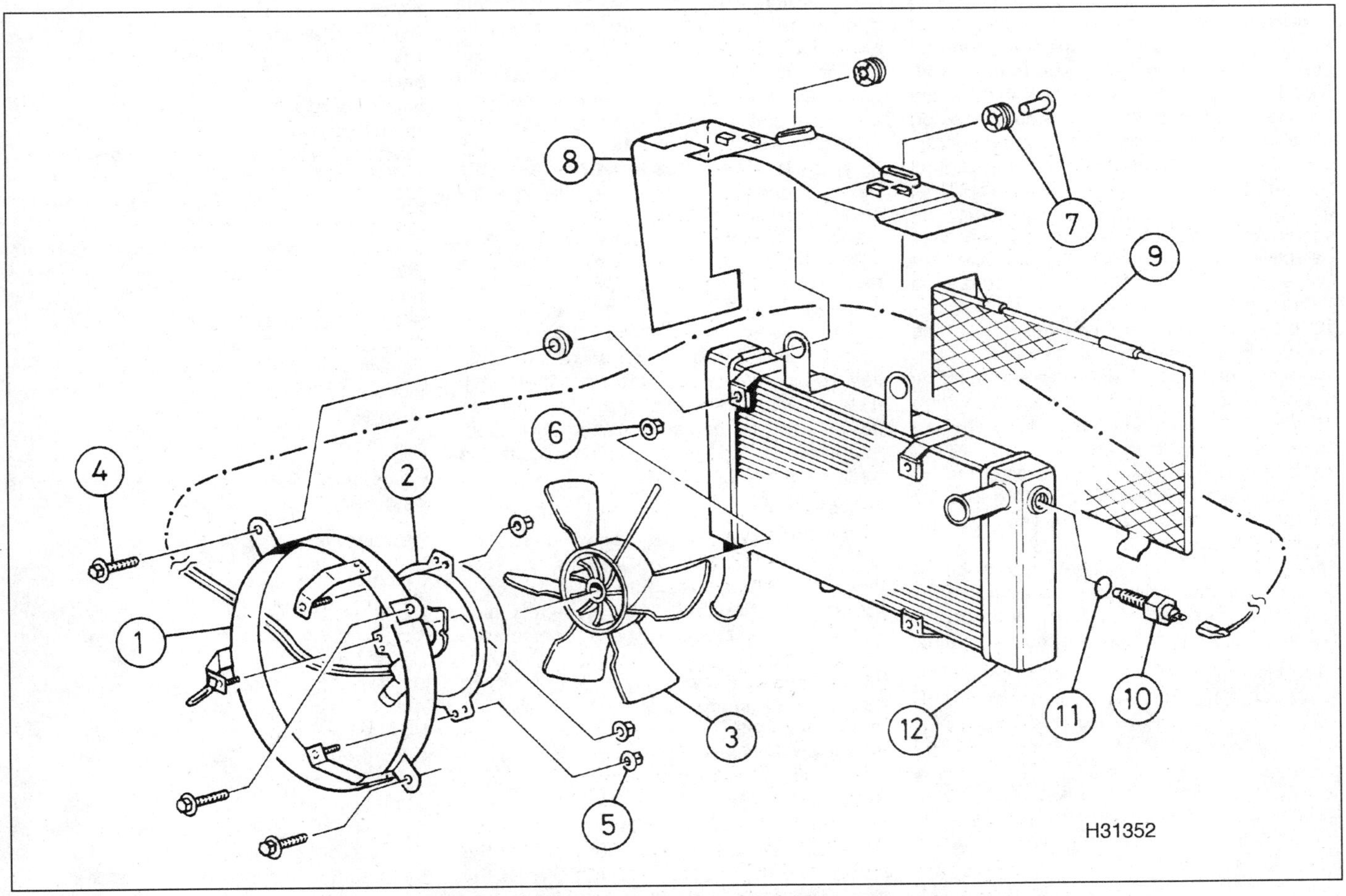

the radiator. Route the motor wiring correctly and secure it in position with all the cover clips. Fit the retaining bolts, not forgetting the earth (ground) lead connector, and tighten them securely.
10 Reconnect the wiring connector to the fan switch then install the radiator (see Section 7).

4.5 Left radiator and cooling fan assembly components

1 Fan shroud	5 Motor nut	9 Radiator grille
2 Fan motor	6 Fan nut	10 Fan switch
3 Fan	7 Radiator mounting rubber and collar	11 O-ring
4 Fan shroud bolt	8 Fan shroud cover	12 Radiator

3•4 Cooling system

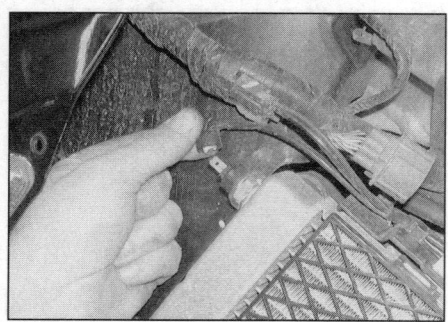

4.12 Disconnect the fan switch wiring connector

Cooling fan switch

Check

11 If the engine is overheating and the cooling fan isn't coming on, first check the cooling fan circuit fuse (see Chapter 9). If the fuse is blown, check the fan circuit for a short to earth (see the wiring diagrams at the end of this book).

12 If the fuse is good, remove the left side lower fairing (see Chapter 8), and disconnect the wiring connector from the fan switch on the top of the left radiator (see illustration). Switch on the ignition and check that battery voltage is present at the fan switch connector; if not check the wiring and connectors as described in Chapter 9. If battery voltage is present, using a jumper wire if necessary, connect the wire to earth (ground). The fan should come on. If it does, the fan switch is defective and must be renewed. If it does not come on, the fan should be tested (see Paragraph 3).

13 If the fan is on all the time, disconnect the switch wiring connector. The fan should stop. If it does, the switch is defective and must be renewed. If it doesn't, check the wiring between the switch and the fan for a short to earth, and the fan itself.

14 If the fan works but is suspected of cutting in at the wrong temperature, a more comprehensive test of the switch can be made as follows.

15 Remove the switch (see Paragraphs 17 to 19). Fill a small heatproof container with oil and place it on a stove. Connect the positive (+) probe of an ohmmeter to the terminal of

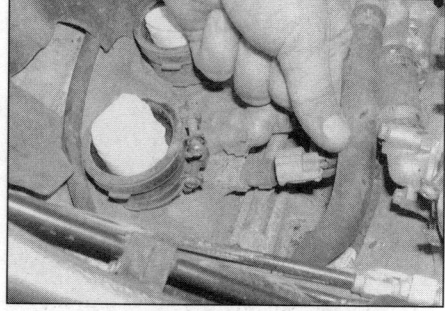

5.2 Disconnect the coolant temperature sensor wiring connector

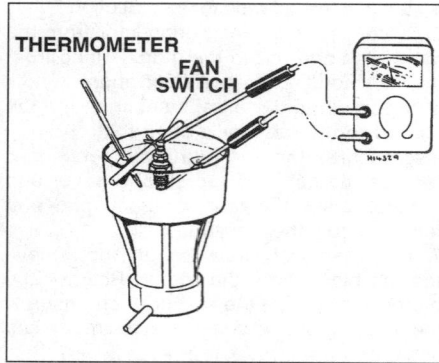

4.15 Cooling fan switch testing set-up

the switch and the negative (–) probe to the switch body, and using some wire or other support suspend the switch in the oil so that just the sensing portion and the threads are submerged (see illustration). Also place a thermometer capable of reading temperatures up to 110°C in the oil so that its bulb is close to the switch. Note: *None of the components should be allowed to directly touch the container.*

16 Initially the ohmmeter reading should show an open circuit (infinite resistance) indicating that the switch is open. Heat the oil, stirring it gently.

 Warning: This must be done very carefully to avoid the risk of personal injury.

When the temperature reaches around 98 to 102°C the switch contacts should close and the meter should show continuity (zero resistance). Now turn the heat off. As the temperature falls below 93 to 97°C the meter reading should show an open circuit again (infinite resistance), indicating that the switch has opened. If the meter readings obtained are different, or they are obtained at different temperatures, then the fan switch is faulty and must be renewed.

Replacement

 Warning: The engine must be completely cool before carrying out this procedure.

17 Remove the lower fairing panels (see Chapter 8) then drain the cooling system (see Chapter 1).

18 Disconnect the wiring connector from the fan switch on the top of the left radiator (see illustration 4.12).

19 Unscrew the switch and remove it from the radiator. Discard the switch O-ring and obtain a new one.

20 Apply a suitable sealant to the switch threads, then fit the new O-ring. Screw the switch into the radiator and tighten it to the specified torque setting. Take care not to overtighten the switch as the radiator could be damaged.

21 Reconnect the switch wiring connector and refill the cooling system (see Chapter 1). Install the lower fairing panels (see Chapter 8).

5 Coolant temperature gauge and sender – check and replacement

Coolant temperature gauge

Check

1 The coolant temperature gauge is part of the instrument cluster LCD unit and is operated by the fuel injection system coolant temperature sensor (see Chapters 4 and 9). If the system malfunctions check first that the battery is fully charged and that the fuses are all good.

2 If the gauge is not working, remove the fuel tank and throttle body assembly (see Chapter 4). Ensure the ignition is switched off then disconnect the wiring connector from the coolant temperature sensor (fitted to the rear of the front cylinder head) (see illustration). Turn the ignition switch on; the gauge should display --°C. Turn the ignition switch off again then earth (ground) the green/blue terminal of the sensor wiring connector. Turn the ignition switch on again; the gauge should display 132°C which flashes continually. If the gauge performs as described, the coolant temperature sensor is proven defective and must be renewed.

Caution: Do not leave the ignition switched on for any longer than is necessary to take the reading, or the gauge may be damaged.

3 If the temperature gauge display does not perform as expected, the fault lies in the wiring or the gauge itself. Check all the relevant wiring and wiring connectors (see Chapter 9). If all appears to be well, the instrument cluster LCD unit or printed circuit board is defective and must be renewed (see Chapter 9).

Replacement

4 The temperature gauge is part of the instrument cluster LCD unit (see Chapter 9).

Temperature gauge sensor

Check

5 The temperature gauge is controlled by the fuel injection system coolant temperature sensor. To check the temperature gauge function of the sensor, first remove the sensor from the engine (see Chapter 4).

6 Fill a small heatproof container with oil and place it on a stove. Connect the positive (+) probe of an ohmmeter to the green/blue terminal of the sensor and the negative (–) probe to the sensor body, and using some wire or other support suspend the switch in the oil so that just the sensing portion and the threads are submerged. Also place a thermometer capable of reading temperatures up to 130°C in the oil so that its bulb is close to the switch. Note: *None of the components should be allowed to directly touch the container.*

Cooling system

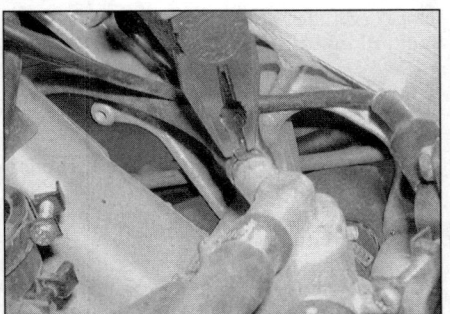

6.6a Disconnect the air bleed hose . . .

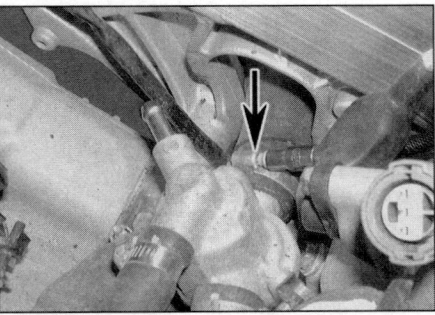

6.6b . . . and the filler neck hose (arrow) from the thermostat housing

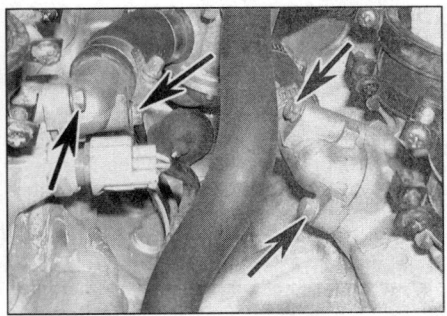

6.7 Unscrew the bolts (arrows) securing the coolant outlet unions to the front and rear cylinders and remove the thermostat housing assembly

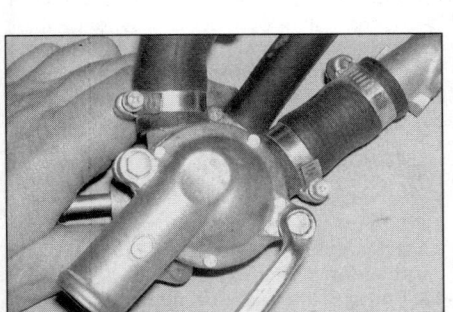

6.8a Unscrew the two bolts and remove the cover and O-ring . . .

6.8b . . . then lift the thermostat out from the housing

7 Heat the oil, stirring it gently. Check that the correct resistance is obtained at the specified temperatures (see Specifications). If the meter readings obtained are significantly different, or they are obtained at different temperatures, then the sensor is faulty and must be renewed. *Note: Keep the temperature of the oil constant for approximately 3 minutes before taking each resistance reading to ensure accurate readings are obtained.*

 Warning: This must be done very carefully to avoid the risk of personal injury.

Replacement

8 Refer to Section 11 of Chapter 4.

6 Thermostat and housing – removal, check and installation

Removal

 Warning: The engine must be completely cool before carrying out this procedure.

1 The thermostat is automatic in operation and should give many years service without requiring attention. In the event of a failure, the valve will probably jam open, in which case the engine will take much longer than normal to warm up. Conversely, if the valve jams shut, the coolant will be unable to circulate and the engine will overheat. Neither condition is acceptable, and the fault must be investigated promptly.

2 Remove the left and right lower fairing panels (see Chapter 8).
3 Drain the cooling system (see Chapter 1).
4 Remove the fuel tank and throttle body assembly (see Chapter 4). The thermostat is fitted to the housing located between the cylinders.
5 Slacken the retaining clip and detach the coolant bypass hose from the water pump.
6 Release the retaining clips and detach the filler neck hose and air bleed hose from the thermostat housing **(see illustrations)**.
7 Unscrew the bolts securing the coolant outlet unions to the front and rear cylinders then manoeuvre the thermostat housing assembly out of position **(see illustration)**. Recover the O-rings from the coolant unions and discard them; new ones must be used on installation.
8 Unscrew the bolts and remove the cover from the base of the thermostat housing then lift out the thermostat **(see illustrations)**. Discard the cover O-ring; a new one must be used on installation.

Check

9 Examine the thermostat visually before carrying out the test. If it remains in the open position at room temperature, it should be renewed.
10 Suspend the thermostat by a piece of wire in a container of cold water. Place a thermometer in the water so that its bulb is close to the thermostat **(see illustration)**. Heat the water, noting the temperature when the thermostat opens, and compare the result with the specifications given at the beginning of the Chapter. Also check the amount the valve opens after it has been heated at 95°C for a few minutes and compare the measurement to the specifications. If the readings obtained differ from those given, the thermostat is faulty and must be renewed.

 Warning: This must be done very carefully to avoid the risk of personal injury.

Installation

11 Install the thermostat in its housing, aligning its body with the housing cut-out. Fit a new O-ring to the housing cover groove then fit the cover, making sure the O-ring remains correctly seated, and securely tighten its bolts **(see illustration)**.

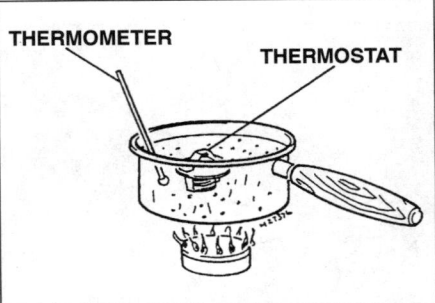

6.10 Thermostat testing set-up

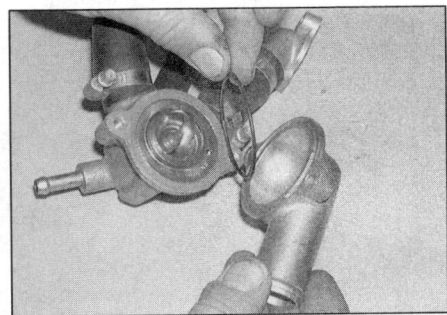

6.11 Ensure the thermostat is correctly located and fit a new O-ring to the cover groove

3•6 Cooling system

12 Fit a new O-ring to the groove of each coolant outlet union then manoeuvre the thermostat housing assembly into position **(see illustration)**. Ensure the union O-rings are correctly positioned then fit the union retaining bolts and tighten them securely.
13 Connect the air bleed hose and filler neck hose to the thermostat housing and secure them in position with the retaining clips.
14 Connect the bypass hose to the water pump and securely tighten its retaining clip.
15 Install the throttle body assembly and fuel tank (see Chapter 4).
16 Refill the cooling system (see Chapter 1).
17 On completion, install the fairing lower panels (see Chapter 8).

7 Radiators – removal and installation

Removal

 Warning: The engine must be completely cool before carrying out this procedure.

1 Remove the lower fairing panels (see Chapter 8) then drain the cooling system (see Chapter 1). The radiators are removed individually as follows.

Left side radiator
2 Trace the wiring back from the radiator cooling fan to its (black) wiring connector, located above the fan switch, and disconnect

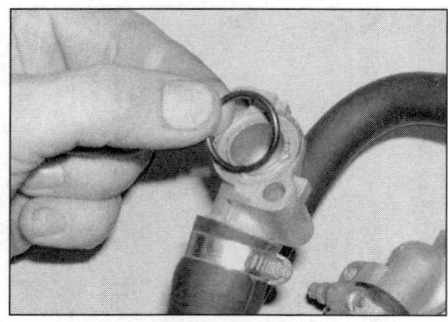

6.12 Fit a new O-ring to each coolant outlet union

it from the main wiring harness **(see illustration 4.3)**.
3 Slacken the retaining clips securing the upper and lower hoses to the radiator.
4 Unscrew the radiator mounting bolt and remove the collar from the rubber mounting **(see illustration)**.
5 Free the radiator from its mountings then detach it from the coolant hoses and remove it from the bike **(see illustrations)**.
6 Check the radiator grille and the radiator for signs of damage and clear any dirt or debris that might obstruct air flow and inhibit cooling. If the radiator fins are badly damaged or broken the radiator must be renewed. Also check the rubber mountings, and renew them if necessary. Refer to Section 4 for information on the cooling fan and switch.

Right side radiator
7 Slacken the retaining clip and detach the filler neck hose from the radiator **(see illustration)**.
8 Slacken the retaining clips securing the upper and lower hoses to the radiator **(see illustration)**.
9 Unscrew the radiator mounting bolt and remove the collar from the rubber mounting **(see illustration)**.
10 Free the radiator from its mountings then detach it from the upper and lower coolant hoses. Remove the radiator from the bike, disconnect the air bleed hose as it becomes accessible.
11 Check the radiator grille and the radiator for signs of damage and clear any dirt or debris that might obstruct air flow and inhibit cooling. If the radiator fins are badly damaged or broken the radiator must be renewed. Also check the rubber mountings, and renew them if necessary.

Installation

Left side radiator
12 Ensure the three mounting rubbers are correctly fitted then reconnect the upper and lower hoses to the radiator.
13 Engage the radiator peg in the lower mounting and seat its upper mounting on the frame peg. Fit the collar to the remaining mounting rubber and install the mounting bolt, tightening it securely.
14 Ensure both coolant hoses are correctly seated then securely tighten the retaining clips.
15 Reconnect the cooling fan wiring connector.

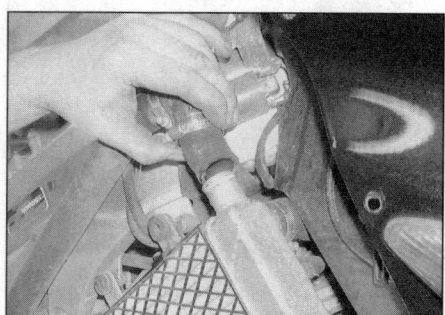

7.4 Unscrew the mounting bolt and remove the collar from the radiator mounting

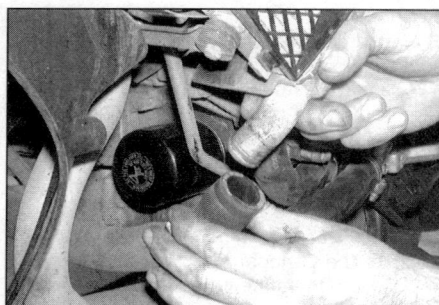

7.5a Disconnect the lower hose from the left radiator . . .

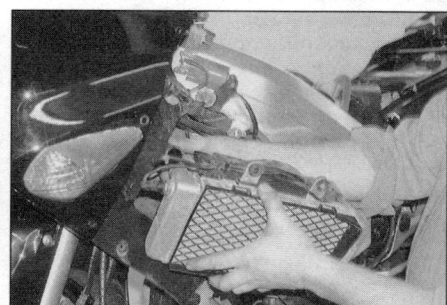

7.5b . . . then free radiator from its mountings and disconnect the upper hose

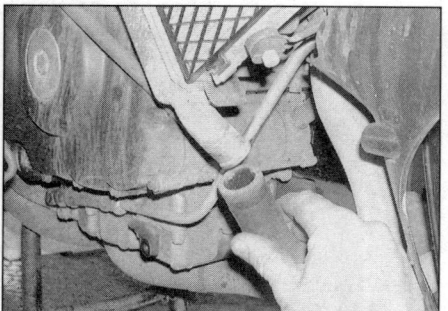

7.7 Disconnect the filler neck hose . . .

7.8 . . . and the lower hose from the right radiator

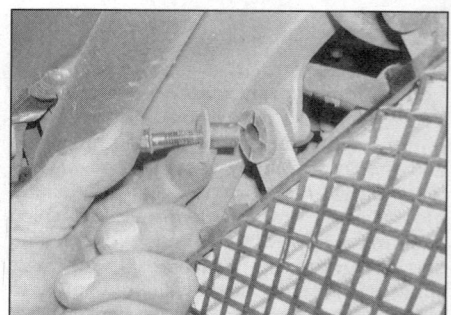

7.9 Unscrew the mounting bolt and remove collar from the radiator mounting

Cooling system 3•7

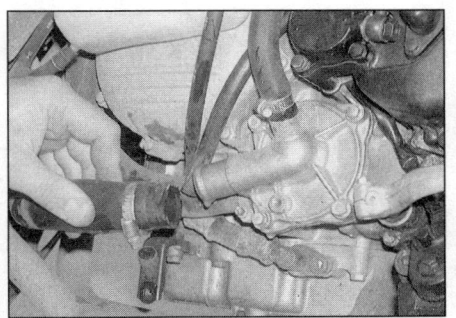

8.5a Slacken the retaining clips and disconnect the radiator lower hose . . .

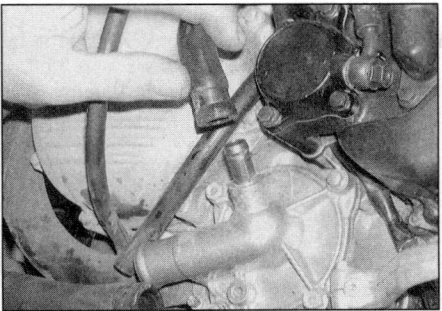

8.5b . . . and the bypass hose from the water pump cover

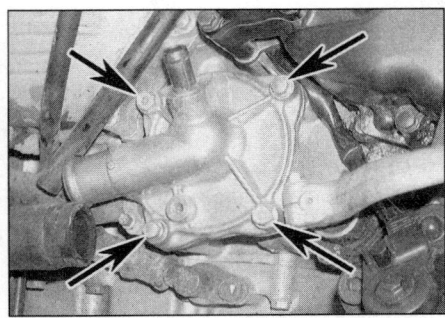

8.6a Unscrew the four bolts (arrows) . . .

16 Refill the cooling system (see Chapter 1).
17 On completion install the fairing panels (see Chapter 8).

Right side radiator

18 Ensure the three mounting rubbers are correctly fitted then offer up the radiator and reconnect the air bleed hose. Secure the hose in position with the retaining clip then connect the upper and lower hoses to the radiator.
19 Engage the radiator peg in the lower mounting and seat its upper mounting on the frame peg. Fit the collar to the remaining mounting rubber and install the mounting bolt, tightening it securely.
20 Ensure the upper and lower coolant hoses are correctly seated then securely tighten their retaining clips.
21 Reconnect the filler neck to the radiator and securely tighten its retaining clip.
22 Refill the cooling system (see Chapter 1).
23 On completion install the fairing panels (see Chapter 8).

8 Water pump – check, removal and installation

Check

1 The water pump is located on the lower left-hand side of the engine. Visually check the area around the pump for signs of leakage.
2 To prevent leakage of water from the cooling system to the lubrication system and vice versa, two seals are fitted on the pump shaft. On the underside of the pump body there is also a drainage hole. If either seal fails, this hole should allow the coolant or oil to escape and prevent the oil and coolant mixing.
3 The seal on the water pump side is of the mechanical type which bears on the rear face of the impeller. The second seal, which is mounted behind the mechanical seal is of the normal feathered lip type. However, neither seal is available as a separate item because the pump is sold as an assembly. Therefore, if on inspection the drainage hole shows signs of leakage, the pump must be removed and renewed.

Removal

4 Drain the coolant (see Chapter 1). Place a suitable container below the water pump to catch any residual oil as the water pump is removed.
5 Slacken their retaining clips and detach the bypass and radiator lower hoses from the pump cover (see illustrations).
6 Unscrew the four bolts and remove the cover from the pump body (see illustration). Discard the cover O-ring; a new one will be needed on installation (see illustration).
7 Slacken the retaining clip and detach the cylinder block hose from the pump body (see illustration).
8 Withdraw the pump from the crankcase. Remove the O-ring from the rear of the pump body and discard it as a new one must be used (see illustration).

8.6b . . . and remove the pump cover and O-ring

Installation

9 Apply a smear of engine oil to the new O-ring and install it onto the rear of the pump body. Install the pump into the crankcase, aligning the slot in the impeller shaft with the tab on the oil pump shaft.
10 Attach the coolant hose to the pump body and securely tighten its retaining clip.
11 Fit the new O-ring to the pump body groove then fit the pump cover. Install the four cover bolts and tighten them to the specified torque.
12 Reconnect the hoses to the pump cover and securely tighten their retaining clips (see illustration).
13 Refill the cooling system (see Chapter 1).
14 On completion install the fairing panels (see Chapter 8).

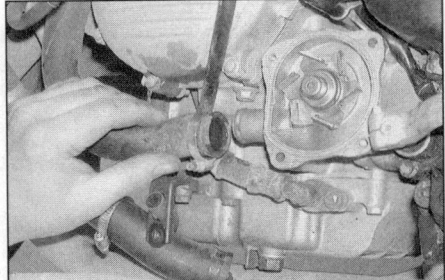

8.7 Slacken the retaining clip and disconnect the cylinder block hose from the pump

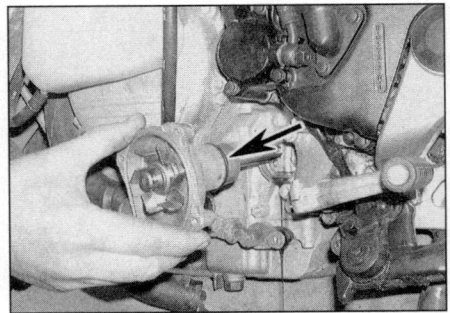

8.8 Withdraw the pump from the crankcase and discard its O-ring (arrow)

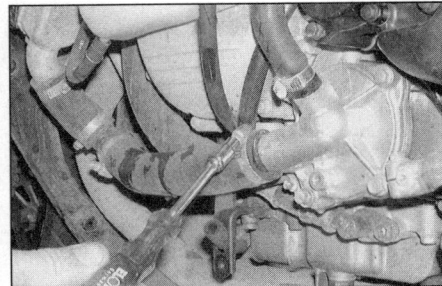

8.12 Ensure all hose clips are securely tightened before refilling the cooling system

3•8 Cooling system

9 Coolant hoses – removal and installation

Removal

1 Before removing a hose, drain the coolant (see Chapter 1).
2 To disconnect a hose, release the retaining clips and move them along the hose, clear of the relevant union. There are two different types of clip, the standard screw-type clip and spring clips (fitted to the smaller bore hoses) which are released by squeezing their ends together with a pair of pliers. Carefully work the hose free.
Caution: The radiator unions are fragile. Do not use excessive force when attempting to remove the hoses.
3 If a hose proves stubborn, release it by rotating it on its union before working it off. If all else fails, cut the hose with a sharp knife then slit it at each union so that it can be peeled off in two pieces. Whilst this means renewing the hose, it is preferable to buying a new radiator.
4 The water inlet unions on the left side of the cylinder block and the outlet unions located between the cylinders can be removed by unscrewing the retaining bolts. Remove the throttle body assembly (see Chapter 4) for access to the outlet unions. If a union is removed, its O-ring must be renewed.

Installation

5 Slide the clips onto the hose and then work it on to its respective union.

 If the hose is difficult to push on its union, it can be softened by soaking it in very hot water, or alternatively a little soapy water can be used as a lubricant.

6 Rotate the hose on its unions to settle it in position before sliding the clamps into place and tightening them securely.
7 If either the coolant inlet or outlet unions have been removed from the cylinder block, fit a new O-ring to the union groove, then install the union and tighten the mounting bolts securely.
8 On completion refill the cooling system (see Chapter 1).

Chapter 4
Fuel and exhaust systems

Contents

Air filter – renewal	see Chapter 1	Fuel injection system components – removal, check and installation	11
Air filter housing – removal and installation	7	Fuel pump – check, removal and installation	5
Air filter housing variable intake system – general information, testing, removal and installation	8	Fuel supply system – flow rate and pressure check	4
Catalytic converter	19	Fuel system – check	see Chapter 1
Choke cable – removal and installation	16	Fuel tank – cleaning and repair	3
Evaporative emission control (EVAP) system	20	Fuel tank – raising, removal and installation	2
Exhaust system – removal and installation	17	General information and precautions	1
Fast idle wax unit	14	Idle speed – check and adjustment	see Chapter 1
Fuel filter – removal and installation	6	Pulse secondary air (PAIR) system	18
Fuel gauge sender unit check and renewal	see Chapter 9	Starter valves – removal, installation and synchronisation	13
Fuel injection system – fault diagnosis and testing	10	Throttle and choke cables – check and adjustment	see Chapter 1
Fuel injection system – general information	9	Throttle body assembly – removal and installation	12
		Throttle cables – removal and installation	15

Degrees of difficulty

Easy, suitable for novice with little experience   **Fairly easy,** suitable for beginner with some experience   **Fairly difficult,** suitable for competent DIY mechanic   **Difficult,** suitable for experienced DIY mechanic 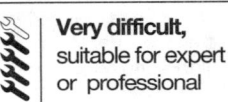   **Very difficult,** suitable for expert DIY or professional

Specifications

Fuel
Grade	Unleaded, minimum 91 RON (Research Octane Number)
Fuel tank capacity	21 litres

Fuel supply system data
Fuel pressure at specified idle speed*	36 psi (2.5 Bar)
Minimum fuel flow rate	150 cc every 10 seconds

Fuel pressure regulator vacuum hose disconnected and plugged

Throttle body assembly
Identification marking	
UK and US (except California) models	GQ30A
California models	GQ30B
Throttle bore diameter	36 mm
Idle speed	see Chapter 1

Component test data
Note: *All values given are only accurate at 20°C (68°F)*

Air filter variable intake solenoid valve resistance	28 to 32 ohms
Atmospheric pressure sensor voltage	4.75 to 5.25 volts
Cam pulse generator peak voltage	0.7 volts minimum
Coolant temperature sensor resistance	2.2 to 2.7 K-ohms
Coolant temperature sensor voltage	4.75 to 5.25 volts
Evaporative emission control (EVAP) system	30 to 34 ohms
Fuel injector resistance	13.0 to 14.4 K-ohms
Ignition pulse generator peak voltage	0.7 volts minimum
Intake air temperature sensor resistance	1 to 4 K-ohms
Intake air temperature sensor voltage	4.75 to 5.25 volts
Manifold pressure	140 to 190 mmHg
MAP sensor voltage	4.75 to 5.25 volts
Oxygen sensor heating element resistance	10 to 40 ohms
Pulse secondary air (PAIR) control valve resistance	20 to 24 ohms
Throttle position sensor voltage	4.75 to 5.25 volts

4•2 Fuel and exhaust systems

Torque settings

Coolant temperature sensor	23 Nm
Exhaust system	
Silencer mounting bolt nut	21 Nm
Front/rear pipe nuts	12 Nm
Fuel hose/pipe union bolt	22 Nm
Fuel hose/pipe union nut	22 Nm
Fuel pump mounting plate nuts	14 Nm
Fuel pressure regulator nut	29 Nm
Fuel rail bolts	10 Nm
Ignition pulse generator bolts	12 Nm
Oxygen sensor	44 Nm
Pulse secondary air (PAIR) valve cover bolts	12 Nm

1 General information and precautions

General information

The fuel system consists of the fuel tank, the fuel pump and filter (which are contained in the tank), the fuel feed and return hose and the throttle body assembly. The fuel pump supplies fuel to the fuel rails on the throttle body assembly, these act as reservoirs for the fuel injectors (one for each cylinder) which inject the fuel into the inlet tracts. The injectors are operated by the Engine Control Module (ECM) using the information obtained from the various sensors it monitors (see Section 9 for information on the fuel injection system operation).

Air is drawn into the throttle body assembly via an air filter which is housed under the fuel tank. The air filter housing incorporates a variable intake system to enable it operate more efficiently (see Section 8 for further information).

The exhaust system is a four-into-one design. On VFR800FI-W and VFR800FI-X German and Swiss models and all VFR800FI-Y and VFR800FI-1 models the exhaust system incorporates a catalytic converter (see Section 19).

Precautions

⚠ **Warning: Petrol (gasoline) is extremely flammable, so take extra precautions when you work on any part of the fuel system. Don't smoke or allow open flames or bare light bulbs near the work area, and don't work in a garage where a natural gas-type appliance is present. If you spill any fuel on your skin, rinse it off immediately with soap and water. When you perform any kind of work on the fuel system, wear safety glasses and have a fire extinguisher suitable for a class B type fire (flammable liquids) on hand.**

Residual pressure will remain in the fuel feed hose and fuel rail assembly long after the motorcycle was last used. Before disconnecting any fuel line, ensure the ignition is switched off then depressurise the fuel system by slackening the union bolt securing the feed hose to the tank (see Section 2 – renew the sealing washers every time the bolt is slackened).

It is vital that no dirt or debris is not allowed to enter the fuel tank or the fuel rail assembly whilst the fuel pipes are disconnected. Any foreign matter in the fuel system components could result in injector damage/malfunction.

Ensure the ignition is switched off before disconnecting/reconnecting any fuel injection system wiring connector. If a connector is disconnected/reconnected with the ignition switched on, the engine control module (ECM) may be damaged.

Always perform service procedures in a well-ventilated area to prevent a build-up of fumes.

Never work in a building containing a gas appliance with a pilot light, or any other form of naked flame. Ensure that there are no naked light bulbs or any sources of flame or sparks nearby.

Do not smoke (or allow anyone else to smoke) while in the vicinity of petrol (gasoline) or of components containing it. Remember the possible presence of vapour from these sources and move well clear before smoking.

Check all electrical equipment belonging to the house, garage or workshop where work is being undertaken (see the Safety first! section of this manual). Remember that certain electrical appliances such as drills, cutters, etc. create sparks in the normal course of operation and must not be used near petrol (gasoline) or any component containing it. Again, remember the possible presence of fumes before using electrical equipment.

Always mop up any spilt fuel and safely dispose of the rag used.

Any stored fuel that is drained off during servicing work must be kept in sealed containers that are suitable for holding petrol (gasoline), and clearly marked as such; the containers themselves should be kept in a safe place. Note that this last point applies equally to the fuel tank if it is removed from the machine; also remember to keep its filler cap closed at all times.

Read the Safety first! section of this manual carefully before starting work.

2 Fuel tank – raising, removal and installation

⚠ **Warning: Refer to the precautions given in Section 1 before starting work.**

Raising (to give access for maintenance)

1 Remove the seat (see Chapter 8).
2 Remove the C-spanner and extension bar from the toolkit supplied with the motorcycle; these will form the prop which will hold the tank in the raised position (see illustration).
3 Unscrew the two front mounting bolts and washers securing the fuel tank to the frame.
4 Lift up the front of the fuel tank, taking care not to lose the collars from the tank mounting rubbers or the tank side mounting rubbers. Screw one of the mounting bolts into the frame and locate the tubular end of the extension bar over the mounting bolt (see illustration). Carefully lower the tank down onto the C-spanner, locating the spanner

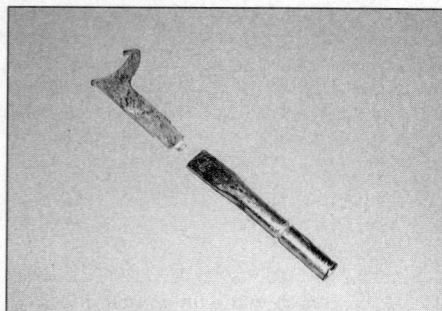
2.2 Use the C-spanner and extension bar in the toolkit to form the fuel tank prop

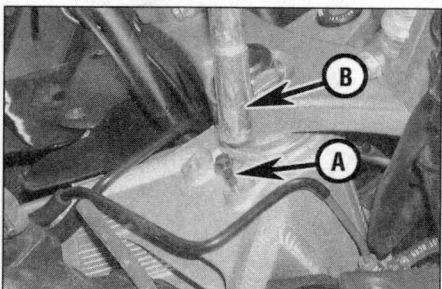
2.4a Screw a mounting bolt (A) into the frame and locate the extension bar (B) over the bolt . . .

Fuel and exhaust systems 4•3

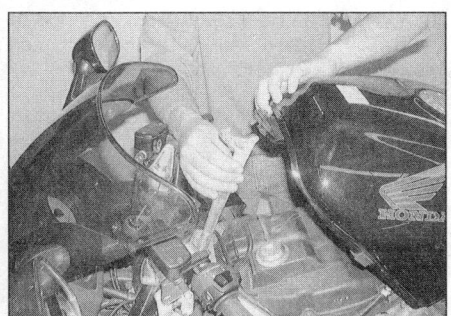

2.4b ... then lower the tank down onto the end of the C-spanner

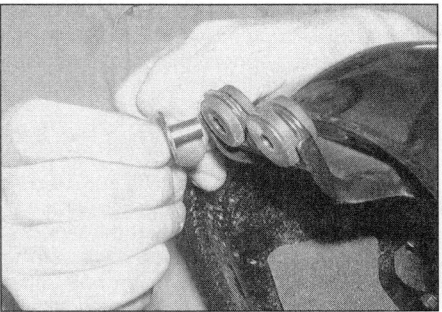

2.5 Ensure the collars are in position in the front mounting rubbers before lowering the tank back into position

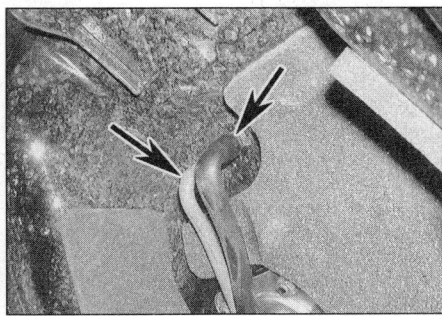

2.7 Disconnect the breather and overflow hoses (arrows) from the underside of the fuel tank

securely in the tank mounting rubber collar **(see illustration)**. Ensure the tank is securely supported before proceeding.

5 On completion, lift the tank and remove the C-spanner arrangement and mounting bolt. Ensure the tank mounting rubbers and collars are in position then lower the tank back down **(see illustration)**. Refit the front mounting bolts and washers and tighten them securely, then install the seat (see Chapter 8).

Removal

6 Raise the tank as described in Paragraphs 1 to 4 then disconnect the battery negative (–) terminal.

7 Disconnect and unclip the breather and overflow hoses from the underside of the tank **(see illustration)**.

8 Disconnect the wiring connectors from the fuel pump and the fuel gauge sender unit and free the wiring from the fuel tank clips **(see illustration)**.

9 Remove the C-spanner arrangement and mounting bolt and lower the tank back down then unscrew the tank rear mounting bolts **(see illustration)**.

10 Have an assistant lift the fuel tank away from the motorcycle and turn it upside down, taking care not to damage the paintwork **(see illustration)**.

11 Wipe clean the area around the tank fuel hose unions then position a wad of rag around the fuel feed hose union bolt, to catch any residual fuel (pressure will still be present in the fuel hose). Slacken and remove the union bolt and disconnect the hose from the tank **(see illustration)**. Remove the sealing washers which are fitted on each side of the hose union and discard them; new ones must be used on installation.

12 Release the retaining clip and disconnect the fuel return hose from the tank then remove the tank from the motorcycle. Store the tank upside down on some soft cloth to prevent damage to the paintwork.

13 Inspect the tank mounting rubbers for signs of damage or deterioration and renew them if necessary. The fuel feed pipe sealing washers must be renewed every time the union bolt is slackened.

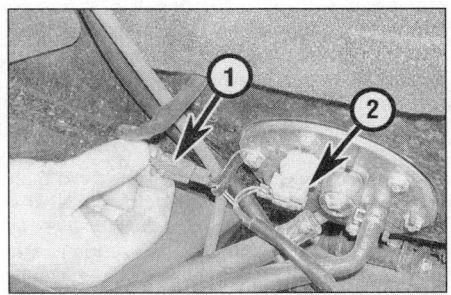

2.8 Disconnect the fuel gauge sender unit (1) and fuel pump (2) wiring connectors and free the wiring from the tank

Installation

14 Position a new sealing washer on each side of the fuel feed hose union and refit the union bolt. Ensure the hose union is correctly seated between the lugs on the tank then tighten the union bolt to the specified torque.

15 Reconnect the return hose to the fuel tank and secure it in position with the retaining clip.

16 Seat the tank in position on the frame then fit the tank rear mounting bolts and tighten them securely.

17 Raise and support the tank then reconnect the pump and sender unit wiring connectors and the breather and overflow hoses to the tank. Ensure the wiring and hoses are correctly routed and retained by all the necessary clips.

18 Make sure the tank side mounting rubbers

2.10 ... then lift the tank away from the bike and turn it upside down

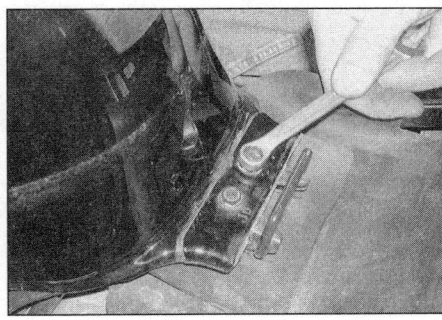

2.9 Unscrew the two rear mounting bolts ...

are correctly fitted and the collars are in position in the front mounting rubbers then lower the tank into position. Install the tank front mounting bolts and washers and tighten them securely.

19 Reconnect the battery then start the engine and check that there is no sign of fuel leakage. If all is well, install the seat (see Chapter 8).

3 Fuel tank – cleaning and repair

1 All repairs to the fuel tank should be carried out by a professional who has experience in this critical and potentially dangerous work.

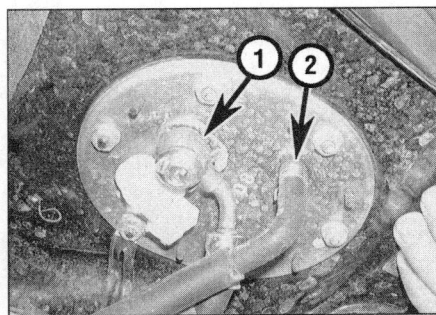

2.11 Unscrew the union bolt and disconnect the fuel feed hose (1) and sealing washers then release the clip and disconnect the return hose (2)

4•4 Fuel and exhaust systems

Even after cleaning and flushing of the fuel system, explosive fumes can remain and ignite during repair of the tank.

2 If the fuel tank is removed from the bike, it should not be placed in an area where sparks or open flames could ignite the fumes coming out of the tank. Be especially careful inside garages where a natural gas-type appliance is located, because the pilot light could cause an explosion.

4 Fuel supply system – flow rate and pressure check

1 If there is thought to be a problem with the fuel supply system, the fuel flow rate and fuel pressure can be checked as follows.

Fuel flow rate check

2 Remove the seat cowling (see Chapter 8).
3 Raise and support the fuel tank (see Section 2).
4 Ensure the ignition is switched off then disconnect the wiring connector from the fuel cut-off relay located on the right side of the motorcycle, just to the rear of the battery (see Section 11). Using an auxiliary wire, bridge the brown and black/white wiring terminals of the connector.
5 Position a wad of rag beneath the fuel return hose union on the fuel tank and have ready a suitable plug to close off the union. Working quickly to minimise fuel loss, disconnect the return hose from the tank and plug the tank union. Mop up all spilt fuel and place the end of the return hose into a suitable container.
6 Turn the ignition on for exactly 10 seconds, catching all the fuel expelled from the return hose in the container, and then turn off the ignition switch. Measure the amount of fuel collected and compare this to the minimum fuel flow amount given in the Specifications.
7 If fuel flow is below the specified minimum there is a problem in the fuel supply system. Likely causes are.
a) *Blocked/restricted fuel feed or return hose.*
b) *Blocked fuel filter.*
c) *Faulty fuel pressure regulator.*
d) *Faulty fuel pump.*

8 On completion, reconnect the return hose to the fuel tank and secure it in position with the retaining clip.
9 Reconnect the fuel cut-off relay wiring connector then start the engine and check that there is no sign of fuel leakage. If all is well, lower the tank back into position (see Section 2) then install the seat cowling (see Chapter 8).

Fuel pressure check

Note: *An adaptor and pressure gauge will be required for this check. Honda specify the use of adaptor (Pt. no. 90008-PP4-E02) and gauge (Pt. No. 07406-0040002) and the necessary sealing washers. New sealing washers for the fuel feed hose union will also be required.*

10 Raise and support the fuel tank (see Section 2) then disconnect the battery negative (–) terminal.
11 Remove the air filter housing (see Section 7).
12 Disconnect the vacuum hose from the fuel pressure regulator and plug the hose end.
13 Wipe clean the area around the tank fuel feed hose union then position a wad of rag around the fuel feed hose union bolt and a container beneath the union to catch all spilt fuel (pressure will still be present in the fuel hose).
14 Slacken and remove the union bolt and replace it with the adaptor and pressure gauge, working quickly to minimise fuel spillage. Ensure the sealing washers are in position on each side of the hose union then securely tighten the adaptor and gauge to ensure there are no fuel leaks. Mop up any spilt fuel.
15 Connect the battery negative (–) lead then start the engine and allow it to idle at the specified speed. Note the pressure present in the fuel system then turn the engine off. Compare the reading obtained to that given in the Specifications.
16 If the fuel pressure is higher than specified, likely causes are.
a) *Blocked/restricted fuel feed or return hose.*
b) *Faulty fuel pressure regulator.*
c) *Faulty fuel pump (although this is unlikely).*
17 If the fuel pressure is lower than specified, likely causes are.
a) *Leaking fuel hose union/injector.*
b) *Blocked fuel filter.*
c) *Faulty fuel pressure regulator.*
d) *Faulty fuel pump.*
18 On completion, disconnect the battery negative (–) lead again. Remove the adaptor, fuel gauge and sealing washers then position a new sealing washer on each side of the hose union and quickly install the union bolt. Ensure the hose union is correctly seated between the lugs on the tank then tighten the union bolt to the specified torque. Reconnect the vacuum hose to the fuel pressure regulator and install the air filter housing (see Section 7). Reconnect the battery then start the engine and check that there is no sign of fuel leakage. If all is well, lower the tank back into position (see Section 2).

5 Fuel pump – check, removal and installation

⚠ **Warning:** *Refer to the precautions given in Section 1 before starting work.*

Check

1 The fuel pump is located inside the fuel tank. The fuel pump runs for a few seconds when the ignition is switched on, to pressurise the fuel system, and then cuts out until the engine is started. If the pump is thought to be faulty, first check the fuses (see Chapter 9). If they are in good condition proceed as follows.
2 Raise and support the fuel tank (see Section 2).
3 Ensure the ignition is switched off then disconnect the fuel pump wiring connector and connect the positive (+) lead of a voltmeter to the brown terminal of the connector and the negative (–) lead to the green terminal of the connector. Turn the ignition on whilst noting the reading obtained on the meter.
4 If battery voltage is present for a few seconds, the fuel pump circuit is operating correctly and the fuel pump is proven to be faulty.
5 If no reading is obtained, check the fuel pump circuit wiring for continuity and make sure all the connectors are free from corrosion and are securely connected. Repair/replace the wiring as necessary and clean the connectors using electrical contact cleaner. If this fails to reveal the fault, check the following components.
a) *Engine stop switch (see Chapter 9).*
b) *Fuel cut-off relay (see Section 11).*
c) *Engine stop relay (see Section 11).*
d) *Bank angle sensor (see Section 11).*
e) *Engine control module (ECM) (see Section 11).*

Removal

6 Remove the fuel tank (see Section 2).
7 With the fuel tank supported upside-down, unscrew the fuel pump mounting plate retaining nuts, noting the correct fitted location of the wiring harness clip. Carefully remove the pump assembly from the tank along with the mounting plate seal; discard the seal – a new seal must be used on installation **(see illustration)**.
8 To remove the pump from its mounting plate, cut the cable tie (where fitted) and remove the wire mesh pre-filter assembly from the intake rubber **(see illustration)**.
9 Unclip and remove the wiring connector locking clip then disconnect the connector from the pump **(see illustrations)**.

5.7 Carefully manoeuvre the fuel pump assembly out from the tank

Fuel and exhaust systems 4•5

5.8 Remove the wire mesh pre-filter from the intake rubber

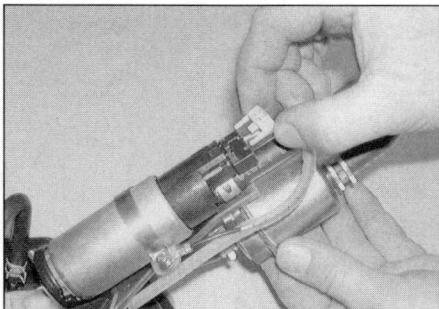

5.9a Slide out the retaining clip . . .

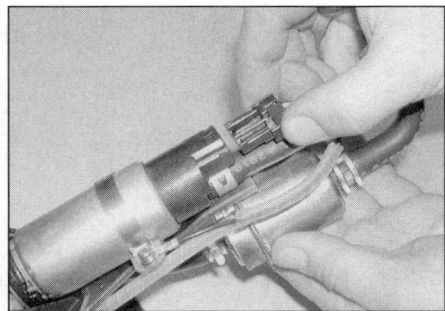

5.9b . . . then disconnect the wiring connector from the fuel pump

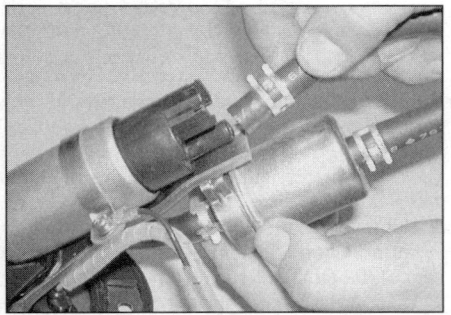

5.10 Release the retaining clip and detach the fuel hose from the pump

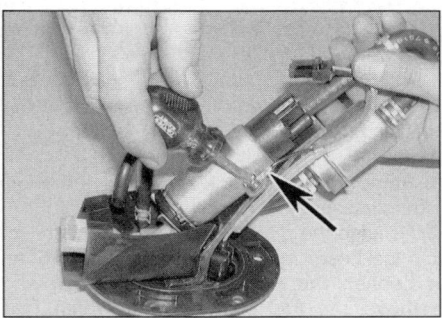

5.11a Remove the clamp screw and detach the earth (ground) lead . . .

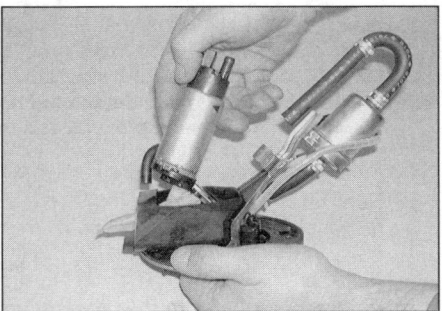

5.11b . . . then remove the clamp and separate the fuel pump from its mounting

10 Release the retaining clip and detach the fuel hose from the pump (see illustration).
11 Undo the retaining clamp screw, noting the correct fitted location of the earth (ground) lead, and remove the clamp (see illustration). The pump can then be separated from its mounting (see illustration). Check the pump intake filter for signs of dirt/damage and clean/renew (as necessary).

Installation

12 If the pump has been dismantled carry out the operations in Paragraphs 13 to 17. If not proceed to Paragraph 18.
13 Where necessary, ensure the intake filter is clean then clip the pump securely onto the mounting plate assembly.
14 Apply a few drops of locking compound to the screw then hook the retaining clamp into position. Fit the earth (ground) lead to the retaining screw then fit the screw and tighten securely.
15 Reconnect the pump wiring connector and secure it in position with the locking clip.
16 Reconnect the fuel hose to the pump and secure it in position with the retaining clip.
17 Fit the wire mesh pre-filter assembly to the intake rubber and secure it in position with a new cable tie.

18 Ensure the mounting plate and tank surfaces are clean and dry, then fit the new seal to the plate making sure its locating pins are all located correctly in the plate holes (see illustration).
19 Manoeuvre the pump assembly into the tank, taking care not to damage the intake rubber and filter, and seat it in position.
20 Fit the wiring harness clip to the correct stud then fit the retaining nuts. Tighten all nuts by hand then go around in the specified sequence and tighten them to the specified torque (see illustration).
21 Install the fuel tank (see Section 2).

5.18 Fit a new rubber seal to the mounting plate ensuring its pins locate correctly in the plate holes

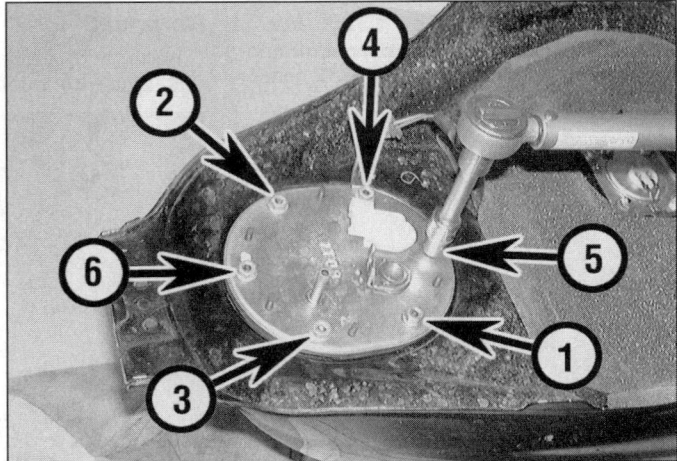

5.20 Tighten the pump nuts to the specified torque in the sequence shown (ensure the wiring harness clip is fitted in location 4)

4•6 Fuel and exhaust systems

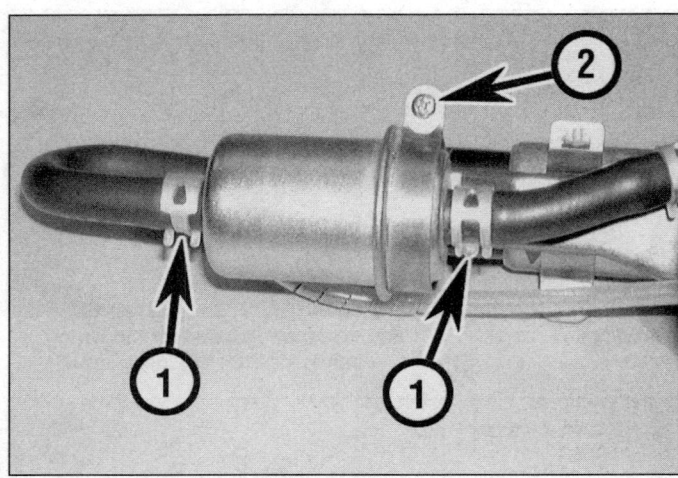

6.3 Release the clips (1) and detach the hoses from the filter then slacken the clamp screw (2) and remove the filter

7.3 Disconnect the crankcase breather hose from the rear of the air filter housing

6 Fuel filter – removal and installation

Warning: Refer to the precautions given in Section 1 before starting work.

Removal

1 The filter is part of the fuel pump assembly which is housed inside the fuel tank.

2 Remove the fuel tank and remove the fuel pump assembly (Sections 2 and 5).
3 Release the retaining clips and disconnect the fuel hoses from the filter **(see illustration)**.
4 Loosen the retaining clamp screw and remove the fuel filter, noting which way around it is installed.

Installation

5 Locate the filter in the retaining clamp, ensuring it is the correct way around, and securely tighten the clamp screw.
6 Reconnect the fuel hoses to the filter and secure them in position with the retaining clips.
7 Install the pump assembly in the fuel tank then fit the tank to the motorcycle (Sections 5 and 2).

7 Air filter housing – removal and installation

Removal

1 Raise and support the fuel tank (see Section 2 – to improve access remove the

tank completely) then remove the air filter element (see Chapter 1).
2 Remove the ignition HT coils for cylinders number 1 and 3 which are mounted on the rear of the air filter housing (see Chapter 5).
3 Disconnect the crankcase breather hose from the rear of the housing **(see illustration)**.
4 Disconnect the wiring connector and vacuum hose from the manifold absolute pressure (MAP) sensor (mounted on the rear of the housing) **(see illustration)**.
5 Undo the screw securing the pulse secondary air (PAIR) valve bracket to the front of the housing **(see illustration)**.
6 On California models disconnect the EVAP system hoses from the five-way connector on the rear of the air filter housing.
7 On all models except California models disconnect the vacuum hose from the left (number 1 and 2 hoses) and right side (number 3 and 4 hoses) of the housing **(see illustration)**.
8 Remove the screws securing the housing intake funnels to the throttle body and remove all four funnels **(see illustrations)**.
9 Lift the air filter housing slightly and disconnect the wiring connector from the intake air temperature sensor (on the left side

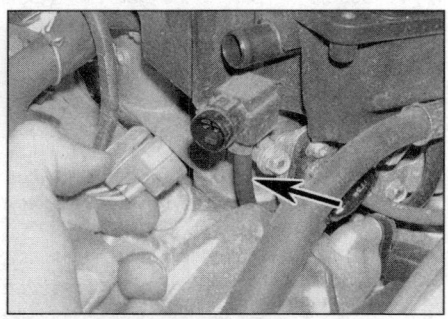

7.4 Disconnect the wiring connector and vacuum hose (arrow) from the MAP sensor

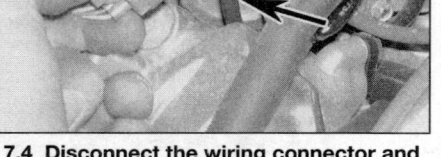

7.5 Remove the screw and free the pulse secondary air (PAIR) valve bracket from the housing

7.7 On UK and US (except California) models disconnect the vacuum hoses from the left and right side of the air filter housing

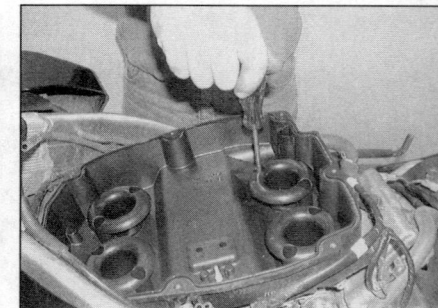

7.8a Undo the retaining screws . . .

Fuel and exhaust systems 4•7

7.8b ... and remove the four intake funnels from the housing

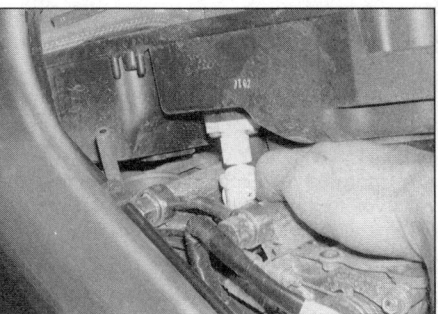
7.9 Lift the housing and disconnect the wiring connector from the intake air temperature sensor

7.10 Disconnecting the vacuum hose from the reservoir on the front of the air filter housing

of the housing) and the breather hoses from the base of the housing (see illustration).
10 Disconnect the wiring connector from the variable intake system solenoid valve (on the right side of the housing). Disconnect the vacuum hose (numbered 12) linking the solenoid to the one-way valve and the hose (numbered 10) from the vacuum reservoir on the front of the housing (numbered 10) (see illustration).
11 Disconnect the PAIR valve hose from the front of the housing and remove the housing from the motorcycle (see illustration).

Installation

12 Installation is the reverse of removal, ensuring all hoses and wiring are correctly routed and securely reconnected. Ensure one of the index marks on each intake funnel is correctly aligned with the index mark on the air filter housing.

8 Air filter housing variable intake system – information, testing, removal and installation

General information

1 The air filter housing incorporates an additional intake opening which is fitted with a flap valve. The valve is controlled by the engine control module (ECM) via a solenoid valve and vacuum diaphragm unit. This additional intake is used to increase engine efficiency by allowing increased airflow through the filter housing at higher engine speeds.
2 At engine speeds below 5000 rpm, the ECM switches on the solenoid valve, allowing vacuum to act on the diaphragm unit. This closes the additional intake flap valve in the filter housing, restricting the airflow through the filter housing. A one-way valve is fitted to the hose linking the solenoid valve and reservoir to the throttle body to ensure enough vacuum is always present in the system, regardless of the engine operating conditions.
3 At engine speeds above 5500 rpm, the ECM switches off the solenoid valve which isolates the vacuum diaphragm unit from the vacuum present in the throttle body/vacuum reservoir. The additional intake flap in the air filter housing is then forced open by the spring in the vacuum diaphragm unit.

Testing

4 Raise and support the fuel tank (see Section 2).
5 Disconnect the wiring connector from the neutral switch (see Chapter 9) then start the engine. Observe the operation of the valve in the air filter housing whilst slowly increasing the engine speed (see illustration). When the engine speed exceeds 5500 rpm the flap should open. Slowly decrease the engine speed and check that the valve closes again when the engine speed drops below 5000 rpm.
6 If the system is operating correctly, switch off the engine then reconnect the neutral switch and lower the fuel tank (see Section 2).
7 If the valve remains open, the most likely cause is that the vacuum diaphragm unit is faulty. To check the valve, disconnect the vacuum hose and connect an auxiliary length of hose to the diaphragm. Suck on the end of the hose and check that the valve closes. If not the vacuum diaphragm is faulty and must be renewed.
8 If the vacuum diaphragm is in good condition, remove the air filter housing (Section 7) and check that all the vacuum hoses and the reservoir are in good condition and the solenoid valve wiring connector is securely connected.
9 Check the operation of the solenoid valve by blowing through the upper hose union; air should flow freely through the valve. Connect battery voltage (12 volts) across the valve terminals and repeat the check; no air should now flow through the valve if it is functioning correctly. If an ohmmeter is available, check the resistance of the control valve windings and compare the reading obtained to that given in the Specifications. Renew the valve if faulty.
10 Check the operation of the one-way valve by blowing through it from the solenoid valve union (hose 12) and then the throttle body union; air should only flow through the valve when blown through the solenoid valve union.

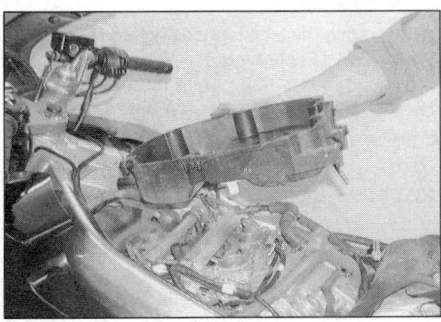
7.11 Removing the air filter housing

If air also passes through the valve when blown through from the throttle body union, the valve is faulty and must be renewed.
11 If all components appear to function correctly, the fault must be in the engine control module (ECM) (which is unlikely if the engine is running normally) or the solenoid valve wiring.

Removal and installation

Vacuum diaphragm unit

12 Raise and support the fuel tank (see Section 2).
13 Disconnect the vacuum hose from the diaphragm unit then undo the retaining screws and lift off the air filter housing cover.
14 Free the diaphragm unit rod from the flap valve then twist the unit to free it from the lid.
15 Installation is the reverse of removal.

8.5 Check the operation of the variable intake valve flap (arrow) as described in text

4•8 Fuel and exhaust systems

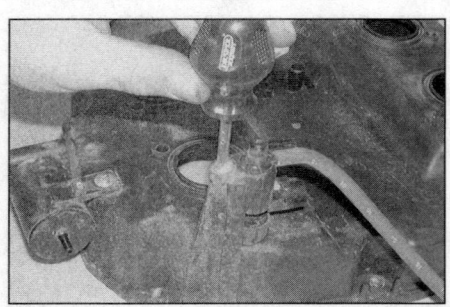

8.17 Undo the screw and remove the PAIR solenoid valve from the base of the air filter housing

Solenoid valve

16 Remove the air filter housing (see Section 7).
17 Undo the retaining screw and remove the valve from the housing **(see illustration)**.
18 Installation is the reverse of removal.

One-way valve

19 Remove the air filter housing (see Section 7).
20 Disconnect the vacuum hoses, noting their correct fitted locations, and remove the valve.
21 Installation is the reverse of removal.

Vacuum reservoir

22 Remove the air filter housing (see Section 7).
23 Undo the retaining screw and remove the reservoir from the housing **(see illustration)**.
24 Installation is the reverse of removal.

9 Fuel injection system – general information

1 All models are equipped with a programmed engine management (PGM-FI) system which operates both the injection and ignition systems on the motorcycle **(see illustration)**. Refer to Chapter 5 for information on the ignition system. The fuel injection side of the system functions as follows.
2 The fuel pump, contained in the fuel tank,

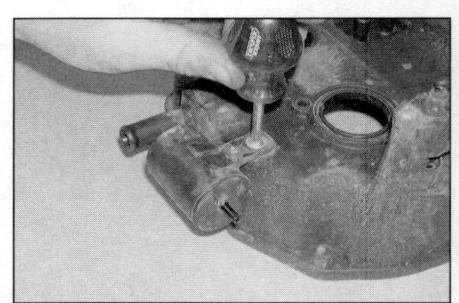

8.23 Undo the screw and remove the vacuum reservoir from the housing

pumps fuel to the fuel rails on the throttle body assembly, via a filter. Fuel supply pressure is controlled by the pressure regulator which keeps the pressure in the fuel rails constant, returning excess fuel to the tank via the return hose. The fuel rails act as reservoirs for the four injectors (one for each cylinder) which are operated by the Engine Control Module (ECM).

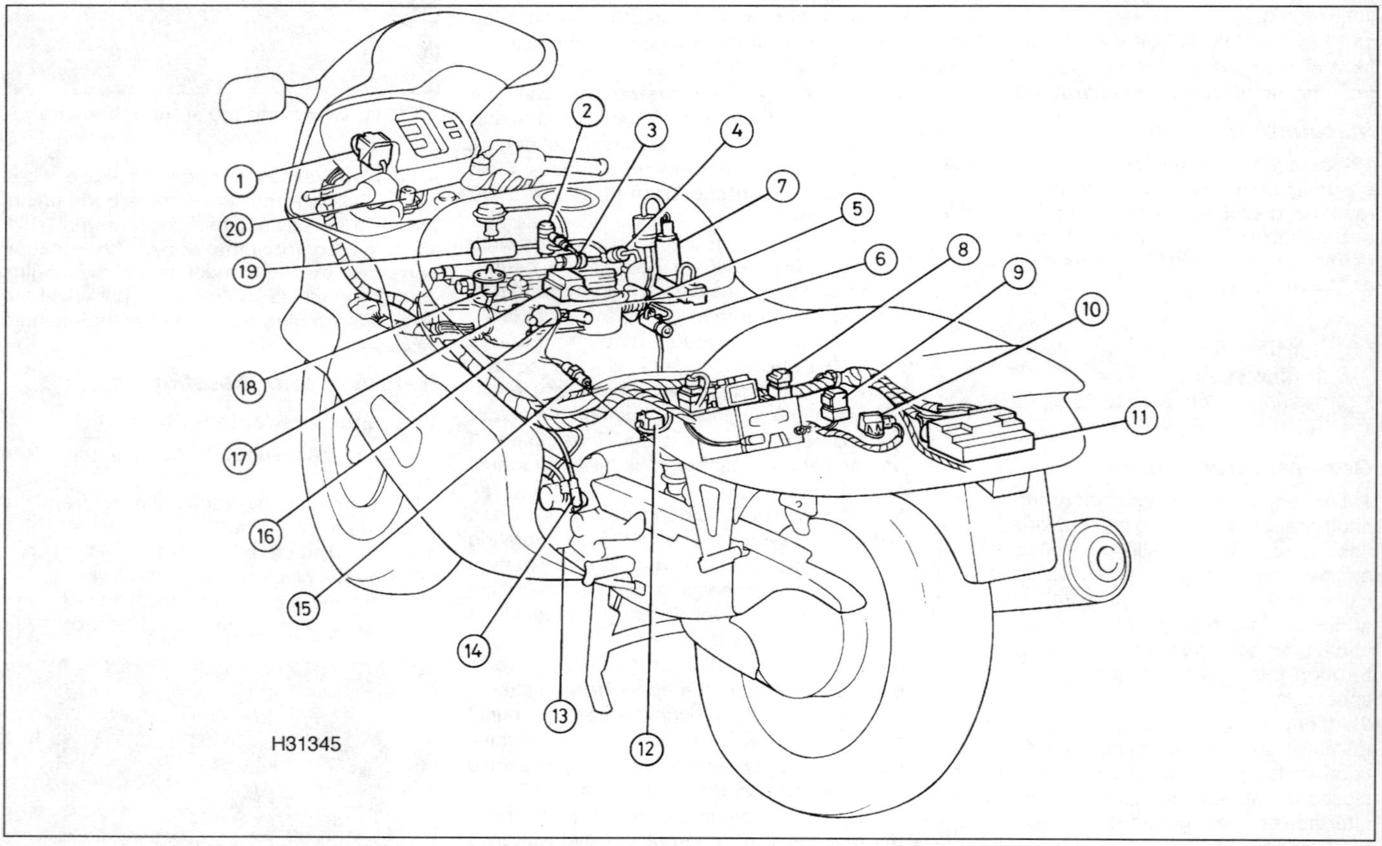

9.1 Engine management system component locations

1 Bank angle sensor
2 Variable intake system solenoid valve
3 Fuel pressure regulator
4 Variable intake system one-way valve
5 Throttle sensor
6 Cam pulse generator
7 Fuel pump assembly
8 Engine stop relay
9 Fuel cut-off relay
10 Atmospheric pressure sensor
11 Engine control module (ECM)
12 Ignition pulse generator
13 Oxygen sensors
14 Speed sensor
15 Coolant temperature sensor
16 Injectors
17 Manifold absolute pressure (MAP) sensor
18 Intake air temperature sensor
19 Variable intake system vacuum reservoir
20 Immobiliser receiver

Fuel and exhaust systems 4•9

3 The ECM monitors signals from the following sensors.
   a) *Throttle sensor – informs the ECM of the throttle position, and the rate of throttle opening or closing.*
   b) *Coolant temperature sensor – informs the ECM of engine temperature.*
   c) *Manifold absolute pressure (MAP) sensor – informs the ECM of the engine load by monitoring the pressure in the throttle body inlet tracts.*
   d) *Intake air temperature sensor – informs the ECM of the temperature of the air entering the throttle body.*
   e) *Cam pulse generator – informs the ECM of engine speed and camshaft position.*
   f) *Ignition pulse generator – informs the ECM of engine speed and crankshaft position.*
   g) *Atmospheric (barometric) pressure sensor – informs the ECM of the atmospheric (barometric) pressure the motorcycle is operating in.*
   h) *Speed sensor – informs the ECM of the motorcycle speed (see Chapter 9).*
   i) *Oxygen sensors (catalytic converter models) – informs the ECM of the oxygen content of the exhaust gases.*

4 All the above information is analysed by the ECM and, based on these signals, it determines the appropriate ignition and fuel requirements for the engine. The ECM controls the fuel injector by varying its pulse width – the length of time the injector is held open – to provide a richer or weaker mixture, as appropriate. The mixture is constantly varied by the ECM, to provide the best setting for starting, warm-up, idle, cruising, and acceleration. Due to the layout of the engine, the fuel needs for each cylinder are slightly different and the ECM is programmed to compensate for this; the injection system is fully sequential, with each injector receiving its own operating signal from the ECM.

5 A choke operates a starter valve arrangement in the throttle body assembly to assist cold starting. When the throttle is closed, the starter valves allow additional air to bypass the throttle valves which increases the engine idle speed.

6 If there is an abnormality in any of the readings obtained from any sensor, the ECM enters its back-up mode. In this event, the ECM ignores the abnormal sensor signal, and assumes a pre-programmed value which will allow the engine to continue running (albeit at reduced efficiency). If the ECM enters this back-up mode, the warning light in the instrument cluster will come on, and the relevant fault code will be stored in the ECM memory. The ECM fault codes can be accessed using the self-diagnosis function (see Section 10).

10.3a The service check wiring connector is taped to the wiring harness on the left side of the subframe

10.3b Bridge the terminals with a jumper wire

10 Fuel injection system – fault diagnosis and checking

Fault diagnosis

1 If the fuel injection system diagnostic warning light on the instrument cluster illuminates when the motorcycle is running, a fault has occurred in the fuel injection/ignition system. The engine control module (ECM) will store the relevant fault code in its memory and this code can be read as follows using the self-diagnostic mode of the ECM.

2 If the engine can be started, place the motorcycle on its sidestand, start the engine and allow it to idle while observing the warning light.

3 If the engine cannot be started, remove the seat cowling (see Chapter 8) to gain access to the fuel injection system service check wiring connector, which is taped to the wiring harness running along the left side of the subframe **(see illustration)**. Ensure the ignition is switched off, then bridge the terminals of the service check connector with a jumper wire **(see illustration)**. With the terminals bridged, turn the ignition on and observe the warning light.

4 The warning light uses a series of long (1.3 second) and short (0.5 second) flashes to identify the fault code(s). The long flashes are the equivalent of ten short flashes eg. a long flash followed by three short flashes indicates fault code 13. If there is more than one fault, the warning light will identify them in order, starting with the lowest and finishing with the highest. Once the code(s) have been identified, the ECM will repeat the process. The fault codes are as shown.

Fault code	Symptoms	Possible causes
No code (warning light constantly on)	Engine does not start	Blown fuse/faulty power supply to ECM Faulty engine stop relay Faulty engine stop switch/open circuit on switch earth (ground) wire Faulty bank angle sensor Faulty ECM
No code (warning light constantly on)	Engine runs normally	Short circuit in warning light wiring Short circuit in service check connector wiring Faulty ECM
No code (warning light off)	Engine runs normally	Blown warning light bulb Open circuit in warning light wiring Faulty ECM
1	Engine runs normally	Faulty MAP sensor
2	Engine runs normally	Faulty MAP sensor/vacuum hose
7	Engine difficult to start at low temperatures	Faulty coolant temperature sensor
8	Poor throttle response	Faulty throttle sensor
9	Engine runs normally	Faulty intake air temperature sensor
10	Engine runs poorly at high altitude	Faulty atmospheric (barometric) pressure sensor
11	Engine operates normally	Speed sensor faulty
12	Engine does not start	Faulty No. 1 injector
13	Engine does not start	Faulty No. 2 injector
14	Engine does not start	Faulty No. 3 injector
15	Engine does not start	Faulty No. 4 injector
18	Engine does not start	Faulty cam pulse generator
19	Engine does not start	Faulty ignition pulse generator
20	Engine operates normally	Faulty EPROM in ECM
21	Engine operates normally	Faulty No.1/2 oxygen sensor
22	Engine operates normally	Faulty No. 3/4 oxygen sensor
23	Engine operates normally	Faulty oxygen sensor heating element

4•10 Fuel and exhaust systems

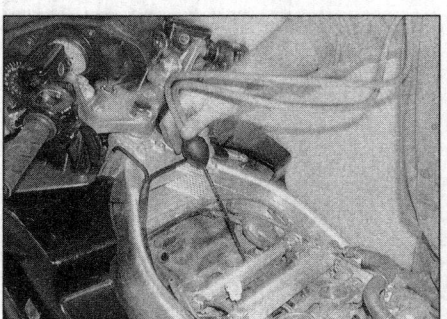

11.2 Using a stethoscope to check each injector

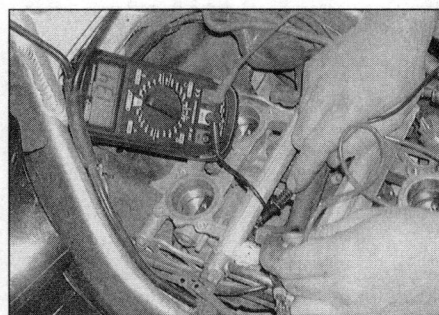

11.3 Checking the resistance of an injector

Once the code(s) have been noted, switch off the ignition and (if necessary) remove the jumper wire from the service check connector. **Note:** *The warning light should come on for a few seconds when the ignition is first switched on as a check of the system.*

5 Once the fault has been corrected, it will be necessary to remove the fault code from the ECM memory. To do this, ensure the ignition is switched off then bridge the terminals of the service check connector (see Paragraph 3). Turn the ignition on, then disconnect the jumper wire from the service check connector and quickly reconnect it. The warning light will illuminate when the wire is disconnected and should start to flash when it is reconnected, indicating that all fault codes have been erased. Turn off the ignition then remove the jumper wire. Turn the ignition on to check the operation of the warning light (in some cases it may be necessary to repeat the erasing procedure more than once) then install the seat.

Checking

6 If a fault appears in the fuel injection system, first ensure that all the system wiring connectors are securely connected and free of corrosion. Ensure that the fault is not due to poor maintenance – check that the air filter element is clean, that the spark plugs are in good condition and correctly gapped, that the valve clearances are correctly adjusted, the cylinder compression pressures are correct, the ignition timing is correct (refer to Chapters 1, 2, 4 and 5).

7 If this fails to correct the fault there are a number of preliminary checks that can be undertaken to isolate the cause (see Section 11).

Note: *The component test data at the beginning of this Chapter is provided by Honda and will be obtained using Honda specified diagnostic equipment. A certain amount of variation is acceptable using aftermarket equipment.*

8 Before purchasing new parts, have any suspect component confirmed faulty by a Honda dealer. Provided the system is in good condition, component faults can be identified and confirmed through the ECM with a plug-in electronic test pin box without the need to remove components from the machine. Faults within the ECM can only be diagnosed with the Honda test pin box.

11 Fuel injection system components – check, removal and installation

Caution: Ensure the ignition is switched off before disconnecting/ reconnecting any fuel injection system wiring connector. If a connector is disconnected/reconnected with the ignition switched on the engine control module (ECM) maybe damaged.

Fuel injectors

 Warning: Refer to the precautions given in Section 1 before starting work.

Check

1 Remove the air filter housing (see Section 7).
2 If the engine runs, start it and allow it to idle. Check the operation of each injector using a stethoscope or sounding rod; an injector will emit a 'clicking' noise when functioning **(see illustration)**. If any injector is silent, either the injector or its wiring harness is faulty. **Note:** *Reconnect the wiring connectors and the vacuum hose to the MAP sensor and intake air temperature sensor (both on the air filter housing) to prevent the ECM detecting a fault while making the check.*

3 If the engine does not run, or if an injector is thought to be faulty (see Section 10), disconnect the wiring connector(s) from the injector(s). Connect an ohmmeter across the terminals of the injector and measure its resistance **(see illustration)**. Compare the reading obtained to that given in the Specifications at the beginning of this Chapter. If the resistance differs greatly from that specified, the injector should be renewed.

4 Check for continuity between the black/white wire terminal of the injector and earth (ground). There should be no continuity; if there is continuity, the injector is faulty and should be renewed.

5 Turn the ignition on and check for battery voltage between the black/white wire terminal in the connector and earth (ground), then turn the ignition off. If there is no voltage, refer to *Wiring Diagrams* (see Chapter 9) and check for a break or short circuit in the black/white wire.

6 If the injector and wiring are good, have the operation of the ECM tested by a Honda dealer.

Removal

7 Remove the throttle body assembly (see Section 12).
8 Disconnect the breather hoses from the throttle body and the vacuum hose from the fuel pressure regulator **(see illustrations)**.
9 Disconnect the wiring connectors from the throttle sensor and injectors. Undo the clamp retaining screw then remove the wiring harness from the throttle body, noting its correct routing **(see illustrations)**.
10 Unscrew the bolts securing both fuel rails to the throttle body then carefully lift off the fuel rail assembly, leaving the injectors in the throttle body **(see illustration)**. Note the correct fitted position of the O-ring and spacer on each injector; these must be renewed.
11 Remove the injectors and lower seals

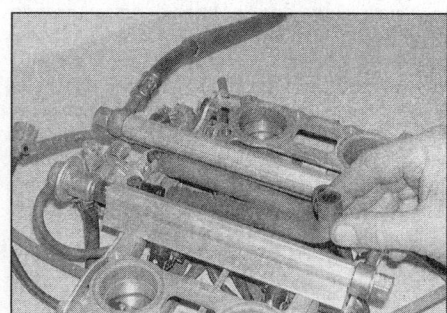

11.8a Disconnect both breather hoses from the throttle body . . .

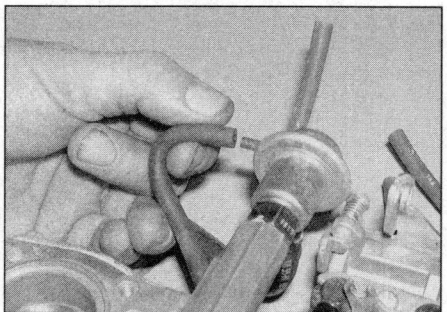

11.8b . . . and the vacuum hose from the fuel pressure regulator

11.9a Disconnect the throttle sensor wiring connector . . .

Fuel and exhaust systems 4•11

11.9b ... and the injector wiring connectors ...

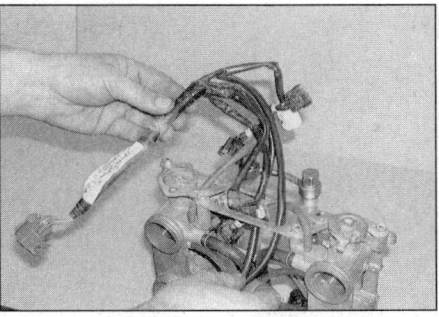

11.9c ... then undo the clamp screw and remove the throttle body wiring harness

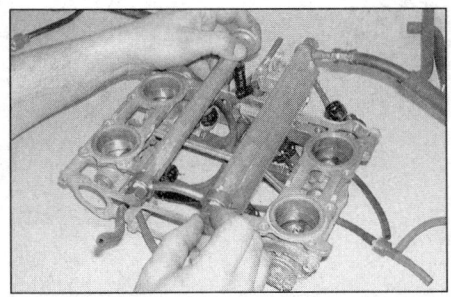

11.10 Unscrew the retaining bolts then lift the fuel rail assembly carefully off the injectors

from the throttle body assembly **(see illustration)**. The seals must all be renewed.

Installation

12 Lubricate the new lower seals, upper spacers and O-rings with a smear of engine oil. Fit the lower seals to the throttle body, ensuring they are all fitted the right way around. Slide the upper spacer onto each injector and fit a new O-ring to each injector making sure each one is correctly seated in the injector recess **(see illustrations)**.
13 Ease the injectors into position in the throttle body, taking care not to damage the lower seals.
14 Ensure the injector wiring connectors are all correctly positioned then, taking care not to damage the O-rings, install the fuel rails fully onto the injectors then fit the retaining bolts and tighten them to the specified torque **(see illustration)**.
15 Fit the wiring harness and reconnect the wiring connectors to the injectors and throttle sensor. Ensure the harness is correctly routed then refit the clamp retaining screw.
16 Reconnect the fuel pressure regulator vacuum hose, then refit the breather hoses and install the throttle body assembly (see Section 12).

Fuel pressure regulator

⚠️ **Warning:** *Refer to the precautions given in Section 1 before starting work.*

Check

17 Check the fuel flow rate and pressure as described in Section 4.

Removal

18 Remove the throttle body assembly (see Section 12).
19 Disconnect the fuel and vacuum hose from the pressure regulator **(see illustration)**.

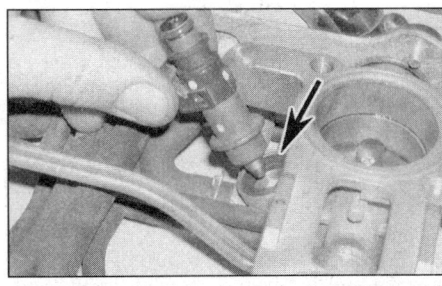

11.11 Ease the injector out of position and remove the lower seal (arrow)

Make alignment marks between the regulator and fuel rail then slacken the regulator nut.
20 Remove the fuel rail from the injectors (see Paragraphs 8 to 10), then unscrew the regulator from the end of the fuel rail **(see illustration)**. Remove the O-ring from the regulator and discard.

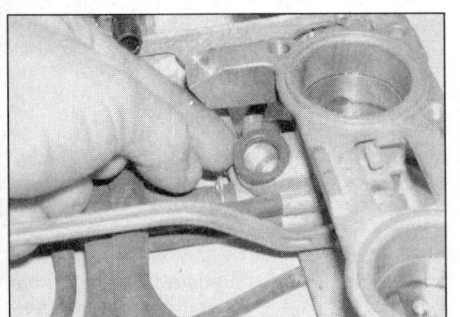

11.12a Fit the new lower seals to the throttle body ...

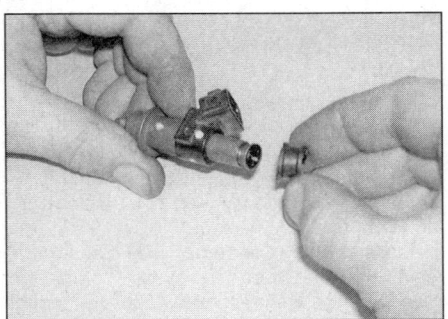

11.12b ... and fit the new upper spacer ...

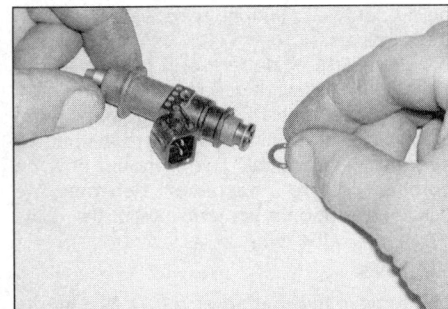

11.12c ... and O-ring to each injector

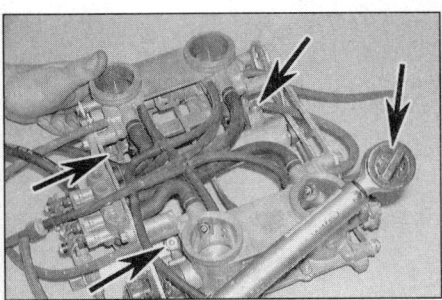

11.14 Ensure the fuel rail assembly is correctly seated then tighten the retaining bolts (arrows) to the specified torque

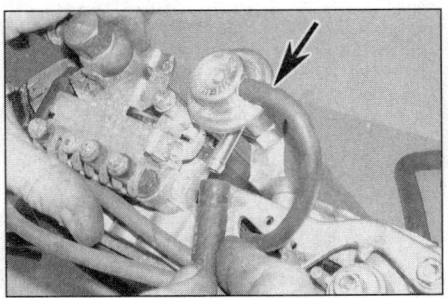

11.19 Disconnect the fuel return hose and the vacuum hose (arrow) from the fuel pressure regulator

11.20 Unscrew the regulator nut and remove the assembly from the end of the fuel rail

4•12 Fuel and exhaust systems

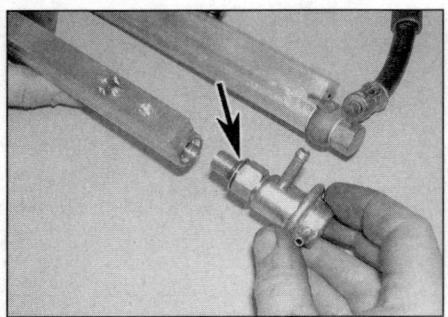

11.21 Fit a new O-ring (arrow) and screw the pressure regulator fully into the fuel rail

11.26 The fuel cut-off relay (arrow) is located behind the rear brake fluid reservoir

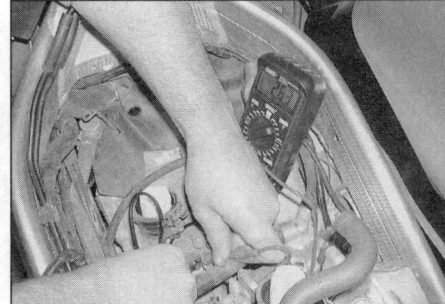

11.33 Checking the coolant temperature sensor resistance

Installation

21 Fit a new O-ring then screw the pressure regulator fully onto the fuel rail and realign the marks made prior to removal (see illustration).
22 Renew all the injector seals and fit the fuel rail assembly (see Paragraphs 9 to 12) and refit the breather hoses to the throttle body assembly.
23 Tighten the pressure regulator nut to the specified torque then reconnect the fuel and vacuum hoses to the regulator before installing the throttle body assembly (see Section 12).

Fuel cut-off relay

Check

24 Remove the relay (see Steps 26 and 27) then connect an ohmmeter to the black/white and brown terminals of the relay. There should be no continuity (infinite resistance) between the terminals.
25 Using a 12V battery and jumper wires, connect the brown/black terminal of the relay to the battery positive (+) terminal and the black/white terminal of the relay to the battery negative (–) terminal. There should now be continuity (zero resistance) between the black/white and brown terminals of the relay. If not, renew the relay.

Removal

26 Remove the seat cowling (see Chapter 8). The fuel cut-off relay is located on the right side of the motorcycle, just behind the rear brake fluid reservoir (see illustration).
27 Disconnect the wiring connector then undo the retaining bolt and remove the relay from the motorcycle.

Installation

28 Installation is the reverse of removal.

Throttle sensor

Check

29 If the throttle sensor is thought to be faulty (see Section 10), disconnect the sensor wiring connector, then turn the ignition on and check for voltage between the pink wire terminal in the connector and earth (ground). The voltage should be as specified at the beginning of this Chapter. If there is no voltage, refer to *Wiring Diagrams* (see Chapter 9) and check for a break or short circuit in the pink wire. Now check for voltage between the pink wire terminal and the green/orange wire terminal in the connector. If there is no voltage, refer to *Wiring Diagrams* (see Chapter 9) and check for a break or short circuit in the green/orange wire. Turn the ignition off.
30 If the power supply to the throttle sensor is good, have the operation of the ECM tested by a Honda dealer.

Removal and installation

31 The throttle sensor is an integral part of the throttle body assembly and is not available separately. If the sensor is faulty, the complete throttle body assembly will have to be renewed (see Section 12).

Coolant temperature sensor

Note: *The sensor also operates the coolant temperature gauge (see Chapter 3).*

Check

32 If the coolant temperature sensor is thought to be faulty (see Section 10), first remove the throttle body assembly to access the temperature sensor which is screwed into the rear of the front cylinder head (see Section 12).
33 With the engine cold, disconnect the sensor wiring connector. Measure the resistance between the yellow/blue and green/orange wiring terminals in the sensor with an ohmmeter and compare the result with the Specifications at the beginning of this Chapter (see illustration). If the resistance reading differs greatly from that specified, the sensor is probably faulty and should be renewed. **Note:** *The specified value is only valid at 20°C (68°F); the sensor resistance will increase at lower temperatures and decrease at higher temperatures.*
34 If the sensor appears to be functioning correctly, check its power supply. Turn the ignition on and check for voltage between the yellow/blue wire terminal in the connector and earth (ground) (see illustration). The voltage should be as specified at the beginning of this Chapter. If there is no voltage, refer to *Wiring Diagrams* (see Chapter 9) and check for a break or short circuit in the yellow/blue wire. Now check for voltage between the yellow/blue wire terminal and the green/orange wire terminal in the connector. If there is no voltage, refer to *Wiring Diagrams* (see Chapter 9) and check for a break or short circuit in the green/orange wire. Turn the ignition off.
35 If the power supply to the temperature sensor is good, have the operation of the ECM tested by a Honda dealer.

Removal

36 Remove the throttle body assembly (see Section 12).
37 Drain the cooling system (see Chapter 1).
38 Disconnect the wiring connector then unscrew the coolant temperature sensor from the rear of the front cylinder head (see illustration). Discard the sealing washer, a new one will be needed for installation.

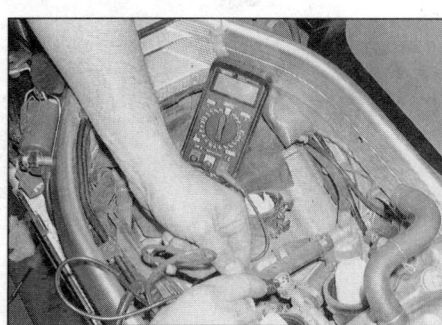

11.34 Measuring the coolant temperature sensor power supply

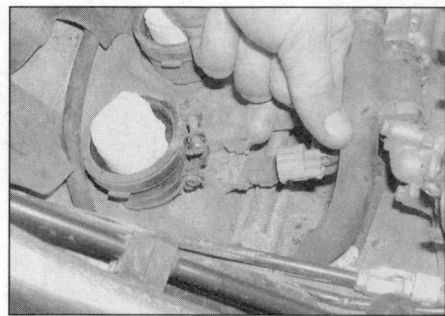

11.38 Disconnect the wiring connector then unscrew the coolant temperature sensor from the cylinder head

Fuel and exhaust systems 4•13

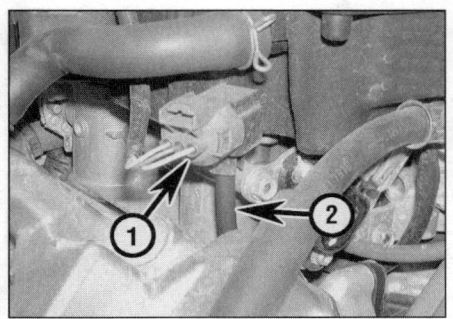

11.41 MAP sensor wiring connector (1) and vacuum hose (2)

Installation

39 Fit the new sealing washer then screw the sensor into the cylinder head. Tighten the sensor to the specified torque then connect the wiring connector securely.
40 Install the throttle body assembly (Section 12) and refill the cooling system (Chapter 1).

Manifold absolute pressure (MAP) sensor

Check

41 If the MAP sensor is thought to be faulty (see Section 10), raise and support the fuel tank (see Section 2) to access the sensor which is mounted on the rear of the air filter housing (see illustration). Note: *Fault code – one flash requires an initial check of the sensor wiring, two flashes requires a check of the vacuum hose.*
42 To check the wiring, disconnect the wiring connector from the sensor, then turn the ignition on and check for voltage between the pink wiring terminal in the connector and earth (ground). The voltage should be as specified at the beginning of this Chapter. If there is no voltage, refer to *Wiring Diagrams* (see Chapter 9) and check for a break or short circuit in the pink wire. Now check for voltage between the pink wire terminal and the green/orange wire terminal in the connector. If there is no voltage, refer to *Wiring Diagrams* (see Chapter 9) and check for a break or short circuit in the green/orange wire. Finally check for voltage between the light green/yellow wire terminal and the green/orange wire terminal in the connector. If there is no voltage, refer to

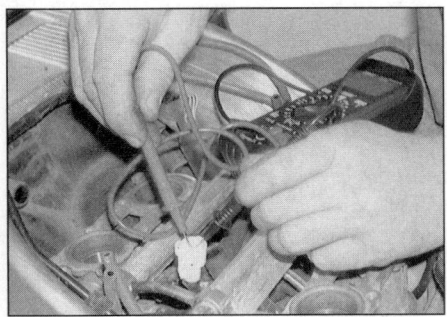

11.51 Checking the intake air temperature sensor power supply

11.44 Using a vacuum gauge to check pressure at the MAP sensor (arrow)

Wiring Diagrams (see Chapter 9) and check for a break or short circuit in the light green/yellow wire. Turn the ignition off.
43 If the power supply to the MAP sensor is good, have the operation of the ECM tested by a Honda dealer.
44 To check the manifold pressure, disconnect the vacuum hose from the sensor and use a three-way adaptor to connect a vacuum gauge between the sensor and the throttle body (see illustration).
45 Start the engine and measure the manifold absolute pressure at idle speed, then compare the result with the Specifications at the beginning of this Chapter. If the pressure differs greatly from that specified, inspect the vacuum hose and its connections and renew the hose if necessary.
46 If the pressure is good, have the operation of the ECM tested by a Honda dealer.

Removal

47 Disconnect the wiring connector and vacuum hose from the MAP sensor. Undo the retaining screw and remove the sensor; access to the retaining screw is awkward (a right-angle screwdriver will be needed) but can be improved by removed the air filter housing (see Section 7).

Installation

48 Installation is the reverse of removal.

Intake air temperature sensor

Check

49 If the intake air temperature sensor is thought to be faulty (see Section 10), remove

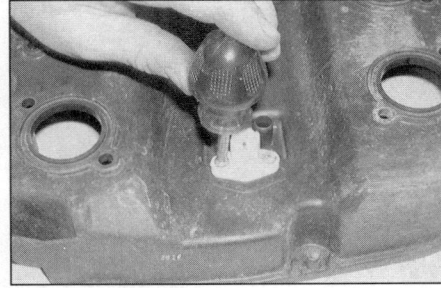

11.53 Undo the screws and remove the intake air temperature sensor from the air filter housing

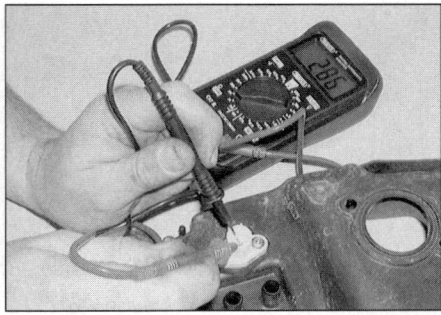

11.50 Measuring the resistance of the intake air temperature sensor

the air filter housing (see Section 7) to gain access to the sensor which is mounted on the left-hand side of the housing.
50 With the sensor cold, measure the resistance between the sensor terminals with an ohmmeter and compare the result with the Specifications at the beginning of this Chapter (see illustration). If the resistance reading differs greatly from that specified, the sensor is probably faulty and should be replaced. Note: *The specified value is only valid at 20°C (68°F); the sensor resistance will increase at lower temperatures and decrease at higher temperatures.*
51 If the sensor appears to be functioning correctly, check its power supply. Turn the ignition on and check for voltage between the grey/blue wire terminal in the connector and earth (ground) (see illustration). The voltage should be as specified at the beginning of this Chapter. If there is no voltage, refer to *Wiring Diagrams* (see Chapter 9) and check for a break or short circuit in the grey/blue wire. Now check for voltage between the grey/blue wire terminal and the green/orange wire terminal in the connector. If there is no voltage, refer to *Wiring Diagrams* (see Chapter 9) and check for a break or short circuit in the green/orange wire. Turn the ignition off.
52 If the power supply to the temperature sensor is good, have the operation of the ECM tested by a Honda dealer.

Removal

53 Remove the air filter housing (see Section 7) then undo the retaining screws and remove the sensor from the housing (see illustration).

Installation

54 Installation is the reverse of removal.

Cam pulse generator

Check

55 If the cam pulse generator is thought to be faulty (see Section 10), remove the air filter housing (see Section 7), then locate the pulse generator wiring connector (white) which is inside the protective cover on the right side of the frame.
56 Disconnect the wiring connector, then use an ohmmeter to check for continuity between

4•14 Fuel and exhaust systems

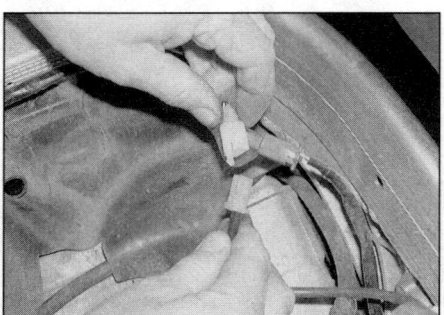

11.56 Disconnect the wiring connector to test the cam pulse generator

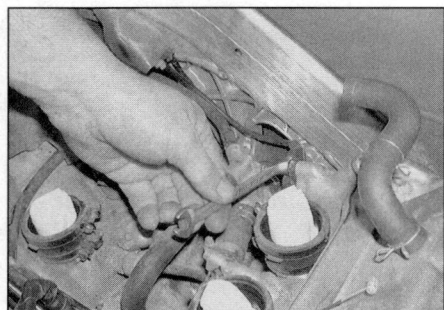

11.60a Unscrew the retaining bolt . . .

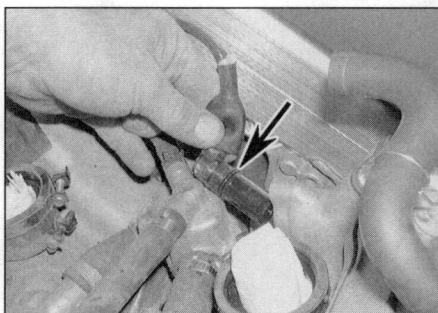

11.60b . . . and ease the cam pulse generator out of position (O-ring arrowed)

the white and grey wire terminals on the pulse generator side of the connector and earth (ground) **(see illustration)**. There should be no continuity (infinite resistance). If there is continuity, renew the cam pulse generator.

57 To measure the cam pulse generator peak voltage (see **Note**) connect the positive (+) lead of the peak voltage adaptor to the white wire terminal on the pulse generator side of the connector and the negative (–) lead to the grey wire terminal. Turn the engine over on the starter motor and note the voltage reading obtained. If this reading is below the specified minimum, the cam pulse generator is faulty.
Note: *Honda specify their own Imrie diagnostic tester (model 625), or the peak voltage adaptor (Pt. No. 07HGJ-0020100) with an aftermarket digital multimeter having an impedance of 10 M-ohm/DCV minimum for this test.*

58 If the cam pulse generator functions correctly then the fault must be in the wiring harness (see *Wiring Diagrams*, Chapter 9) or the ECM.

Removal

59 Remove the throttle body assembly (see Section 12) to gain access to the pulse generator which is fitted to the front of the rear cylinder head assembly.
60 Disconnect the wiring connector **(see illustration 11.56)** then undo the retaining bolt and withdraw the pulse generator from the cylinder head, easing it past the thermostat housing **(see illustrations)**. Remove the O-ring and discard it as new one will be needed for installation.

Installation

61 Fit the new O-ring to the sensor recess and lubricate it with a smear of engine oil.
62 Ease the sensor into position, taking care not to damage the O-ring, and tighten the retaining bolt securely.
63 Connect the wiring connector then install the throttle body assembly (Section 12).

Ignition pulse generator

Check

64 If the ignition pulse generator is thought to be faulty (see Section 10), remove the air filter housing (see Section 7), then locate the pulse generator wiring connector (red) which is inside the protective cover on the right side of the frame **(see illustration)**.
65 Disconnect the wiring connector, then use an ohmmeter to check for continuity between the white/yellow and yellow wire terminals on the pulse generator side of the connector and earth (ground) **(see illustration)**. There should be no continuity (infinite resistance). If there is continuity, renew the ignition pulse generator.
66 To measure the ignition pulse generator peak voltage (see **Note**, Paragraph 57) connect the positive (+) lead of the peak voltage adaptor to the yellow wire terminal on the pulse generator side of the connector and the negative (–) lead to the white/yellow wire terminal. Turn the engine over on the starter motor and note the voltage reading obtained. If this reading is below the specified minimum, the ignition pulse generator is faulty.
67 If the ignition pulse generator functions correctly then the fault must be in the wiring harness (see *Wiring Diagrams*, Chapter 9) or the ECM.

Removal

68 Remove the right crankcase cover from the engine (see Chapter 2, Section 13).
69 Undo the retaining bolts then free the wiring grommet from the cover and remove the pulse generator **(see illustration)**.

Installation

70 Remove all traces of sealant from the sensor wiring grommet and crankcase cover and apply a smear of fresh sealant to the grommet.
71 Locate the grommet and pulse generator correctly in the crankcase cover and tighten the retaining bolts to the specified torque. Install the cover on the engine (see Chapter 2).

Atmospheric (barometric) pressure sensor

Check

72 If the atmospheric pressure sensor is thought to be faulty (see Section 10), remove

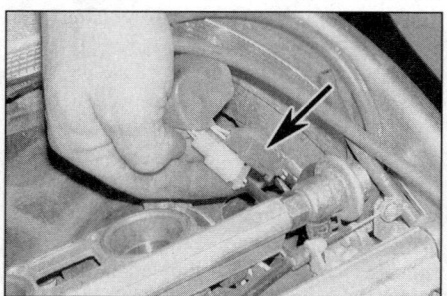

11.64 Locate the ignition pulse generator (red) wiring connector which is on the right side of the frame

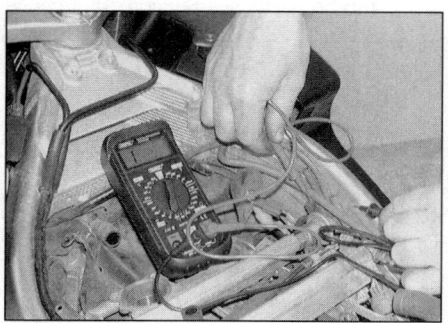

11.65 Checking ignition pulse generator for a short circuit to earth (ground)

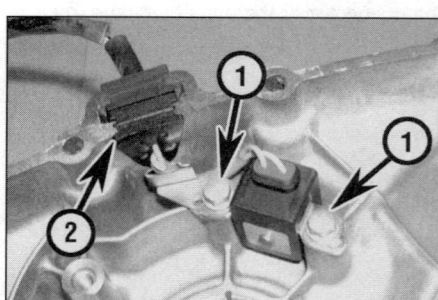

11.69 Undo the retaining bolts (1) then free the sealing grommet (2) and remove the ignition pulse generator from the crankcase cover

Fuel and exhaust systems 4•15

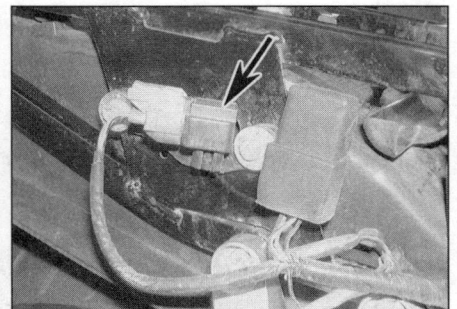

11.72 The atmospheric pressure sensor (arrow) is located on the right side of the subframe

11.76a Undo the retaining screw . . .

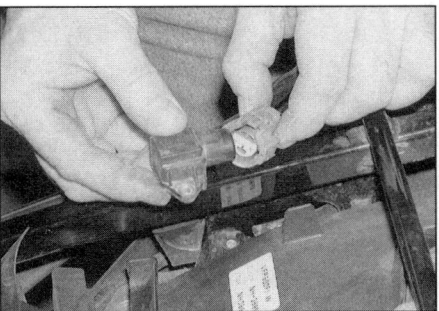

11.76b . . . then disconnect the atmospheric pressure sensor from the wiring connector and remove it from the bike

the seat cowling (see Chapter 8) to access the sensor **(see illustration)**.

73 To check the wiring, disconnect the wiring connector from the sensor, then turn the ignition on and check for voltage between the pink wiring terminal in the connector and earth (ground). The voltage should be as specified at the beginning of this Chapter. If there is no voltage, refer to *Wiring Diagrams* (see Chapter 9) and check for a break or short circuit in the pink wire. Now check for voltage between the pink wire terminal and the green/orange wire terminal in the connector. If there is no voltage, refer to *Wiring Diagrams* (see Chapter 9) and check for a break or short circuit in the green/orange wire. Finally check for voltage between the light green/black wire terminal and the green/orange wire terminal in the connector. If there is no voltage, refer to *Wiring Diagrams* (see Chapter 9) and check for a break or short circuit in the light green/black wire. Turn the ignition off.

74 If the power supply to the sensor is good, have the operation of the ECM tested by a Honda dealer.

Removal

75 Remove the seat cowling (see Chapter 8). **Note:** *The sensor can be accessed with just the seat removed, but removal is a lot easier with the cowling removed as well.*

76 Undo the retaining screw, then disconnect the wiring connector and remove the sensor from the motorcycle **(see illustrations)**.

Installation

77 Installation is the reverse of removal.

Speed sensor

78 See Chapter 9, Section 16.

Bank angle sensor

Check

79 Position the motorcycle on its centrestand and remove the upper fairing (see Chapter 8) to access the bank angle sensor wiring connector **(see illustration)**.

80 With the ignition switch on and the wiring still connected, connect the negative (–) lead of a voltmeter to the green wire terminal. Connect the voltmeter positive (+) lead to the white wire terminal to check that battery voltage (approximately 12 volts) is present, then connect the positive (+) lead to the red/green (or red/white) terminal and check that between 0 to 1 volt is present.

81 Switch the ignition off then remove the retaining nuts and bolts and free the bank angle sensor from its mounting. Hold the sensor horizontal and switch the ignition on; the engine stop relay (located behind the right side of the seat cowling) should click. Slowly tilt the sensor to the left whilst listening to the engine stop relay; once the sensor reaches an angle of approximately 60° the relay should be heard to click (indicating power supply is open). Switch the ignition off and return the sensor to the horizontal, then switch the ignition back on again (engine stop relay should click again) and tilt the sensor to the right. The engine stop relay should be heard to click again once the sensor reaches an angle of around 60°.

82 If voltage readings/relay performance are not as given it is likely the bank angle sensor is faulty and should be replaced.

Removal

83 Remove the upper fairing (see Chapter 8).
84 Disconnect the bank angle sensor wiring connector then slacken and remove the retaining nuts and bolts and remove the sensor from its mounting **(see illustration 11.79)**.

Installation

85 Installation is the reverse of removal ensuring the sensor is fitted with its 'UP' mark facing upwards.

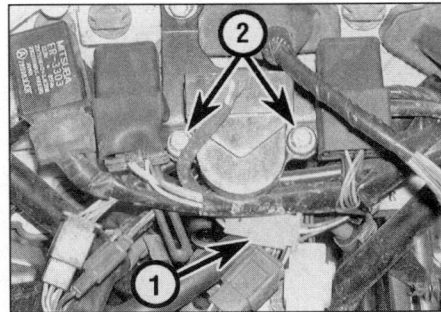

11.79 Bank angle sensor wiring connector (1) and retaining nuts and bolts (2)

Engine stop relay

Check

86 Remove the relay (see below).
87 Connect an ohmmeter across the black/pink and brown (or black/white) wire terminals of the relay. Using a battery and jumper wires, connect the battery positive (+) terminal to the red/orange wire terminal of the relay and the negative (–) terminal to the black wire terminal of the relay and note the reading obtained. If the relay is operating correctly there should be continuity (zero resistance) when the battery is connected and no continuity (infinite resistance) when the battery is disconnected. If this is not the case, renew the relay.

Removal

88 Remove the seat (see Chapter 8). The engine stop relay is located on the right side of the motorcycle, just in front of the rear brake fluid reservoir **(see illustration)**.
89 Free the relay from its mounting and manoeuvre it out from behind the cowling, then disconnect the wiring connector and remove the relay from the motorcycle.

Installation

90 Installation is the reverse of removal.

Engine control module (ECM)

Check

91 If the ECM is thought to be faulty (see Section 10), remove the seat (see Chapter 8)

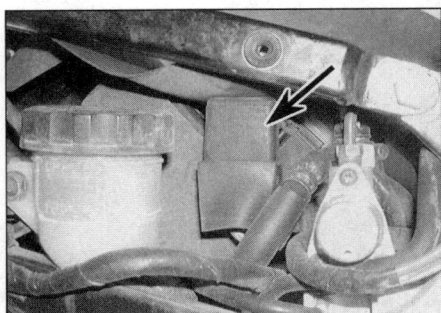

11.88 The engine stop relay (arrow) is located in front of the rear brake fluid reservoir (shown with seat cowling removed)

4•16 Fuel and exhaust systems

11.98a Remove the four trim clips . . .

11.98b . . . and lift off the cover to gain access to the ECM

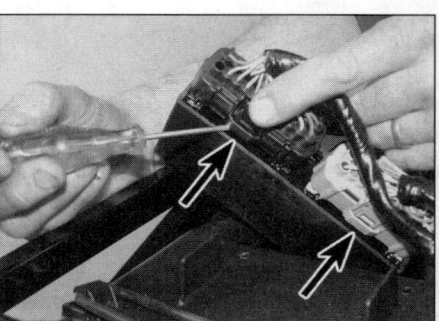

11.98c Wiring connectors are secured by clips (arrows)

to access the module **(see illustrations 11.98a and 11.98b)**.

92 Disconnect the wiring connectors and check that the terminals are clean and free from corrosion **(see illustration 11.98c)**. Reconnect the connectors.

93 Ensure the ignition is switched off, then bridge the terminals of the service check connector (see Section 10, Paragraph 3). Turn the ignition on, and count how many times the warning light flashes.

94 If the light gives two long (1.3 second) flashes (equivalent to fault code 20), turn the ignition off and reset the ECM memory (see Section 10, Paragraph 5). Now turn the ignition on and count how many times the warning light flashes. If the light gives two long (1.3 second) flashes the ECM is probably faulty and should be tested by a Honda dealer using the Honda diagnostic test pin box (see Section 10).

95 If the light does not give two long flashes, turn the ignition off and remove the jumper wire. Disconnect the atmospheric pressure sensor (see Paragraph 72), turn the ignition on and count how many times the warning light flashes.

96 The light should indicate a fault with the atmospheric sensor (see Section 10). If so, reset the ECM memory (see Section 10, Paragraph 5). If the light gives two long flashes, turn the ignition off, bridge the terminals of the service check connector (see Section 10, Paragraph 3) and turn the ignition on. If the light gives two long flashes, reset the ECM memory. If the light still gives two long flashes the ECM is probably faulty and should

be tested by a Honda dealer using the Honda diagnostic test pin box (see Section 10).

Removal

97 Remove the seat (see Chapter 8) and disconnect the battery negative (–) terminal.

98 Remove the four trim clips and lift off the ECM cover from the rear of the mudguard, then disconnect the wiring connectors from the ECM **(see illustrations)**.

99 Undo the retaining screw and remove the ECM.

Installation

100 Installation is the reverse of removal.

Oxygen sensors – models with a catalytic converter

Check

101 The operation of the oxygen sensors can only be checked using the Honda diagnostic test pin box (see Section 10).

102 To check the sensor heating element, remove the left side lower fairing lower panel and the coolant reservoir (see Chapters 3 and 8). Trace the wiring back from the sensors to the protective cover on the left side of the engine and disconnect both wiring connectors **(see illustration)**.

103 Connect an ohmmeter across the black wiring terminals of each sensor wiring connector and measure the resistance. Compare the reading to that given in the Specifications. If the resistance differs greatly from that specified, the sensor should be renewed.

Removal

Note: *Ensure the exhaust system is cold before proceeding. The oxygen sensor is delicate and will not work if it is dropped or knocked, if its power supply is disrupted, or if any cleaning materials are used on it.*

104 Remove the left side fairing lower panel (see Chapter 8).

105 Remove the coolant reservoir (see Chapter 3).

106 Trace the wiring back from the relevant sensor to the protective cover on the left side of the engine then disconnect the sensor wiring connector.

107 If the lower sensor is being removed, slacken the retaining bolts and remove the sensor heat shield **(see illustration)**.

108 Using a special oxygen sensor socket wrench, Honda service tool (Pt. No. 07LAA-PT50101) or 22 mm ring spanner, unscrew the oxygen sensor and remove it from the exhaust front pipe **(see illustration)**.

Installation

109 Installation is the reverse of removal, tightening the sensor to the specified torque.

12 Throttle body assembly – removal and installation

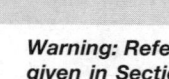

Warning: *Refer to the precautions given in Section 1 before starting work.*

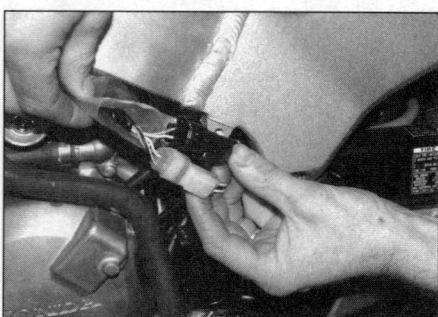

11.102 Disconnect the oxygen sensor wiring connectors

11.107 Lower oxygen sensor is protected by a heat shield (arrow)

11.108 Using and special socket wrench to unscrew the oxygen sensor

Fuel and exhaust systems 4•17

12.6 Disconnect the water hoses (arrows)

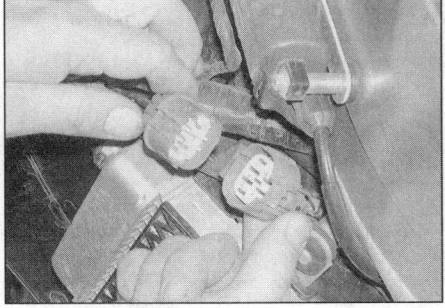

12.7 Disconnect the throttle body wiring harness connector

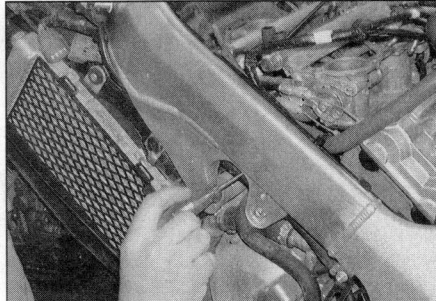

12.8 All the intake rubber retaining clips can be slackened from the left side of the bike

Removal

1 Remove the fairing left side lower panel (see Chapter 8).
2 Remove the coolant reservoir (see Chapter 3).
3 Remove the fuel tank (Section 2).
4 Remove the air filter housing (Section 7).
5 On VFR800FI-W and X models, unclip the choke outer cable from its bracket then free the inner cable from the starter valve lever.
6 On VFR800FI-Y and 1 models, loosen the clips and disconnect the water hoses from the fast idle wax unit (see illustration).
7 Trace the wiring back from the throttle body assembly to its connector on the left side of the motorcycle, directly above the radiator. Disconnect the connector so the harness is free to be removed with the throttle body (see illustration).
8 Slacken the four retaining clips securing the intake rubbers to the throttle body assembly, then ease the throttle body assembly out of position (see illustration).
9 Remove the screws securing the throttle cable bracket to the throttle body. Free both inner cables from the throttle cam (see illustrations).
Caution: Do not snap the throttle cam/valves from fully open to fully closed once the cables have been disconnected because this can lead to engine idle speed problems.
10 Remove the throttle body assembly from the motorcycle (see illustration). Whilst the throttle body assembly is removed, tape over/plug the intake rubbers to prevent dirt/debris from entering the intake ports. If the intake rubbers shown signs of damage or deterioration they must be renewed.
Caution: The throttle body assembly must be treated as a sealed unit. NEVER loosen any of the white-painted bolts/screws on the assembly as these are pre-set at the factory to ensure correct synchronisation of the throttle valves and idle circuit. The only components on the assembly which can be serviced are the starter valves (see Section 13), fast idle wax unit (see Section 14) if fitted, and the various vacuum hoses.

Installation

11 Prior to installation, check the throttle body vacuum hoses for signs of damage or deterioration and renew any suspect hoses.

12.9a Undo the throttle cable bracket screws (arrows) . . .

12 Remove the tape/plugs from the intake rubbers and make sure the retaining clip screws are correctly position so they are all accessible from the left side of the machine.
13 Lubricate the rubbers with a smear of engine oil to ease installation then rest the throttle body assembly in position.
14 Connect the throttle inner cables to the cam then fit the cable bracket to the throttle body and securely tighten its retaining screws.
15 Securely reconnect the wiring harness connector then ease the throttle body assembly into the intake rubbers. Ensure the throttle body is fully engaged in the intake rubbers then securely tighten all the retaining clips.
15 On VFR800FI-W and X models, connect the choke inner cable to the starter valve lever then clip the outer cable into its bracket.
16 On VFR800FI-Y and 1 models, connect the water hoses to the fast idle wax unit and secure them with the clips.

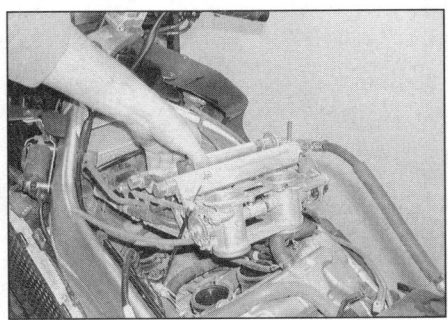

12.10 Removing the throttle body assembly

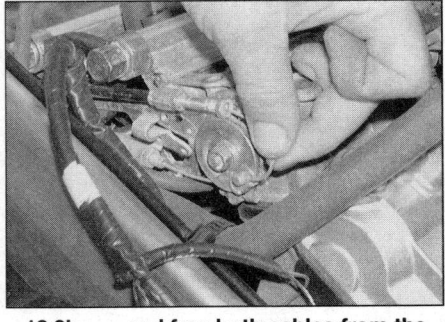

12.9b . . . and free both cables from the throttle cam

17 Install the air filter housing and the fuel tank (Sections 7 and 2).
18 Install the coolant reservoir (Chapter 3) then fit fairing lower panel (Chapter 8).

13 Starter valves – removal, installation and synchronisation

Warning: Refer to the precautions given in Section 1 before starting work.

Removal

1 Remove the throttle body assembly (see Section 12).
2 Disconnect the breather hoses from the throttle body assembly.
3 Locate the starter valves on the right side of the throttle body assembly (see illustration).

13.3 The starter valves (arrows) are located on the right side of the throttle body assembly

4•18 Fuel and exhaust systems

13.4 Fast idle wax unit mounting screws (arrows)

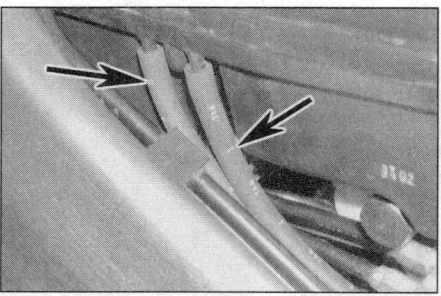

13.18a On UK and US (except California) models disconnect the vacuum hoses 1 and 2 (arrows) from the left side of the air filter housing . . .

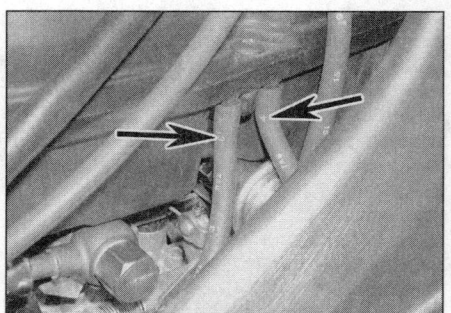

13.18b . . . and hoses 3 and 4 (arrows) from the right side of the housing

Screw each starter valve adjustment screw (only three of the four valves are adjustable) in until it seats lightly and note the amount of turns needed to do so on a piece of paper.

4 On VFR800FI-Y and 1 models, undo the fast idle wax unit mounting screws and remove them **(see illustration)**.

5 Slacken and remove the nut and washer from the starter valve lever pivot bolt. Withdraw the bolt, washer and return spring and remove the lever, noting the correct fitted location of all components. Prise the pivot shaft bushes out from the body bracket. **Note:** *On VFR800FI-Y and 1 models, do not separate the fast idle wax unit from the starter valve lever – the position of the double nuts at the end of the link rod is pre-set at the factory and should not be disturbed.*

6 Loosen the starter valve lock nuts and remove the four valve assemblies from the throttle body, keeping them in their correct fitted order.

7 Check all components for wear and damage and renew as necessary. If required, disassemble the fast idle wax unit (see Section 14).

Installation

8 Clean the starter valves and throttle body passages using only compressed air.

Caution: NEVER use a solvent-based carburettor cleaner to clean the throttle body assembly. The throttle bores are covered with a molybdenum coating which could be removed by the cleaner.

9 Install each starter valve in its original location making sure the cut-out on the valve body aligns correctly with the guide in the valve bore. Securely tighten the valve lock nuts then check that each valve moves smoothly and easily in its bore.

10 Fit the pivot bushes to the body then install the starter valve lever. Ensure the lever is correctly engaged with all the valves then fit the return spring, washer and pivot bolt. Refit the washer and nut to the pivot bolt and tighten securely. Make sure the return spring ends are correctly located then check the operation of the starter valve lever before continuing; the lever should move smoothly and easily and return to the fully closed position under return spring pressure.

11 On VFR800FI-Y and 1 models, install the fast idle wax unit mounting screws and tighten them securely.

12 Turn the starter valve adjustment screws in until they seat lightly then back each one out by the exact amount of turns noted prior to removal.

13 Fit the breather hoses and install the throttle body assembly (Section 12). On completion check the starter valve synchronisation as follows.

Synchronisation

14 If the starter valves have been disturbed, it will be necessary to synchronise them as follows. This is done by measuring the vacuum produced in each cylinder. If the valves are out of synchronisation, the engine will idle unevenly resulting in higher then normal vibration levels.

15 To synchronise the starter valves, you will need a set of vacuum gauges or calibrated tubes (manometer) to indicate engine vacuum. The equipment used should be suitable for a four cylinder engine and come complete with the necessary adapters and hoses to connect to the engine. **Note:** *Because of the nature of the synchronisation procedure and the need for special instruments, most owners leave the task to a Honda dealer.*

16 Start the engine and let it run until it reaches normal operating temperature, then shut it off.

17 Raise and support the fuel tank (see Section 2).

18 On UK and US (except California) models, disconnect the vacuum hoses (numbered 1 to 4) from the left and right sides of the air filter housing and connect them to the vacuum gauges/manometer according to the manufacturer's instructions **(see illustrations)**. On California models disconnect the four vacuum hoses from the union on the rear of the air filter housing and connect them to the gauges/manometer. Make sure there are no leaks in the set-up, as false readings will result.

19 Disconnect the pulse secondary air (PAIR) hoses from the front and rear valve covers and plug the cover unions.

 Warning: Take great care not to burn your hands on the hot engine unit.

20 Start the engine and make sure the idle speed is correct. If it isn't, adjust it (see Chapter 1). If the gauges are fitted with damping adjustment, set this so that the needle flutter is just eliminated but so that they can still respond to small changes in pressure.

21 Adjustment is made via the three screws fitted to the ends of number 2 to 4 starter valves; number 1 starter valve is not adjustable. The screws are accessible from above and can be adjusted using a 7 mm spanner **(see illustrations)**.

22 Using number 1 starter valve vacuum reading as the base setting, adjust the screws so that all the vacuum readings for cylinders 1 to 4 are the same. From this point, adjust

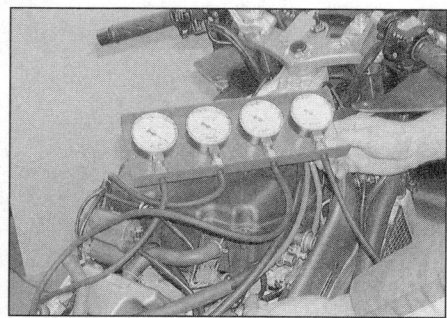

13.21a Using a set of vacuum gauges to synchronise the starter valves

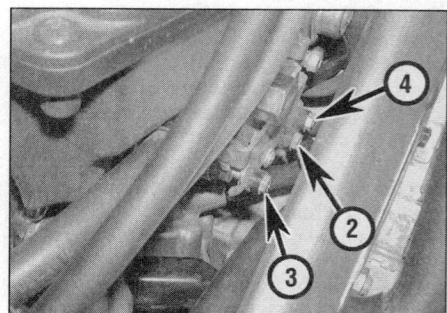

13.21b Adjustment is via the screws (arrows) on the starter valves

Fuel and exhaust systems 4•19

number 3 valve until its reading is 20 ± 5 mmHg below the base setting then adjust the number 4 valve until its reading is 10 ± 5 mmHg below the base setting.
23 When the adjustment is complete, recheck the vacuum readings and idle speed, then stop the engine. Remove the vacuum gauge or manometer and securely reconnect the vacuum hoses to the air filter housing.
24 Remove the plugs then securely connect the PAIR hoses to the valve covers.
25 Recheck the idle speed then lower the fuel tank back down into position (see Section 2).

14 Fast idle wax unit – removal, inspection and installation

Note: *This information only applies to VFR800FI-Y and VFR800FI-1 models*

⚠ **Warning:** *Refer to the precautions given in Section 1 before starting work.*

Removal

1 Follow Steps 1 to 5 in Section 13 to remove the wax unit and starter valve lever.

Inspection

2 Undo the three screws securing the wax unit cover and remove it. Discard the O-ring as a new one must be used on reassembly.
3 Remove the spring, spring seat and wax element from the body (or in reverse order from the cover, depending where they are). Discard the wax element O-rings, noting how they fit.
4 Visually inspect all components for signs of wear and damage and renew them as required.
5 The wax element expands with heat. If you suspect it is not working correctly, place it first in a cold place and check that it is fully retracted. Now gently heat it using a hairdryer or equivalent and check that it expands.
6 Check that the pushrod moves smoothly in and out of the body.
7 Check the spring for fatigue and distortion.
8 Fit new O-rings in the grooves in the wax element, then install the spring, spring seat and element into the cover. Fit a new O-ring into the groove in the cover, then install the cover onto the body, ensuring the O-rings remain in place, and tighten the cover screws evenly in a criss-cross pattern.

Installation

9 Follow Steps 10 to 13 in Section 13 to install the wax unit.

15 Throttle cables – removal and installation

⚠ **Warning:** *Refer to the precautions given in Section 1 before proceeding.*

Removal

1 Raise and support the fuel tank (see Section 2).
2 Slacken the locknuts securing the cables to the mounting bracket on the throttle body assembly; the upper cable in the bracket opens the throttle cam and the lower cable closes it **(see illustration)**.
3 Remove the retaining screws and free the cable bracket from the throttle body assembly **(see illustrations 12.9a and 12.9b)**. Free the inner cables from the throttle cam then unscrew the locknuts and separate both cables from the mounting bracket.
4 Work back along the cables, freeing them from the retaining clamps whilst noting their correct routing.
5 Unscrew the right handlebar switch screws then free the switch from the handlebar and disconnect the throttle cables from the throttle grip **(see illustration)**.
6 Pull the cables out from between the instrument cluster and frame then unscrew the nut securing each cable to the switch assembly. Free both cables from the switch (the opening cable end fitting is a screw-fit) and remove them from the machine.

Installation

7 Attach both cables to the handlebar switch assembly (ensure the opening cable is screwed fully into position) and lightly tighten the securing nuts.
8 Pass both cables between the instrument cluster and frame, making sure they are correctly routed, and secure them in position with the necessary clamps. The cables must not interfere with any other component and should not be kinked or bent sharply.
9 Lubricate the upper end of each cable with multi-purpose grease and attach the cables to the throttle grip.
10 Fit the switch lower half to the handlebar, locating its peg in the handlebar hole. Fit the top half of the switch and securely tighten the screws. Check that both cable end fittings are correctly positioned then tighten their securing nuts.
11 Locate the lower end of the cables correctly in their mounting bracket. Securely tighten the lower cable locknut but tighten the upper cable locknuts only lightly at this stage. Lubricate the end of each cable with multi-purpose grease and attach them to the throttle cam. Locate the cable mounting bracket correctly on the throttle body assembly and securely tighten its retaining screws.
12 Adjust the cables as described in Chapter 1. Turn the handlebars back and forth to make sure the cables don't cause the steering to bind.
13 Start the engine and turn the handlebars back and forth to make sure the idle speed doesn't rise as the bars are turned. If it does, the cables are incorrectly routed and the problem must be sorted out before the motorcycle is ridden.
14 On completion lower the fuel tank back into position (Section 2).

16 Choke cable – removal and installation

Note: *This information only applies to VFR800FI-W and VFR800FI-X models*

Removal

1 Raise and support the fuel tank (see Section 2).
2 Unclip the choke outer cable from its mounting bracket then free the inner cable from the starter valve lever **(see illustrations)**.

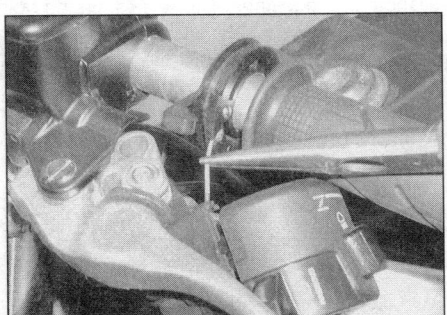

15.2 Slacken the locknuts securing the cables to the throttle body bracket

15.5 Separate the right handlebar switch assembly and detach the throttle cables from the grip

16.2a Unclip the choke outer cable from the bracket . . .

4•20 Fuel and exhaust systems

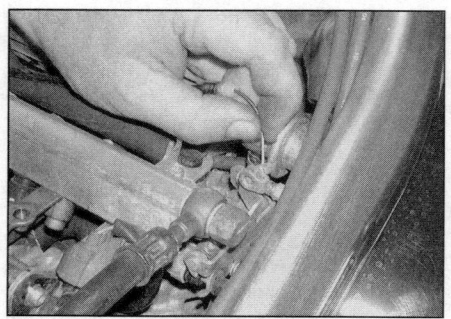

16.2b ... then free the inner cable from the starter valve lever

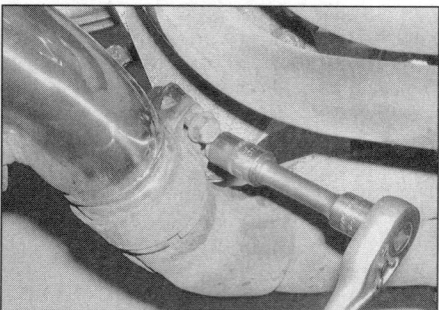

17.2 Slacken the silencer clamp bolts ...

17.3a ... then remove the nut, mounting bolt and washer ...

3 Unscrew the left handlebar switch screws then free the switch from the handlebar and detach the upper end of the choke cable from the choke lever.
4 Work back along the choke cable, freeing it from the retaining clamp(s) whilst noting its correct routing, and pull the cable out from between the instrument cluster and frame.
5 Unscrew the securing nut and cable from the lower half of the switch and remove the cable from the machine.

Installation

6 Screw the cable fully into the handlebar switch assembly and lightly tighten its securing nut.
7 Pass the cable between the instrument cluster and frame, making sure it is correctly routed. The cable must not interfere with any other component and should not be kinked or bent sharply.
8 Lubricate the upper end of the cable with multi-purpose grease and attach it to the choke lever.
9 Engage the switch lower half with the choke lever then locate its peg in the handlebar hole. Fit the top half of the switch and securely tighten the screws. Check that cable end fitting is correctly positioned then securely tighten the securing nut.
10 Lubricate the cable lower end with multi-purpose grease then attach the inner cable to the starter valve lever and clip the outer cable securely into its bracket.
11 Check the operation of the choke lever as described in Chapter 1 then lower the fuel tank back into position (see Section 2).

17 Exhaust system – removal and installation

⚠ **Warning:** *If the engine has been running the exhaust system will be very hot. Allow the system to cool before carrying out any work.*

Silencer

Removal

1 Remove the fairing right lower panel (see Chapter 8).
2 Loosen the bolts of the clamp securing the silencer to the front pipe assembly **(see illustration)**.
3 Slacken and remove the silencer mounting nut, bolt and washer then free the silencer from the front pipe assembly and remove it from the motorcycle **(see illustrations)**. Remove the silencer gasket and discard it.
4 Remove the collar from the rear of the silencer mounting and check the mounting rubber for damage or deterioration. Renew the mounting if damaged.

Installation

5 Remove all traces of gasket and dirt from the front pipe and silencer joint. Check that the silencer clamp is correctly fitted then fit the new gasket **(see illustration)**.
6 Ensure the collar is in position in the rear of the mounting rubber then manoeuvre the silencer into position. Insert the mounting bolt

17.3b ... and free the silencer from the front pipe

and washer then fit the mounting nut and tighten it to the specified torque.
7 Securely tighten the silencer clamp bolts then run the engine and check that there are no exhaust gas leaks.

Front pipe assembly

Removal

8 Remove the lower fairing panels (see Chapter 8).
9 Remove the silencer (see above).
10 Remove the swingarm pivot bolt cap from the right side rider's footrest bracket then slacken and remove the nut from the pivot bolt **(see illustration)**.
11 Slacken and remove the right side rider's footrest bracket heatshield upper bolt **(see illustration)**. Free the complete footrest assembly from the swingarm pivot bolt, taking

17.5 On installation fit a new gasket to the silencer

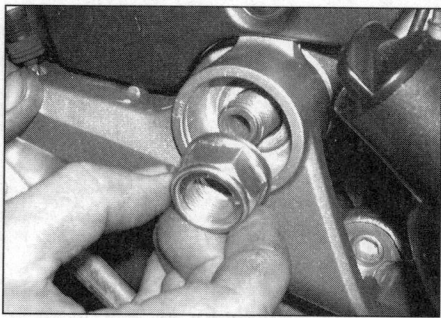

17.10 Unscrew the nut from the swingarm pivot bolt ...

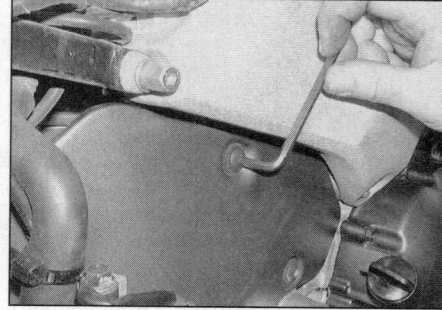

17.11a ... then unscrew the heatshield upper bolt

Fuel and exhaust systems 4•21

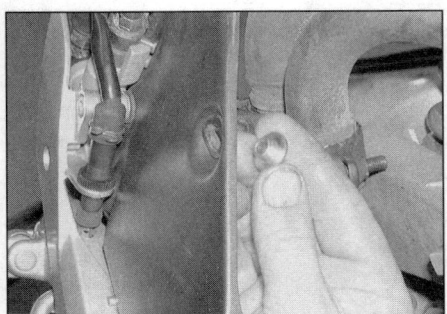

17.11b Free the rider's right footrest bracket assembly from the swingarm pivot bolt and recover the collar fitted between the heatshield and frame

17.12 Front pipe is secured to the centrestand bracket by nut and bolt (arrow)

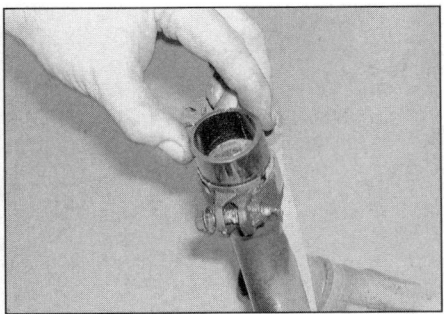

17.16 Fit a new gasket to each front pipe clamp joint . . .

care not to lose the collar from the rear of the heatshield **(see illustration)**.
12 On models fitted with a catalytic converter, remove the coolant reservoir (see Chapter 3). Trace the wiring back from the oxygen sensors and disconnect both wiring connectors so the sensors are free to be removed with the front pipe **(see illustration 11.102)**. Slacken and remove the nut and bolt securing the front pipe to the centrestand mounting bracket **(see illustration)**.
13 Loosen both clamps securing the front pipe assembly to the rear pipes.
14 Unscrew the nuts securing the front pipes to the front cylinder head and free the collars from their studs.
15 Free the front pipe assembly from the rear pipes and remove it from the motorcycle. Remove the gaskets from the cylinder head ports and the rear pipe joints and discard them. Recover the two collars, insulating washers and mounting rubbers from the mounting bracket on the centrestand, where fitted.

Installation

16 Remove all traces of gasket and dirt from the front pipe and rear pipe joints. Check that the rear pipe clamps are correctly fitted, then fit the new gaskets to the front pipe **(see illustration)**. Where fitted, ensure the collars, insulating washers and mounting rubbers are correctly installed in the mounting bracket on the centrestand.
17 Ensure the front cylinder head ports are

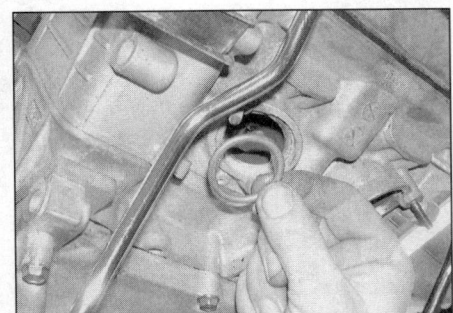

17.17 . . . and install new gaskets in the front cylinder head ports

free of dirt and debris then fit a new gasket to each port, using a dab of grease to hold the gaskets in position **(see illustration)**.
18 Manoeuvre the front pipe assembly into position and engage it with the rear pipes and cylinder head ports, taking care not damage the gaskets. Where fitted, insert the front pipe mounting bolt and fit the mounting nut.
19 Locate the collars on the cylinder head studs and fit the retaining nuts. Ensure the front pipe assembly and rear pipes are correctly joined then tighten the nuts evenly and progressively to the specified torque.
20 Ensure the rear pipe clamp lugs are correctly located in the front pipe cut-outs then securely tighten the clamp bolts.
21 Where fitted, tighten the front pipe mounting bolt nut securely. On models fitted with a catalytic converter, ensure the oxygen sensor wiring is correctly routed, then reconnect both wiring connectors and install the protective cover. Install the coolant reservoir (see Chapter 3).
22 Install the silencer (see Paragraphs 5 to 7) then start the engine and check that there are no exhaust gas leaks before proceeding further.
23 If all is well, locate the footrest assembly on the swingarm bolt. Ensure the collar is in position between the heatshield and frame, then fit the heatshield mounting bolt. Fit the swingarm pivot bolt nut and tighten to the specified torque (see Chapter 6) then securely tighten the heatshield bolt. Fit the pivot bolt cap to the bracket.

Rear pipes

Removal

24 Remove the front pipe assembly (see above).
25 Unscrew the retaining nuts then free the collars from their studs and remove both rear pipes. Remove the gaskets from the cylinder head ports and discard them. **Note:** *The rear pipes are different and are not interchangeable, make identification markings to avoid confusion on installation.*

Installation

26 Ensure the rear cylinder head ports are free of dirt and debris then fit a new gasket to each port, using a dab of grease to hold the gaskets in position.

27 Fit each rear pipe to its relevant port and locate the collars on the cylinder head studs. Screw on the retaining nuts, tightening them lightly at this stage.
28 Install the front pipe as described in Paragraphs 16 to 19.
27 Tighten the rear pipe retaining nuts evenly and progressively to the specified torque then complete the installation procedure as described in Paragraphs 20 to 23.

18 Pulse secondary air (PAIR) system

General information

1 To reduce the amount of unburned hydrocarbons released in the exhaust gases, a pulse secondary air (PAIR) system is fitted. The system consists of the control valve (mounted on the front of the air filter housing), the check valves (fitted to the front and rear cylinder head valve covers) and the hoses linking them. The control valve is connected to the air filter housing and is controlled by the engine control module (ECM).
2 Under certain operating conditions, the ECM opens up the PAIR control valve which then allows filtered air to be drawn through the check valves and cylinder head passages and into the exhaust ports. The air mixes with the exhaust gases, causing any unburned particles of the fuel in the mixture to be burnt in the exhaust port/pipes. This process changes a considerable amount of hydrocarbons and carbon monoxide into relatively harmless carbon dioxide and water. The check valves in the valve covers are fitted to prevent the flow of exhaust gases back up the cylinder head passages and into the air filter housing.

Testing

Control valve

3 Remove the valve from the motorcycle (see below).
4 Check the operation of the control valve by blowing through the lower (air filter housing)

4•22 Fuel and exhaust systems

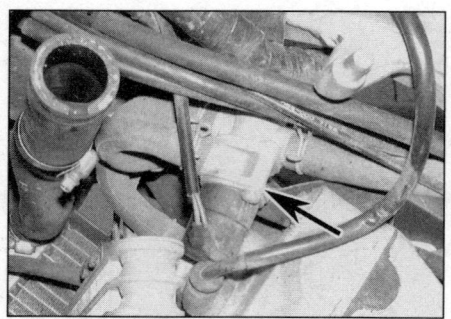

18.11 The pulse secondary air (PAIR) system solenoid valve (arrow) is located above the front cylinder head

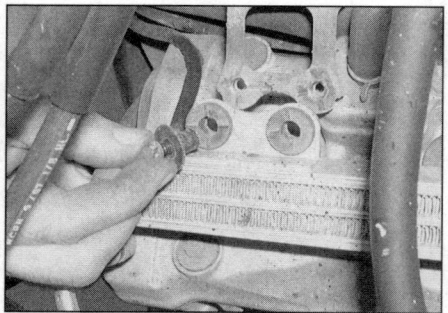

18.15 Remove the oil cooler mounting bolts and collars and move the cooler forwards . . .

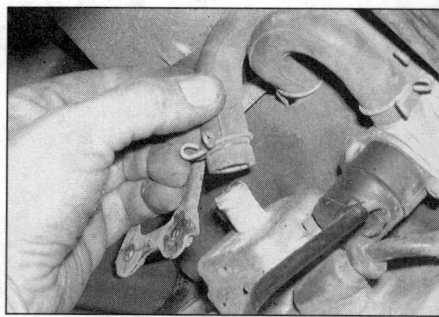

18.16a . . . to gain access to the front check valve assembly

hose union; air should flow freely through the upper (check valve) hose unions. Connect battery voltage (12 volts) across the valve terminals and repeat the check; no air should now flow through the valve if it is functioning correctly. If an ohmmeter is available, check the resistance of the control valve windings and compare the reading obtained to that given in the Specifications. Renew the valve if faulty.

Check valve

5 Disconnect the check valve hoses from the control valve (see below).
6 Check each valve by blowing and sucking on the hose end. Air should flow through the hose only when blown down the hose and not when sucked back up. If this is not the case the check valve is faulty.

Removal and installation
Control valve

7 Remove the fairing left side lower panel (see Chapter 8).
8 Raise and support the fuel tank (see Section 2) then remove the air filter element (see Chapter 1).
9 Remove the left side radiator (see Chapter 3).
10 Undo the retaining screw and free the control valve bracket from the front of the air filter housing.
11 Trace the wiring back from the control valve and disconnect it at the connector **(see illustration)**.
12 Release the retaining clips and disconnect the air hoses then remove the control valve from the motorcycle.
13 Installation is the reverse of removal.

Check valve

14 To gain access to the rear cylinder check valve, raise and support the fuel tank (see Section 2).
15 To gain access to the front cylinder check valve, remove the fairing lower panels and inner panel (see Chapter 8). Unscrew the oil cooler mounting bolts and position the cooler clear of its mounting, taking care not to lose the collars from the mounting rubbers **(see illustration)**.
16 To remove either valve, first release the retaining clip and disconnect the air hose. Undo the retaining bolts and remove the cover and reed valve assembly components, noting which way around they are fitted **(see illustrations)**.
17 Installation is the reverse of removal making sure the reed valve assembly components are correctly fitted and the cover bolts are tightened to the specified torque **(see illustrations)**.

19 Catalytic converter

Note: *This information applies to VFR800FI-W and VFR800FI-X German and Swiss models, all VFR800FI-Y and VFR800FI-1 models*

General information

1 A catalytic converter is incorporated in the exhaust front pipe assembly to minimise the level of exhaust pollutants released into the atmosphere **(see illustration)**.

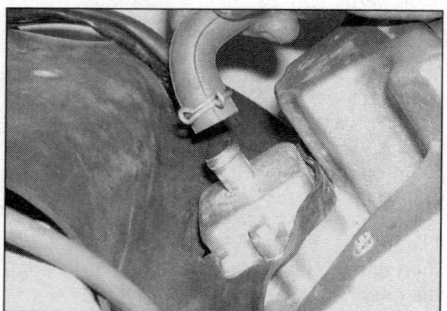

18.16b Disconnecting the air hose from the rear check valve

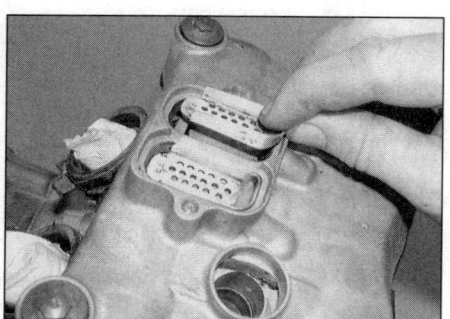

18.17a Install the baffle plates . . .

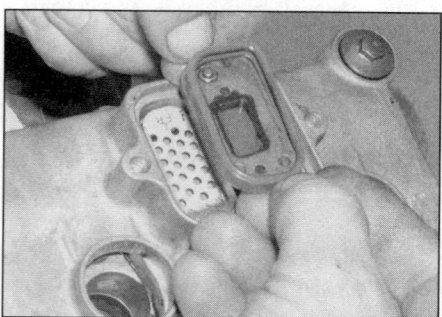

18.17b . . . then fit the rubber seals to the reed valves . . .

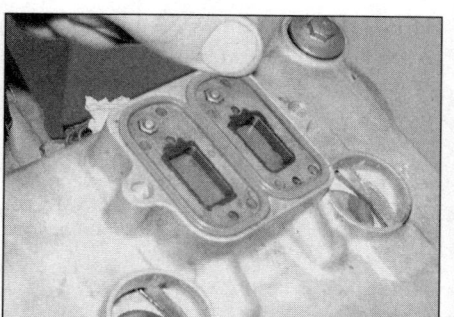

18.17c . . . and seat both valves correctly in the cylinder head cover (shown with cover removed)

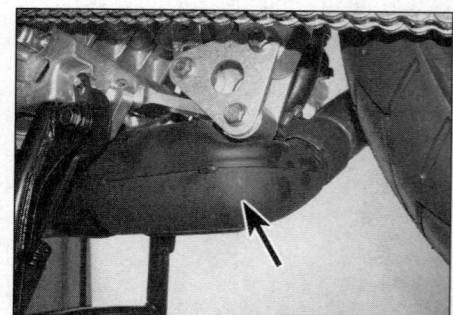

19.1 Location of the catalytic converter

Fuel and exhaust systems 4•23

2 The catalytic converter consists of a canister containing a fine mesh impregnated with a catalyst material, over which the hot exhaust gases pass. The catalyst speeds up the oxidation of harmful carbon monoxide, unburned hydrocarbons and soot, effectively reducing the quantity of harmful products released into the atmosphere via the exhaust gases.

3 The catalytic converter is of the closed-loop type with exhaust gas oxygen content information being fed back to the fuel injection system engine control module (ECM) by the oxygen sensors. There are two sensors which are both screwed into the exhaust front pipe, just in front of the catalytic converter; one sensor monitors the gases from cylinder number 1 and 2 downpipe and the other sensor monitors cylinder number 3 and 4 gases **(see illustration 11.107)**.

4 Both oxygen sensors contain heating elements which are controlled by the ECM. When the engine is cold, the ECM switches on these heating elements which warm the exhaust gases as they pass over each sensor. This brings the catalytic converter quickly up to its normal operating temperature and decreases the level of exhaust pollutants emitted whilst the engine warms up. Once the engine is sufficiently warmed up, the ECM switches off both oxygen sensor heating elements.

5 Refer to Section 17 for information on exhaust front pipe (which incorporates the catalytic converter) removal and installation, and Section 11 for oxygen sensor removal and installation.

Precautions

6 The catalytic converter is a reliable and simple device which needs no maintenance in itself, but there are some facts of which an owner should be aware if the converter is to function properly for its full service life.

a) *DO NOT use leaded or lead replacement petrol (gasoline) – the additives will coat the precious metals, reducing their converting efficiency and will eventually destroy the catalytic converter.*

b) *Always keep the ignition and fuel systems well-maintained in accordance with the manufacturer's schedule – if the fuel/air mixture is suspected of being incorrect have it checked on an exhaust gas analyser.*

c) *If the engine develops a misfire, do not ride the bike at all (or at least as little as possible) until the fault is cured.*

d) *DO NOT use fuel or engine oil additives – these may contain substances harmful to the catalytic converter.*

e) *DO NOT continue to use the bike if the engine burns oil to the extent of leaving a visible trail of blue smoke.*

f) *Remember that the catalytic converter is FRAGILE – do not strike it with tools during servicing work.*

20 Evaporative emission control (EVAP) system

Note: *This information applies to California models only*

General information

1 An evaporative emissions control system is fitted to minimise the escape of fuel vapour from the fuel tank into the atmosphere. The tank filler cap is sealed and a charcoal canister is mounted on the front of the engine unit. The canister collects the fuel vapours generated in the tank when the motorcycle is parked and stores them until they can be cleared from the canister, via the control valve, into the throttle body inlet tracts to be burned by the engine during normal combustion. The system is controlled by the engine control module (ECM).

2 To ensure that the engine runs correctly when it is cold and/or idling, the control valve is not opened by the ECM until the engine has warmed up, and the engine is under load; the control valve is then modulated on and off to allow the stored vapour to pass into the inlet tracts.

Testing

Control valve

3 Remove the valve from the motorcycle (see below).

4 Check the operation of the control valve by blowing through the inlet (canister hose) union; air should flow freely through the valve and out the outlet hose union. Connect battery voltage (12 volts) across the valve terminals and repeat the check; no air should now flow through the valve if it is functioning correctly. If an ohmmeter is available, check the resistance of the control valve windings and compare the reading obtained to that given in the Specifications. Renew the valve if faulty.

Charcoal canister

5 No testing of the canister is possible, if it is thought to be faulty it must be renewed.

Removal and installation

Control valve

6 Remove the fairing right side lower panel (see Chapter 8).

7 Disconnect the wiring connector and vacuum hoses from the valve then undo the mounting bracket bolt and remove the valve assembly from the right side of the frame.

8 Installation is the reverse of removal.

Charcoal canister

9 Remove the fairing left and right side lower panels (see Chapter 8).

10 Disconnect the hoses from the canister, noting their correct fitted locations.

11 Unscrew the mounting bolts and remove the canister from the front of the engine, taking care not to lose the collars and spacer from the mounting rubbers.

12 Installation is the reverse of removal.

Notes

Chapter 5
Ignition system

Contents

General information	1	Ignition timing – general information and check	4
Ignition (main) switch – check, removal and installation	see Chapter 9	Immobiliser (HISS) system	5
Ignition HT coils – check, removal and installation	3	Neutral switch – check and replacement	see Chapter 9
Ignition pulse generator – check, removal and installation	see Chapter 4	Sidestand switch – check and replacement	see Chapter 9
Ignition system – check	2	Spark plugs – gap check and renewal	see Chapter 1

Degrees of difficulty

Easy, suitable for novice with little experience	Fairly easy, suitable for beginner with some experience	Fairly difficult, suitable for competent DIY mechanic	Difficult, suitable for experienced DIY mechanic 	Very difficult, suitable for expert DIY or professional

Specifications

General information
Cylinder identification
 Number 1 .... Left rear cylinder
 Number 2 .... Left front cylinder
 Number 3 .... Right rear cylinder
 Number 4 .... Right front cylinder
Firing order and spacing .... 1 (180°) – 3 (270°) – 2 (180°) – 4 (90°)

Spark plugs
Type and electrode gap .... see Chapter 1

Ignition HT coils
Initial voltage (see text) .... Battery voltage (approximately 12 volts)
Minimum peak voltage (see text) .... 100 volts

Ignition timing
At specified idle speed
 UK and US (except California) models .... 15° BTDC
 California models .... 10° BTDC

Torque setting
Ignition timing inspection plug .... 18 Nm

1 General information

The ignition system is integrated with the fuel injection system to form a combined engine management system which is operated by the engine control module (ECM) (see Chapter 4 for further information). The ECM uses its inputs from the various sensors to calculate the required ignition advance setting and ignition HT coil charging time. The only components which solely operate the ignition side of the system are the four separate ignition HT coils, one for each cylinder.

A safety interlock circuit will cut the ignition if the sidestand is put down whilst the engine is running and in gear, or if a gear is selected whilst the engine is running and the sidestand is down. It also prevents the engine from being started if the sidestand is down and the engine is in gear unless the clutch lever is pulled in.

VFR800FI-Y and VFR800FI-1 models are equipped with Honda's HISS immobiliser system which will not allow the engine to be started unless the correct ignition key is used. The immobiliser system has its own fault-diagnosis function.

Because of their nature, the individual ignition system components can be checked but not repaired. If ignition system troubles occur, and the faulty component can be isolated, the only cure for the problem is to replace the part with a new one. Keep in mind that most electrical parts, once purchased, cannot be returned. To avoid unnecessary expense, make very sure the faulty component has been positively identified before buying a replacement part.

Note that there is no provision for adjusting the ignition timing on these models.

2 Ignition system – check

Warning: *The energy levels in electronic systems can be very high. On no account should the ignition be switched on whilst the plugs or plug caps are being held. Shocks from the HT circuit can be most unpleasant.*

5•2 Ignition system

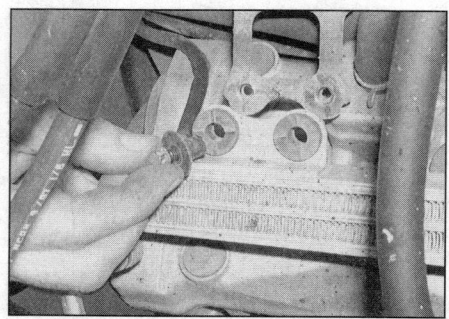

2.3 Remove the mounting bolts and collars and move the oil cooler forwards to gain access to the front cylinder spark plug caps

Secondly, it is vital that the engine is not turned over or run with any of the plug caps removed, and that the plugs are soundly earthed (grounded) when the system is checked for sparking. The ignition system components can be seriously damaged if the HT circuit becomes isolated.

1 As no means of adjustment is available, any failure of the system can be traced to failure of a system component or a simple wiring fault. Of the two possibilities, the latter is by far the most likely. In the event of failure, check the system in a logical fashion, as described below.

2 To gain access to the rear cylinder spark plugs, raise and support the fuel tank (see Chapter 4).

3 To gain access to the front cylinder plugs, remove the lower and inner fairing panels (see Chapter 8). Unscrew the oil cooler mounting bolts and position the cooler clear of the spark plug caps, taking care not to lose the collars from the mounting rubbers **(see illustration)**.

4 Ensure the ignition is switched off then disconnect the fuel pump wiring connector from the base of the fuel tank.

5 Disconnect the plug caps from the spark plugs. Connect each cap to a spare spark plug and lay each plug on the engine with the threads contacting the engine. If necessary, hold each spark plug with an insulated tool.

2.8 With the tester securely connected between a plug cap and earth, sparks should jump between the nail ends when the engine is cranked over

A simple spark gap testing tool can be made from a block of wood, a large alligator clip and two nails, one of which is fashioned so that the spark plug cap or bare HT lead end can be connected to it. Make sure the gap between the two nail ends is the same as specified

⚠️ **Warning:** Do not remove any of the spark plugs from the engine to perform this check – atomised fuel being pumped out of the open spark plug hole could ignite, causing severe injury!

6 Having observed the above precautions, check that the kill switch is in the RUN position and the transmission is in neutral, then turn the ignition switch ON and turn the engine over on the starter motor. If the system is in good condition a regular, fat blue spark should be evident at each plug electrode. If the spark appears thin or yellowish, or is non-existent, further investigation will be necessary. Before proceeding further, turn the ignition OFF and remove the key as a safety measure.

7 The ignition system must be able to produce a spark which is capable of jumping a particular size gap. Honda do not provide a specification, but a healthy system should produce a spark capable of jumping at least 6 mm. A simple testing tool can be made to test the minimum gap across which the spark will jump (see **Tool Tip**) or alternatively it is possible to buy an ignition spark gap tester tool and some of these tools are adjustable to alter the spark gap.

8 Connect the plug cap of the number 1 coil to the protruding electrode on the test tool, and clip the tool to a good earth (ground) on the engine or frame **(see illustration)**. Check that the transmission is in neutral and the kill switch is in the RUN position, then turn the ignition switch ON and turn the engine over on the starter motor. If the system is in good condition a regular, fat blue spark should be seen to jump the gap between the nail ends. Repeat the test for the number 2 to 4 coils. If the test results are good the entire ignition system can be considered good. If the spark appears thin or yellowish, or is non-existent, further investigation is required.

9 Ignition faults can be divided into two categories, namely those where the ignition system has failed completely, and those which are due to a partial failure. The likely faults are listed below, starting with the most probable source of failure. Work through the list systematically, referring to the subsequent sections for full details of the necessary checks and tests. **Note:** *Before checking the following items ensure that the battery is fully charged and that all fuses are in good condition.*

  a) Loose, corroded or damaged wiring connections, broken or shorted wiring between any of the component parts of the ignition system (see Chapter 9).
  b) Faulty HT lead or spark plug cap, faulty spark plug, dirty, worn or corroded plug electrodes, or incorrect gap between electrodes.
  c) Faulty ignition (main) switch or engine kill switch (see Chapter 9).
  d) Faulty neutral, clutch or sidestand switch (see Chapter 9).
  e) Faulty ignition pulse generator coil or damaged rotor (see Chapter 4 – if this is the problem the fuel injection system warning light will be illuminated).
  f) Faulty ignition HT coil(s).
  g) Faulty engine control module (see Chapter 4).

10 If the above checks don't reveal the cause of the problem, have the ignition system tested by a Honda dealer.

11 On completion, securely reconnect the plug caps to the plugs then reconnect the fuel pump wiring connector and lower the fuel tank back down into position (see Chapter 4). Ensure the collars are in position then refit the oil cooler mounting bolts and tighten securely before installing the fairing panels (see Chapter 8).

3 Ignition HT coils – check, removal and installation

Check

1 Carry out the operations described in Paragraphs 2 to 5 of Section 2.

2 Remove the fairing left and right side lower panels (see Chapter 8) to gain access to number 2 and 4 ignition coils.

3 Starting with number 1 coil (see below for coil location details), connect the positive (+) lead of the voltmeter and peak voltage adaptor arrangement* to the blue/black terminal of the coil (wiring connector still securely connected) and connect the negative (–) lead to a suitable earth (ground) point. *****Note:** *Honda specify their own Imrie diagnostic tester (model 625), or the peak voltage adaptor (Pt. No. 07HGJ-0020100) with an aftermarket digital multimeter having an impedance of 10 M-ohm/DCV minimum, for this test.*

4 Check that the kill switch is in the RUN position and the transmission is in neutral, then turn the ignition switch on. Note the initial

Ignition system 5•3

3.11 Disconnecting number 1 cylinder plug cap

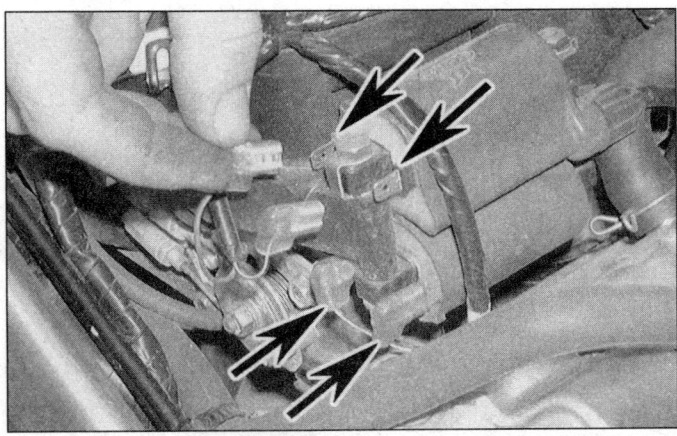

3.12 Note the correct locations of the wiring connectors before disconnecting them from the coil terminals (arrows)

voltage reading on the meter then turn the engine over on the starter motor and note the ignition coil peak voltage reading obtained. Once both readings have been noted, turn the ignition switch off and disconnect the meter.

5 Connect the positive (+) lead of the meter and peak voltage adaptor arrangement to the yellow/white terminal of the number 2 coil (wiring connector still securely connected) and connect the negative (–) lead to a suitable earth (ground) point. Repeat the check in Paragraph 4.

6 Connect the positive (+) lead of the meter and peak voltage adaptor arrangement to the red/blue terminal of the number 3 coil (wiring connector still securely connected) and connect the negative (–) lead to a suitable earth (ground) point. Repeat the check in Paragraph 4.

7 Connect the positive (+) lead of the meter and peak voltage adaptor arrangement to the red/yellow terminal of the number 4 coil (wiring connector still securely connected) and connect the negative (–) lead to a suitable earth (ground) point. Repeat the checks in Paragraph 4.

8 Compare the results with the Specifications at the beginning of this Chapter. If the initial voltage reading is not as specified or any of the peak voltage readings are lower than the specified minimum, then there is a fault in the ignition system circuit (see Section 2). **Note:** *The peak voltage readings for each coil may vary but each one must exceed the specified minimum.*

9 If the initial and peak voltage readings are as specified and the plug does not spark, then the ignition HT coil, HT lead or plug cap are faulty (the plug caps and leads are available separately). In order to determine conclusively that an ignition coil is defective, it should be tested by a Honda dealer. If the coil is confirmed to be faulty, it must be renewed; the coil is a sealed unit and cannot therefore be repaired.

Removal

Note: *The spark plug caps and leads are available separately and can be renewed without removing the coil. The spark plug cap is a screw-fit in the end of the lead and the lead is secured to the coil by a knurled retaining ring as well as being screwed onto the coil output terminal.*

Number 1 and 3 cylinder coils

10 Raise and support the fuel tank (see Chapter 4) to gain access to the coils which are mounted on the rear of the air filter housing; the lower coil is number 1 coil and the upper coil number 3.

11 Disconnect the spark plug caps from the number 1 and 3 plugs **(see illustration)**.

12 Note the correct fitted locations of the primary wires (note them on a piece of paper if necessary to avoid confusion on installation), then disconnect them from the ignition coil terminals **(see illustration)**.

13 Slacken and remove the nuts, mounting bolts and spacers and remove both coils from the air filter housing **(see illustrations)**.

Number 2 and 4 cylinder coils

14 Remove the fairing lower panels and the inner panel (see Chapter 8). Number 2 coil is mounted on the left side of the frame and number 4 on the right side.

15 Unscrew the oil cooler mounting bolts and position the cooler clear of its mounting **(see illustration 2.3)**, taking care not to lose the collars from mounting rubbers. Disconnect the spark plug cap from its plug.

3.13a Unscrew the mounting nuts . . .

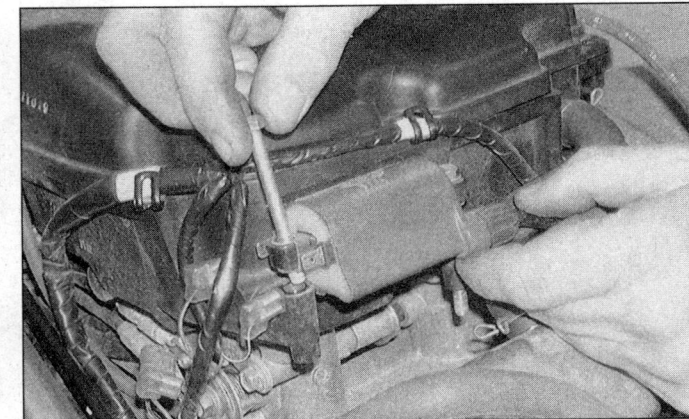

3.13b . . . then withdraw the bolts and remove number 1 and 3 HT coils from the rear of the air filter housing

5•4 Ignition system

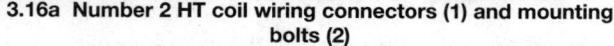

3.16a Number 2 HT coil wiring connectors (1) and mounting bolts (2)

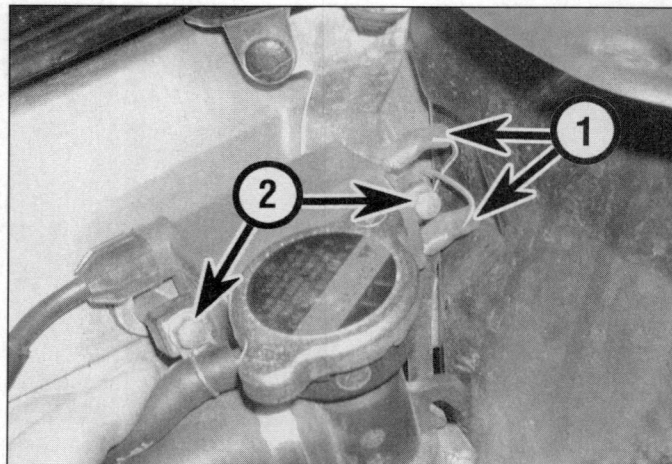

3.16b Number 4 HT coil wiring connectors (1) and mounting bolts (2)

16 Note the correct fitted locations of the primary wires (note them on a piece of paper if necessary to avoid confusion on installation), then disconnect them from the ignition coil terminals (see illustrations).

17 Slacken and remove the mounting bolts and spacers and remove the coil from the frame.

Installation

18 Installation is the reverse of removal. Make sure the wiring connectors and HT leads are securely connected.

4 Ignition timing – general information and check

General information

1 Since no provision exists for adjusting the ignition timing and since no component is subject to mechanical wear, there is no need for regular checks; only if investigating a fault such as a loss of power or a misfire, should the ignition timing be checked.

2 The ignition timing is checked dynamically

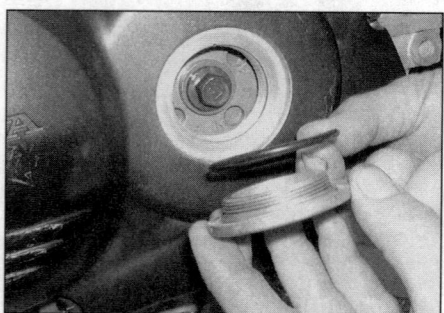

4.5 Unscrew the inspection plug from the right crankcase cover and discard the O-ring, as new one should be used on installation

(engine running) using a stroboscopic lamp. The inexpensive neon lamps should be adequate in theory, but in practice may produce a pulse of such low intensity that the timing mark remains indistinct. If possible, one of the more precise xenon tube lamps should be used, powered by an external source of the appropriate voltage. Note: Do not use the machine's own battery as an incorrect reading may result from stray impulses within the machine's electrical system.

Check

Note: The ignition timing can be checked on each individual cylinder as marks are provided for all four cylinders. The following check describes the check for number 1 cylinder; if necessary repeat the check on the other cylinders.

3 Raise and support the fuel tank (see Chapter 4) and remove the fairing right-side lower panel (See Chapter 8).

4 Warm the engine up to normal operating temperature then stop it.

5 Unscrew the timing inspection plug from the right crankcase cover. Discard the cover O-ring; a new one should be used on installation (see illustration).

6 The timing marks are stamped on the ignition pulse generator rotor. There are four sets of marks, one for each cylinder. On European and US (except California) models the timing mark is the line next to the F mark of the relevant cylinder. On California models the timing mark is the set of punch marks located between the F and T marks of the relevant cylinder (see illustration).

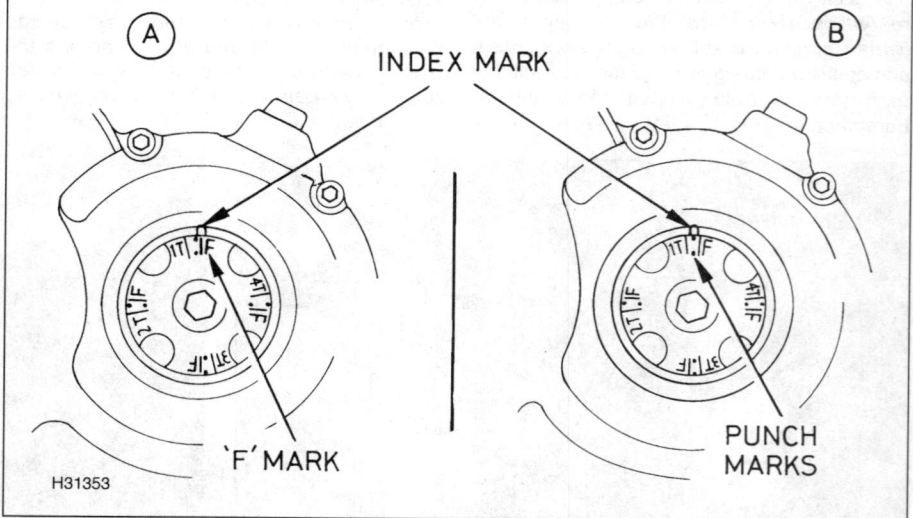

4.6 Timing mark details

A UK and US (except California) models - the line next to the F mark must be aligned with the cover index mark (arrows)

B California models - the punch marks in between the F and T marks must be aligned with the cover index mark (arrows)

Ignition system 5•5

 HAYNES HiNT *The timing marks can be highlighted with white paint to make them more visible under the stroboscope light.*

7 Connect the timing light to the number 1 cylinder HT lead as described in the manufacturer's instructions.
8 Start the engine and aim the light at the static timing mark.
9 With the machine idling at the specified speed, the rotor timing mark should align with the index mark, in the form of a cut-out or line, on the top of the crankcase cover aperture **(see illustration)**.
10 Slowly increase the engine speed whilst observing the timing mark. The timing mark should move anti-clockwise, increasing in relation to the engine speed until it reaches full advance (no identification mark).
11 As already stated, there is no means of adjusting the ignition timing on these machines. If the ignition timing is incorrect, or suspected of being incorrect, there is a fault in the ignition system, and the system must be tested as described in the preceding Sections of this Chapter.
12 When the check is complete, fit a new O-ring to the inspection plug. Lubricate the plug threads with a smear of multi-purpose grease then fit it to the crankcase cover and tighten to the specified torque **(see illustration)**.
13 Disconnect the timing light, then lower the fuel tank back down and install the fairing lower panel (see Chapters 4 and 8).

5 Immobiliser (HISS) system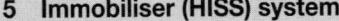

General information

1 The Honda Ignition Security System (HISS) will only allow the machine to be started if the correct registered key is used to turn the ignition on. The system consists of a transponder which is part of the ignition key, a receiver which is fitted around the ignition switch, and the electronic control module (ECM) **(see illustration)**.
2 When the ignition is switched on, the ECM sends power through the receiver to the

5.1 Location of the immobiliser receiver

4.9 On UK and US (except California) models point the timing light at the rotor and check the F mark line is aligned with the index mark (arrows)

transponder. The transponder sends a coded signal back through the receiver to the ECM. If the signal sent by the transponder matches the signal stored in the ECM memory, the immobiliser indicator light in the speedometer (marked by a key symbol) comes on for two seconds, then goes out, and the ECM allows the engine to be started. If the key code signal is not recognised, or if there is a fault in the system, the indicator light stays on. If the light stays on, refer to the *Fault diagnosis* and *Troubleshooting* sections below. Likewise if the light does not come on at all.
3 The ECM can store the codes for up to four registered keys. They keys should be kept separately (ie. not on the same key-ring) as the proximity of another key to the one being used in the switch can lead to the signal from it being jammed, and the bike will not start. The key has a built in transponder which can be damaged if the key is dropped or knocked, gets too hot, is too close to a magnetic object, or is submerged in water for too long. If all the keys are lost, the ECM will have to be replaced with a new one, so always make sure you have at least one spare key. If a new key is obtained, it must be registered into the system before the bike can be started.

Key registration procedure

To register a new key

Note: *To do this you will need the Honda special tool (Part No. 07XMZ-MBW0100) which is a wiring loom adapter that plugs into the ignition pulse generator wiring connector. If this tool is not available, registration must be carried at a Honda dealer with the special tool.*
4 Obtain a new key from a Honda dealer, then have it cut to match the original key.
5 Raise or remove the fuel tank (see Chapter 4), then disconnect the ignition pulse generator (red) wiring connector which is inside the protective cover on the right side of the frame (see Chapter 4, Section 11). Connect the special tool to the ECM side of the connector, then connect the red clip on the tool to the battery positive (+) terminal and the green clip to the battery negative (–) terminal.
6 Turn the ignition on using your original key.

4.12 On completion, fit a new O-ring and tighten the inspection plug to the specified torque

The immobiliser indicator light should come on and stay on. **Note:** *If the light starts to flash after ten seconds, there is a fault in the system which will have gone into diagnostic mode. The sequence of flashes is the fault code which should be matched with the fault (see Paragraph 37).*
7 Disconnect the red clip from the battery positive terminal, leave it disconnected for at least two seconds, then reconnect it. The indicator should now come on for two seconds, then begin to flash four times repeatedly. This indicates that the system is in registration mode. At this point the registrations of all keys except the one in the switch will have been cancelled, so if you have another spare apart from the new one you want to register, this will also have to be registered.
8 Turn the ignition off and remove the original key, placing it well away from the receiver.
9 Insert the new key into the switch and turn it on. The indicator should now come on for two seconds, then begin to flash four times repeatedly. This indicates that the system has registered the new key. Turn the ignition off and remove the key.
10 To register any other spare keys that will have been cancelled, repeat Steps 8 and 9. Up to four keys can be registered.
11 On completion turn the ignition off, then remove the special tool and reconnect the ignition pulse generator wiring connector. Now turn the ignition on using any of the registered keys to return the system to normal mode.
12 Check that all registered keys can start the motorcycle.

To register new keys with a new ignition switch

Note: *To do this you will need the Honda special tool (Part No. 07XMZ-MBW0100) which is a wiring loom adapter that plugs into the ignition pulse generator wiring connector. If this tool is not available, registration must be carried at a Honda dealer with the special tool.*
13 Obtain a new switch and two (or more if required) new keys.
14 Remove the faulty switch (see Chapter 9), but retain the receiver to fit with the new switch.

5•6 Ignition system

15 Raise or remove the fuel tank (see Chapter 4), then disconnect the ignition pulse generator wiring connector (see Paragraph 5). Connect the special tool to the ECM side of the connector, then connect the red clip on the tool to the battery positive (+) terminal and the green clip to the battery negative (–) terminal.

16 Place one of the original registered keys from the faulty switch next to the receiver.

17 Connect the new ignition switch to the connector in the wiring loom, but keep it away from the receiver. Turn the new switch on with one of the new keys. The immobiliser indicator light should come on and stay on, which means the ECM recognises the old key that is next to the receiver. If there is a fault in the system the light will start to flash (see Paragraph 37).

18 Disconnect the red clip from the battery positive terminal and leave it disconnected for at least two seconds, then reconnect it. The indicator should now come on for two seconds, then begin to flash four times repeatedly. This indicates that the system is in registration mode. At this point the registrations of all keys except the one near the receiver will have been cancelled.

19 Turn the ignition off and remove the new key.

20 Install the new ignition switch and fit the receiver onto it (see Chapter 9).

21 Insert the new key and turn the switch on. The indicator should now come on for two seconds, then begin to flash four times repeatedly. This indicates that the system has registered the new key. Turn the ignition off and disconnect the red clip from the battery positive terminal.

22 Turn the ignition on using the newly registered key. The indicator light should come on for two seconds, then go off.

23 Turn the ignition off and reconnect the red clip to the battery positive terminal.

24 Turn the ignition on using the newly registered key. The indicator light should come on and stay on. Now disconnect the red clip from the battery positive terminal and leave it disconnected for at least two seconds, then reconnect it. The indicator should now come on for two seconds, then begin to flash four times repeatedly. This indicates that the system is in registration mode. At this point the registrations of all old keys for the faulty switch are cancelled.

25 Turn the ignition off and remove the new key, placing it well away from the receiver.

26 Insert the second new unregistered key and turn the ignition on. The indicator should now come on for two seconds, then begin to flash four times repeatedly. This indicates that the system has registered the second new key. Turn the ignition off and remove the key.

27 To register any other new keys, repeat Step 26. Up to four keys can be registered.

28 On completion turn the ignition off, disconnect the special tool and reconnect the ignition pulse generator wiring connector. Now turn the ignition on using any of the registered keys to return the system to normal mode.

29 Check that all newly registered keys can start the motorcycle.

To register new keys with a new ECM

30 Obtain a new ECM along with two (or more if required) new keys. Install the ECM (see Chapter 4). Have the keys cut to match the original key for your ignition switch.

31 Insert a new key into the switch and turn it on. The indicator should now come on for two seconds, then begin to flash four times repeatedly. This indicates that the system has registered the new key. If there is a fault in the system the light will start to flash (see Paragraph 37).

32 Turn the ignition off and remove the key.

33 Insert the second new key and turn the ignition on. The indicator should now come on for two seconds, then begin to flash four times repeatedly. This indicates that the system has registered the second new key.

34 Turn the ignition off and remove the key.

35 The new ECM will only register two new keys at this stage. If you have a third key that you want to register, refer to Steps 4 to 11 to register it, noting that you will need the special tool mentioned.

36 Check that both newly registered keys can start the motorcycle.

Fault diagnosis

Note: *To enter the fault diagnosis mode of the system you will need the Honda special tool (Part No. 07XMZ-MBW0100) which is a wiring loom adapter that plugs into the ignition pulse generator wiring connector. If this tool is not available, fault diagnosis must be carried at a Honda dealer with the special tool.*

37 There are two fault diagnosis modes, one for a fault which occurs during normal use, and one for a fault that occurs when registering a new key. Make sure you refer to the correct table when matching the fault code.

38 If the indicator light has come on and stayed on during normal use, raise or remove the fuel tank (see Chapter 4). Disconnect the ignition pulse generator (red) wiring connector (see Paragraph 5). Connect the special tool to the ECM side of the connector, then connect the red clip on the tool to the battery positive (+) terminal and the green clip to the battery negative (–) terminal.

39 Turn the ignition switch on. The indicator light in the speedometer will come on for ten seconds, then start to flash. This means it has entered diagnostic mode, and the sequence of the flashes indicates the fault that has occurred. The pattern repeats continuously. Match the pattern with the fault below, making sure you refer to the relevant table. If the indicator stays on after ten seconds and does not flash, then there is no fault logged in the system.

Troubleshooting procedure

Indicator light does not come on when ignition switched on

40 Check the fuses (see Chapter 9).

41 If the fuses are good, remove the fairing (see Chapter 8) to access the instrument cluster grey 10-pin connector (see Chapter 9, Section 15).

42 Pull back the boot on the connector but do not disconnect it. Turn the ignition on and use a voltmeter to check for battery voltage between the brown/blue and green wire

Fault is indicated during normal use		
Flash pattern	Fault	Solution
Two short, one long, one short	Faulty ECM	Install new ECM
Two short, two long	Faulty receiver or wiring	Follow Troubleshooting procedure below
One long, three short	Signal jammed by other key	Place other key well away from receiver
One long, two short, one long	Signal jammed by other key	Place other key well away from receiver

Fault is indicated during key registration		
Flash pattern	Fault	Solution
One short, one long, one short, one long	Key already registered	Use a new or cancelled key
Two short, two long	Faulty receiver or wiring	Follow Troubleshooting procedure below
One short, one long, two short	Key already registered on old ECM	Use a new key

Ignition system 5•7

5.42 Checking for battery voltage in the instrument cluster wiring connector

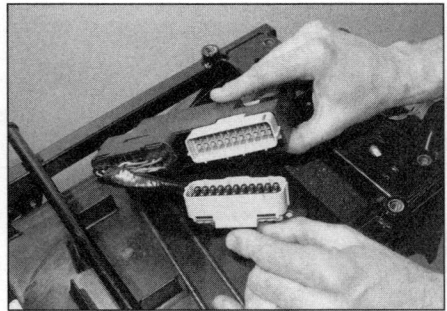

5.50 Check the ECM and connector terminals for damage and corrosion

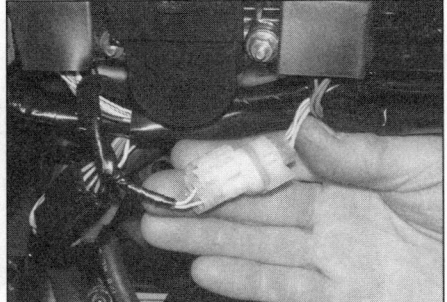

5.52 Immobiliser receiver wiring connector

terminals in the connector **(see illustration)**. Turn the ignition off.

43 If there is no voltage, check for continuity in the wiring (see *Wiring Diagrams*, Chapter 9). If there is voltage, test the operation of the ECM (see Paragraph 49). If the ECM signal is good, the indicator light unit is faulty and must be replaced with a new one (see Chapter 9).

Indicator light stays on when ignition switched on

44 The indicator light should go out after approximately two seconds. If the light stays on, check that none of the other registered keys are close to the receiver. If they are, remove them and try the ignition again.

45 Turn the ignition on with a spare key and check the indicator light, which should come on for two seconds, then go out. If it does, the first key is faulty. If it doesn't, perform the fault diagnosis procedure described above (see Steps 37 to 39). If a fault code is displayed, use the appropriate table to determine the fault and the solution.

46 If no fault code is displayed, or the system does not go into fault diagnosis mode, remove the fairing (see Chapter 8) to access the instrument cluster grey 10-pin connector (see Chapter 9, Section 15).

47 Pull back the boot on the connector but do not disconnect it. Connect the voltmeter probes between the red/white and green wire terminals and turn the ignition on **(see illustration 5.42)**. There should be no voltage for approximately two seconds, then the meter should register battery voltage.

48 If there is no voltage at all, check for continuity in the wiring (see *Wiring Diagrams*, Chapter 9), then check the operation of the ECM (see Paragraph 46).

Operation of the ECM

49 Disconnect the ECM wiring connectors (see Chapter 4, Section 11). Turn the ignition on and using a voltmeter, check for battery voltage between the white/red wire terminal in the black connector and earth (ground). Turn the ignition off. Now turn the ignition on and check for battery voltage between the black/white wire terminal in the grey connector and earth (ground). Finally check for continuity between the green wire terminal and earth (ground).

50 If voltage is present and the wiring is good, check the ECM connectors for loose, damaged or corroded terminals **(see illustration)**. If the connectors are good, then the ECM could be faulty, and should be tested by a Honda dealer.

Ignition pulse generator wiring

51 Disconnect the ignition pulse generator (red) wiring connector (see Chapter 4, Section 11) then check for continuity in the yellow and white/yellow wires between the ECM and the ignition pulse generator (see *Wiring Diagrams*, Chapter 9). If there is no continuity, trace the fault and repair or replace the wiring as necessary. If there is continuity, the ECM could be faulty and should be tested by a Honda dealer.

Faulty immobiliser receiver

52 If the 'two short, two long' flash pattern has been indicated during the fault diagnosis procedure, remove the fairing to access the receiver (see Chapter 8). Trace the wiring from the receiver and disconnect it at the white 4-pin connector **(see illustration)**. Using a voltmeter, connect the positive (+) probe to the yellow/red wire terminal on the loom side of the connector and the negative (–) probe to earth (ground). Turn the ignition on – there should be approximately 5 volts present. If there is no voltage, check for continuity in the yellow/red wire between the ECM and the receiver, and repair or replace the wiring if there is no continuity.

53 If there is voltage, check for continuity to earth (ground) in the green/orange wire, and repair or replace the wiring if there is no continuity.

54 If the wiring is good, using a voltmeter, connect the positive (+) probe to the pink wire terminal on the loom side of the receiver connector and the negative (–) probe to earth (ground). Turn the ignition on – there should be approximately 5 volts present. If there is, the receiver is faulty.

55 If there is no voltage, check for continuity in the orange/blue and pink wires between the ECM and the receiver, and repair or replace the wiring if there is no continuity between the connectors, or if there is continuity in either to earth (ground). If the wiring is good, the receiver is faulty.

Renewal

56 To renew the receiver, remove the fairing (see Chapter 8). Trace the wiring from the receiver and disconnect it at the white 4-pin connector **(see illustration 5.52)**. Undo the screws securing the receiver to the ignition (main) switch and remove the receiver, noting how it fits **(see illustration)**.

57 To renew the ECM, see Chapter 4, Section 11.

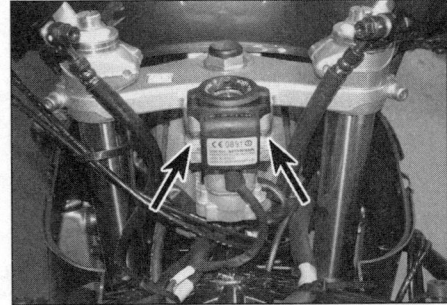

5.56 Receiver is secured by two screws (arrows)

Chapter 6
Frame, suspension and final drive

Contents

Drive chain – removal, cleaning and installation	15
Drive chain and sprockets – check, adjustment and lubrication	see Chapter 1
Footrests and brackets – removal and installation	3
Forks – disassembly, inspection and reassembly	7
Forks – removal and installation	6
Frame – inspection and repair	2
General information	1
Handlebar switches – check	see Chapter 9
Handlebar switches – removal and installation	see Chapter 9
Handlebars – removal and installation	5
Rear shock absorber – removal, inspection and installation	11
Rear sprocket coupling – inspection and overhaul	17
Rear suspension linkage – removal, inspection and installation	10
Sidestand – check	see Chapter 1
Sidestand – lubrication	see Chapter 1
Sidestand switch – check and replacement	see Chapter 9
Sprockets – check and renewal	16
Stands – removal and installation	4
Steering head bearings – freeplay check and adjustment	see Chapter 1
Steering head bearings – inspection and renewal	9
Steering head bearings – lubrication	see Chapter 1
Steering stem – removal and installation	8
Suspension – adjustments	12
Suspension – check	see Chapter 1
Swingarm – inspection and bearing renewal	14
Swingarm – removal and installation	13
Swingarm and suspension linkage bearings – lubrication	see Chapter 1

Degrees of difficulty

Easy, suitable for novice with little experience 	**Fairly easy,** suitable for beginner with some experience	**Fairly difficult,** suitable for competent DIY mechanic	**Difficult,** suitable for experienced DIY mechanic 	**Very difficult,** suitable for expert DIY or professional

Specifications

Front forks
Fork oil type
    UK models .... Honda fork fluid
    US models .... Pro Honda suspension fluid SS-8
Fork oil capacity .... 457 ± 2.5 cc
Fork oil level* .... 130 mm
Fork spring free length
    Standard .... 382.7 mm
    Service limit .... 375.0 mm
Fork tube runout limit .... 0.2 mm
*Oil level is measured from the top of the tube with the fork spring removed, the leg fully compressed and the damper rod fully inserted.

Final drive
Chain type .... RK 50MF0Z3 or DID 50VA7 (108 links, split)
Joining link pin projection from side plate (unstaked) .... 1.2 to 1.4 mm
Joining link staked ends diameter .... 5.55 to 5.85 mm

6•2 Frame, suspension and final drive

Torque settings

Bottom yoke fork clamp bolt	49 Nm
Centrestand pivot bolts	54 Nm
Fork damper assembly bolt	20 Nm
Fork top cap damper rod locknut	20 Nm
Fork top cap	23 Nm
Front sprocket bolt	51 Nm
Handlebar clamp bolt	26 Nm
Handlebar end weight screw	10 Nm
Master cylinder mounting clamp bolts	12 Nm
Passenger footrest bracket bolts	
Left side bracket	31 Nm
Right side bracket	32 Nm
Rear shock absorber	
Upper mounting nut	42 Nm
Upper mounting bracket pivot bolt nut	42 Nm
Lower mounting bolt nut	42 Nm
Rear sprocket nuts	34 Nm
Rear sub-frame to main frame bolts	44 Nm
Sidestand	
Pivot bolt	10 Nm
Pivot bolt locknut	29 Nm
Bracket throughbolt nut	39 Nm
Switch mounting bolt	10 Nm
Silencer mounting bolt	21 Nm
Steering head bearing adjuster nut	25 Nm
Steering stem nut	103 Nm
Suspension linkage pivot bolts	
Linkage to swingarm bolt	44 Nm
Linkage to shock absorber bolt	42 Nm
Linkage to connecting link bolt	44 Nm
Connecting link to mounting bracket bolt	59 Nm
Swingarm air guide bolts	9 Nm
Swingarm drive chain slider bolts	9 Nm
Swingarm pivot bolt nut	93 Nm
Top yoke fork clamp bolt	23 Nm

1 General information

All models use a twin spar box-section aluminium frame which uses the engine as a stressed member.

Front suspension is by a pair of oil-damped telescopic forks which have a conventional damper system and are adjustable for spring pre-load.

At the rear, an alloy swingarm acts on a single shock absorber via a three-way linkage. The shock absorber is adjustable for spring pre-load and rebound damping. The swingarm is mounted onto the rear of the engine.

The drive to the rear wheel is by chain.

2 Frame – inspection and repair

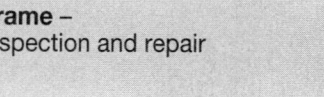

1 The frame should not require attention unless accident damage has occurred. In most cases, frame renewal is the only satisfactory remedy for such damage. A few frame specialists have the jigs and other equipment necessary for straightening the frame to the required standard of accuracy, but even then there is no simple way of assessing to what extent the frame may have been over stressed.

2 After the machine has accumulated a lot of miles, the frame should be examined closely for signs of cracking or splitting at the welded joints. Loose engine mount bolts can cause ovaling or fracturing of the mounting tabs. Minor damage can often be repaired by specialist welding, depending on the extent and nature of the damage.

3 Remember that a frame which is out of alignment will cause handling problems. If misalignment is suspected as the result of an accident, it will be necessary to strip the machine completely so the frame can be thoroughly checked.

3 Footrests and brackets – removal and installation

Rider's footrest

Removal

1 Remove the split pin and washer from the footrest pivot pin, then withdraw the pivot pin and remove the footrest, noting the fitting of the return spring.

Installation

2 Installation is the reverse of removal. Use a new split pin and bend its ends securely.

Passenger footrests

Removal

3 Remove the split pin and washer from the footrest pivot pin.

4 Withdraw the pivot pin and carefully remove the footrest from the bracket, noting the correct fitted position of the footrest detent plate, ball and spring.

Installation

5 Installation is the reverse of removal ensuring the detent plate, ball and spring are all correctly positioned. Check the detent mechanism functions correctly then secure the pivot pin in position with a new split pin.

Rider's right side footrest bracket assembly

Removal

6 Remove the seat cowling (see Chapter 8).

7 Trace the wiring back from the rear brake light switch and disconnect it at the wiring connectors (see illustration). Free the switch

Frame, suspension and final drive 6•3

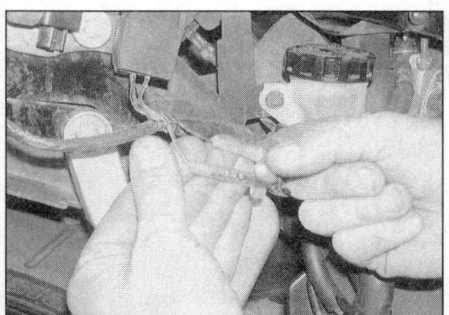

3.7 Disconnecting the rear brake light switch wiring connectors

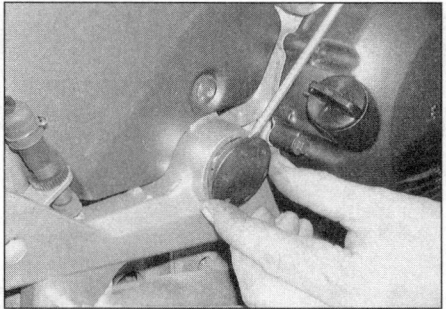

3.10a Prise out the cap from the right side rider's footrest bracket . . .

3.10b . . . and unscrew the nut from the swingarm pivot bolt

wiring from any relevant clips or ties so the switch is free to be removed with the bracket.
8 Remove the split pin and withdraw the clevis pin securing the rear brake master cylinder pushrod to the brake pedal.
9 Slacken and remove the mounting bolts and nut and free the master cylinder from the footrest bracket. **Note:** *There is no need to disconnect the hydraulic hoses from the cylinder.*
10 Remove the swingarm pivot bolt cap from the right side rider's footrest bracket then slacken and remove the nut from the pivot bolt **(see illustrations)**.
11 Slacken and remove the right side rider's footrest bracket heatshield upper bolt **(see illustration)**. Free the complete footrest assembly from the swingarm pivot bolt, taking care not to lose the collar from the rear of the heatshield, and remove it from the bike **(see illustration)**.
12 If necessary, unhook the return spring and brake light switch spring from the brake pedal then remove the circlip and washer and remove the brake pedal from the bracket.

Installation

13 If the brake pedal was removed, apply a smear of multi-purpose grease to the bracket pivot pin then install the pedal and washer and secure them in position with the circlip. Ensure the circlip is correctly located in its groove then hook the return spring and brake light switch spring back into position.
14 Ensure the collar is fitted to the rear of the heatshield upper mounting then locate the footrest bracket on the swingarm pivot bolt and its lower locating bolt. Fit the swingarm pivot bolt nut and tighten it to the specified torque then fit the pivot bolt cap to the bracket. Refit the heatshield upper bolt and tighten securely.
15 Locate the master cylinder on the bracket and fit its mounting bolts and nut, tightening them securely. Align the pushrod with the pedal then fit the clevis pin and secure it in position with a new split pin.
16 Ensure the rear brake light switch wiring is correctly routed and retained by all the necessary clips then reconnect it to the main wiring harness.

17 Refit the seat cowling (see Chapter 8). Check the operation of the rear brake and brake light before riding the bike on the road.

Rider's left side footrest bracket assembly

Removal

18 Position the motorcycle on its centrestand then support the rear wheel with a block of wood positioned between the tyre and ground; this will prevent the swingarm moving when the pivot bolt is withdrawn.
19 Remove the swingarm pivot bolt caps from the rider's footrest brackets.
20 Slacken and remove the swingarm pivot bolt nut then carefully withdraw the pivot bolt, complete with the left footrest bracket assembly. Slide off the footrest bracket then insert the pivot bolt back into position to securely locate the swingarm whilst the bracket is removed.

Installation

21 Withdraw the pivot bolt and apply a light coating of multi-purpose grease to its shaft.
22 Slide the bracket onto the pivot bolt then insert the bolt back through the swingarm and crankcase. Ensure the footrest bracket is correctly seated on its lower locating bolt then push the pivot bolt fully home.
23 Fit the nut to the swingarm pivot bolt and tighten to the specified torque. Fit the caps to both footrest brackets then remove the support from under the rear wheel.

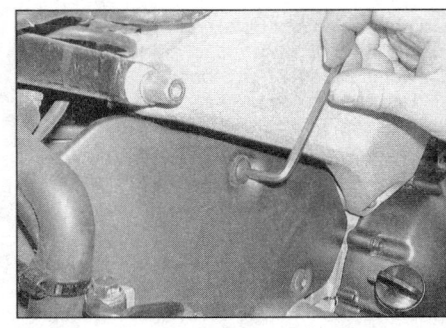

3.11a Unscrew the heatshield upper bolt . . .

Passenger right side footrest bracket assembly

Removal

24 Remove the seat cowling (see Chapter 8).
25 Slacken and remove the silencer mounting bolt nut then withdraw the bolt and washer.
26 Unscrew the two mounting bolts and remove the footrest bracket from the rear subframe, taking care not to lose the collar from the rear of the silencer mounting

Installation

27 Installation is the reverse of removal tightening the bracket and silencer mounting bolts to the specified torque.

Passenger left side footrest bracket assembly

Removal

28 Remove the seat cowling (see Chapter 8).
29 Unhook the seat lock outer cable from the rear of the subframe and detach the inner cable from the lock. Note the correct routing of the cable then free it from the subframe.
30 Unscrew the two mounting bolts and remove the footrest bracket and lock cable assembly from the subframe.

Installation

31 Installation is the reverse of removal tightening the bracket mounting bolts to the specified torque. Check the operation of the seat lock before installing the cowling.

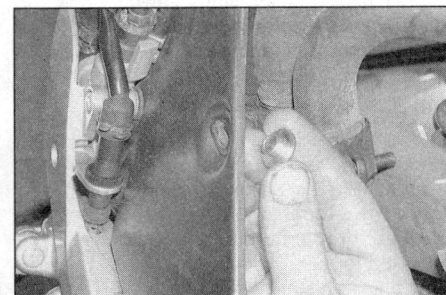

3.11b . . . then free the rider's right footrest bracket from the swingarm pivot bolt and recover the collar

6•4 Frame, suspension and final drive

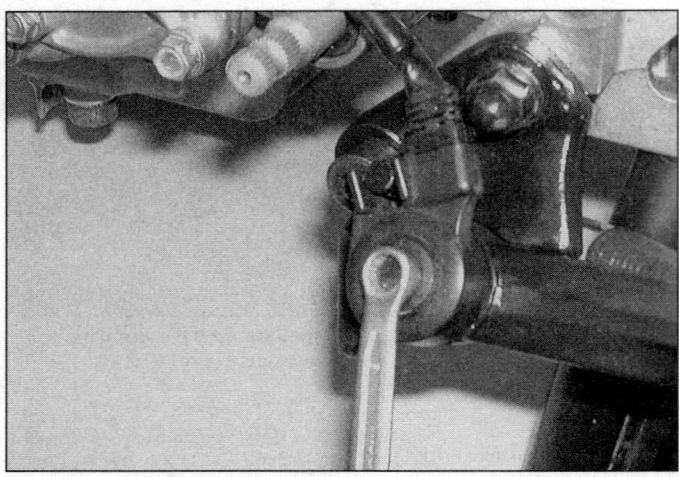

4.2 Unscrew the bolt and free the switch from the sidestand pivot

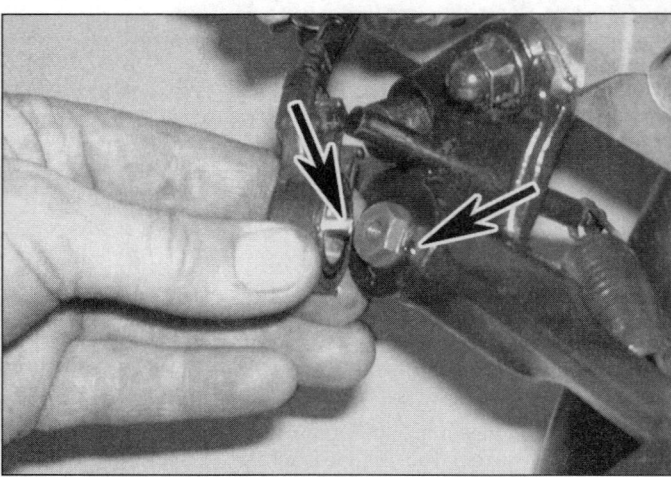

4.9 Ensure the sidestand switch pin engages correctly with the hole (arrows) on installation

4 Stands –
removal and installation

Sidestand

Note: *Honda recommend that a new sidestand switch retaining bolt is used on installation (the bolt is supplied with its threads pre-coated with locking compound). If the old bolt is to be reused, remove all original locking compound from its threads and apply fresh compound before installing it*

Removal

1 The sidestand is attached to a bracket which is mounted to the rear of the engine. A spring anchored to the stand ensures that it is held in the retracted or extended position.
2 Support the bike on the centrestand then unscrew the sidestand switch retaining bolt and free the switch from the stand **(see illustration)**. **Note:** *There is no need to disconnect the switch wiring.*
3 Carefully unhook the stand spring and remove it from the stand and bracket.
4 Unscrew the nut from the sidestand pivot bolt then remove the pivot bolt and stand from the mounting bracket.
5 If necessary, unscrew the retaining bolt and nut and remove the sidestand bracket from the centrestand/suspension linkage mounting bracket.

Installation

6 Where necessary, fit the sidestand bracket to the centrestand/suspension linkage bracket. Fit the bracket retaining bolt and tighten it securely, then fit the nut to the bracket throughbolt and tighten it to the specified torque.
7 Apply multi-purpose grease to the pivot bolt shank then fit the sidestand to its bracket and insert the pivot bolt. Tighten the pivot bolt to the specified torque then fit the nut to the bolt and tighten it to the specified torque.

8 Reconnect the sidestand spring and check that it holds the stand securely up when not in use – an accident is almost certain to occur if the stand extends while the machine is in motion.
9 Locate the switch onto the sidestand, making sure the pin locates in the hole in the sidestand, and the lug on the stand bracket locates into the cutout in the switch body **(see illustration)**. Fit the new retaining bolt and tighten it to the torque setting specified. Check the operation of the sidestand switch (see Chapter 1).

Centrestand

Note: *Honda recommend that new pivot bolts should be used on installation of the centrestand (the bolts are supplied with their threads pre-coated with locking compound). If the old bolts are to be reused, remove all original locking compound from their threads and apply fresh compound before installing them.*

Removal

10 Support the motorcycle securely on the sidestand.
11 Carefully unhook the stand springs and remove them. The spring plate can then be removed from its pivot.
12 Support the stand then unscrew the left and right side pivot bolts (noting that the bolts are of different length) and recover the plain and spring washer from each bolt.
13 Remove the centrestand and recover the large spacer which is fitted between the left stand pivot and the bracket and the smaller spacer and washer positioned between the right stand pivot and bracket.

Installation

14 Lubricate the stand pivots and spacers with multi-purpose grease. Fit the large spacer to the left side pivot and the smaller spacer and washer to the right side pivot then locate the stand on its bracket. Fit the plain washer and spring washer to each pivot bolt

then fit the bolts and tighten them to the specified torque.
15 Check the stand pivots freely then hook the spring plate onto its pivot. Install the centrestand springs and check that they hold the stand securely up when not in use – an accident is almost certain to occur if the stand falls down while the machine is in motion.

5 Handlebars –
removal and installation

Right handlebar

Note: *Honda recommend that a new end weight retaining screw is used on installation (the screw is supplied with its threads pre-coated with locking compound). If the old screw is to be reused, remove all original locking compound from its threads and apply fresh compound before installing it.*

Removal

1 Undo the retaining screw and remove the end weight from the handlebar.
2 Disconnect the wires from the brake light switch on the master cylinder assembly.
3 Unscrew the two master cylinder assembly clamp bolts and position the assembly clear of the handlebar, making sure no strain is placed on the hydraulic hose. Keep the master cylinder reservoir upright to prevent possible fluid leakage.
4 Unscrew the two handlebar switch screws then free the switch from the handlebar.
5 Carefully prise off the circlip from the top of the right fork tube.
6 Slacken the clamp bolt then free the handlebar from the top of the fork tube and the throttle twistgrip and remove it from the bike.
7 To remove the inner weight assembly from the handlebar, unhook the retaining ring tab from the handlebar hole and spray a silicone-

Frame, suspension and final drive 6•5

based aerosol lubricant around the inside of the handlebar tube. Refit the end weight securely to the inner weight and use it to withdraw the inner weight assembly, complete with its rubber seats. Discard the retaining ring; a new one should be used on refitting.

Installation

8 Where the inner weight has been removed, ensure the rubber seats are correctly positioned in the weight grooves, then fit the new retaining ring to the outer end of the assembly. Fit the end weight to the inner weight and lubricate the rubber seats with a silicone-based lubricant spray. Insert the weight assembly into the handlebar tube until one of the retaining ring tabs locates correctly in the handlebar hole, then remove the end weight **(see illustration)**.
9 Lubricate the bearing surfaces of the throttle twistgrip and handlebar with a smear of multi-purpose grease.
10 Insert the handlebar into the throttle twistgrip then locate it on the fork tube. Ensure the handlebar lug is correctly located in the upper yoke slot, then tighten the handlebar clamp bolt to the specified torque.
11 Fit the circlip to the top of the fork tube making sure it is correctly seated in the tube groove.
12 Ensure the locating pin in the lower half of the switch is correctly seated in the hole in the underside of the handlebar then assemble the switch and securely tighten its screws, tightening the front screw first.
13 Locate the master cylinder on the handlebar and fit its mounting clamp with its UP mark facing upwards. Align the outer corner of the master cylinder mounting with the punch mark on the top of the handlebar, then tighten the mounting clamp upper bolt to the specified torque followed by the lower bolt.
14 Locate the end weight in the handlebar then fit the new retaining screw and tighten it to the specified torque.
15 Reconnect the wiring connectors to the front brake light switch. Check the operation of the switch and throttle twistgrip before using the bike.

Left handlebar

Note: *If the handlebar end weight is to be removed, note that Honda recommend that a new retaining screw should be used on installation (the screw is supplied with its threads pre-coated with locking compound). If the old screw is to be reused, remove all original locking compound from its threads and apply fresh compound before installing it. The choke lever is fitted to VFR800FI-W and X models only.*

Removal

16 Disconnect the wires from the clutch switch on the master cylinder assembly.
17 Unscrew the two master cylinder assembly clamp bolts and position the assembly clear of the handlebar, making sure no strain is placed on the hydraulic hose. Keep the master cylinder reservoir upright to prevent possible fluid leakage.
18 Unscrew the two handlebar switch screws and position the switch clear of the handlebar.
19 Detach the choke cable (where fitted) from the lever. On VFR800FI-Y and 1 models a spacer is fitted in place of the choke lever **(see illustration)**.
20 Carefully prise off the circlip from the top of the left fork tube. Slacken the clamp bolt then free the handlebar from the top of the fork tube and remove it from the bike.
21 If necessary, undo the retaining screw and remove the end weight from the handlebar then remove the grip (cut it off if necessary) and slide off the choke lever or spacer (as fitted). The inner weight can be removed as described in Paragraph 7.

Installation

22 Where necessary, clean the handlebar surface then slide on the choke lever or spacer (as fitted). Apply a suitable adhesive to the inside of the grip then slide the grip into position, making sure it is correctly positioned before the adhesive sets. Fit the inner weight (where necessary – see Paragraph 8) then install the end weight, tightening the new retaining screw to the specified torque.
23 Locate the handlebar on the fork tube, making sure its lug is correctly located in the upper yoke slot, and tighten its clamp bolt to the specified torque.
24 Fit the circlip to the top of the fork tube making sure it is correctly seated in the tube groove.
25 Where fitted, attach the choke cable to the lever and engage the lever correctly with the switch lower half. Align the locating pin in the lower half of the switch with the hole in the underside of the handlebar, then assemble the switch and securely tighten its screws, tightening the front screw first.
26 Locate the master cylinder on the handlebar and fit its mounting clamp with its UP mark facing upwards. Align the outer corner of the master cylinder mounting with the punch mark on the top of the handlebar, then tighten the mounting clamp upper bolt to the specified torque followed by the lower bolt.
27 Reconnect the wiring connectors to the clutch switch and check the operation of the switch and choke lever (where fitted) before using the bike.

6 Forks – removal and installation

Removal

Caution: *Although not strictly necessary, before removing the forks it is recommended that the fairing lower and inner panels are removed (see Chapter 8). This will prevent accidental damage to the paintwork.*

1 Remove the front wheel (see Chapter 7). Remove each fork as described under the relevant sub-heading.

Right fork

2 Unscrew the bolt securing the braking system hose fitting to the delay valve, then undo the two bolts securing the delay valve to the slider (see Chapter 6) **(see illustrations)**. **Note:** *There is no need to disconnect any of the hydraulic unions.*

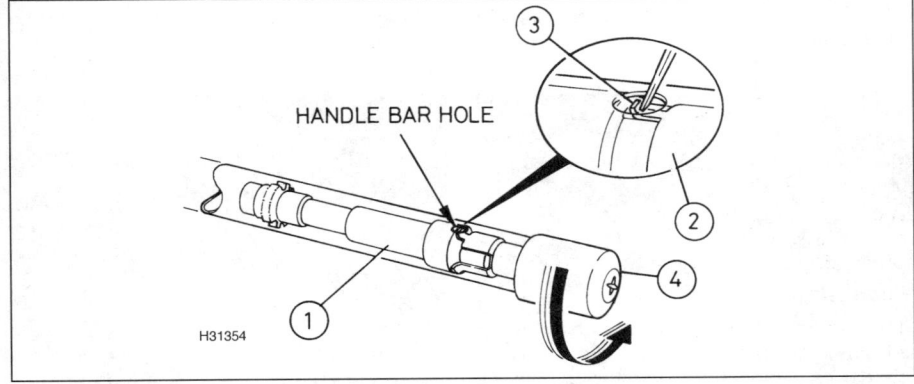

5.8 When installing the inner weight (1) and retaining ring (2), ensure a retaining ring tab (3) is correctly located in the handlebar hole. Rotate the end weight (4) to position the tab correctly

5.19 A spacer (arrow) is fitted inside the switch housing on VFR800FI-Y and 1 models in place of the choke lever

6•6 Frame, suspension and final drive

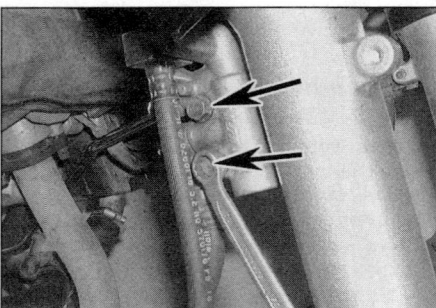

6.2a Unscrew the bolt securing the hose fitting to the delay valve . . .

6.2b . . . then unbolt the delay valve from the right fork slider

6.3a Prise off the circlip from the top of the fork tube . . .

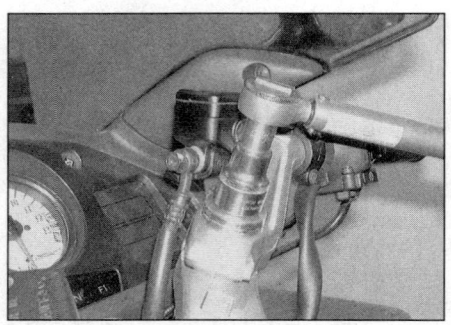

6.3b . . . and release the throttle cable tie from the tube

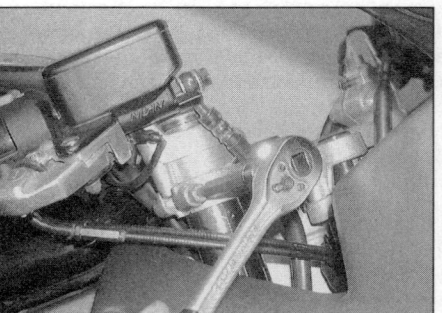

6.4 If the fork is to be dismantled, slacken the top cap by a couple of turns whilst it is still clamped in the yokes

6.5a Slacken the top yoke . . .

3 Remove the circlip from the top of the fork tube then slacken the handlebar clamp bolt **(see illustration)**. Release the tie securing the throttle cables to the fork tube **(see illustration)**.
4 If the fork is to be dismantled, unscrew the top cap from the fork tube by a few turns **(see illustration)**.
5 Slacken the top and bottom yoke clamp bolts and remove the fork by twisting it and pulling it downwards **(see illustrations)**. As the fork is removed, support the handlebar assembly so that no strain is placed on the master cylinder hose. Whilst the fork is removed, support the handlebar assembly so that the master cylinder fluid reservoir is upright, to prevent possible fluid spillage.

 *If the fork legs are seized in the yokes, spray the area with penetrating oil and allow time for it to soak in before trying again.*

Left fork
Note: *Honda recommend that a new brake caliper bracket bolt and secondary master cylinder pushrod bolt should be used on installation (the bolts are supplied with their threads pre-coated with locking compound). If the old bolts are to be reused, remove all original locking compound from its threads and apply fresh compound before installing it*
6 Slacken and remove the bolts securing the brake caliper bracket, the secondary master cylinder pushrod and the hydraulic union assembly to the left fork slider. Position the caliper assembly clear of the fork and support it to avoid placing any strain on the hydraulic pipes and hoses. **Note:** *There is no need to disconnect any of the hydraulic unions.*
7 Remove the fork as described in Paragraphs 3 to 5.

Installation
Right fork
8 Remove all traces of corrosion from the fork tube and the yokes.
9 Slide the fork up through the bottom and top yokes and locate the handlebar on the top of the fork tube **(see illustration)**. Lightly tighten the top yoke clamp bolt to hold the fork in position

6.5b . . . and bottom yoke clamp bolts (arrow) . . .

6.5c . . . and slide the fork out of the yokes

6.9 Slide the fork up through the yokes and locate the handlebar on the fork tube

Frame, suspension and final drive 6•7

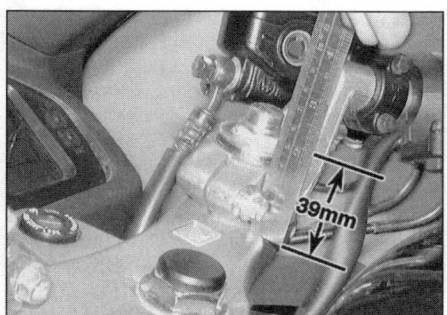

6.10a Position the fork tube so the upper surface of the top cap is exactly 39 mm above the top yoke . . .

6.10b . . . then tighten the yoke clamp bolts to the specified torque

6.11 Ensure the handlebar is correctly seated then tighten its clamp bolt to the specified torque

10 Ensure the fork top cap is tightened to the specified torque (applies if the fork has been dismantled) then position the fork so that the upper surface of the top cap is exactly 39 mm above the upper surface of the top yoke **(see illustration)**. Once the fork tube is correctly positioned tighten both the top and bottom yoke clamp bolts to their specified torque settings **(see illustration)**.
11 Ensure the handlebar lug is correctly located in the top yoke cutout then tighten its clamp bolt to the specified torque **(see illustration)**. Secure the handlebar in position with the circlip, making sure it is correctly seated in the fork tube groove.
12 Fit the braking system delay valve bolts, tightening them to the specified torque then fit the hose fitting retaining bolt and tighten it to the specified torque (see Chapter 7).
13 Install the front wheel (see Chapter 7) and the front mudguard (see Chapter 8). Check the operation of the front forks and brakes before taking the machine out on the road.

Left fork

14 Install the fork as described in Paragraphs 8 to 11.
15 Align the braking system hydraulic union with the slider and tighten its mounting bolt to the specified torque (see Chapter 7). Engage the brake caliper mounting bracket and secondary master cylinder pushrod correctly with the slider, then fit the **new** retaining bolts and tighten them to the specified torque (see Chapter 7).
16 Install the front wheel (see Chapter 7) and the front mudguard (see Chapter 8). Check the operation of the front forks and brakes before taking the machine out on the road.

7 Forks – disassembly, inspection and reassembly

Disassembly

1 Always dismantle the fork legs separately to avoid interchanging parts and thus causing an accelerated rate of wear. Store all components in separate, clearly marked containers **(see illustration)**.

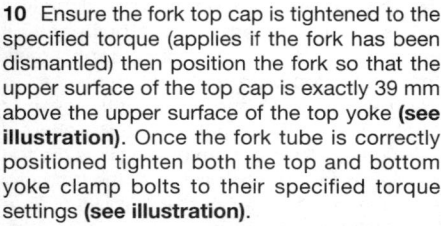

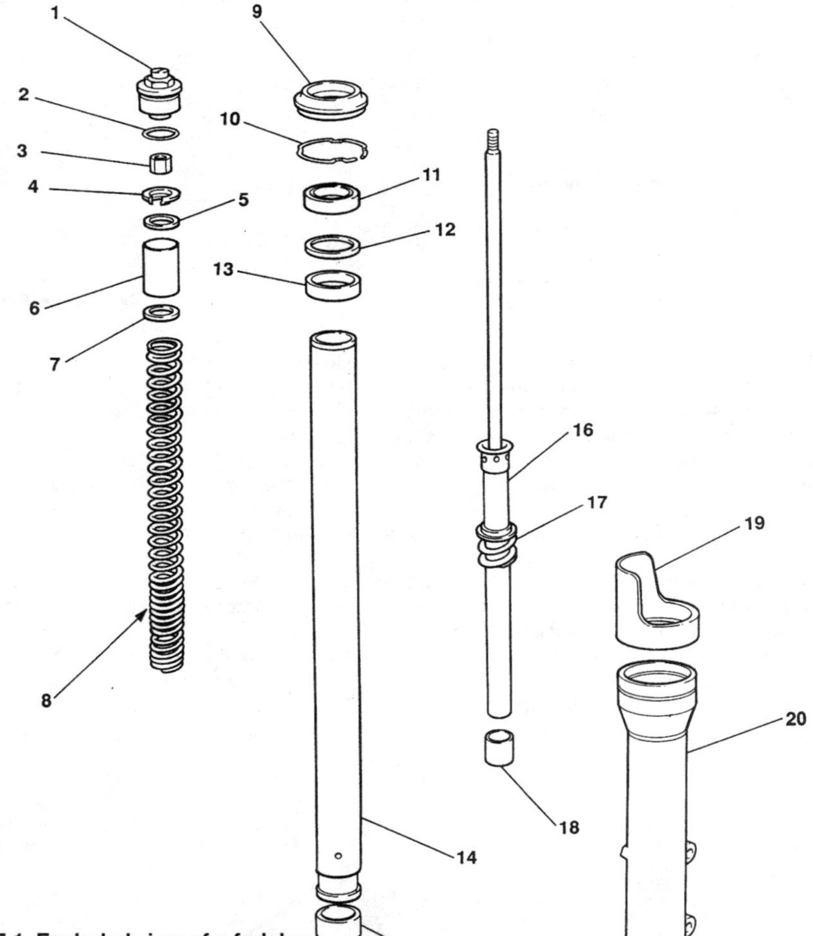

7.1 Exploded view of a fork leg

1 Top cap	12 Washer
2 O-ring	13 Upper bush
3 Locknut	14 Fork tube
4 Slotted collar	15 Bottom bush
5 Washer	16 Damper assembly
6 Spacer	17 Spring
7 Spring seat	18 Damper assembly seat
8 Spring	19 Fork protector
9 Dust seal	20 Fork slider
10 Retaining clip	21 Damper bolt
11 Oil seal	

6•8 Frame, suspension and final drive

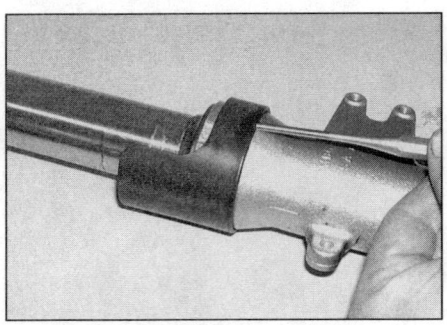

7.3 Remove the protector from the top of the fork slider

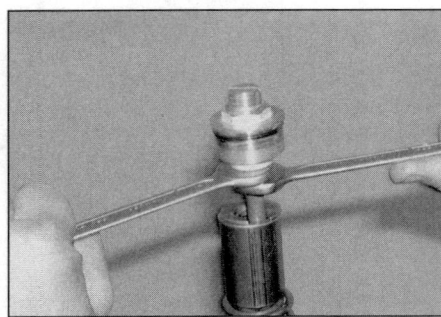

7.5 Slacken the locknut then unscrew the top cap from the damper rod

7.12 Prise out the dust seal from the top of the slider . . .

7.13 . . . then carefully remove the oil seal retaining clip

2 Before dismantling the fork, it is advisable to slacken the damper bolt in the base of the slider. If necessary, to prevent the damper rotating, compress the fork tube in the slider so that the spring exerts maximum pressure on the damper, then have an assistant slacken the bolt.
3 Carefully ease off the fork tube protector from the top of the slider and remove it. If the protector is a tight-fit, tap it off using a hammer and punch **(see illustration)**.
4 Support the fork in an upright position then fully unscrew the top cap from the tube.
5 Hold the top cap with an open-ended spanner then slacken the top cap locknut **(see illustration)**. The top cap assembly can then be unscrewed from the damper rod.
6 Slide the fork tube down into the slider and remove the slotted collar from the damper rod, noting which way around it is fitted.
7 Lift the washer off the damper rod followed by the spacer and the spring seat.
8 Withdraw the fork spring, noting which way up it is fitted.
9 Invert the fork leg over a suitable container and pump both the fork tube and damper rod to expel as much fork oil as possible.
10 Remove the previously slackened damper bolt and its copper sealing washer from the bottom of the slider. Discard the sealing washer as a new one must be used on reassembly. If the damper bolt was not slackened before dismantling the fork, it may be necessary to re-install the spring, spring seat, spacer, washer and top cap to prevent the damper from turning.
11 Invert the fork and withdraw the damper assembly complete with its seat.
12 Carefully prise out the dust seal from the top of the slider to gain access to the oil seal retaining clip **(see illustration)**. Discard the dust seal as a new one must be used.
13 Slide the fork tube fully into the slider then carefully remove the retaining clip, taking care not to scratch the surface of the tube **(see illustration)**.
14 To separate the tube from the slider it is necessary to displace the top bush and oil seal. The bottom bush should not pass through the top bush, and this can be used to good effect. Push the tube gently inwards until it stops then pull the tube sharply outwards until the bottom bush strikes the top bush. Repeat this operation until the top bush and seal are tapped out of the slider **(see illustrations)**.
15 With the tube removed, slide off the oil seal and its washer, noting which way up they fit. Discard the oil seal as a new one must be used. The top bush can then also be slid off its upper end.
Caution: Do not remove the bottom bush from the tube unless it is to be renewed.

Inspection
16 Clean all parts in solvent and blow them dry with compressed air, if available. Check the fork tube for score marks, scratches, flaking of the chrome finish and excessive or abnormal wear. Look for dents in the tube and renew the tube in both forks if any are found. Check the fork seal seat for nicks, gouges and scratches. If damage is evident, leaks will occur. Also check the oil seal washer for damage or distortion and renew it if necessary.

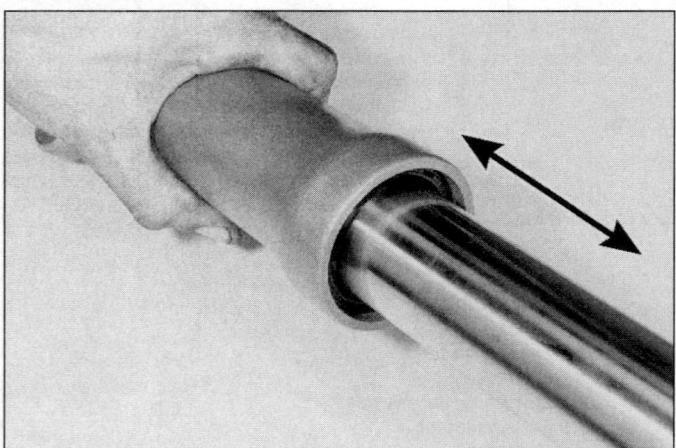

7.14a To separate the fork tube and slider, pull them apart firmly several times - the slide hammer effect will force the oil seal and top bush out of position

7.14b Slide off the oil seal (1), washer (2) and top bush (3) from the fork tube. Do not remove the bottom bush (4) unless it is to be renewed

Frame, suspension and final drive 6•9

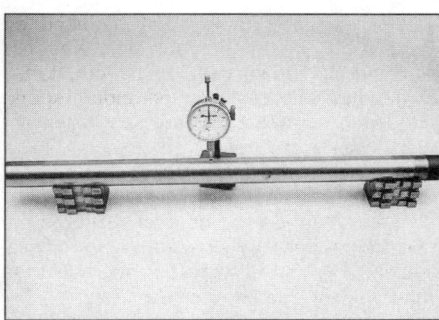

7.17 Check the fork tube for runout using V-blocks and a dial gauge

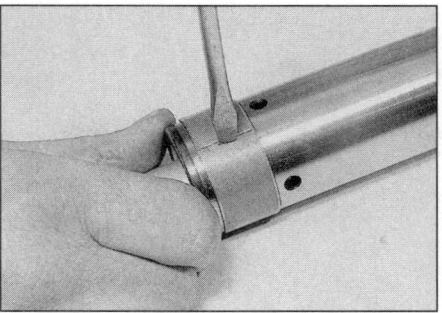

7.19 Prise off the bottom bush using a flat-bladed screwdriver

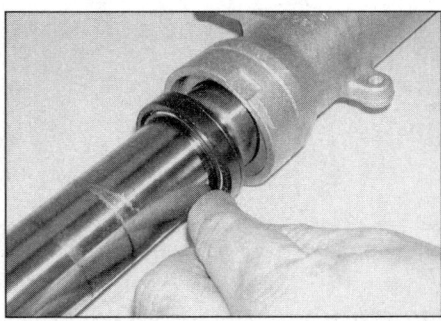

7.24 Slide the oil seal down the fork tube and tap it fully into the fork slider . . .

17 Check the fork tube for runout using V-blocks and a dial gauge, or have it done by a dealer **(see illustration)**. If the amount of runout exceeds the service limit specified, the tube should be renewed.

 Warning: If the tube is bent or exceeds the runout limit, it should not be straightened; renew it.

18 Check the spring for cracks and other damage. Measure the spring free length and compare the measurement to the specifications at the beginning of the Chapter. If it is defective or sagged below the service limit, renew the springs in both forks. Never renew only one spring.

19 Examine the working surfaces of the two bushes and the washer fitted between the top bush and oil seal; if worn or scuffed they must be renewed. To remove the bottom bush from the fork tube, prise it apart at the slit using a flat-bladed screwdriver and slide it off **(see illustration)**. Make sure the new one seats properly.

20 Check the damper assembly and its rebound spring for damage and wear, and renew the assembly if necessary. The rebound spring is not available separately.

21 If work is being carried out on the left fork leg, slide out the inner sleeve from the brake caliper mounting and inspect the bearing arrangement for signs of wear. If there is any sign of wear, renew the bearings, the inner sleeve and both dust seals. Prise out the seals and note the correct fitted location of each bearing before pressing/drifting them out of position. The new bearings should be pressed or drawn into position rather than driven into

position to prevent possible damage. In the absence of a press, a suitable drawbolt arrangement can be made up as described in *Tools and Workshop Tips* in the Reference section. Ensure both bearings are correctly positioned (there should be a 3.5 mm gap between the outer edge of each bearing and the slider) then press the new dust seals into position. Lubricate the bearings with multi-purpose grease then slide in the inner sleeve.

Reassembly

22 Oil the fork tube and lower bush with the specified fork oil then carefully insert the fork tube fully into the slider.

23 Oil the top bush and slide it down over the tube. Press the bush squarely into its recess in the slider as far as possible, then install the oil seal washer. Tap the bush squarely into the slider using a suitable piece of tubing as a slide-hammer, taking great care not to scratch the fork tube. In the absence of the Honda service tool, a suitable piece of tubing slightly larger in diameter than the fork tube and slightly smaller in diameter than the oil seal recess in the slider can be used. Ensure the bush is correctly seated before continuing.

 *Ensure the fork tube remains fully pushed into the slider during assembly. Any accidental damage should then be confined to the area of the fork tube which remains above the oil seal during normal use.*

24 Lubricate the new oil seal with fork oil then carefully slide the seal onto the fork tube, making sure its marked surface is facing upwards **(see illustration)**. Drive the seal into place as described in Paragraph 23 until the retaining clip groove is visible above the seal.

 Wrap some insulating tape around the circlip groove in the top of the fork tube – this will prevent the possibility of damage to the oil seal lips as it is slid down the tube.

25 Once the seal is correctly seated, fit the retaining clip, making sure it is correctly located in its groove **(see illustration)**.

26 Lubricate the lips of the new dust seal then slide it down the fork tube and press it into position **(see illustration)**.

27 Slide the fork tube protector along the fork tube and clip it securely onto the slider, aligning its locating lug with the slider cutout **(see illustration)**.

28 Fit the seat to the base of the damper assembly then insert the assembly fully into the fork tube **(see illustration)**. Fit a new copper sealing washer to the damper bolt then apply a few drops of a suitable non-permanent thread-locking compound to the bolt threads **(see illustration)**. Install the bolt into the bottom of the slider and tighten it to the specified torque setting **(see illustration)**. If the damper rod rotates inside the tube, temporarily install the fork spring, spring seat, spacer, washer and top cap (see Paragraphs 30 to 35) and compress the fork to hold the damper rod.

7.25 . . . then secure it in position with the retaining clip

7.26 Ensure the retaining clip is correctly seated in its groove then fit the new dust seal

7.27 Align the fork protector lug with the slider cutout (arrows) and fit the protector

6

6•10 Frame, suspension and final drive

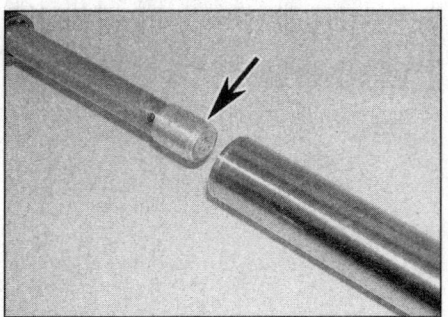

7.28a Fit the seat (arrow) to the base of the damper and insert the assembly into the tube

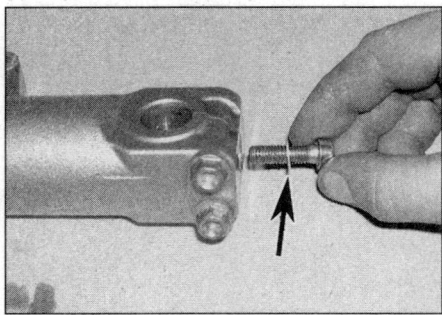

7.28b Fit a new sealing washer (arrow) then apply locking compound to the damper bolt threads . . .

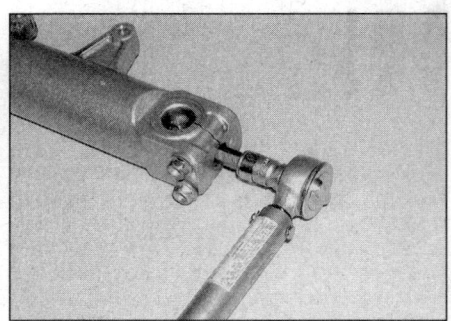

7.28c . . . before fitting the bolt and tightening it to the specified torque

29 Slowly pour in the specified amount and type of fork oil whilst pushing the damper rod piston up and down **(see illustration)**. Once the oil has been added, slowly pump the fork tube in and out at least five times, and pump the damper rod at least another 10 times. This will ensure that the fork oil is evenly distributed. Fully insert both the tube and damper rod then check the fork oil level from the top of the tube **(see illustration)**. Add or subtract fork oil until the oil is at the level specified at the beginning of the Chapter.

30 Clamp the slider securely in the padded jaws of a vice and fully extend the damper rod.

31 Insert the fork spring, ensuring that its tighter-pitched coils are at the bottom **(see illustration)**.

32 Ensure the damper rod is still fully extended then fit the spring seat, the spacer and the washer **(see illustrations)**. Ensure all components are correctly fitted then slide the slotted collar into the damper rod, ensuring it is fitted the correct way up (concave face upwards).

33 Fit a new O-ring to the top cap and lubricate it with a smear of fork oil.

34 Ensure the locknut is screwed fully onto the damper rod then screw on the top cap until it contacts the locknut **(see illustration)**. Hold the top cap stationary and tighten the locknut to the specified torque (where possible).

35 Check that all components are correctly seated then carefully screw the top cap into the fork tube, making sure it is not cross-threaded. **Note:** *The top cap can be tightened to its specified torque setting at this stage if the tube can be held firmly enough, but do not risk distorting the tube by overtightening it. A better method is to tighten the top cap when the fork has been reinstalled and is securely held in the yokes.*

36 Install the forks and set the spring pre-load as required (see Sections 6 and 12).

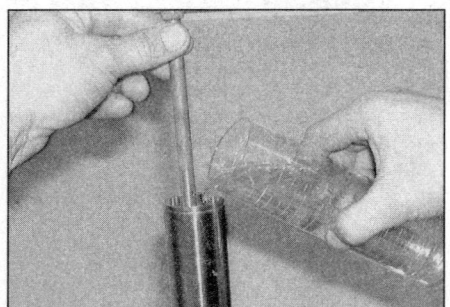

7.29a Pour in the correct amount and type of oil whilst pumping the damper rod . . .

7.29b . . . and check the oil level as described in text

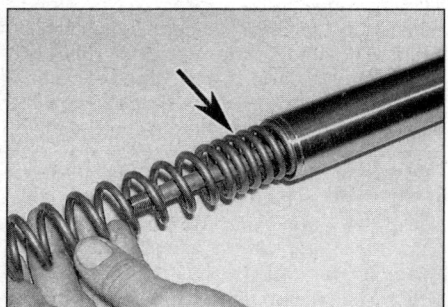

7.31 Fit the spring with its tighter-pitched coils (arrow) at the bottom

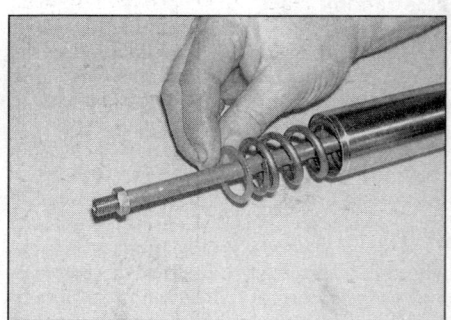

7.32a Fit the spring seat . . .

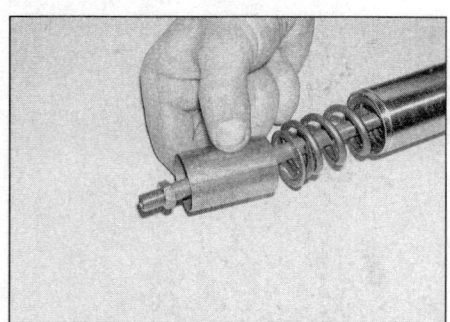

7.32b . . . followed by the spacer . . .

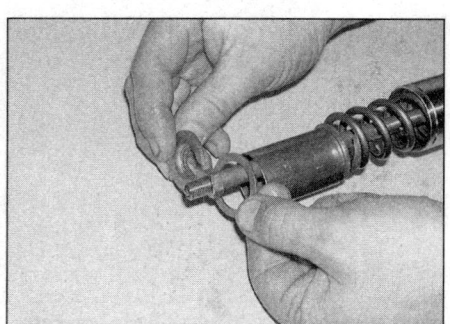

7.32c . . . then fit the washer and slotted collar

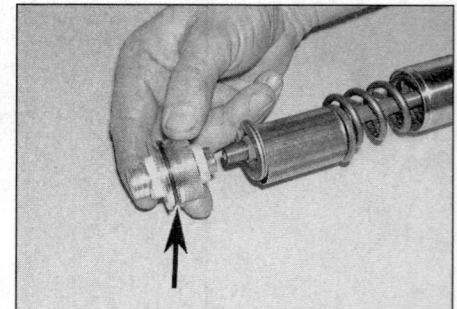

7.34 Fit a new O-ring (arrow) then screw the top cap onto the damper rod

Frame, suspension and final drive 6•11

8 Steering stem – removal and installation

Caution: *Although not strictly necessary, before removing the steering stem it is recommended that the fuel tank is removed or at least raised. This will prevent accidental damage to the paintwork. Access can also be improved by removing the upper fairing.*

Removal

1 Remove the front forks (see Section 6).
2 Undo the bolts and free the horn mounting bracket from the bottom yoke.
3 Remove the cap from the top of the steering stem then unscrew the steering stem nut **(see illustration)**.
4 Lift the top yoke off the steering stem and place it aside, making sure no strain is placed on the ignition switch wiring. Use a rag to protect other components. If the fairing has been removed, disconnect the wiring connector for the ignition switch and (where fitted) the immobiliser receiver, and remove the top yoke assembly.
5 Prise the lockwasher tabs out of the notches in the locknut. Unscrew the locknut using either a C-spanner or a suitable drift located in one of the notches.
6 Remove the lockwasher, bending up the remaining tabs to release it from the adjuster nut if necessary. Discard the lockwasher, a new one should be used on installation.
7 Support the bottom yoke then slacken the adjuster nut using either a C-spanner, a peg-spanner or a drift located in one of the notches. Unscrew the adjuster nut from the steering stem then gently lower the bottom yoke and steering stem out of the frame. Remove the lower bearing from the steering stem.
8 Remove the dust seal and the upper bearing and its inner race from the top of the steering head.
9 Remove all traces of old grease from the bearings and races and check them for wear or damage as described in Section 9. **Note:** *Do not attempt to remove the races from the frame or the steering stem unless the bearings are to be renewed.*

Installation

10 Smear a liberal quantity of multi-purpose grease on the bearing races in the frame and steering stem. Work the grease well into both the upper and lower bearings and oil the threads of the adjuster nut.
11 Ensure the lower bearing is correctly seated on its inner race then carefully lift the steering stem/bottom yoke up through the frame. Install the upper bearing and its inner race in the top of the steering head, then install the dust seal. Lubricate the adjuster nut threads with engine oil then screw the nut onto the steering stem.
12 To preload the bearings to the torque specified by the manufacturer (see Specifications) it will be necessary to use the Honda service tool (Pt. No. 07916-3710101) which is a socket that fits the adjuster nut. Using the service tool, tighten the adjuster nut to the specified torque setting then turn the steering stem from lock to lock approximately 5 times to settle the bearings and races in position. After pre-loading the bearings, slacken the adjuster nut one full turn then tighten it again to the specified torque setting.
Note: *It is important to check the feel of the steering afterwards as described below; if it is too tight readjust the bearings as described below.*
13 If the service tool is not available, tighten the adjuster nut hard using a conventional C-spanner to preload the bearings then adjust them as follows.
14 Slacken the adjuster nut slightly until pressure is just released, then turn it slowly clockwise until resistance is just evident. The object is to set the adjuster nut so that the bearings are under a very light loading, just enough to remove any freeplay.
Caution: Take great care not to apply excessive pressure because this will cause premature failure of the bearings.
15 With the bearings correctly adjusted, fit a new lockwasher to the adjuster nut. Bend down two opposite lockwasher tabs into the grooves of the adjuster nut.
16 Install the locknut and tighten it finger-tight only.
17 Hold the adjuster nut, to prevent it from moving, and tighten the locknut approximately 90° more until its slots align with the remaining lock washer tabs. Recheck the bearing adjustment before securing the locknut in position by bending up the other pair of lockwasher tabs into its slots.
18 Fit the top yoke to the steering stem and install the stem nut, tightening lightly only at this stage. Where necessary, reconnect the ignition switch and immobiliser receiver wiring connectors, making sure the wiring is correctly routed.
19 Fit the horn mounting bracket to the bottom yoke and securely tighten its bolts.
20 Install the forks as described in Section 6 of this Chapter.
21 Tighten the steering stem nut to the specified torque then install the cap in the top of the stem.
22 On completion check that the steering stem bearing adjustment as described in Chapter 1.

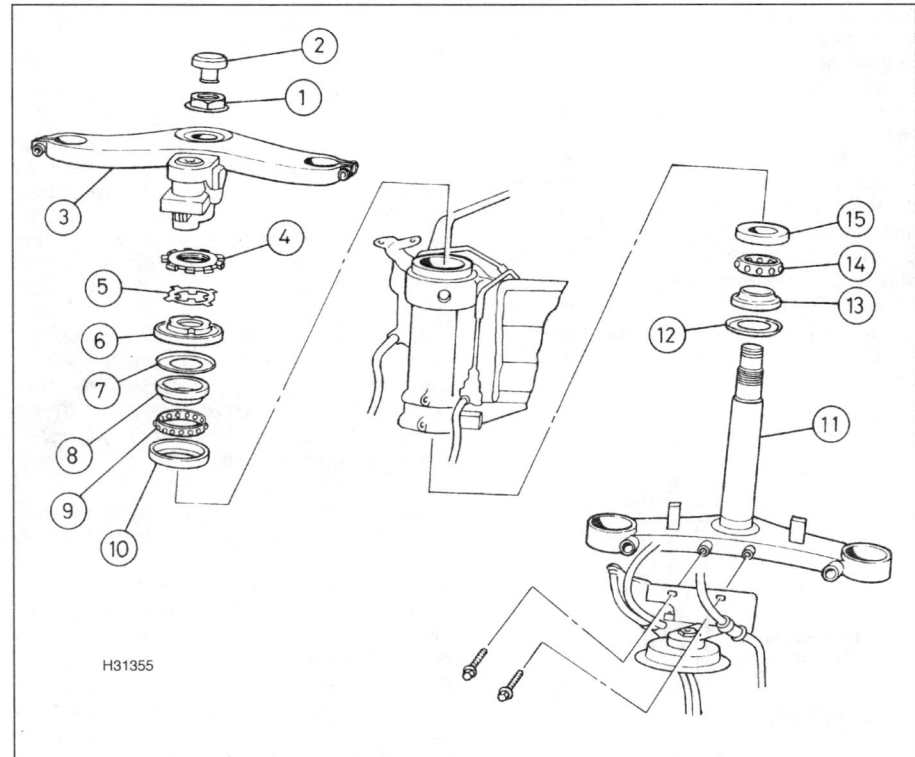

8.3 Exploded view of steering stem components

1 Steering stem nut	6 Adjuster nut	11 Steering stem
2 Cap	7 Upper dust seal	12 Lower dust seal
3 Top yoke	8 Upper bearing inner race	13 Lower bearing inner race
4 Locknut	9 Upper bearing	14 Lower bearing
5 Lockwasher	10 Upper bearing outer race	15 Lower bearing outer race

6•12 Frame, suspension and final drive

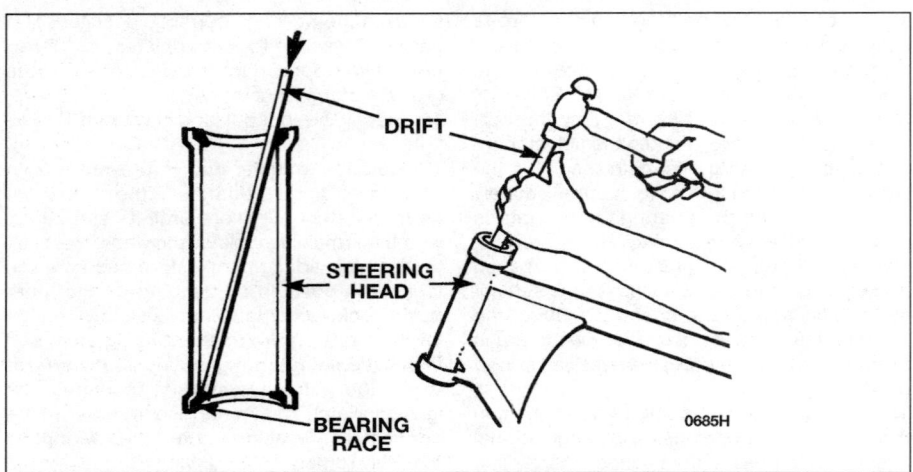

9.4 Drive the bearing outer races out with a hammer and drift as shown

9 Steering head bearings – inspection and renewal

Inspection

1 Remove the steering stem (see Section 8).
2 Remove all traces of old grease from the bearings and races and check them for wear or damage.
3 The outer races should be polished and free from indentations. Inspect the bearing rollers for signs of wear, damage or discoloration, and examine the bearing ball retainer cage for signs of cracks or splits. Spin the bearings by hand. They should spin freely and smoothly. If there are any signs of wear on any of the above components both upper and lower bearing assemblies must be renewed as a set. Only remove the outer races and the lower bearing inner race if they need to be renewed – do not re-use them once they have been removed.

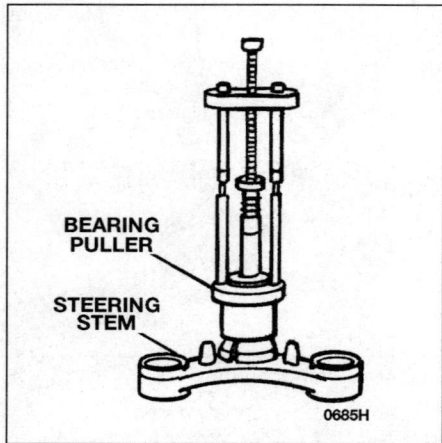

9.7 Using a bearing puller to remove the lower bearing inner race from the steering stem

Renewal

Note: *Obtain new bearings and new dust seals before proceeding.*

4 The outer races are an interference fit in the frame headstock and can be tapped from position with a suitable drift **(see illustration)**. Tap firmly and evenly around each race to ensure that it is driven out squarely. It may prove advantageous to curve the end of the drift slightly to improve access.
5 Alternatively, the races can be removed using a slide-hammer type bearing extractor; these can often be hired from tool shops.
6 The new outer races can be pressed into the head using a drawbolt arrangement, or by using a large diameter tubular drift **(see illustration)**. Ensure that the drawbolt washer or drift (as applicable) bears only on the outer edge of the race and does not contact the race bearing surface.

> **HAYNES HiNT** *Installation of new bearing outer races is made much easier if the races are left overnight in the freezer. This causes them to contract slightly making them a looser fit.*

7 To remove the lower bearing inner race from the steering stem, use two screwdrivers placed on opposite sides of the race to work it free. If the bearing is firmly in place it will be necessary to use a bearing puller **(see illustration)**. With the inner race removed, lift off the dust seal and discard it.
8 Fit the new dust seal to the steering stem then slide on the new inner race. A length of tubing with an internal diameter slightly larger than the steering stem will be needed to tap the new race into position **(see illustration)**. Ensure that the drift bears only on the inner edge of the race and does not contact the bearing surface.
9 Ensure the lower bearing inner race and both outer races are correctly seated then install the steering stem as described in Section 8.

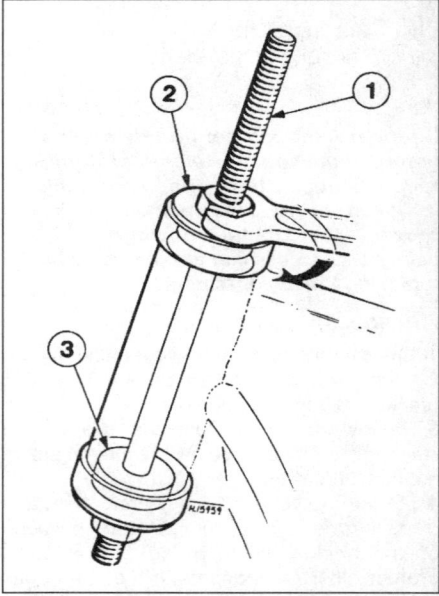

9.6 Drawbolt arrangement for fitting steering stem bearing outer races

1 Long bolt or threaded rod
2 Thick washer
3 Guide for lower race

10 Rear suspension linkage – removal, inspection and installation

Removal

1 Place the motorcycle on its centrestand then support the rear wheel with a block of wood positioned between the tyre and ground; this will prevent the swingarm dropping as the suspension linkage bolts are withdrawn.
2 Unscrew the nut and remove the shock absorber lower mounting bolt.
3 Unscrew the nuts and remove the bolts securing the linkage plates to the swingarm

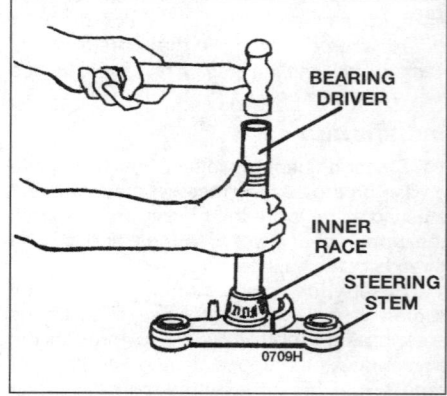

9.8 Drive the new inner race onto the steering stem using a tubular drift which bears only on the inner edge of the race

Frame, suspension and final drive 6•13

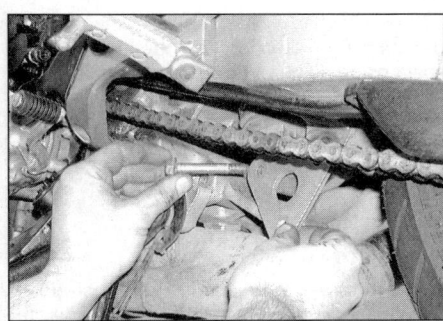

10.3 Withdraw the pivot bolts and manoeuvre the suspension linkage plates out of position

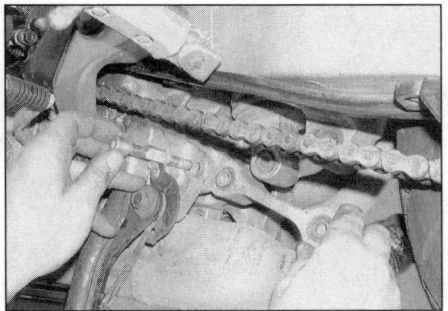

10.4 Remove the front pivot bolt and remove the connecting link from the bike

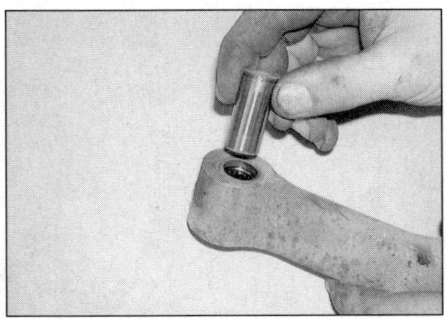

10.5 Check the connecting link inner sleeves and bearings for signs of wear or damage

and connecting link. Note the correct fitted location of each plate and remove both plates from the bike **(see illustration)**.
4 Slacken and remove the nut and bolt securing the connecting link to its mounting bracket and remove the link from the bike **(see illustration)**.

Inspection

5 Withdraw the inner sleeves from the connecting link and thoroughly clean all components, removing all traces of dirt, corrosion and grease **(see illustration)**.
6 Inspect all components closely, looking for obvious signs of wear such as heavy scoring, or for damage such as cracks or distortion. Renew as necessary.
7 If the connecting link dust seals are damaged, the old seals can be levered out of position using a flat-bladed screwdriver. Ensure the sealing lip of the new seal is facing inwards then press the seal squarely into position until it is flush with the link.
8 If the connecting link bearings are damaged, it will be necessary to renew the bearings, inner sleeves and dust seals. Remove the dust seals and note the correct fitted location of each bearing before pressing/drifting them out of position. The new bearings should be pressed or drawn into their bores rather than driven into position to prevent possible damage. In the absence of a press, a suitable drawbolt arrangement can be made up as described in *Tools and Workshop Tips* in the Reference section.

Ensure both bearings are centrally positioned in the link (there should be a 5.2 to 5.7 mm gap on each side of the bearing) then press the new dust seals into position.
9 Lubricate the needle roller bearings, the inner sleeves and the dust seal lips with multi-purpose grease and slide both sleeves carefully into position.

Installation

10 If not already done, lubricate the seals, needle roller bearings, inner sleeves and the pivot bolts with multi-purpose grease.
11 Align the connecting link with its mounting bracket and insert the pivot bolt from the left side. Fit the nut to the bolt tightening it lightly only.
12 Offer up both suspension linkage plates making sure the ←FR marking on each plate is facing forwards (both plates are the same) **(see illustration)**. Insert the bolts (from the left side of the machine) securing the plates to the connecting link, shock absorber and swingarm and screw on the nuts; tighten the nuts lightly at this stage.
13 Retain the bolt and tighten the connecting link to mounting bracket bolt nut to the specified torque setting.
14 Tighten the suspension linkage plate to swingarm and connecting link bolt nuts to the specified torque, then tighten the shock absorber lower mounting bolt nut to its specified torque.
15 Remove the wood from underneath the rear wheel then take the motorcycle off its

centrestand and check the operation of the rear suspension before taking the machine on the road.

11 Rear shock absorber – removal, inspection and installation

Removal

1 Remove the suspension linkage plates and connecting link (Section 10).
2 Remove the seat (see Chapter 8).
3 Unscrew the fuel tank front and rear mounting bolts (see Chapter 4). Lift the rear of the fuel tank and support it securely to allow access to the shock absorber upper mounting nut. **Note:** *There is no need to disconnect the fuel hoses from the tank.*
4 Lift the rubber insulating cover to gain access to the shock absorber upper mounting which is located in front of the tank rear hinge assembly **(see illustration)**. If necessary, to improve access, slacken and remove the pivot bolt and nut and remove the hinge and its pivot bushes from the mounting bracket; the rubber covers can then be positioned clear of the nut.
5 Have an assistant support the shock absorber then unscrew the upper mounting nut. Free the shock absorber upper mounting from the frame and manoeuvre it out from underneath the bike **(see illustration)**.

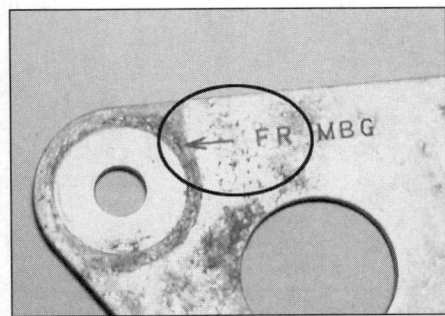

10.12 Ensure both linkage plates are fitted with the ←FR marking facing forwards

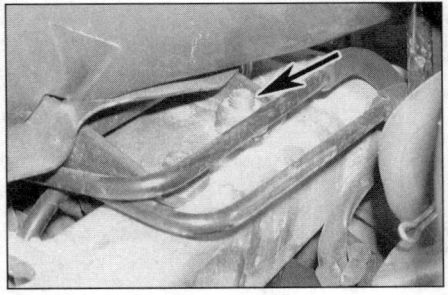

11.4 Lift the rubber cover to gain access to the shock absorber upper mounting nut (arrow)

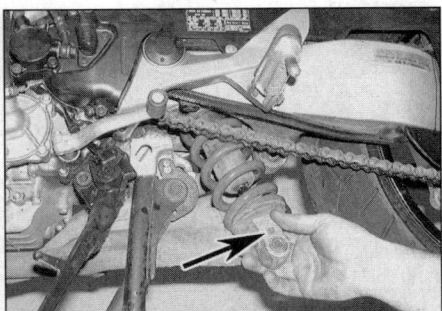

11.5 Removing the rear shock absorber (rebound damping adjuster arrowed)

6•14 Frame, suspension and final drive

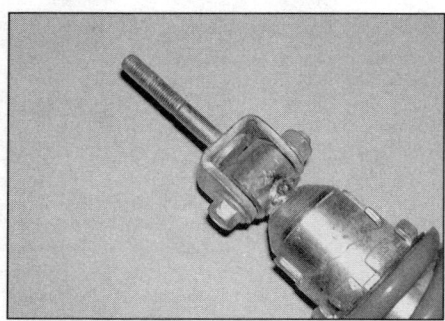

11.6 If necessary, slacken and remove the nut and bolt and separate the shock absorber from the upper mounting bracket

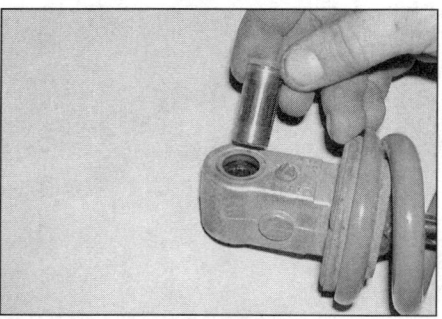

11.7 Check the shock absorber lower mounting inner sleeve and bearings for signs of wear or damage

6 If necessary, slacken and remove the nut and bolt and separate the shock absorber from the upper mounting bracket **(see illustration)**.

Inspection

7 Withdraw the inner sleeve from the lower mounting then inspect the shock absorber for obvious physical damage and oil leakage, and the coil spring for looseness, cracks or signs of fatigue **(see illustration)**. The shock absorber is a sealed unit and cannot be dismantled; the only components available separately are the upper mounting bracket and the lower mounting bearing components.
8 If the lower bearing dust seals are damaged, the old seals can be levered out of position using a flat-bladed screwdriver. Ensure the sealing lip of the new seal is facing inwards then press the seal squarely into position until it is flush with the shock absorber.
9 If the bearing is damaged, it will be necessary to renew the bearing, inner sleeve and dust seals. Remove the dust seals and note the correct fitted location of the bearing before pressing/drifting it out of position. The new bearing should be pressed or drawn into its bore rather than driven into position to prevent possible damage. In the absence of a press, a suitable drawbolt arrangement can be made up as described in *Tools and Workshop Tips* in the Reference section. Ensure the bearing is centrally positioned in the link then press the new dust seals into position.

Installation

10 Where necessary, fit the upper mounting bracket to the shock absorber and fit the mounting bolt and nut. Position the bracket so that its stud is parallel to the shock absorber then tighten the mounting bolt nut to the specified torque.
11 Ensure the rebound damping adjuster is positioned on the left side **(see illustration 11.5)** then manoeuvre the shock absorber into position. Fit the upper mounting nut, tightening it to the specified torque, then locate the rubber covers correctly on the frame/engine and (where necessary) install the fuel tank hinge arrangement **(see illustration)**.
12 Ensure all the fuel tank mounting rubbers are correctly fitted (see Chapter 4) and the collars are in position in the front mountings then seat the fuel tank back in position. Fit the tank front and rear mounting bolts and tighten them securely.

13 Install the suspension linkage as described in Section 10 then adjust the suspension as required (see Section 12).

12 Suspension – adjustments

Front forks

1 The forks are adjustable for spring pre-load. The pre-load is adjusted using the adjuster fitted to the centre of each fork top cap. The standard setting recommended by Honda is checked by measuring the distance from the top of the adjuster to the upper surface of the top cap; this should be 9 mm. Turn the adjuster clockwise to increase the pre-load and anti-clockwise to decrease it **(see illustration)**. The amount of pre-load is indicated by lines on the body of the adjuster.
Caution: Always make sure both left and right adjusters are adjusted equally to ensure the motorcycle handles predictably.

Rear shock absorber

2 The rear shock absorber features both spring pre-load adjustment and rebound damping adjustment.
3 Spring pre-load adjustment is made using a suitable C-spanner (one is provided in the toolkit) to turn the spring seat on the top of the shock absorber **(see illustration)**. There are seven positions, position 1 is the softest setting and position 7 the hardest. Align the setting required with the adjustment stopper. Honda recommend position 2 as the standard setting.
4 Rebound damping adjustment is made by turning the adjuster on the left side of the shock absorber lower end **(see illustration)**. To establish the standard setting, rotate the adjuster fully clockwise (in the direction of the H arrow) until it stops; do not force it. From this point, rotate the adjuster anti-clockwise

11.11 Ensure the shock absorber is correctly fitted then tighten its upper mounting nut to the specified torque

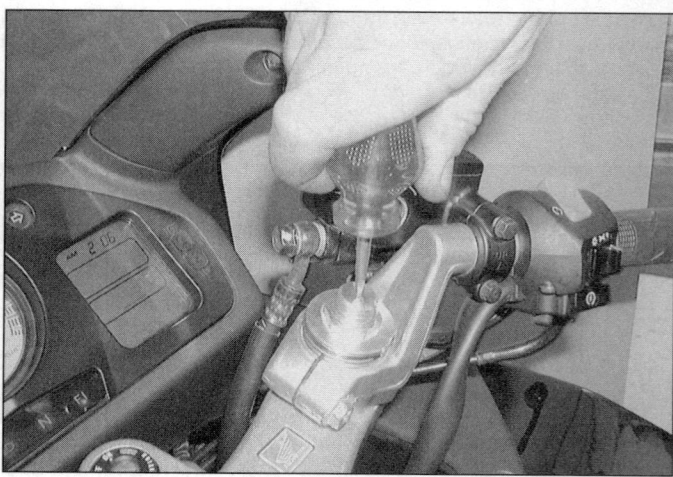

12.1 Adjusting the front fork pre-load

Frame, suspension and final drive 6•15

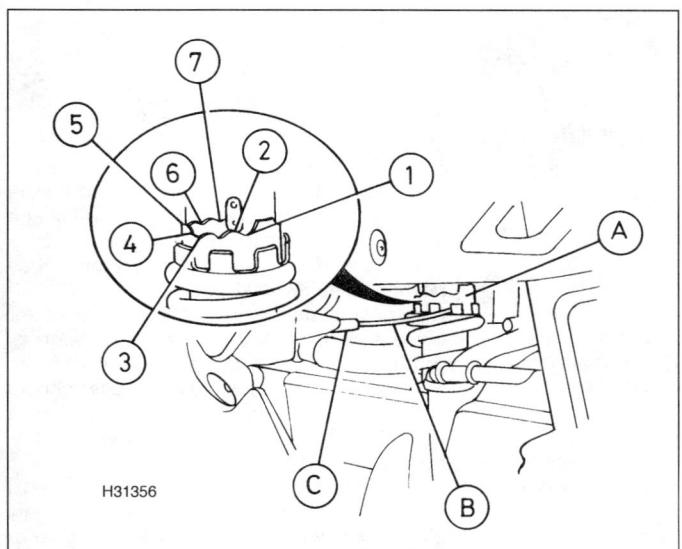

12.3 Rear shock absorber pre-load adjustment details. Rotate the spring seat (A) with the C-spanner (B) and extension bar (C) supplied in the toolkit (inset shows pre-load positions 1 to 7)

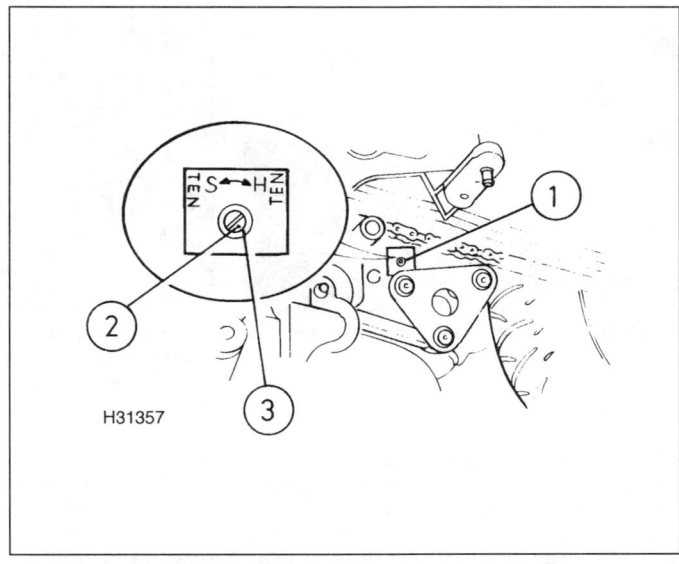

12.4 The rebound damping adjuster (1) is located on the left side of the shock absorber. The standard setting can be set by aligning the punch marks (2 and 3) as described in text

through approximately one and a half turns until the punch mark on the adjuster aligns with the reference mark on the shock absorber. This is the standard setting recommended by Honda. To soften the damping, rotate the adjuster further anti-clockwise (in the direction of the S arrow) and to stiffen the damping rotate the adjuster clockwise (in the direction of the H arrow).

13 Swingarm – removal and installation

Removal

Note: *Honda recommend that new brake caliper mounting bolts should be used on installation (the bolts are supplied with their threads pre-coated with locking compound). If the old bolts are to be reused, remove all original locking compound from their threads and apply fresh compound before installing them.*

1 Remove the rear wheel (see Chapter 7).
2 If the swingarm is to be renewed, remove the rear wheel bearing holder assembly (see Chapter 7).
3 If the swingarm is to be fitted again, remove the chainguard retaining bolts and inner retaining clips and free it from the swingarm. Slacken and remove the retaining bolts and collars and remove the air guide from the base of the swingarm (see illustration). Unscrew the bearing holder clamp bolt from the rear of the swingarm and free the brake hose guide from the bolt. Remove the bolts securing the rear brake caliper to its mounting bracket then slide the caliper off the disc and tie it to the subframe to avoid any strain being placed on the hydraulic hoses (see illustration). Release the drive chain tension and unhook the chain from the rear sprocket.
Caution: Do not operate the rear brake whilst the caliper is removed.
4 Prise out the swingarm pivot bolt caps from the footrest brackets then slacken and remove the nut from the pivot bolt (see illustrations).
5 Slacken and remove the right side rider's footrest bracket heatshield upper bolt then free the complete footrest assembly from the pivot bolt, taking care not to lose the collar from the rear of the heatshield (see illustrations).

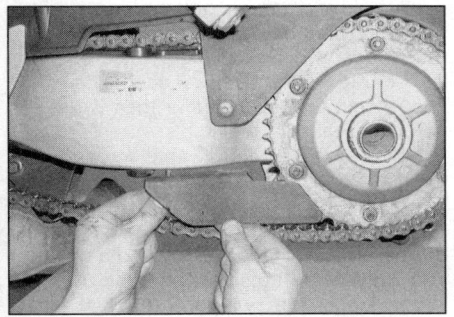

13.3a Unbolt the air guide from the base of the swingarm

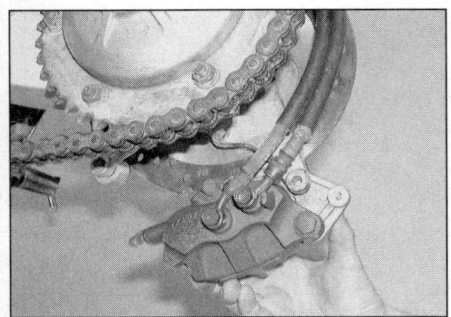

13.3b Undo the two bolts and slide the rear brake caliper assembly off the disc

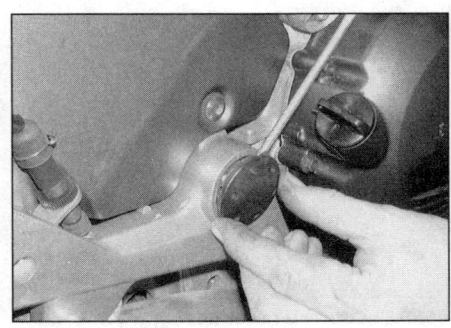

13.4a Prise out the cap from the right side rider's footrest bracket . . .

13.4b . . . and unscrew the nut from the swingarm pivot bolt

6

6•16 Frame, suspension and final drive

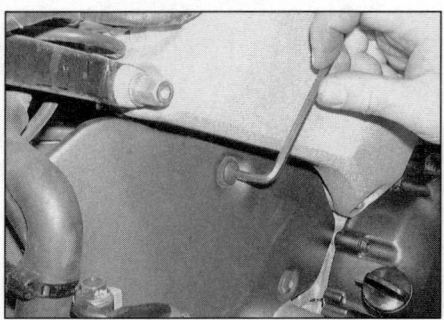

13.5a Unscrew the heatshield upper bolt . . .

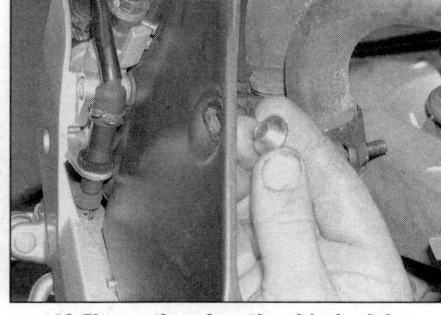

13.5b . . . then free the rider's right footrest bracket from the swingarm pivot bolt and recover the collar

14 Swingarm – inspection and bearing renewal

Inspection

1 Remove the inner sleeves from the swingarm left pivot and suspension linkage needle roller bearings and the spacers from each side of the swingarm right pivot ball bearings **(see illustrations)**.
2 Inspect all components closely, looking for obvious signs of wear such as heavy scoring, or for damage such as cracks or distortion. The right pivot ball bearing inner races should rotate freely and easily without any sign of freeplay or roughness. Renew worn components as necessary.
3 If the dust seals are damaged, the old seals can be levered out of position using a flat-bladed screwdriver. Ensure the sealing lip of the new seal is facing inwards then press the seal squarely into position until it is flush with the swingarm.
4 If the left pivot bearing or suspension linkage bearing is damaged, it will be necessary to renew the bearing, its inner sleeve and the dust seals. Remove the dust seals and note the correct fitted location of the bearing before pressing/drifting it out of position. The new bearing should be pressed or drawn into its bore rather than driven into position to prevent possible damage. In the absence of a press, a suitable drawbolt arrangement can be made up as described in *Tools and Workshop Tips* in the Reference section. Ensure the bearing is correctly positioned in its bore (the left pivot bearing should be positioned 4.0 mm in from the swingarm outer surface and there should be a 5.5 to 6.0 mm gap on each side of the suspension linkage bearing) then press the new dust seals into position.
5 To renew the right pivot bearings, prise out the dust seals then extract the bearing circlip from inside the swingarm bore. Support the swingarm and tap both bearings squarely out of position simultaneously, using a hammer and suitable drift inserted through from the inner side of the swingarm. Pack the new bearings with multi-purpose grease then

6 Slacken and remove the nut and withdraw the bolt securing the suspension linkage plates to the base of the swingarm.
7 Support the swingarm then withdraw the pivot bolt, complete with left side footrest bracket **(see illustration)**. Manoeuvre the swingarm out of position taking care not to lose the inner and outer spacers from the right side pivot. Inspect the assembly as described in Section 14.

Installation

8 If not already done, lubricate the seals, bearings, inner sleeves and the pivot bolts with multi-purpose grease.
9 Ensure the inner sleeves are correctly fitted to the left pivot and suspension linkage bearings and the outer (shorter) and inner (longer) spacers are fitted to the right pivot then manoeuvre the swingarm into position, passing it through the drive chain **(see illustration)**.
10 Slide the left footrest bracket onto the pivot bolt then insert the bolt back through the swingarm and crankcase. Make sure the drive chain passes between the footrest bracket and swingarm then seat the bracket on its lower locating bolt and push the pivot bolt fully home.
11 Locate the right footrest assembly on the swingarm bolt, making sure the collar is in position between the heatshield and frame. Fit the swingarm pivot bolt nut and tighten to the

specified torque then securely tighten the heatshield bolt. Fit both pivot bolt caps to the footrest brackets.
12 Align the swingarm with the suspension linkage and insert the pivot bolt from the left side. Fit the nut to the bolt and tighten it to the specified torque.
13 Where necessary, fit the rear bearing holder and associated components (see Chapter 7).
14 Hook the drive chain onto the rear sprocket then slide the brake caliper into position, ensuring the pads pass each side of the disc. Fit the **new** caliper mounting bolts and tighten them to the specified torque (see Chapter 7).
15 Locate the chainguard on the swingarm and secure it in position with the retaining clips and bolts, tightening them securely. Fit the brake hose guide to the rear bearing holder clamp bolt and loosely install the bolt in the swingarm.
16 Apply a few drops of locking compound to the threads of the air guide bolts. Install the guide on the base of the swingarm then fit the collars and retaining bolts, tightening them to the specified torque.
17 Fit the rear wheel (see Chapter 7) then adjust the drive chain slack as described in Chapter 1.
18 Check the operation of the rear brake and suspension before taking the machine on the road.

13.7 Withdraw the pivot bolt and left footrest bracket then manoeuvre the swingarm out of position

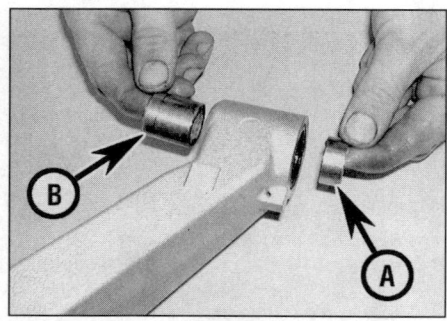

13.9 Ensure the outer spacer (A) and inner spacer (B) are correctly fitted to the right pivot before installing the swingarm

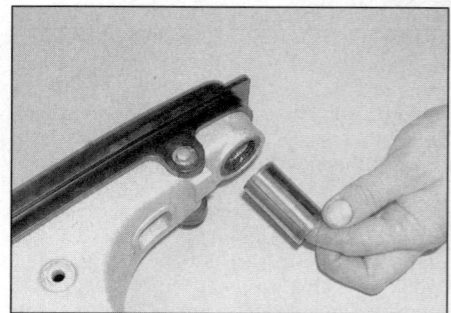

14.1a Withdraw the inner sleeves and check the needle roller bearings for wear or damage

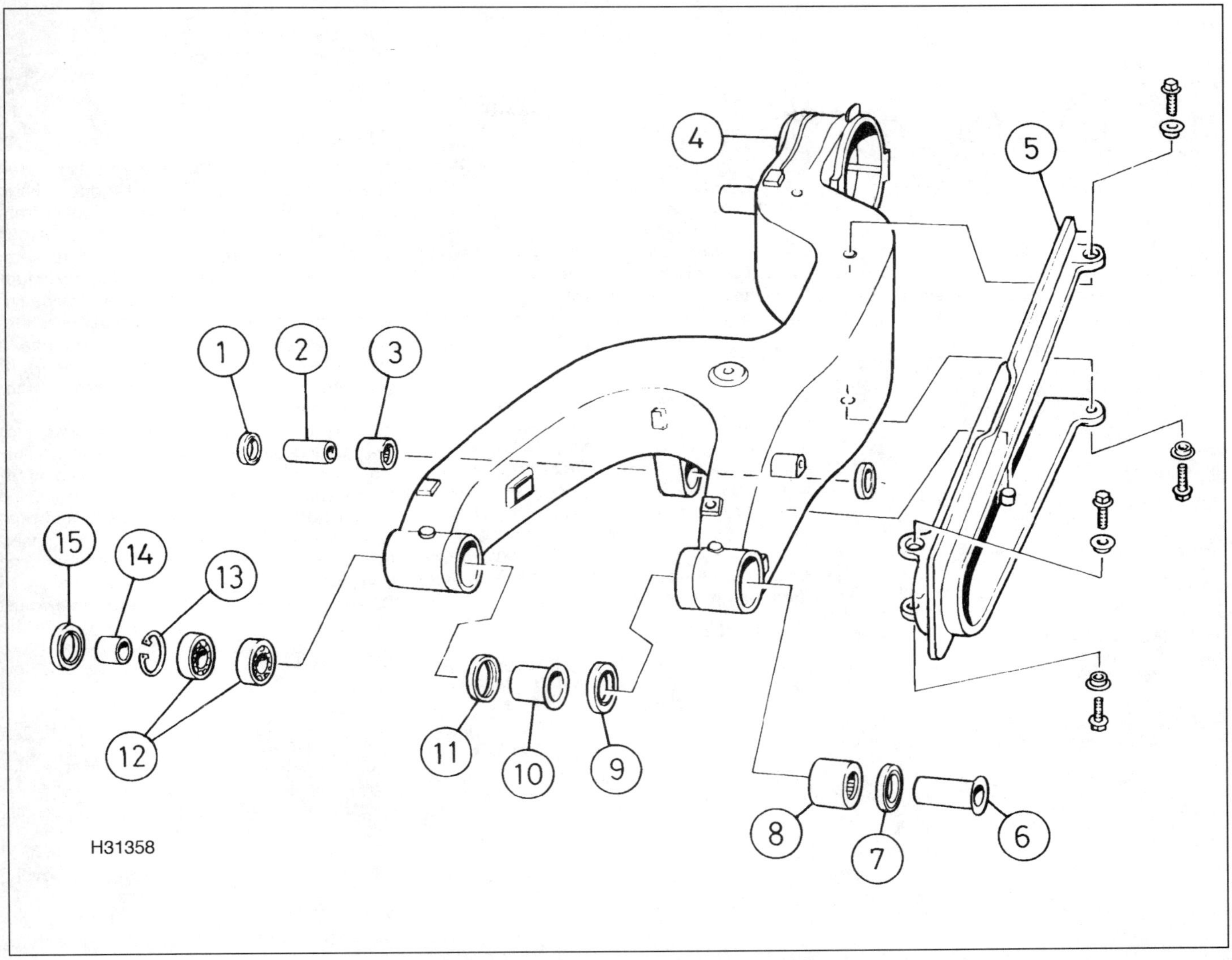

Frame, suspension and final drive 6•17

14.1b Exploded view of swingarm assembly

1 Dust seal	4 Swingarm	8 Left pivot needle roller bearing
2 Suspension linkage pivot bearing inner sleeve	5 Drive chain slider	9 Dust seal
3 Suspension linkage pivot needle roller bearing	6 Left pivot bearing inner sleeve	10 Right pivot inner spacer
	7 Dust seal	

11 Dust seal
12 Right pivot ball bearings
13 Circlip
14 Right pivot outer spacer
15 Dust seal

securely support the inner side of the swingarm right pivot. Locate the inner bearing in its bore, with its marked surface facing outwards, and tap it squarely into position using a tubular spacer (such as a socket) which bears only on the bearing outer race. Ensure the inner bearing is firmly against its stop then tap the outer bearing (marked surface facing outwards) squarely into position until it is firmly in contact with the inner bearing. Secure both bearings in position with the circlip, making sure it is correctly located in its groove, then press the new dust seals into position.

6 Lubricate all the bearings and seals with multi-purpose grease then fit the inner sleeves to the left pivot and suspension linkage bearings and install the outer (shorter) and inner (longer) spacer in the right pivot.

7 Inspect the swingarm drive chain slider for signs of wear or damage along its complete length. If it is worn down to below the wear marking on the front of the slider (see Chapter 1) it must be renewed. To renew the slider, slacken and remove the retaining bolts and collars and remove the slider from the swingarm. Install the new slider, ensuring its locating pins are correctly seated in the swingarm holes **(see illustration)**. Apply fresh locking compound to each retaining bolt then install the collars and retaining bolts, tightening them to the specified torque.

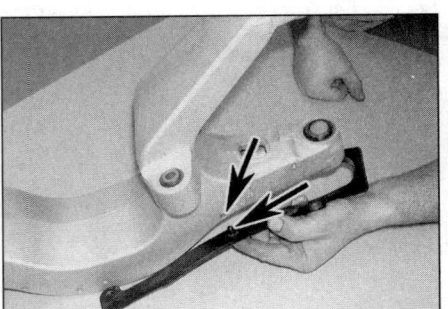

14.7 Ensure the drive chain slider pins are correctly located in the swingarm holes (arrows)

6•18 Frame, suspension and final drive

15.8 Ensure that the ends of the joining link are properly staked

15 Drive chain –
removal, cleaning and installation

Removal

Note: *The original equipment drive chain fitted to these models has a staked-type master (joining) link which can be disassembled using either Honda service tool, Pt. No. 07HMH-MR10103, or one of several commercially-available drive chain cutting/staking tools. Such chains can be recognised by the master link side plate's identification marks (and usually its different colour), as well as by the staked ends of the link's two pins which look as if they have been deeply centre-punched, instead of peened over as with all the other pins.*

⚠ **Warning: NEVER install a drive chain which uses a clip-type master (split) link. Use ONLY the correct service tools to secure the staked-type of master link – if you do not have access to such tools, have the chain renewed by a dealer to be sure of having it securely installed.**

1 Position the bike on its centrestand and locate the joining link in a suitable position to work on by rotating the back wheel.
2 Slacken the drive chain as described in Chapter 1.
3 Split the chain at the joining link using the chain cutter, following carefully the manufacturer's operating instructions (see also Section 8 in *Tools and Workshop Tips* in the

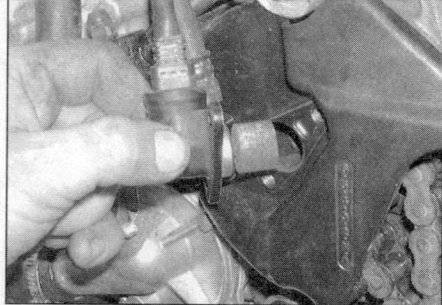

16.6 Undo the two bolts and free the speed sensor from the sprocket cover

Reference Section). Free the chain from the rear sprocket then, with the transmission in neutral, pull it off the front sprocket.

> **HAYNES HINT** *If a spare length of chain is available, hook this onto one end of the fitted chain before pulling the chain off of the front sprocket. Pull the fitted chain off the front sprocket and detach it, leaving the spare chain in position. The spare chain can then be used to draw the fitted chain back into position with the sprocket cover in position.*

Cleaning

4 Soak the chain in paraffin (kerosene) for approximately five or six minutes.
Caution: Don't use gasoline (petrol), solvent or other cleaning fluids which might damage its internal sealing properties. Don't use high-pressure water. Remove the chain, wipe it off, then blow dry it with compressed air immediately. The entire process shouldn't take longer than ten minutes – if it does, the O-rings in the chain rollers could be damaged.

Installation

⚠ **Warning: If you do not have access to a chain riveting tool, have the chain fitted by a dealer.**

5 If necessary, remove the front sprocket cover from the engine as described in Paragraphs 5 to 8 of Section 16.
6 Install the drive chain around the front sprocket, making sure it passes between the swingarm and footrest bracket, and leave the two ends in a convenient position to work on.
7 Refer to Section 8 in *Tools and Workshop Tips* in the Reference Section. Install the new joining link from the inside with the four O-rings correctly located between the link plates. Install the new side plate with its identification marks facing out. Measure the amount that the joining link pins project from the side plate and check they are within the measurements specified at the beginning of the Chapter. Stake the new link using the drive chain cutting/staking tool, following carefully the instructions of both the chain manufacturer and the tool manufacturer. DO NOT re-use old joining link components.
8 After staking, check the joining link and staking for any signs of cracking **(see illustration)**. If there is any evidence of cracking, the joining link, O-rings and side plate must be renewed. Measure the diameter of the staked ends in two directions and check that it is evenly staked and within the measurements specified at the beginning of the Chapter.
9 Install the sprocket cover (see Section 16) then adjust and lubricate the chain following the procedures described in Chapter 1.
Caution: Use only the recommended lubricant.

16 Sprockets –
check and renewal

Check

1 To gain access to the front sprocket, drive chain guide plate and swingarm drive chain slider, remove the sprocket cover as described in Paragraphs 5 to 8.
2 Check the wear pattern on both sprockets (see Chapter 1, Section 1). If the sprocket teeth are worn excessively, renew the chain and both sprockets as a set. Whenever the sprockets are inspected, the drive chain should be inspected also (see Chapter 1). If you are renewing the chain, renew the sprockets as well.
3 Also check the drive chain guide plate and the drive chain slider on the swingarm for signs of wear or damage and renew if necessary (see Chapter 1).
4 Install the sprocket cover (see Paragraphs 13 to 16) then adjust and lubricate the chain following the procedures described in Chapter 1.
Caution: Use only the recommended lubricant.

Renewal

Front sprocket

5 Remove the left side fairing lower panel (see Chapter 8).
6 Unscrew the mounting bolts and free the speed sensor from the sprocket cover **(see illustration)**. **Note:** *There is no need to disconnect its wiring.*
7 Unscrew the clutch slave cylinder bolts and free the cylinder from the sprocket cover, taking care not to lose the locating dowels. Remove the gasket and discard it; a new one should be used on refitting. **Note:** *There is no need to disconnect the hydraulic hose from the cylinder.*
Caution: Do not operate the clutch lever whilst the cylinder is detached. To prevent the piston being accidentally expelled, retain it with a cable passed through the cylinder mounting bolts holes and securely tightened around the cylinder (see illustration).

16.7 Prevent the clutch slave cylinder piston from being expelled by retaining it with a cable tie (arrow)

Frame, suspension and final drive 6•19

8 Unscrew the sprocket cover bolts and remove the cover from the engine (see illustration). Remove the drive chain guide plate from the crankcase along with the cover locating dowels (see illustration).
9 Have an assistant apply the rear brake hard, to prevent rotation, then unscrew the sprocket bolt and remove the washer.
10 Slide the sprocket and chain off the shaft and slip the sprocket out of the chain. If there is not enough slack on the chain to remove the sprocket, slacken the bearing holder clamp bolt and adjust the drive chain until it is fully slack (see Chapter 1).
11 Engage the new sprocket with the chain, making sure the sprocket '530' marking is facing outwards, then slide it on the shaft. Take up any slack in the chain.
12 Install the sprocket bolt with its washer and tighten it to the torque setting specified at the beginning of the Chapter, using the method employed on removal to prevent the sprocket from turning.
13 Fit the locating dowels to the crankcase and locate the drive chain guide on the dowels. Fit the sprocket cover and securely tighten its retaining bolts. **Note:** *The longer cover retaining bolt is slightly shorter than the two clutch cylinder retaining bolts (90 as opposed to 95 mm); do not get these bolts mixed up.*
14 Ensure the cover and clutch slave surfaces are clean and dry and apply a smear of silicone grease to the tip of the clutch pushrod. Ensure the locating dowels are correctly positioned and fit a new gasket to the cover. Remove the cable tie (where fitted) then refit the slave cylinder, tightening its retaining bolts securely.
15 Align the speed sensor drive with the front

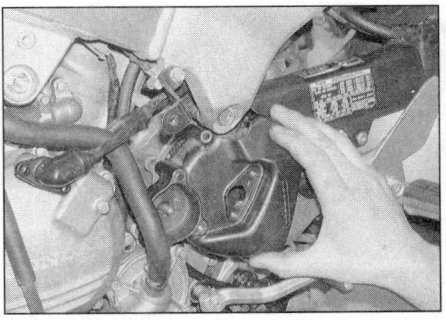

16.8a Remove the sprocket cover from the engine . . .

sprocket bolt and seat the sensor on the sprocket cover. Ensure the sensor is correctly located then securely tighten its retaining bolts.
16 Install the lower fairing panel (see Chapter 8) then adjust and lubricate the chain following the procedures described in Chapter 1.
Caution: Use only the recommended lubricant.

Rear sprocket

17 Remove the retaining bolts and inner retaining clips and free the chainguard from the swingarm.
18 Slacken and remove the retaining bolts and collars and remove the air guide from the base of the swingarm.
19 Slacken the bearing holder clamp bolt and adjust the drive chain until it is fully slack (see Chapter 1) then unhook the chain from the rear sprocket.
20 Slacken and remove the nuts and washers securing the sprocket to the coupling

16.8b . . . and lift off the drive chain guide plate

whilst retaining the sprocket bolts with an Allen key. Remove the sprocket noting which way round it fits.
21 Inspect the sprocket bolts and renew any which show signs of wear or damage.
22 Install the sprocket onto the coupling with the stamped tooth number facing out. Fit the washers and nuts to the sprocket bolts and tighten them to the specified torque whilst retaining the bolts with an Allen key.
23 Hook the chain onto the rear sprocket then refit the chainguard bolts, tightening them securely, and retaining clips.
24 Apply a few drops of locking compound to the threads of the air guide bolts. Install the guide on the base of the swingarm then fit the collars and retaining bolts, tightening them to the specified torque.
25 Adjust and lubricate the chain following the procedures described in Chapter 1.
Caution: Use only the recommended lubricant.

17 Rear sprocket coupling – inspection and overhaul

1 Remove the rear sprocket coupling assembly from the stub axle as described in Section 18 of Chapter 7, noting that there is no need to remove the rear wheel (see illustration).
2 Ease out the driven flange from the rear of the coupling hub, leaving the rubber dampers in position inside, and remove the spacer (see illustrations).

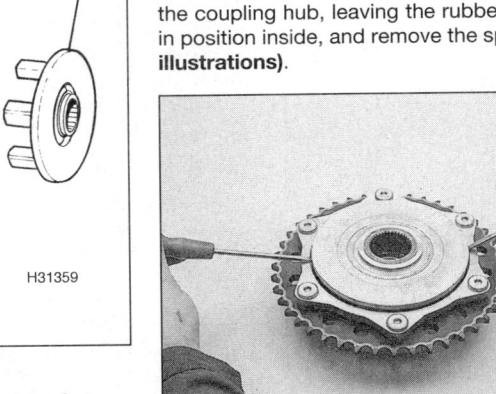

17.2a Ease the driven flange out from the rear of the coupling hub . . .

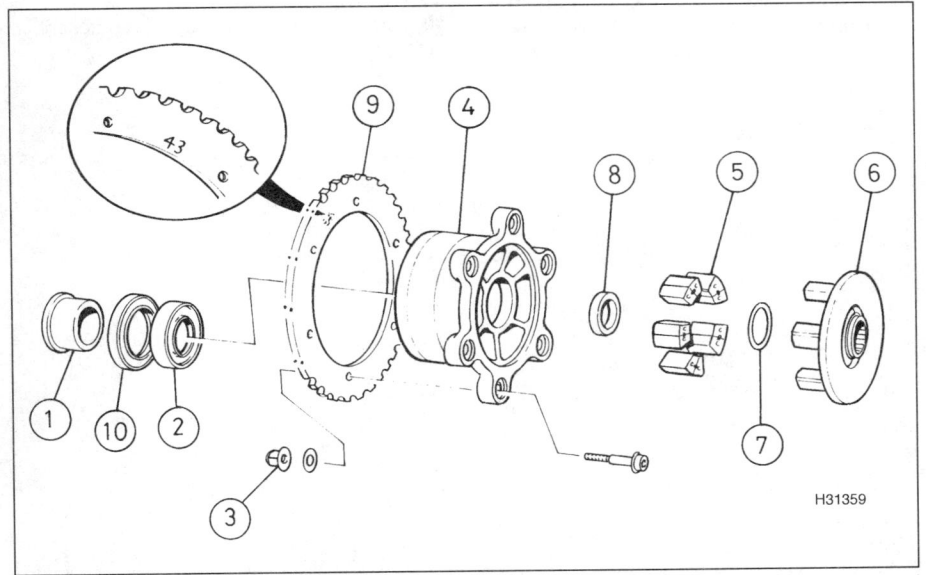

17.1 Exploded view of the rear sprocket coupling assembly

1 Flanged spacer	4 Coupling hub	6 Driven flange	9 Rear sprocket
2 Bearing	5 Rubber damper	7 O-ring	10 Dust seal
3 Sprocket nut	segment	8 Spacer	

6•20 Frame, suspension and final drive

17.2b ... and recover the spacer

17.3 Lift out the rubber damper segments and inspect them for wear or damage

17.5 The dust seal can be levered out from the hub with a large, flat-bladed screwdriver

3 Lift out the rubber damper segments and check them for cracks, hardening and general deterioration **(see illustration)**. Renew the rubber dampers as a set if necessary. Check the driven flange O-ring and renew it if it shows signs of damage or deterioration.
4 Inspect the coupling assembly components closely, looking for obvious signs of wear such as heavy scoring, or for damage such as cracks or distortion. The coupling bearing inner race should rotate freely and easily without any sign of freeplay or roughness. Renew worn components as necessary.
5 If the dust seal is damaged, the old seal can be levered out of position using a flat-bladed screwdriver **(see illustration)**. Ensure the sealing lip of the new seal is facing inwards then press the seal squarely into position until it is flush with the hub.
6 To renew the bearing, prise out the dust seal then securely support the hub outer face and drift/press the bearing squarely out of position **(see illustration)**. Turn the bearing holder over and locate the new bearing in its bore then tap/press the bearing squarely into position using a tubular spacer (such as a socket) which bears only on the bearing outer race. Ensure the bearing is firmly against its stop then press the new dust seal into position.
7 On reassembly, ensure the spacer is correctly fitted to the rear of the coupling bearing and all the damper segments are correctly installed.
8 Ensure the O-ring is correctly seated in the driven flange recess and lubricate it with a smear of multi-purpose grease. Align the flange with the rubber dampers and press it fully into the coupling **(see illustration)**.
9 Lubricate the dust seal lip with multi-purpose grease then fit the flanged spacer to the coupling dust seal.
10 Install the sprocket coupling as described in Chapter 7.

17.6 Support the hub and drive the bearing out from the inside

17.8 Ensure the rubber dampers and the O-ring are correctly fitted then fit the driven flange to the coupling hub

Chapter 7
Brakes, wheels and tyres

Contents

Brake fluid level check . . . . see *Daily (pre-ride) checks*	
Brake hoses, pipes and unions – inspection and renewal 11	
Brake light switches – check and replacement . . . . see Chapter 9	
Brake pad wear check . . . . see Chapter 1	
Brake system bleeding 12	
Brake system check . . . . .see Chapter 1	
Dual combined brake system (DCBS) components – removal, overhaul and installation 10	
Front brake calipers – removal, overhaul and installation 3	
Front brake disc – inspection, removal and installation 4	
Front brake master cylinder – removal, overhaul and installation . . . 5	
Front brake pads – renewal 2	
Front wheel – removal and installation 15	
Front wheel bearings – renewal 16	
General information 1	
Rear brake caliper – removal, overhaul and installation 7	
Rear brake disc – inspection, removal and installation 8	
Rear brake master cylinder – removal, overhaul and installation . . . 9	
Rear brake pads – renewal 6	
Rear wheel – removal and installation 17	
Rear wheel bearing holder – removal, overhaul and installation 18	
Tyres – general information and fitting 19	
Tyres – pressure, tread depth and condition see *Daily (pre-ride) checks*	
Wheel bearings – check . . . . .see Chapter 1	
Wheels – alignment check 14	
Wheels – general check . . . . .see Chapter 1	
Wheels – inspection and repair 13	

Degrees of difficulty

Easy, suitable for novice with little experience	**Fairly easy,** suitable for beginner with some experience	**Fairly difficult,** suitable for competent DIY mechanic	**Difficult,** suitable for experienced DIY mechanic 	**Very difficult,** suitable for expert DIY or professional

Specifications

Brakes

Brake fluid type . . . . DOT 4

Front caliper bore ID

	New	Service limit
Left caliper		
Upper piston	25.400 to 25.450 mm	25.460 mm
Centre and lower pistons	22.650 to 22.700 mm	22.710 mm
Right caliper		
Upper piston	27.000 to 27.050 mm	27.060 mm
Centre piston	22.650 to 22.700 mm	22.710 mm
Lower piston	25.400 to 25.450 mm	25.460 mm

Front caliper piston OD

Left caliper		
Upper piston	25.318 to 25.368 mm	25.310 mm
Centre and lower pistons	22.585 to 22.618 mm	22.560 mm
Right caliper		
Upper piston	26.916 to 26.968 mm	26.910 mm
Centre piston	22.585 to 22.618 mm	22.560 mm
Lower piston	25.318 to 25.368 mm	25.310 mm

Rear caliper bore ID

Outer pistons	22.650 to 22.700 mm	22.710 mm
Centre piston	27.000 to 27.050 mm	27.060 mm

Rear caliper piston OD

Outer pistons	22.585 to 22.618 mm	22.560 mm
Centre piston	26.916 to 26.968 mm	26.910 mm
Front master cylinder bore ID	12.700 to 12.743 mm	12.760 mm
Front master cylinder piston OD	12.657 to 12.684 mm	12.650 mm
Rear master cylinder bore ID	17.460 to 17.503 mm	17.515 mm
Rear master cylinder piston OD	17.417 to 17.444 mm	17.405 mm
Secondary master cylinder bore ID	12.700 to 12.743 mm	12.760 mm
Secondary master cylinder piston OD	12.657 to 12.684 mm	12.650 mm

7•2 Brakes, wheels and tyres

Brakes (continued)

Front disc minimum thickness
- Standard .... 4.5 mm
- Service limit .... 3.5 mm

Front disc maximum runout .... 0.3 mm

Rear disc minimum thickness
- Standard .... 6.0 mm
- Service limit .... 5.0 mm

Rear disc maximum runout .... 0.3 mm

Wheels

Maximum wheel runout (front and rear)
- Axial (side-to-side) .... 2.0 mm
- Radial (out-of-round) .... 2.0 mm

Maximum axle runout (front and rear) .... 0.2 mm

Tyres

Tyre pressures and tread depth .... see *Daily (pre-ride) checks*

Tyre sizes
- Front 120/70 ZR17 (58W)
- Rear 180/55 ZR17 (73W)

Torque settings

Bleed valve .... 6 Nm
Brake hose banjo bolts .... 34 Nm
Brake hose fitting mounting bolt .... 12 Nm
Brake pad retaining pin .... 18 Nm
Brake pipe union nut .... 17 Nm
Delay valve mounting bolts .... 12 Nm
Front brake caliper
  Left caliper
    Bracket pivot bolt .... 31 Nm
    Secondary master cylinder pushrod bolt .... 31 Nm
    Caliper half joining bolts .... 32 Nm
  Right caliper
    Bracket mounting bolts .... 31 Nm
    Caliper half joining bolts .... 32 Nm
Front brake disc bolts .... 20 Nm
Front brake master cylinder
  Mounting clamp bolt .... 12 Nm
  Brake lever pivot bolt nut .... 6 Nm
Front wheel
  Axle bolt .... 59 Nm
  Axle clamp bolts .... 22 Nm
Proportional control valve mounting bolts .... 12 Nm
Rear brake caliper
  Bracket mounting bolts .... 31 Nm
  Caliper half joining bolts .... 32 Nm
  Mounting plate torque rod bolts .... 34 Nm
Rear brake disc bolt nuts .... 34 Nm
Rear brake master cylinder
  Mounting bolts .... 12 Nm
  Fluid reservoir mounting bolt .... 12 Nm
  Pushrod clevis locknut .... 18 Nm
Rear wheel bearing holder clamp bolt .... 74 Nm
Rear wheel nuts .... 108 Nm
Silencer mounting bolt nut .... 21 Nm
Stub axle nut .... 201 Nm
Swingarm air guide bolts .... 9 Nm

Brakes, wheels and tyres 7•3

1 General information

All models covered in this manual are fitted with cast alloy wheels designed for tubeless tyres only. Both front and rear brakes are hydraulically operated disc brakes. The machine is fitted with Honda's Dual Combined Braking System (DCBS), where the front and rear brakes are applied simultaneously.

Each front brake is operated by a three-piston caliper. The upper and lower piston of each caliper are hydraulically linked to the front master cylinder and the centre pistons are linked to the rear master cylinder. The right side caliper is fixed to the fork slider but the left side caliper is mounted onto a bracket which is allowed to pivot on the left fork slider. This mounting bracket incorporates the secondary master cylinder which is connected to the inner piston of the rear brake caliper, via the proportional control valve.

The rear brake is also operated by a three-piston caliper. The outer pistons of the caliper are linked to the rear master cylinder and the centre piston to the secondary master cylinder (via the proportional control valve). The rear master cylinder is also connected to the centre piston of each front brake caliper, via a delay valve. The system operates as follows.

Front brake operation

When the front brake lever is applied, hydraulic pressure from the front master cylinder is applied to the upper and lower pistons of each front brake caliper (as in a conventional braking system) **(see illustration)**. As the calipers grip the discs, the torque causes the left side caliper mounting bracket to pivot around its lower mounting. The forward motion of the mounting bracket transfers this force to the secondary master cylinder pushrod which then hydraulically applies the centre piston of the rear brake caliper.

The hydraulic pressure to the rear brake caliper centre piston is controlled by the proportional control valve (located under the seat). The valve is effectively a pressure regulating valve and is fitted to prevent the rear wheel locking under extreme braking.

Rear brake operation

When the rear brake pedal is applied, hydraulic pressure from the master cylinder is applied to the outer pistons of the rear brake caliper (as in a conventional braking system) and also to the centre piston of each front brake caliper **(see illustration)**.

A delay valve is incorporated in the hydraulic supply to the front brake calipers. When the rear brake pedal is applied, the delay valve allows hydraulic pressure to act immediately only on the left side caliper centre piston and isolates the right side caliper. Only when the pressure in the hydraulic system rises above a preset amount does the delay valve open and also apply the right side caliper centre piston. The action of the delay valve prevents the front forks 'diving' when the rear brake is applied hard and ensures a more natural feel to the braking system.

Caution: *Disc brake components rarely require disassembly. Do not disassemble components unless absolutely necessary. If a hydraulic brake line is loosened, the entire system must be disassembled, drained, cleaned and then properly filled and bled upon reassembly. Do not use solvents on internal brake components. Solvents will cause the seals to swell and distort. Use only clean brake fluid or denatured alcohol for cleaning. Use care when working with brake fluid as it can injure your eyes and it will damage painted surfaces and plastic parts.*

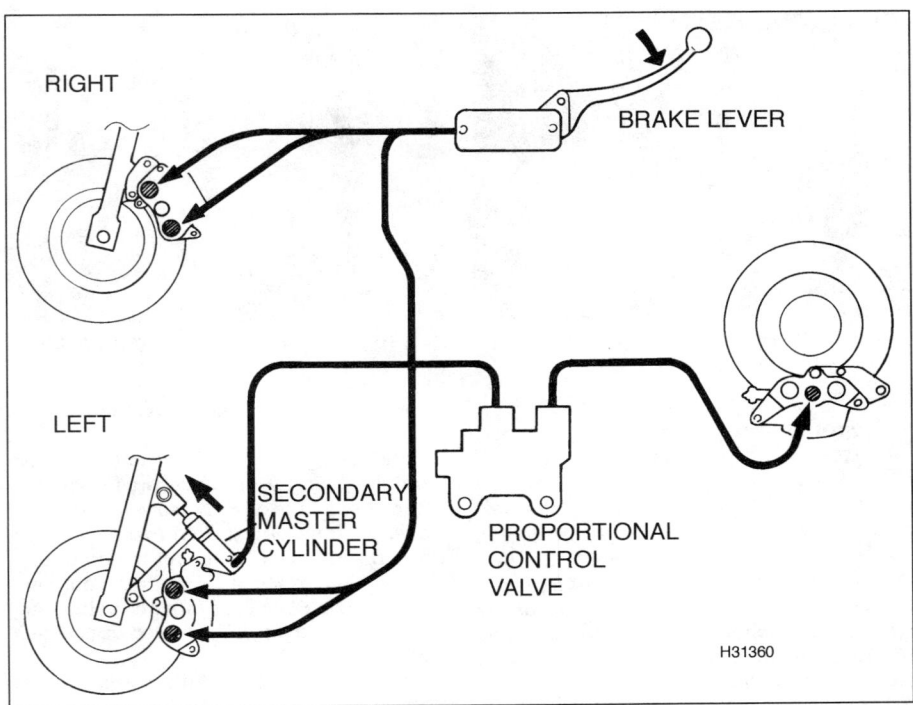

1.1 Dual combined braking system (CBS) front brake system

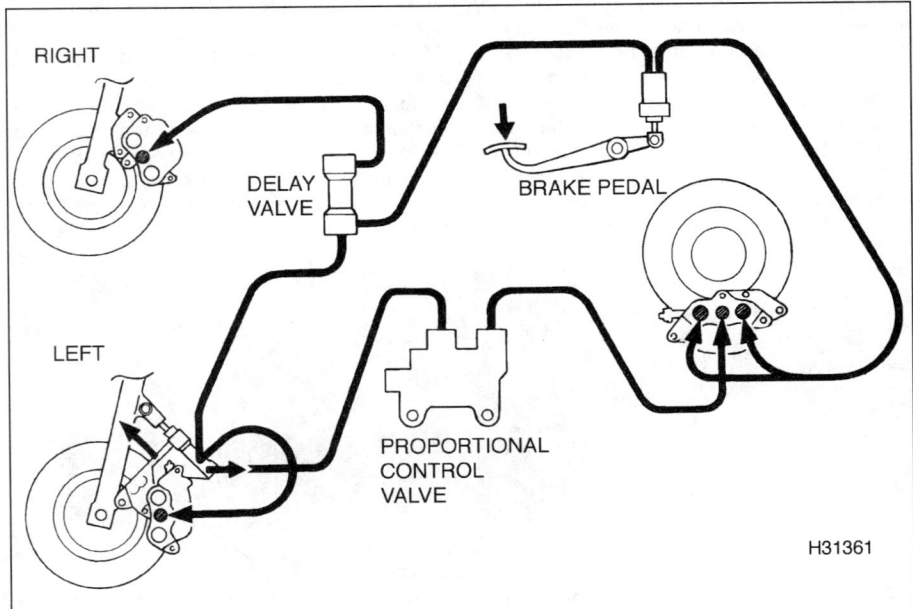

1.2 Dual combined braking system (CBS) rear brake system

7•4 Brakes, wheels and tyres

2.1a Remove the pad retaining pin plug from the caliper . . .

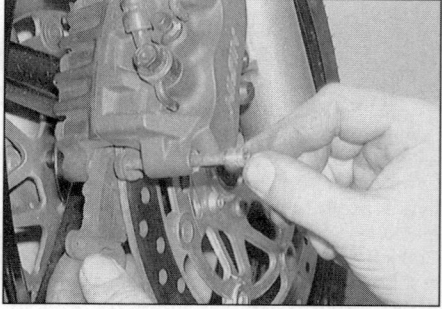

2.1b . . . then slacken and remove the retaining pin . . .

2.1c . . . and withdraw both pads from the caliper

2 Front brake pads – renewal

⚠️ **Warning:** *The dust created by the brake system may contain asbestos, which is harmful to your health. Never blow it out with compressed air and don't inhale any of it. An approved filtering mask should be worn when working on the brakes.*

1 Unscrew or prise out the pad retaining pin plug followed by the pad retaining pin and withdraw the pin, noting how it fits **(see illustrations)**. Slide the pads out of the caliper, taking care not to displace the pad retainer from the top of the caliper bracket **(see illustration)**.

2 Inspect the surface of each pad for contamination and check that the friction material has not worn down level with or beyond the wear limit grooves in the pad edge **(see illustration)**. If either pad is worn down to, or beyond, the wear limit, is fouled with oil or grease, or heavily scored or damaged by dirt and debris, both pads must be renewed as a set. Note that it is not possible to degrease the friction material; if the pads are contaminated in any way they must be renewed. Also check that each pad has worn evenly at each end, and that each has the same amount of wear as the other. If uneven wear is noticed, one or more of the pistons are probably sticking in the caliper, in which case the calipers must be overhauled (see Section 3).

3 If the pads are in good condition clean them carefully, using a fine wire brush which is completely free of oil and grease to remove all traces of road dirt and corrosion. Using a pointed instrument, clean out the grooves in the friction material and dig out any embedded particles of foreign matter. Any areas of glazing may be removed using emery cloth.

4 Check the condition of the brake disc (see Section 4).

5 Remove all traces of corrosion from the pad pin. Inspect the pin for signs of damage and renew it if necessary.

6 If new pads are being installed, push the pistons as far back into the caliper as possible. A good way of doing this is to insert one of the old pads between the outside of the disc and the piston, then push the caliper against the pad and disc using hand pressure. Due to the increased friction material thickness of new pads, it may be necessary to remove the master cylinder reservoir cover and diaphragm and remove some fluid.

7 Smear the backs of the pads and the shank of the pad pin with copper-based grease, making sure that none gets on the front or sides of the pads.

8 Installation of the pads is the reverse of removal. Make sure the pad spring is correctly positioned in the caliper body and the pad retainer is in place on the caliper bracket. Insert the pads into the caliper so that the friction material faces the disc, making sure they locate correctly in the pad retainer **(see illustration)**. Push up on the pads to align the holes and slide the pad retaining pin through. Make sure the pin passes through the hole in each pad. Tighten the pad retaining pin to the torque setting specified then install the pad pin plug **(see illustration)**.

9 Operate the brake lever and brake pedal several times to bring the pads into contact with the disc then check the master cylinder reservoir fluid level (see *Daily (pre-ride) checks*). Once the fluid level is correct, check the operation of the brake before riding the motorcycle.

3 Front brake calipers – removal, overhaul and installation

⚠️ **Warning:** *If a caliper indicates the need for an overhaul (usually due to leaking fluid or sticky operation), all old brake fluid should be flushed from the system. Also, the dust created by the brake system may contain asbestos, which is harmful to your health. Never blow it out with compressed air and don't inhale any of it. An approved filtering mask should be worn when working on the brakes. Do not, under any circumstances, use petroleum-based solvents to clean brake parts. Use clean brake fluid, brake cleaner or denatured alcohol only.*

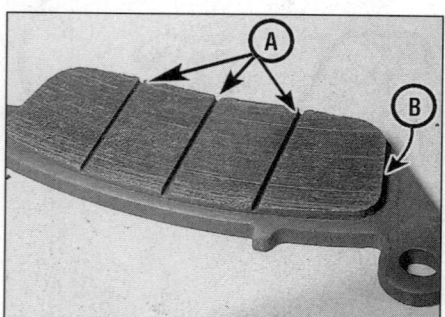

2.2 Brake pads must be replaced when the wear grooves (A) are no longer visible or when the cutout (B) is exposed

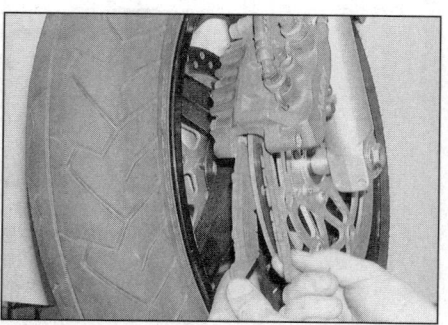

2.8a Slide the pads into position ensuring each pads friction material is against the disc

2.8b Tighten the pad retaining pin to the specified torque then securely refit the pin plug

Brakes, wheels and tyres 7•5

3.3 Unscrew the banjo bolts (arrows) and detach the brake hoses from the caliper

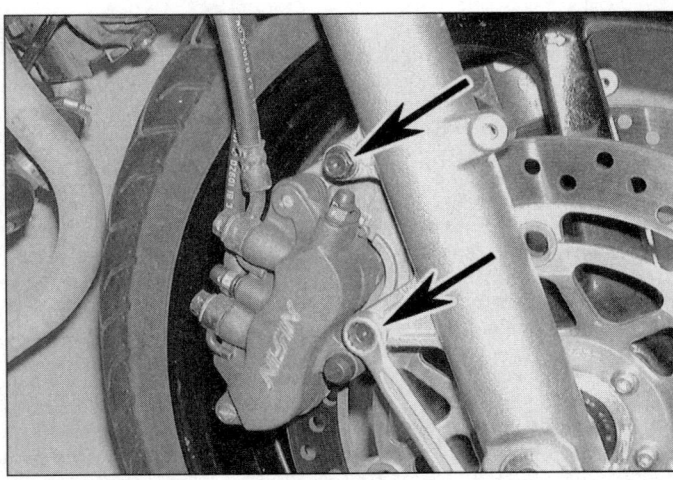

3.4 Unscrew the caliper mounting bolts (arrows) and remove the right caliper from the bike

Removal

1 Remove the brake pads from the caliper (see Section 2).

2 If the caliper pistons are to be removed, loosen the three bolts securing the inner half of the caliper body to the main body.

3 Note the correct fitted locations of the brake hose end fittings on the caliper then unscrew the brake hose banjo bolts and separate both hoses from the caliper (see illustration). Plug the hose ends or wrap a plastic bag tightly around each hose to minimise fluid loss and prevent dirt entering the system. Discard the sealing washers as new ones must be used on installation. Note: If you are planning to overhaul the caliper and don't have a source of compressed air to blow out the pistons, just loosen the banjo bolts at this stage and retighten them lightly. The bike's hydraulic system can then be used to force the pistons out of the body once the caliper has been freed from the fork slider. Disconnect the hoses once the pistons have been sufficiently displaced.

Right caliper

4 Slacken and remove the caliper bracket mounting bolts then remove the assembly from the fork slider (see illustration). The caliper and bracket can then be separated. Take care not to lose the pad spring from the caliper body or the pad retainer from the bracket. Note: Honda recommend that new brake caliper bracket bolts should be used on installation (the **new** bolts are supplied with their threads pre-coated with locking compound). If the old bolts are to be reused, remove all original locking compound from their threads and apply fresh compound before installing them.

Left caliper

5 Unscrew the caliper mounting bracket pivot bolt and the secondary master cylinder pushrod bolt and free the caliper and mounting bracket assembly from the fork slider (see illustration). Slide off the caliper and remove it from the mounting bracket assembly. Take care not to lose the pad spring from the caliper body or the pad retainer from the bracket. Note: Honda recommend that a new bracket pivot bolt and secondary master cylinder pushrod bolt should be used on installation (the bolts are supplied with their threads pre-coated with locking compound). If the old bolts are to be reused, remove all original locking compound from their threads and apply fresh compound before installing them.

6 Withdraw the inner sleeve from the caliper bracket mounting and inspect the bearing arrangement for signs of wear (see illustration). If there is any sign of wear, renew the bearings, the inner sleeve and both dust seals (see Section 7 of Chapter 6).

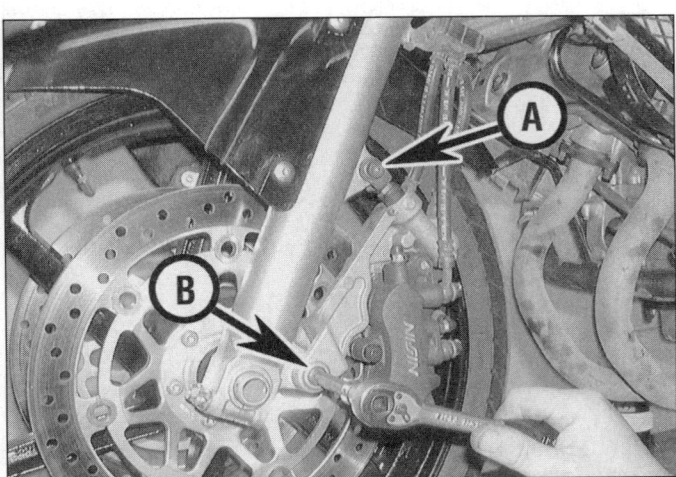

3.5 Unscrew the secondary master cylinder pushrod bolt (A) and the mounting bracket pivot bolt (B) and remove the left brake caliper

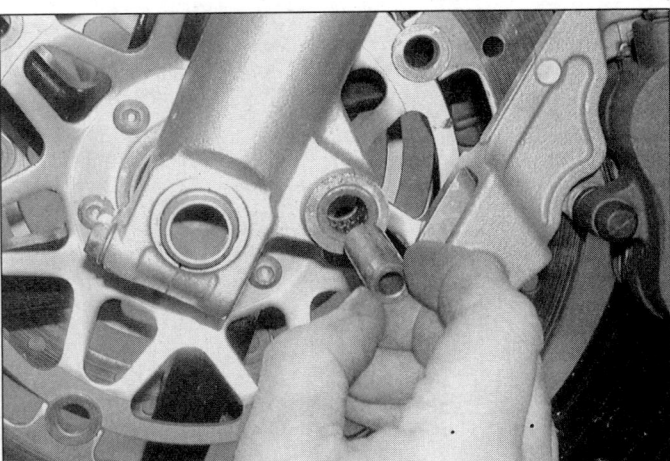

3.6 Withdraw the left caliper mounting bracket pivot sleeve and check the bearing for wear or damage

Overhaul

7 Remove the pad spring from inside the caliper, noting which way around it is fitted **(see illustration)**. Clean the exterior of the caliper with denatured alcohol or brake system cleaner. Make sure all old grease is removed from the slider pins.

8 Unscrew the three bolts and separate the caliper inner half from the main body. **Note:** *Honda recommend that new bolts should be used on reassembly (the bolts are supplied with their threads pre-coated with locking compound). If the old bolts are to be reused, remove all original locking compound from their threads and apply fresh compound before installing them.*

9 Withdraw the partially ejected pistons from the caliper body. **Note:** *On the left side caliper note that the centre and lower pistons are the same diameter; the centre piston being the shorter of the two.*

HAYNES HINT *If the pistons cannot be withdrawn by hand, they can be pushed out by applying compressed air to the brake hose union hole. Only low pressure should be required, such as is generated by a foot pump. Cover the caliper with a wad of rag to prevent the pistons being forcibly expelled.*

10 Using a wooden or plastic tool, remove the dust seals from the caliper bores. Discard them as new ones must be used on installation. If a metal tool is being used, take great care not to damage the caliper bores.

11 Remove and discard the piston seals in the same way.

12 Clean the pistons and bores with denatured alcohol, clean brake fluid or brake system cleaner. If compressed air is available, use it to dry the parts thoroughly (make sure it's filtered and unlubricated).

Caution: Do not, under any circumstances, use a petroleum-based solvent to clean brake parts.

13 Inspect the caliper bores and pistons for signs of corrosion, nicks and burrs and loss of plating. If surface defects are present, the caliper assembly must be renewed. If the necessary measuring equipment is available, compare the dimensions of the pistons and bores to those given in the Specifications Section of this Chapter, renewing any component that is worn beyond the service limit. If the caliper is in bad shape the master cylinder should also be checked.

14 Clean off all traces of corrosion from the slider pins and their bores in the caliper and bracket. Renew the rubber dust boots if they are damaged or deteriorated. The slider pins are a screw-fit and can be renewed if damaged. On installation, apply a suitable non-permanent thread locking compound and tighten each pin securely.

15 Lubricate the new piston seals with clean brake fluid and install them in their grooves in the caliper bores.

⚠ **Warning: Ensure all seals are fitted only to their relevant caliper bore. As a precaution, have the dealer where you purchased the seals mark each seal set with its correct fitted location. This will avoid confusion on reassembly; the seal sizes are only marginally different but failure to install them correctly could lead to faulty operation/brake failure.**

16 Lubricate the new dust seals with silicone-based grease and install them in their grooves in the caliper bores.

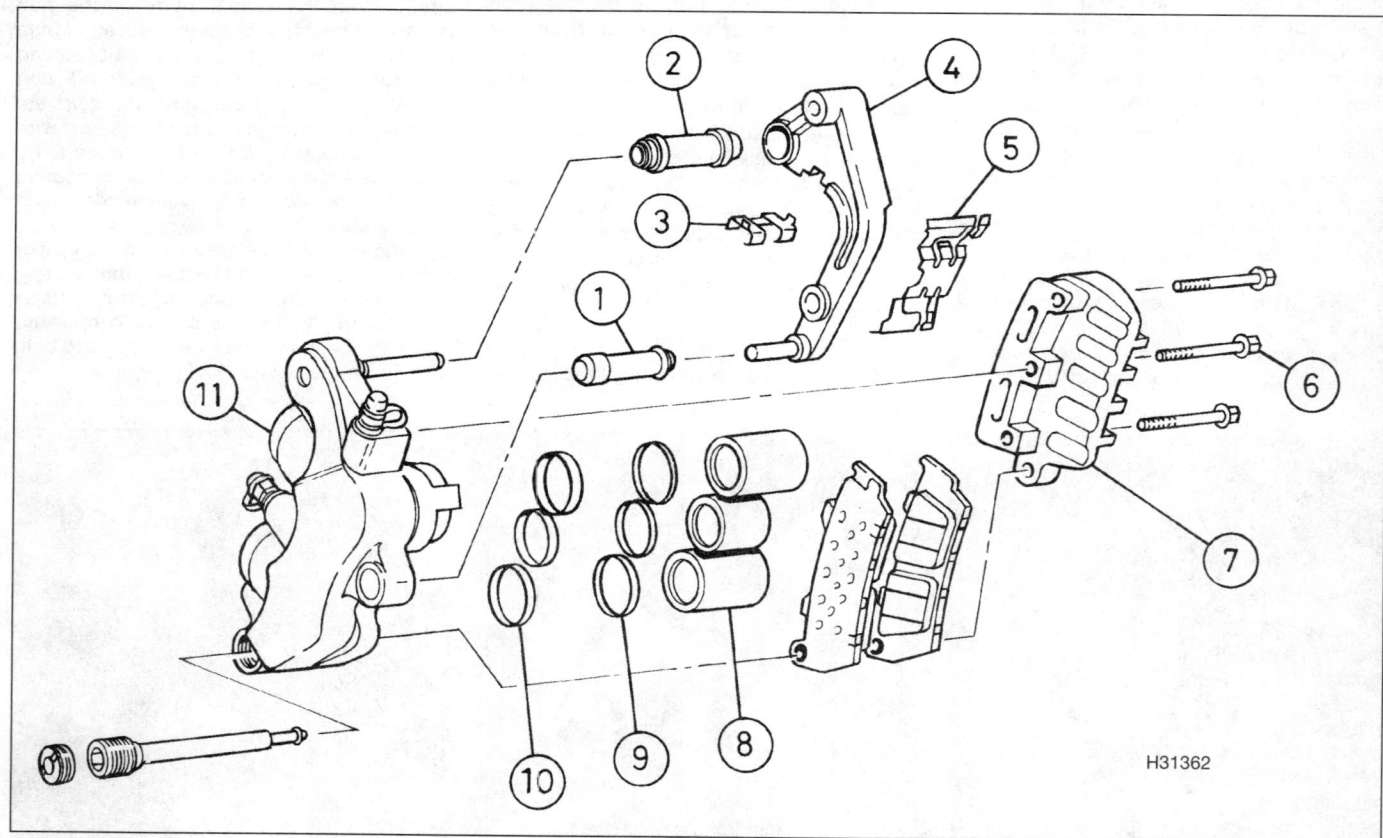

3.7 Exploded view of the right side front brake caliper (left side similar)

1 Slider pin rubber boot	4 Mounting bracket	7 Caliper inner half	10 Piston seals
2 Slider pin rubber boot	5 Pad spring	8 Pistons	11 Caliper main body
3 Pad retainer	6 Caliper body bolts	9 Dust seals	

17 Lubricate the pistons with clean brake fluid and install them closed-end first into their original caliper bores. Using your thumbs, push the pistons all the way in, making sure they enter the bore squarely.
18 Ensure the mating surfaces are clean and dry then fit the inner half of the caliper to the main body. Fit the three new retaining bolts and tighten them to the specified torque. **Note:** *The bolts can be tightened to the specified torque setting at this stage if the caliper can be held firmly enough, but do not risk damaging it. A better method is to tighten the bolts when the caliper has been reinstalled and is securely bolted to the slider.*
19 Clip the pad spring securely into the caliper body, making sure it is fitted the correct way around as noted on removal.

Installation

Right caliper

20 Clean all old grease from the slider pins and check the pin rubber boots for damage and deterioration, renewing them if necessary. Ensure both boots are correctly seated then apply a smear of silicone-based grease to each pin.
21 Ensure the pad retainer is securely fitted to the mounting bracket then align the slider pins and slide the caliper fully into position. Make sure the rubber boot lips are correctly seated then check that the caliper is able to slide freely.
22 Locate the caliper assembly on the fork slider and fit the new mounting bolts, tightening them to the specified torque **(see illustration)**.
23 Position a new sealing washer on each side of each hose union and screw in the banjo bolts. Ensure both hose end fitting pins are correctly located against the caliper body then tighten both banjo bolts to the specified torque.
24 Ensure the pad spring and retainer are still correctly positioned then install the brake pads as described in Section 2.
25 Bleed both the front brake lever hydraulic circuit and the rear brake pedal hydraulic circuit as described in Section 12.
26 Check that there are no fluid leaks and thoroughly test the operation of the brake before riding the motorcycle.

Left caliper

27 Slide out the inner sleeve from the caliper bracket mounting in the fork slider and lubricate the bearings with multi-purpose grease before sliding the inner sleeve back into position.
28 Carry out the operation described in Paragraph 20 and join the caliper and mounting bracket.
29 Engage the mounting bracket with its mounting then locate the secondary master cylinder pushrod on the fork slider and fit the original bolts. Do not fit the new bolts yet as it will be necessary to remove the caliper again to bleed the rear brake pedal hydraulic circuit.

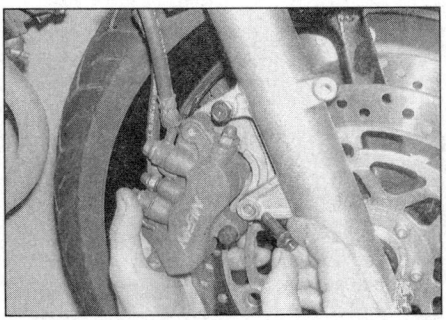

3.22 Secure the caliper in position with new mounting bolts, tightening them to the specified torque

30 Position a new sealing washer on each side of each hose union and screw in the banjo bolts. Ensure both hose end fittings are correctly located against the stops on the caliper body then tighten both banjo bolts to the specified torque.
31 Carry out the operations described in Paragraphs 24 to 26.

4 Front brake discs – inspection, removal and installation

Inspection

1 Visually inspect the surface of the disc for score marks and other damage. Light scratches are normal after use and won't affect brake operation, but deep grooves and heavy score marks will reduce braking efficiency and accelerate pad wear. If a disc is badly grooved it must be machined or renewed.
2 To check disc runout, position the bike on the centrestand and support it so that the front wheel is raised off the ground. Mount a dial gauge on one of the fork sliders, with the plunger on the gauge touching the surface of the disc about 10 mm from the outer edge **(see illustration)**. Rotate the wheel and watch the gauge needle, comparing the reading with the limit listed in the Specifications at the beginning of the Chapter. If the runout is greater than the service limit, check the wheel bearings for play (see Chapter 1). If the bearings are worn, renew them (see Section 16) and repeat this check. If

4.3a The minimum disc thickness is marked on each disc

4.2 Set up a dial gauge with its probe contacting the brake disc then rotate the wheel to check for runout

the disc runout is still excessive, it will have to be renewed, although machining by an engineer may be possible.
3 The disc must not be machined or allowed to wear down to a thickness less than the service limit listed in this Chapter's Specifications and as marked on the disc itself **(see illustration)**. The thickness of the disc can be checked with a micrometer **(see illustration)**. If the thickness of the disc is less than the service limit, it must be renewed.

Removal

4 Remove the front wheel (see Section 15).
Caution: *Do not lay the wheel down and allow it to rest on the disc – the disc could become warped. Set the wheel on wood blocks so the disc doesn't support the weight of the wheel.*
5 Mark the relationship of the disc to the wheel, so it can be installed in the same position. Unscrew the disc retaining bolts, loosening them a little at a time in a criss-cross pattern to avoid distorting the disc, then remove the disc from the wheel **(see illustration)**. **Note:** *Honda recommend that new bolts should be used on reassembly (the bolts are supplied with their threads pre-coated with locking compound). If the old bolts are to be reused, remove all original locking compound from their threads and apply fresh compound before installing them.*
6 If both discs are to be removed, note that the left and right discs are different and must not be swapped. Make identification marks on removal to avoid confusion on installation.

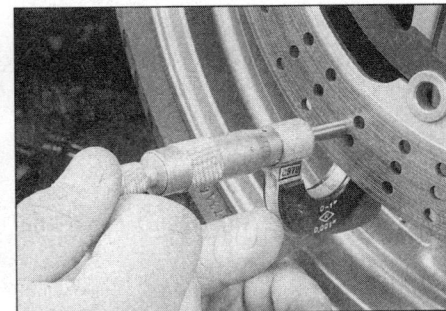

4.3b Using a micrometer to measure disc thickness

7•8 Brakes, wheels and tyres

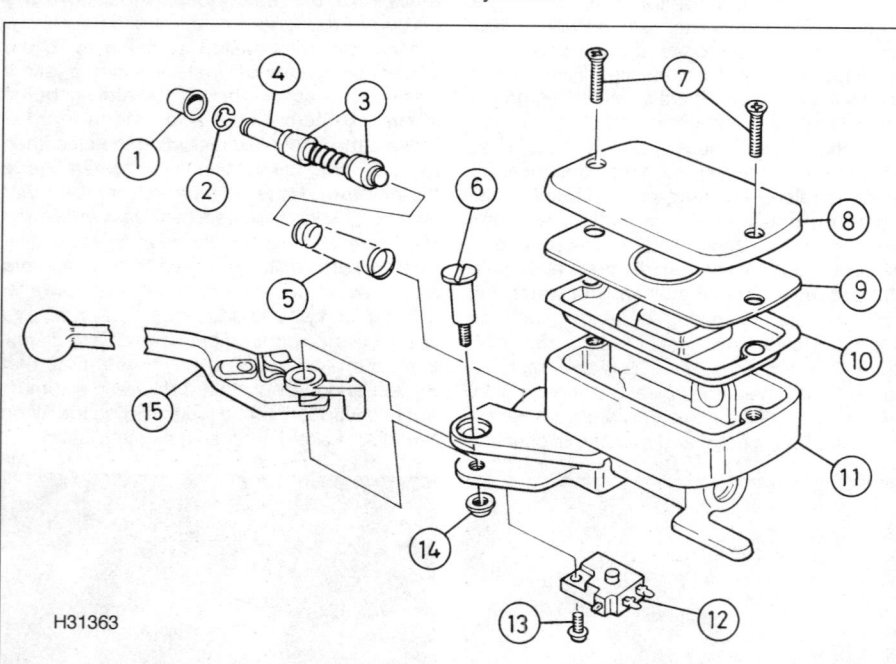

4.5 Evenly and progressively slacken the retaining bolts (arrows) to prevent the disc being warped

Installation

7 Install the disc on the wheel, making sure the marked side is on the outside and the arrow stamped on the disc is pointing in the direction of normal wheel rotation. If the original disc is being installed, align the previously applied matchmarks.
8 Fit the new disc retaining bolts and tighten them in a criss-cross pattern evenly and progressively to the specified torque.
9 Clean off all grease from the brake discs using acetone or brake system cleaner. If a new brake disc has been installed, remove any protective coating from its working surfaces.
10 Install the wheel (see Section 15).
11 Operate the brake lever several times to bring the pads into contact with the disc. Check the operation of the brake carefully before riding the bike.

5 Front brake master cylinder – removal, overhaul and installation

1 If the master cylinder is leaking fluid, or if the lever does not produce a firm feel when the brake is applied, and bleeding the brakes does not help (see Section 12), and the hydraulic hoses are all in good condition, then master cylinder overhaul is recommended.
2 Before disassembling the master cylinder, read through the entire procedure and make sure that you have obtain all new parts required. Also, you will need some new brake fluid, some clean rags and internal circlip pliers.
Caution: Disassembly, overhaul and reassembly of the brake master cylinder must be done in a spotlessly clean work area to avoid contamination and possible failure of the brake hydraulic system components. To prevent damage to the paint from spilled brake fluid, always cover the fuel tank and fairing panels when working on the master cylinder.

Removal

3 Disconnect the wiring connectors from the brake light switch on the base of the master cylinder.
4 Note the correct fitted locations of the brake hose end fitting then unscrew the banjo bolt and separate the hose from the master cylinder. Plug the hose end or wrap a plastic bag tightly around it to minimise fluid loss and prevent dirt entering the system. Discard the sealing washers as new ones must be used on installation.
5 Unscrew the mounting clamp bolts and remove the master cylinder from the handlebar.
6 Remove the reservoir cover retaining screws and lift off the cover, the diaphragm plate and the rubber diaphragm and empty the reservoir contents.

Overhaul

7 Undo the screw and remove the brake light switch from the base of the master cylinder **(see illustration)**.
8 Unscrew the locknut from the brake lever pivot screw then remove the screw and lever from the master cylinder.
9 Carefully remove the dust boot from the end of the piston **(see illustration)**.
10 Using circlip pliers, remove the circlip and slide out the piston assembly and the spring, noting how they fit **(see illustrations)**. Lay the parts out in the correct fitted order to prevent confusion during reassembly.
11 Clean all parts with clean brake fluid or denatured alcohol. If compressed air is available, use it to dry the parts thoroughly (make sure it's filtered and unlubricated).

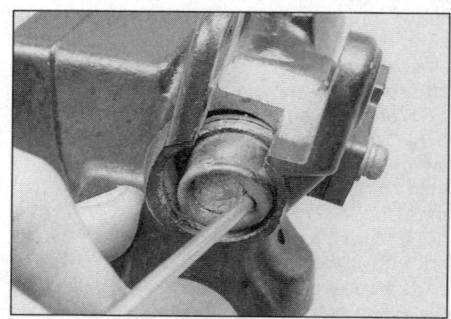

5.9 Remove the rubber boot from the end of the master cylinder piston . . .

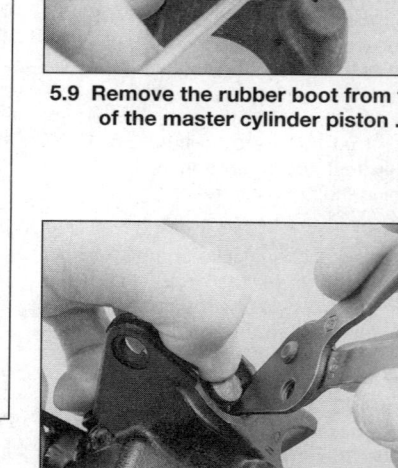

5.10a . . . then depress the piston and remove the circlip

5.7 Exploded view of the front brake master cylinder assembly

1 Dust boot	6 Pivot screw	11 Body
2 Circlip	7 Cover screws	12 Brake light switch
3 Seals	8 Cover	13 Brake light switch screw
4 Piston assembly	9 Diaphragm plate	14 Pivot screw locknut
5 Spring	10 Rubber diaphragm	15 Brake lever

Brakes, wheels and tyres 7•9

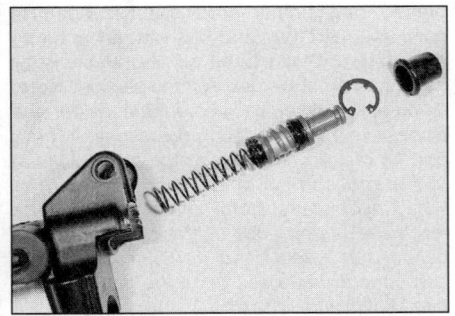

5.10b Layout the master cylinder components in their correct fitted order

Caution: Do not, under any circumstances, use a petroleum-based solvent to clean brake parts.

12 Check the master cylinder bore for corrosion, scratches, nicks and score marks. If the necessary measuring equipment is available, compare the dimensions of the piston and bore to those given in the Specifications Section of this Chapter. If damage or wear is evident, the master cylinder must be renewed. If the master cylinder is in poor condition, then the calipers should be checked as well. Check that the fluid inlet and outlet ports in the master cylinder are clear. Inspect the reservoir cover rubber diaphragm and renew it if it is damaged or deteriorated.

13 The dust boot, circlip, piston, seals and spring are included in the new piston/seal kit. Use all of the new parts, regardless of the apparent condition of the old ones. If the seals are not already on the piston, fit them according to the layout of the old piston assembly ensuring both are installed the correct way around.

14 Install the new spring in the master cylinder so that its tapered end faces the piston.

15 Lubricate the new piston and seal assembly with clean brake fluid. Ensure the assembly is the correct way around then carefully ease it into the master cylinder, making sure the seal lips do not turn inside out as they slip into the bore. Depress the piston and install the new circlip, making sure that it locates in the master cylinder groove.

16 Install the new rubber dust boot, making sure the lip is seated correctly in the piston groove and bore.

17 Fit the lever and install the pivot screw. Securely tighten the pivot screw then fit the locknut and tighten it to the specified torque.

18 Fit the brake light switch to the master cylinder and securely tighten its retaining screw.

Installation

19 Locate the master cylinder on the handlebar and fit the mounting clamp with its UP mark facing upwards. Align the outer corner of the master cylinder mounting with the punch mark on the top of the handlebar then tighten the mounting clamp upper bolt to the specified torque followed by the lower bolt.

20 Position a new sealing washer on each side of the brake hose end fitting then connect the hose to the master cylinder. Ensure the hose end fitting is correctly positioned against its stop then tighten the banjo bolt to the specified torque.

21 Connect the wiring connector to the brake light switch wiring.

22 Fill the fluid reservoir with specified brake fluid as described in *Daily (pre-ride) checks*. Refer to Section 12 of this Chapter and bleed the air from the front brake lever hydraulic system.

23 Ensure the fluid level is correct then fit the reservoir rubber diaphragm, diaphragm plate and cover and securely tighten its retaining screws.

24 Thoroughly check the operation of the front brake before riding the motorcycle.

6 Rear brake pads – renewal

 Warning: The dust created by the brake system may contain asbestos, which is harmful to your health. Never blow it out with compressed air and don't inhale any of it. An approved filtering mask should be worn when working on the brakes.

1 Unscrew or prise out the pad retaining pin plug, then unscrew the retaining pin and remove it, noting how it fits **(see illustrations)**. Slide the pads out of the caliper, taking care not to displace the pad retainer from the front of the caliper bracket.

2 Inspect the surface of each pad for contamination and check that the friction material has not worn down level with or beyond the wear limit grooves in the pad edge **(see illustration 2.2)**. If either pad is worn down to, or beyond, the wear limit, is fouled with oil or grease, or heavily scored or damaged by dirt and debris, both pads must be renewed as a set. Note that it is not possible to degrease the friction material; if the pads are contaminated in any way they must be renewed. Also check that each pad has worn evenly at each end, and that each has the same amount of wear as the other. If uneven wear is noticed, one or more of the pistons are probably sticking in the caliper, in which case the caliper must be overhauled (see Section 7).

3 If the pads are in good condition clean them carefully, using a fine wire brush which is completely free of oil and grease to remove all traces of road dirt and corrosion. Using a pointed instrument, clean out the grooves in the friction material and dig out any embedded particles of foreign matter. Any areas of glazing may be removed using emery cloth.

4 Check the condition of the brake disc (see Section 8).

6.1a Remove the plug from the caliper . . .

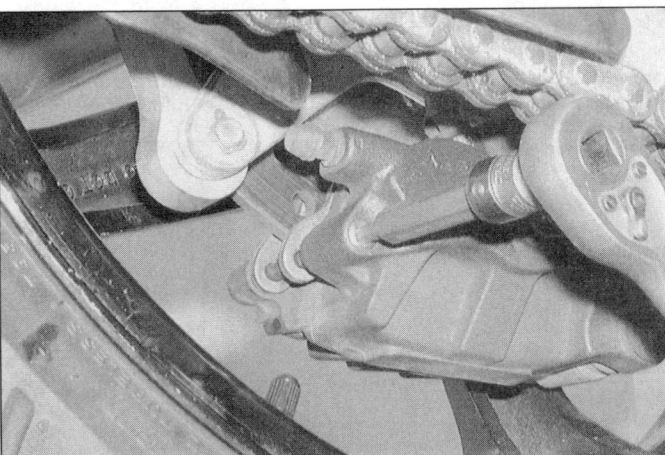

6.1b . . . then slacken and remove the pad retaining pin and withdraw both brake pads

7•10 Brakes, wheels and tyres

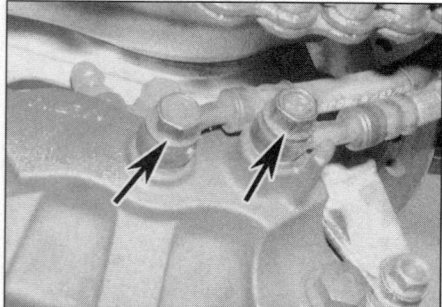

6.8a Slide the pads into position, ensuring the friction material of each pad is against the disc . . .

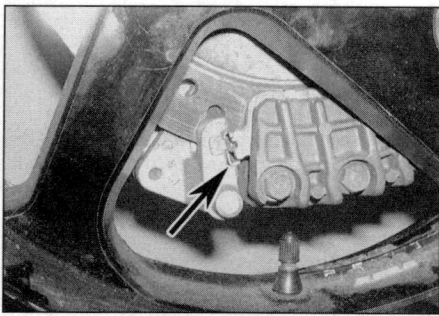

6.8b . . . and locate them both correctly with the pad retainer

5 Remove all traces of corrosion from the pad pin. Inspect the pin for signs of damage and renew it if necessary.
6 If new pads are being installed, push the pistons as far back into the caliper as possible. A good way of doing this is to insert one of the old pads between the outside of the disc and the piston, then push the caliper against the pad and disc using hand pressure. Due to the increased friction material thickness of new pads, it may be necessary to remove the master cylinder reservoir cover and diaphragm and remove some fluid.
7 Smear the backs of the pads and the shank of the pad pin with copper-based grease, making sure that none gets on the front or sides of the pads.
8 Installation of the pads is the reverse of removal. Make sure the pad spring is correctly positioned in the caliper body and the pad retainer is in place on the caliper bracket. Insert the pads into the caliper so that the friction material faces the disc, making sure they locate correctly in the pad retainer (see illustrations). Push down on the pads to align the holes and slide the pad retaining pin through. Make sure the pin passes through the hole in each pad. Tighten the pad retaining pin to the torque setting specified then install the pad pin plug.
9 Operate the brake pedal several times to bring the pads into contact with the disc then check the master cylinder reservoir fluid level (see *Daily (pre-ride) checks*). Once the fluid level is correct, check the operation of the brake before riding the motorcycle.

7 Rear brake caliper – removal, overhaul and installation

Warning: *If a caliper indicates the need for an overhaul (usually due to leaking fluid or sticky operation), all old brake fluid should be flushed from the system. Also, the dust created by the brake system may contain asbestos, which is harmful to your health. Never blow it out with compressed air and don't inhale any of it. An approved filtering mask should be worn when working on the brakes. Do not, under any circumstances, use petroleum-based solvents to clean brake parts. Use clean brake fluid, brake cleaner or denatured alcohol only.*

Removal

1 Remove the brake pads from the caliper (see Section 6).
2 Remove the rear wheel (see Section 17).
3 If the caliper pistons are to be removed, loosen the three bolts securing the inner half of the caliper body to the main body.
4 Note the correct fitted locations of the brake hose end fittings on the caliper then unscrew the brake hose banjo bolts and separate both hoses from the caliper (see illustration). Plug the hose ends or wrap a plastic bag tightly around each hose to minimise fluid loss and prevent dirt entering the system. Discard the sealing washers as new ones must be used on installation. **Note:** *If you are planning to overhaul the caliper and don't have a source of compressed air to blow out the pistons, just loosen the banjo bolts at this stage and retighten them lightly. The bike's hydraulic system can then be used to force the pistons out of the body once the caliper has been freed from the swingarm. Disconnect the hoses once the pistons have been sufficiently displaced.*
5 Slacken and remove the caliper bracket mounting bolts then remove the assembly from the swingarm bracket (see illustration). **Note:** *Honda recommend that new brake caliper bracket bolts should be used on installation (the bolts are supplied with their threads pre-coated with locking compound). If the old bolts are to be reused, remove all original locking compound from their threads and apply fresh compound before installing them.*
6 The caliper and bracket can then be separated. Take care not to lose the pad spring from the caliper body or the pad retainer from the bracket.

Overhaul

7 Remove the pad spring from inside the caliper, noting which way around it is fitted (see illustration). Clean the exterior of the caliper with denatured alcohol or brake system cleaner. Make sure all old grease is removed from the slider pins.
8 Unscrew the three bolts and separate the caliper inner half from the main body. **Note:** *Honda recommend that new bolts should be used on reassembly (the bolts are supplied with their threads pre-coated with locking compound). If the old bolts are to be reused, remove all original locking compound from their threads and apply fresh compound before installing them.*
9 Withdraw the partially ejected pistons from the caliper body. **Note:** *The outer pistons are both the same diameter; mark the head and caliper bore with a felt marker to ensure that they are matched to their original bores on reassembly.*

> **HAYNES HINT** *If the pistons cannot be withdrawn by hand, they can be pushed out by applying compressed air to the brake hose union hole. Only low pressure should be required, such as is generated by a foot pump. Cover the caliper with a wad of rag to prevent the pistons being forcibly expelled.*

10 Using a wooden or plastic tool, remove the dust seals from the caliper bores. Discard them as new ones must be used on installation. If a metal tool is being used, take great care not to damage the caliper bores.
11 Remove and discard the piston seals in the same way.

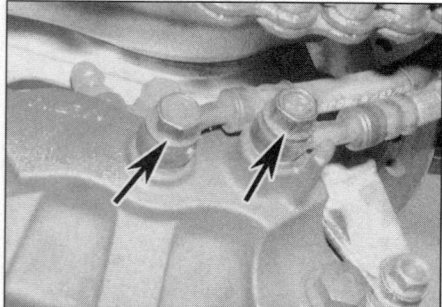

7.4 Rear brake caliper hose banjo bolts (arrows)

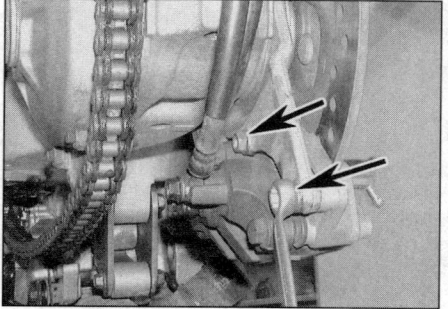

7.5 Unscrew the caliper bracket mounting bolts (arrows) and slide the caliper off the rear disc

12 Clean the pistons and bores with denatured alcohol, clean brake fluid or brake system cleaner. If compressed air is available, use it to dry the parts thoroughly (make sure it's filtered and unlubricated).
Caution: Do not, under any circumstances, use a petroleum-based solvent to clean brake parts.
13 Inspect the caliper bores and pistons for signs of corrosion, nicks and burrs and loss of plating. If surface defects are present, the caliper assembly must be renewed. If the necessary measuring equipment is available, compare the dimensions of the pistons and bores to those given in the Specifications Section of this Chapter, renewing any component that is worn beyond the service limit. If the caliper is in bad shape the master cylinder should also be checked.
14 Clean off all traces of corrosion from the slider pins and their bores in the caliper and bracket. Renew the rubber dust boots if they are damaged or deteriorated. The slider pins are a screw-fit and can be renewed if damaged. On installation, apply a suitable non-permanent thread locking compound and tighten each pin securely.
15 Lubricate the new piston seals with clean brake fluid and install them in their grooves in the caliper bores. Ensure the larger seal is fitted to the centre bore and the smaller seals to the outer bores.
16 Lubricate the new dust seals with silicone-based grease and install them in their grooves in the caliper bores.
17 Lubricate the pistons with clean brake fluid and install them closed-end first into their original caliper bores. Using your thumbs, push the pistons all the way in, making sure they enter the bore squarely.
18 Ensure the mating surfaces are clean and dry then fit the inner half of the caliper to the main body. Fit the three new retaining bolts and tighten them to the specified torque.
Note: *The bolts can be tightened to the specified torque setting at this stage if the caliper can be held firmly enough, but do not risk damaging it. A better method is to tighten the bolts when the caliper has been reinstalled and is securely bolted to the swingarm.*
19 Clip the pad spring securely into the caliper body, making sure it is fitted the correct way around as noted on removal.

Installation

20 Clean all old grease from the slider pins and check the pin rubber boots for damage and deterioration, renewing them if necessary. Ensure both boots are correctly seated then apply a smear of silicone-based grease to each pin.
21 Ensure the pad retainer is securely fitted to the mounting bracket then align the slider pins and slide the caliper fully into position. Make sure the rubber boot lips are correctly seated then check that the caliper is able to slide freely.
22 Locate the caliper assembly on the swingarm and fit the original mounting bolts; do not fit the new bolts yet, it will be necessary to remove the caliper again to bleed it.
23 Position a new sealing washer on each side of each hose union and screw in the banjo bolts. Ensure both hose end fitting are correctly located against the caliper body stops then tighten both banjo bolts to the specified torque.
24 Ensure the pad spring and retainer are still correctly positioned then install the brake pads as described in Section 6.
25 Top up the master cylinder reservoir with the specified brake fluid (see *Daily (pre-ride) checks*) and bleed the rear brake pedal hydraulic system as described in Section 12.
26 Once brake has been correctly bled, install the caliper assembly correctly making sure the brake pads pass each side of the disc. Fit the new caliper bracket bolts and tighten them to the specified torque.
27 Refit the rear wheel (Section 17) and check that there are no fluid leaks and thoroughly test the operation of the brake before riding the motorcycle.

8 Rear brake disc – inspection, removal and installation

Inspection

1 Visually inspect the surface of the disc for score marks and other damage. Light scratches are normal after use and won't affect brake operation, but deep grooves and heavy score marks will reduce braking efficiency and accelerate pad wear. If a disc is badly grooved it must be machined or renewed.
2 To check disc runout, position the bike on the centrestand so that the wheel is raised off the ground. Mount a dial gauge on the swingarm, with the plunger on the gauge touching the surface of the disc about 10 mm from the outer edge. Rotate the wheel and watch the gauge needle, comparing the reading with the limit listed in the Specifications at the beginning of the Chapter. If the runout is greater than the service limit, check the wheel bearings for play (see Chapter 1). If the bearings are worn, renew them (see Section 18) and repeat this check. If the disc runout is still excessive, it will have to be renewed, although machining by an engineer may be possible.
3 The disc must not be machined or allowed to wear down to a thickness less than the service limit listed in this Chapter's Specifications and as marked on the disc itself. The thickness of the disc can be checked with a micrometer. If the thickness of the disc is less than the service limit, it must be renewed.

Removal

4 Remove the rear wheel stub axle as described in Section 18.
5 Mark the relationship of the disc to the stub axle, so it can be installed in the same position.

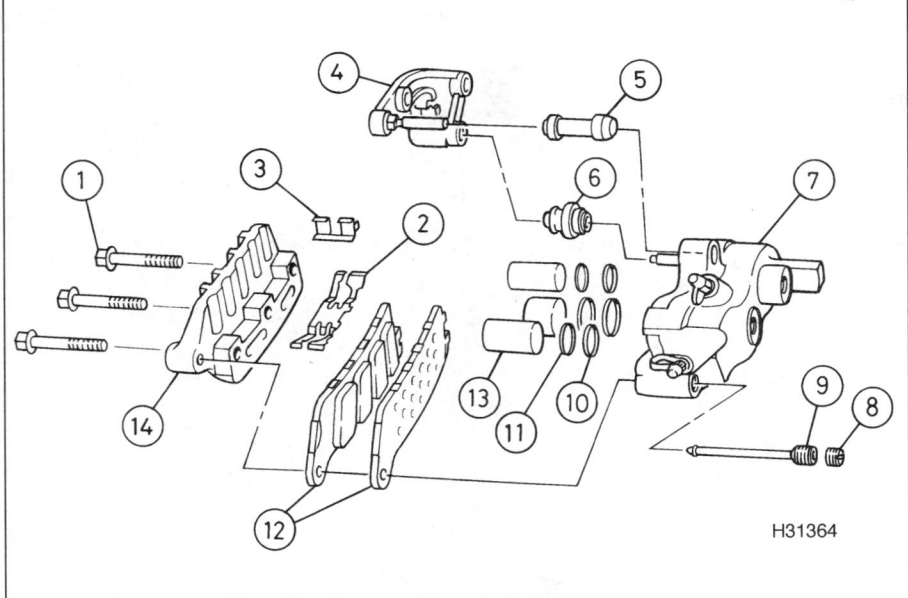

7.7 Exploded view of the rear brake caliper

1 Caliper body bolts	6 Slider pin rubber boot	10 Piston seals
2 Pad spring	7 Caliper main body	11 Dust seals
3 Pad retainer	8 Pad retaining pin plug	12 Brake pads
4 Mounting bracket	9 Pad retaining pin	13 Pistons
5 Slider pin rubber boot		14 Caliper inner half

7•12 Brakes, wheels and tyres

8.6 Unscrew the retaining nuts and bolts (arrows) and remove the disc

6 Slacken and remove the disc retaining nuts and bolts, then lift the disc and its spacer plate (if fitted) off the stub axle **(see illustration)**.

Installation

7 Where fitted, install the spacer plate on the stub axle, aligning its holes with the rear wheel stud heads.
8 Install the disc on the stub axle, making sure its marked side is facing away from the axle flange/spacer plate, and the arrow on the disc is pointing in the normal direction of wheel rotation. If the original disc is being installed, align the previously applied matchmarks.
9 Fit the disc retaining bolts and nuts and tighten them in a criss-cross pattern evenly and progressively to the specified torque setting.

10 Clean off all grease from the brake disc using acetone or brake system cleaner. If a new brake disc has been installed, remove any protective coating from its working surfaces.
11 Install the rear wheel stub axle (see Section 18).
12 Operate the brake pedal several times to bring the pads into contact with the disc. Check the operation of the brake carefully before riding the bike.

9 Rear brake master cylinder – removal, overhaul and installation

1 If the master cylinder is leaking fluid, or if the pedal does not produce a firm feel when the brake is applied, and bleeding the brake does not help (see Section 12), and the hydraulic hoses are all in good condition, then master cylinder overhaul is recommended.
2 Before disassembling the master cylinder, read through the entire procedure and make sure that you obtain all parts required. Also, you will need some new brake fluid, some clean rags and internal circlip pliers. *Note: To prevent damage to the paint from spilled brake fluid, always cover the surrounding components when working on the master cylinder.*
Caution: Disassembly, overhaul and reassembly of the brake master cylinder must be done in a spotlessly clean work area to avoid contamination and possible failure of the brake hydraulic system components.

Removal

3 Remove the seat cowling (see Chapter 8) to gain access to the rear brake fluid reservoir.
4 Note the correct fitted locations of the brake hose and pipe end fittings then unscrew the banjo bolt and separate them from the master cylinder. Plug the end fittings or wrap a plastic bag tightly around them to minimise fluid loss and prevent dirt entering the system. Discard the three sealing washers as new ones must be used on installation.
5 Remove the split pin from the clevis pin securing the brake pedal to the master cylinder pushrod and withdraw the clevis pin. Discard the split pin as a new one must be used.
6 Unscrew the retaining bolts (the lower bolt has a nut fitted) and free the master cylinder from the footrest bracket and heatshield.
7 Undo the retaining screw and free the fluid reservoir union from the master cylinder body. Remove the master cylinder from the bike and allow the fluid reservoir contents to drain into a container. Discard the union O-ring, a new one must be used on installation.

Overhaul

8 If required, mark the position of the clevis locknut on the pushrod, then slacken the locknut and unscrew the clevis and its locknut off from the pushrod **(see illustration)**.
9 Dislodge the rubber dust boot from the base of the master cylinder to reveal the pushrod retaining circlip.
10 Depress the pushrod and, using circlip pliers, remove the circlip. Slide out the piston/pushrod assembly, primary cup and spring from the master cylinder. If they are difficult to remove, apply low pressure compressed air to the fluid outlet. Lay the parts out in the correct fitted order to prevent confusion during reassembly.
11 Clean all of the parts with clean brake fluid or denatured alcohol.
Caution: Do not, under any circumstances, use a petroleum-based solvent to clean brake parts. If compressed air is available, use it to dry the parts thoroughly (make sure it's filtered and unlubricated).
12 Check the master cylinder bore for corrosion, scratches, nicks and score marks. If the necessary measuring equipment is available, compare the dimensions of the piston and bore to those given in the Specifications Section of this Chapter. If damage is evident, the master cylinder must be renewed. If the master cylinder is in poor condition, then the caliper should be checked as well.
13 Inspect the fluid reservoir and its joining hose for cracks or splits and renew it if necessary.

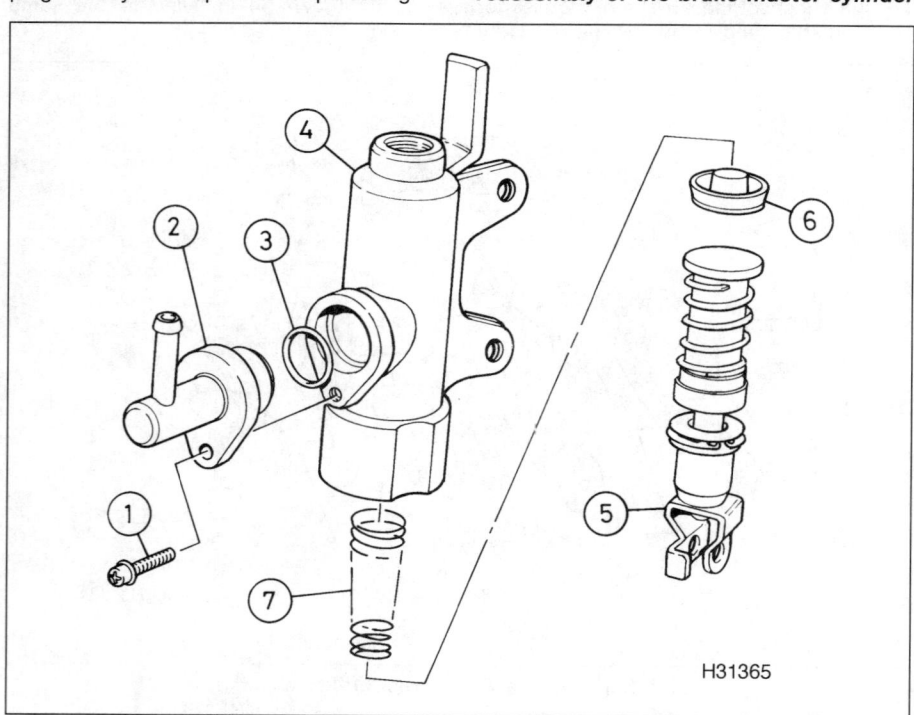

9.8 Exploded view of the rear brake master cylinder

1 Screw
2 Fluid reservoir union
3 O-ring
4 Master cylinder body
5 Piston/pushrod assembly
6 Primary cup
7 Spring

Brakes, wheels and tyres 7•13

14 The dust boot, circlip, piston/pushrod assembly, primary cup, seals and spring are included in the piston/seal kit. Use all of the new parts, regardless of the apparent condition of the old ones. If the seals are not already on the piston, fit them according to the layout of the old piston assembly, ensuring both are fitted the correct way around.
15 Fit the primary cup to the tapered end of the new spring then fit the assembly to the master cylinder so that the tapered end/primary cup faces the piston.
16 Lubricate the new piston/pushrod and seal assembly with clean brake fluid. Carefully ease the assembly into the master cylinder, making sure the seal lips do not turn inside out as they slip into the bore.
17 Depress the piston/pushrod and install the new circlip, making sure that it locates in the master cylinder groove.
18 Install the new rubber dust boot, making sure the lip is seated properly in the bore.
19 If removed, install the locknut and clevis onto the pushrod end. Position the clevis as noted on removal, then tighten the clevis locknut to the specified torque. Honda recommend that the distance between the centres of the master cylinder lower mounting bolt hole and the clevis pin hole should be 67.5 mm. However the pedal height can be altered to suit individual tastes. This is best done after the assembly is installed, and is done by slackening the clevis locknut, then turning the pushrod itself using a spanner on the flats on the top of the rod, until the desired pedal height is obtained. Tighten the clevis locknut to the specified torque (where possible) on completion.

Installation

20 Apply a drop of thread locking compound to the threads of the fluid reservoir union screw. Fit a new O-ring to the union then join it to the master cylinder and tighten its retaining screw securely.
21 Slide the master cylinder into position between the footrest and heatshield ensuring the pushrod engages correctly with the brake pedal. Fit the master cylinder retaining bolts and nut and tighten them securely.
22 Align the pushrod with the brake pedal then fit the clevis pin and secure it in position with a new split pin.
23 Position a new sealing washer on each side of the brake hose and pipe end fittings (there are three washers in total) and insert the banjo bolt. Ensure both end fittings are located against the master cylinder stop then tighten the banjo bolt to the specified torque.
24 Fill the fluid reservoir with the specified brake fluid (see *Daily (pre-ride) checks*) and bleed the rear brake pedal hydraulic system following the procedure in Section 12.
25 Install the seat cowling (see Chapter 8). Check the operation of the brake carefully before riding the motorcycle.

10 Dual combined brake system (DCBS) components – removal, overhaul and installation

Secondary master cylinder

Removal

1 The secondary master cylinder is an integral part of the left side front brake caliper mounting bracket and is removed as follows.
2 Remove the left side front brake caliper as described in Section 3.
3 Note the correct fitted locations of the brake hose fittings on the mounting bracket then unscrew the banjo bolts and detach the bracket from the hoses. Plug the hose fittings or wrap a plastic bag tightly around each hose to minimise fluid loss and prevent dirt entering the system. Discard the sealing washers as new ones must be used on installation.

Overhaul

4 Dislodge the rubber dust boot from the mounting bracket to reveal the pushrod retaining circlip (see illustration).
5 Depress the pushrod and, using circlip pliers, remove the circlip and pushrod assembly. **Note:** *Do not attempt to dismantle the pushrod.*
6 Withdraw the piston assembly, noting which way around it is fitted, and spring from the master cylinder. If they are difficult to remove, apply low pressure compressed air to the fluid outlet. Lay the parts out in the correct fitted order to prevent confusion during reassembly.
7 Clean all of the parts with clean brake fluid or denatured alcohol.
Caution: Do not, under any circumstances, use a petroleum-based solvent to clean brake parts. If compressed air is available, use it to dry the parts thoroughly (make sure it's filtered and unlubricated).

8 Check the master cylinder bore for corrosion, scratches, nicks and score marks. If the necessary measuring equipment is available, compare the dimensions of the piston and bore to those given in the Specifications Section of this Chapter. If damage is evident, the mounting bracket will have to be renewed.
9 The dust boot, circlip, piston assembly, seals and spring are included in the piston/seal kit. Use all of the new parts, regardless of the apparent condition of the old ones. If the seals are not already on the piston, fit them according to the layout of the old piston assembly, ensuring both are fitted the correct way around.
10 Fit the new spring to the master cylinder bore so that its tapered end is facing the piston.
11 Lubricate the new piston and seal assembly with clean brake fluid. Ensure the piston is the correct way around (smaller diameter end towards the spring) and carefully ease it into position, making sure the seal lips do not turn inside out as they slip into the bore.
12 Fit the new dust boot to the pushrod and insert the pushrod assembly.
13 Depress the pushrod and install the new circlip. Ensure the circlip is correctly located in its groove then seat the rubber dust boot lip correctly in the cylinder bore.

Installation

14 Position a new sealing washer on each side of each brake hose fitting then attach both hoses to the caliper mounting bracket. Ensure the hose fittings are correctly positioned then tighten both banjo bolts to the specified torque.
15 Fit the caliper as described in Section 3.
16 Fill the fluid reservoir with the specified brake fluid (see *Daily (pre-ride) checks*) and bleed the front brake lever hydraulic system and the rear brake pedal hydraulic system

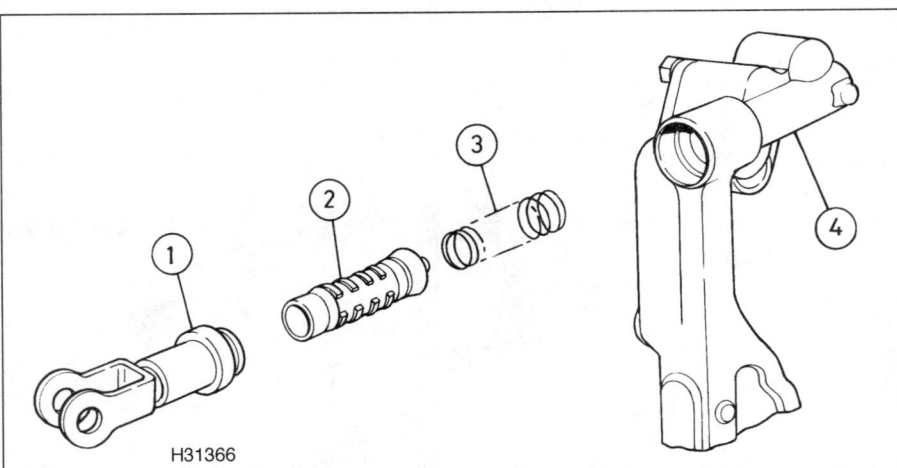

10.4 Exploded view of the secondary master cylinder

1 Pushrod
2 Piston assembly
3 Spring
4 Caliper mounting bracket

7•14 Brakes, wheels and tyres

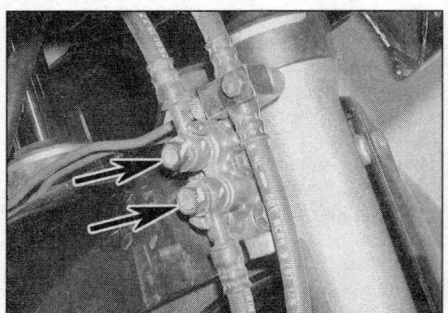

10.18 Delay valve banjo bolts (arrows)

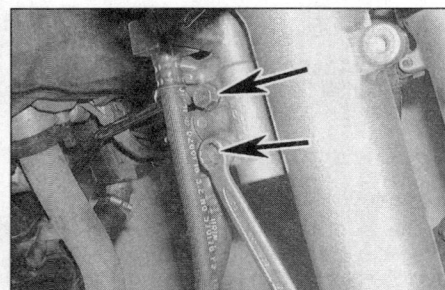

10.21 Unscrew the mounting bolts (arrows) and remove the delay valve from the fork slider

following the procedures in Section 12. Check the operation of both front and rear brakes carefully before riding the motorcycle.

Delay valve

Removal

17 Unscrew its four mounting bolts and remove the front mudguard. The delay valve is mounted on the right fork slider and can be removed as follows.
18 Note the correct fitted locations of the brake hose end fittings on the delay valve then unscrew the banjo bolts and detach both hoses (see illustration). Plug the hose fittings or wrap a plastic bag tightly around each hose to minimise fluid loss and prevent dirt entering the system. Discard the sealing washers as new ones must be used on installation.
19 Unscrew the union nut securing the brake pipe to the inner face of the delay valve.
20 Unscrew the bolt securing the brake pipe fitting to the delay valve.
21 Unscrew the two mounting bolts and remove the delay valve from the fork slider (see illustration).

Overhaul

22 Overhaul of the valve is not possible. If the delay valve is faulty it must be renewed.

Installation

23 Manoeuvre the delay valve into position and engage it with the metal brake pipe, tightening the union nut lightly at this stage.
24 Fit the delay valve mounting bolts and tighten them to the specified torque.

25 Tighten the brake pipe union nut to the specified torque (where possible).
26 Position a new sealing washer on each side of each brake hose end fitting then attach both hoses to the valve. Ensure the hose end fittings are located against the valve stops then tighten both banjo bolts to the specified torque.
27 Fit the brake pipe fitting retaining bolt and tighten it to the specified torque (see illustration).
28 Fill the fluid reservoir with the specified brake fluid (see *Daily (pre-ride) checks*) and bleed the rear brake pedal hydraulic system following the procedure in Section 12. Check the operation of both front and rear brakes carefully before riding the motorcycle.

Proportional control valve

Removal

29 The proportional control valve is located underneath the seat, just to the rear of the fuel tank (see illustration). Removal is as follows.
30 Place the motorcycle on its centrestand then support the rear wheel with a block of wood positioned between the tyre and ground; this will prevent the swingarm dropping when the shock absorber upper nut is unscrewed.
31 Remove the seat cowling (see Chapter 8).
32 Unscrew the fuel tank front and rear mounting bolts (see Chapter 4). Lift the rear of the fuel tank and support it securely or to allow access to the fuel tank hinge. **Note:** *There is no need to disconnect the fuel hoses from the tank.*
33 Slacken and remove the pivot bolt and nut

and remove the fuel tank hinge and pivot bushes from its mounting bracket. Release the rubber insulating covers from the mounting bracket then unscrew the shock absorber upper mounting nut and the retaining bolt and remove the hinge mounting bracket from the frame.
34 To improve access to the valve, free the engine stop relay from its mounting and unclip the fusebox.
35 Note the correct fitted location of the brake hose end fitting on the control valve then unscrew the banjo bolt (which incorporates the bleed valve) and detach the hose. Plug the hose fitting or wrap a plastic bag tightly around the hose to minimise fluid loss and prevent dirt entering the system. Discard the sealing washers as new ones must be used on installation.
36 Unscrew the union nut and detach the brake pipe from the control valve. Plug the pipe end or wrap a plastic bag tightly around the pipe to minimise fluid loss and prevent dirt entering the system.
37 Unscrew the mounting bolts and manoeuvre the control valve out of position.

Overhaul

38 Overhaul of the valve is not possible. If the control valve is faulty it must be renewed.

Installation

39 Manoeuvre the control valve into position and engage it with the metal brake pipe, tightening the union nut lightly only at this stage.
40 Fit the control valve mounting bolts and tighten them to the specified torque.
41 Tighten the brake pipe union nut to the specified torque.
42 Position a new sealing washer on each side of the brake hose end fitting then attach the hose to the valve. Ensure the hose end fitting is securely located against the valve stop then tighten the banjo bolt to the specified torque.
43 Clip the fusebox and engine stop relay back into position and install the fuel tank hinge bracket. Securely tighten the bracket retaining bolt then fit the shock absorber upper mounting nut and tighten to the specified torque (see Chapter 6). Seat the rubber covers over the mounting bracket then fit the fuel tank hinge, ensuring the pivot bushes are in position. Insert the hinge pivot bolt and nut and tighten securely.
44 Fill the fluid reservoir with the specified brake fluid (see *Daily (pre-ride) checks*) and bleed the rear brake pedal hydraulic system following the procedure in Section 12. Check the operation of both front and rear brakes carefully before riding the motorcycle.
45 Ensure all the fuel tank mounting rubbers are correctly fitted (see Chapter 4) and the collars are in position in the front mountings then seat the fuel tank back in position. Fit the tank front and rear mounting bolts and tighten them securely.
46 Install the seat cowling as described in Chapter 8.

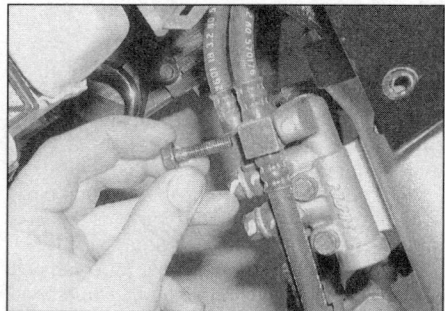

10.27 Install the fitting retaining bolt and tighten to the specified torque

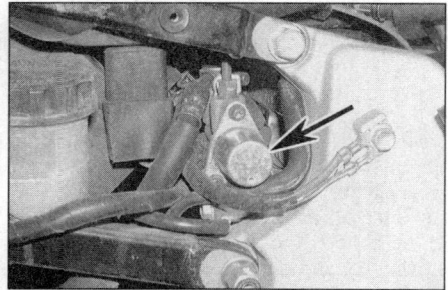

10.29 The proportional control valve (arrow) is located just to the rear of the fuel tank, on the right side

Brakes, wheels and tyres 7•15

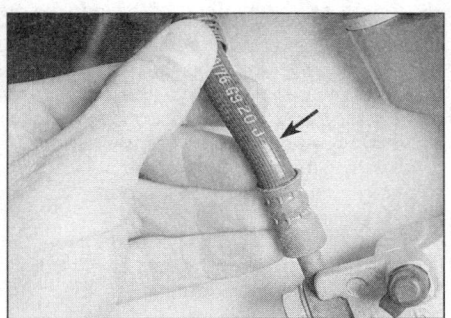

11.2 Flex the brake hoses and check for cracks, bulges or fluid leaks

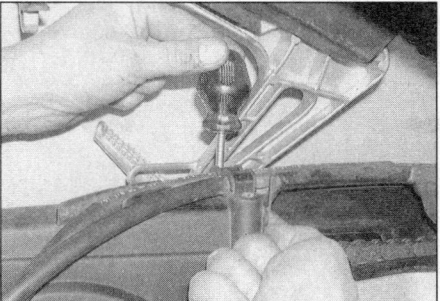

11.16a Undo the retaining screws . . .

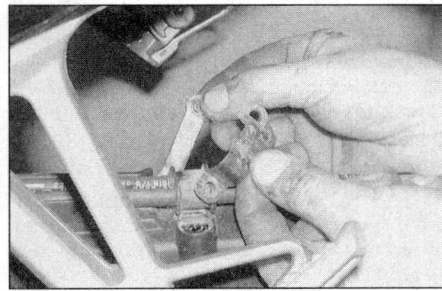

11.16b . . . and remove the brake hose clamps and retaining plates from the chainguard

11 Brake hoses, pipes and unions – inspection and renewal

Inspection

1 The brake hoses, pipes and their associated fittings should be checked regularly (see Chapter 1).
2 Twist and flex the rubber hoses while looking for cracks, bulges and seeping fluid **(see illustration)**. Check extra carefully around the areas where the hoses connect with the end fittings, as these are common areas for hose failure.
3 Inspect the metal brake pipes which link the front brake hoses between the fork sliders and the pipes, located underneath the fuel tank, which connect the rear master cylinder and proportional control valve to the front hoses. If the pipes or their fittings are rusted, scratched or cracked, renew them.

Renewal

4 The brake hoses and pipes are all linked either by banjo union fittings or union nuts. Note that some of the hoses also have a union along their lengths as well as at each end.
5 Before disconnecting a hose/pipe union, remove all traces of dirt then cover the surrounding area with plenty of rags to catch all spilt fluid.

Front brake hoses

6 As a precaution against damage, remove the front mudguard (see Chapter 8). Depending on which hose is being renewed, it may also be necessary to remove the upper fairing to improve access.
7 Where a hose union is secured with a banjo bolt, note the correct positioning of the hose fitting then unscrew the bolt and discard the sealing washers.
8 Where the hose is joined to a pipe, unscrew the pipe union nut then slacken and remove the hose fitting mounting bolt.
9 Note the correct routing of the hose then free it from its retaining clips and guides and remove it from the bike.
10 Fit the new hose, making sure it is correctly routed and isn't twisted or otherwise strained, and secure it in position with the necessary clips and guides.
11 Where the hose is joined to a pipe, screw the union nut into the hose fitting by a few turns then install the fitting mounting bolt and tighten it to the specified torque. The pipe union nut can then be tightened to the specified torque.
12 Where a banjo fitting is used, position a new sealing washer on each side of the fitting then screw in the banjo bolt. Ensure the fitting is correctly positioned against its stop or between the lugs (as applicable) then tighten the banjo bolt to the specified torque setting.
13 Fill the brake system with new fluid of the specified type (see *Daily (pre-ride) checks*) and bleed the air from the relevant hydraulic system (see Section 12). Check the operation of the brakes carefully before riding the motorcycle.

Rear brake hoses

14 Remove the seat cowling (see Chapter 8) and the rear wheel (see Section 17).
15 If the hose linking the proportional control valve to the caliper is to be renewed, peel back the rubber insulating cover to gain access to the valve. If better access is required, carry out the operations described in Paragraphs 30 to 34 of Section 10.
16 Reach in through the cut-out in the rear of the chainguard and support the brake hose clamp retaining plates. Undo the retaining screws and remove both clamps and recover the clamp retaining plates from inside the chainguard **(see illustrations)**. **Note:** *Honda recommend that new clamp screws should be used on installation (the screws are supplied with their threads pre-coated with locking compound). If the old screws are to be reused, remove all original locking compound from their threads and apply fresh compound before installing them.*
17 Undo the screw and free the front retaining clamp from the hoses.
18 Unscrew the banjo bolt securing the hose to the caliper and master cylinder/valve (as applicable) and discard the sealing washers. On the master cylinder, note that it will also be necessary to renew the sealing washer fitted between the pipe union and cylinder.
19 Free the hose from the guide on the bearing holder clamp bolt and remove it from the bike, noting its correct routing.
20 If the hose is being connected to the master cylinder, position a new sealing washer between the pipe union and cylinder and one on each side of the hose union, then screw in the banjo bolt. Ensure both end fittings are located against the master cylinder stop then tighten the banjo bolt to the specified torque.
21 If the hose being connected to the control valve, position a new sealing washer on each side of hose end fitting then screw in the banjo bolt. Ensure the hose end fitting is securely located against the valve stop then tighten the banjo bolt to the specified torque.
22 Locate both hoses in the front retaining clamp then securely tighten the clamp screw.
23 Pass both hoses through their guide on the bearing holder clamp bolt then fit both chainguard clamps to them. Slide the retaining plates into position inside the chainguard then fit the new clamp retaining screws and tighten them securely.
24 Position a new sealing washer on each side of the caliper end fitting then screw in the banjo bolt. Ensure the end fitting is correctly positioned against the caliper stop then tighten the banjo bolt to the specified torque.
25 Fill the fluid reservoir with the specified brake fluid (see *Daily (pre-ride) checks*) and bleed the rear brake pedal hydraulic system following the procedure in Section 12.
26 Install the seat cowling (see Chapter 8). Check the operation of the brake carefully before riding the motorcycle.

Front brake hose joining pipes

27 As a precaution against damage, remove the front mudguard (see Chapter 8).
28 Unscrew the union nuts securing the pipe to each hose fitting.
29 Unscrew the retaining bolt securing the hose end fittings to the bracket on the top of the left fork slider then remove the brake pipe.
30 Manoeuvre the new pipe into position and lightly tighten both its union nuts.
31 Fit the brake hose end fitting retaining bolt and tighten it to the specified torque then tighten both pipe union nuts to the specified torque.
32 Fill the fluid reservoir with the specified

7•16 Brakes, wheels and tyres

brake fluid (see *Daily (pre-ride) checks*) and bleed the relevant hydraulic system following the procedure in Section 12.
33 Install the mudguard (see Chapter 8). Check the operation of the brake carefully before riding the motorcycle.

Rear brake/front brake system pipes

34 Remove the fuel tank (see Chapter 4).
35 If the master cylinder pipe is to be renewed, remove the seat cowling (see Chapter 8).
36 Unscrew the union nut securing the pipe to the hose end fitting on the headstock.
37 Free the pipe from its clips on the inside of the frame then unscrew the union nut securing it to the control valve/master cylinder fitting and remove it from the bike.
38 Manoeuvre the new pipe into position, ensuring it is correctly routed and clip it securely into the frame clips.
39 Connect the pipe to the hose end fitting and control valve/master cylinder and tighten the union nuts to the specified torque.
40 Fill the fluid reservoir with the specified brake fluid (see *Daily (pre-ride) checks*) and bleed the rear brake pedal hydraulic system following the procedure in Section 12.
41 Install the seat cowling (where removed) and fuel tank (see Chapters 4 and 8).
42 Check the operation of the brake carefully before riding the motorcycle.

12 Brake system – bleeding

Note: *Honda specify the use of a vacuum-type brake bleeder to simplify the bleeding process. The bleeder is connected to the bleed valve, which is then opened, and is used to draw the air/fluid through the system and out of the bleed valve. Once all the air has been removed, the bleeder is removed and the brakes are then bled in the conventional way (described below) to complete the bleeding procedure. If you find that the conventional method of bleeding fails to remove the trapped air from the system, it will be necessary to obtain a vacuum-type brake bleeder kit from a motor accessory shop or entrust the work to a Honda dealer.*

1 Bleeding the brakes is simply the process of removing all the air bubbles from the brake fluid reservoirs, the hoses/pipes, the brake calipers and associated components. Bleeding is necessary whenever a brake system hydraulic connection is loosened, when a component or hose is renewed, or when the master cylinder or caliper is overhauled. Leaks in the system may also allow air to enter, but leaking brake fluid will reveal their presence and warn you of the need for repair.
2 To bleed the brakes using the conventional method, you will need some new DOT 4 brake fluid, a length of clear vinyl or plastic tubing, a

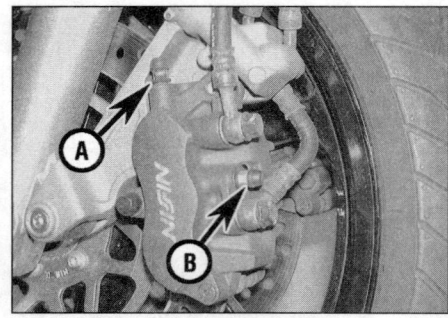

12.7 Each front brake caliper upper bleed valve (A) is for the front brake lever circuit and the lower bleed valve (B) for the rear brake pedal circuit

small container partially filled with clean brake fluid, some rags and a spanner to fit the brake caliper bleed valves.
3 The dual combined braking system (CBS) can be split into two separate hydraulic circuits; the front brake lever circuit and the rear brake pedal circuit. The front brake lever circuit consists of the hoses and pipes linking the front master cylinder to the front brake calipers. The rear brake pedal circuit consists of the hoses and pipes linking the rear master cylinder to the rear caliper and delay valve, the hoses from the delay valve to the front brake calipers and secondary master cylinder and the hoses and pipe from the secondary master cylinder to the proportional control valve and rear brake caliper.
4 If bleeding is being carried out after a hose/pipe has been disconnected identify which circuit the hose is part of, then carry out the relevant bleeding procedure described below.

Front brake lever hydraulic circuit

5 Cover the fuel tank and other painted components to prevent damage in the event that brake fluid is spilled.
6 Undo the retaining screws and remove the master cylinder reservoir cover, diaphragm plate and diaphragm and slowly pump the brake lever a few times, until no air bubbles can be seen floating up from the holes in the bottom of the reservoir. Doing this bleeds the

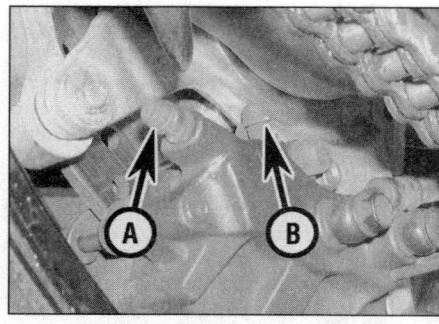

12.15 Access to the rear brake caliper outer (A) and inner (B) bleed valves is impossible with the caliper in position

air from the master cylinder end of the line. Loosely refit the reservoir cover.
7 Starting on the left side caliper, pull the dust cap off the **upper** bleed valve (the lower bleed valve is for the rear brake pedal circuit) **(see illustration)**. Attach one end of the clear vinyl or plastic tubing to the bleed valve and submerge the other end in the brake fluid in the container.
8 Remove the reservoir cover and check the fluid level. Do not allow the fluid level to drop below the lower mark during the bleeding process.
9 Carefully pump the brake lever three or four times and hold it in while opening the caliper bleed valve. When the valve is opened, brake fluid will flow out of the caliper into the clear tubing and the lever will move toward the handlebar.
10 Retighten the bleed valve, then release the brake lever gradually. Repeat the process until no air bubbles are visible in the brake fluid leaving the valve and the lever is firm when applied. Disconnect the bleeding equipment, then tighten the bleed valve to the specified torque and install the dust cap.
11 Repeat the procedure described in Paragraphs 7 to 10 on the right side front caliper, until all air is removed from the system and the brake lever feels firm again.
12 Top the fluid level up to the upper level mark (see *Daily (pre-ride) checks*) then install the diaphragm, diaphragm plate and reservoir cover assembly and securely tighten the retaining screws. Wipe up any spilled brake fluid and check the entire system for leaks.

> **Haynes Hint:** If it's not possible to produce a firm feel to the lever the fluid may be aerated. Let the brake fluid in the system stabilise for a few hours and then repeat the procedure when the tiny bubbles in the system have settled out. Failure to bleed satisfactorily after a reasonable repetition of the bleeding procedure may be due to worn master cylinder seals.

Rear brake pedal hydraulic circuit

13 Remove the seat cowling (see Chapter 8) to gain access to the rear brake master cylinder fluid reservoir.
14 The brake pedal hydraulic circuit must always be bled in the following sequence.
  a) *Right side front brake caliper (lower) bleed valve.*
  b) *Left side front brake caliper (lower) bleed valve.*
  c) *Proportional control valve bleed valve.*
  d) *Rear brake caliper inner bleed valve.*
  e) *Rear brake caliper outer bleed valve.*

15 Access to the rear brake caliper bleed valves is impossible with the caliper in position **(see illustration)**. To get around this remove the rear wheel (see Section 17), then undo the bolts securing the caliper bracket to the swingarm bracket. Slide the caliper off the brake disc then turn it around, so the bleed valves are facing away from the swingarm, and slide it back onto the disc. **Note:** *Honda recommend that new brake caliper bracket bolts should be used on installation (the bolts are supplied with their threads pre-coated with locking compound). If the old bolts are to be reused, remove all original locking compound from their threads and apply fresh compound before installing them.*

16 Unscrew the fluid reservoir cover and lift out the diaphragm plate and diaphragm. Slowly pump the brake pedal a few times, until no air bubbles can be seen floating up from the holes in the bottom of the reservoir. Doing this bleeds the air from the master cylinder end of the line. Loosely refit the reservoir cover.

17 Starting with the right side front caliper **lower** bleed valve (the upper bleed valve is for the front brake lever pedal circuit), pull off the dust cap then attach one end of the clear vinyl or plastic tubing to the bleed valve. Submerge the other end of the tubing in the brake fluid in the container.

18 Check the fluid level. Do not allow the fluid level to drop below the lower mark during the bleeding process. If the fluid level is allowed to drop air will enter the hydraulic circuit and the bleeding process will have to be restarted.

19 Carefully pump the brake pedal three or four times and hold it down while opening the bleed valve. When the valve is opened, brake fluid will flow out into the clear tubing and the pedal will move fully downwards.

20 Retighten the bleed valve, then release the brake pedal gradually. Repeat the process until no air bubbles are visible in the brake fluid leaving the valve. Disconnect the bleeding equipment, then tighten the bleed valve to the specified torque and install the dust cap.

21 Remove the dust cap from the left side front caliper **lower** bleed valve (the upper bleed valve is for the front brake lever pedal circuit) **(see illustration 12.7)**. Attach one end of the clear vinyl or plastic tubing to the bleed valve and submerge the other end in the brake fluid in the container.

22 Repeat the procedure described in Paragraphs 18 to 20, until all air is removed from the left side front caliper.

23 Unscrew the left side front caliper mounting bracket pivot bolt and the secondary master cylinder pushrod bolt then slide the caliper assembly off of the disc. Insert a wooden wedge between the brake pads (to prevent the pistons being expelled) then have an assistant support the caliper/mounting bracket assembly so the secondary master cylinder axis is positioned at angle of approximately 15° above the horizontal, with its pushrod pointing downwards **(see illustration)**. This will cause any trapped air in the secondary master cylinder to work its way

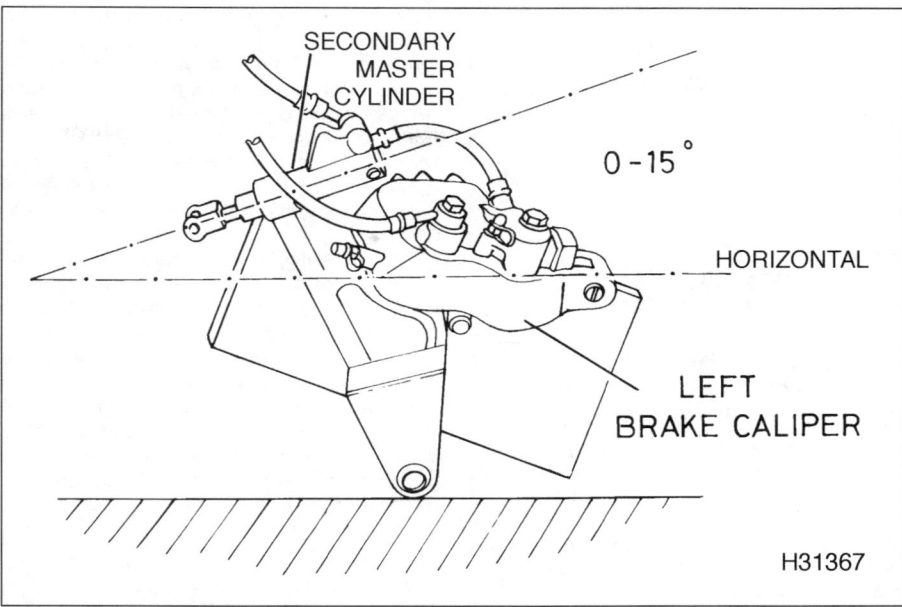

12.23 Bleed the secondary master cylinder of air by positioning the mounting bracket so the master cylinder axis is approximately 15° above the horizontal with the pushrod pointing downwards

up to the proportional control valve. **Note:** *Honda recommend that a new bracket pivot bolt and secondary master cylinder pushrod bolt should be used on installation (the bolts are supplied with their threads pre-coated with locking compound). If the old bolts are to be reused, remove all original locking compound from their threads and apply fresh compound before installing them.*

24 Remove the dust cap from the proportional control valve bleed valve which is located at the rear of the fuel tank, on the right side (lift the rubber insulating cover to gain access to the valve) **(see illustration)**. Attach one end of the clear vinyl or plastic tubing to the bleed valve and submerge the other end in the brake fluid in the container.

25 With the secondary master cylinder correctly positioned, repeat the procedure described in Paragraphs 18 to 20 until all traces of air are removed from the valve.

26 Remove the dust cap from the rear caliper **inner** bleed valve **(see illustration 12.15)**.

12.24 The proportional control valve bleed valve (arrow) is located underneath the seat

Attach one end of the clear vinyl or plastic tubing to the bleed valve and submerge the other end in the brake fluid in the container.

27 Repeat the procedure described in Paragraphs 18 to 20, until all air is removed.

28 Remove the dust cap from the rear caliper **outer** bleed valve. Attach one end of the clear vinyl or plastic tubing to the bleed valve and submerge the other end in the brake fluid in the container.

29 Repeat the procedure described in Paragraphs 18 to 20, until all air is removed.

30 The bleeding procedure is now complete and the brake pedal should feel firm.

> **HAYNES HiNT**
> *If it's not possible to produce a firm feel to the pedal the fluid may be aerated. Let the brake fluid in the system stabilise for a few hours and then repeat the procedure when the tiny bubbles in the system have settled out. Failure to bleed the system satisfactorily (even with a vacuum-type brake bleeder) may be due to worn master cylinder seals.*

31 Top the fluid level up to the upper level mark (see *Daily (pre-ride) checks*) then install the diaphragm, diaphragm plate and reservoir cover assembly. Wipe up any spilled brake fluid.

32 Remove the wooden wedge from the left side front caliper then slide the assembly into position, making sure the pads pass each side of the disc. Engage the mounting bracket with its mounting and locate the secondary master cylinder pushrod on the fork slider. Fit the new bolts and tighten them both to the specified torque setting.

7•18 Brakes, wheels and tyres

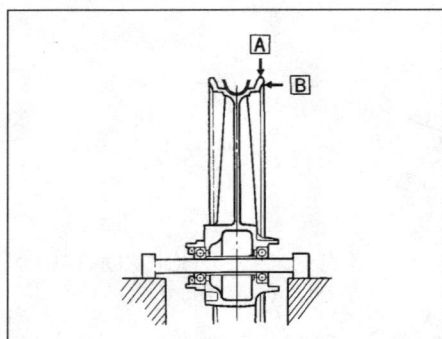

13.2 Check the wheel for radial (out-of-round) runout (A) and axial (side-to-side) runout (B)

33 Install the rear caliper assembly correctly, making sure the brake pads pass each side of the disc. Fit the new caliper bracket bolts and tighten them to the specified torque setting.
34 Refit the rear wheel (Section 17) and check that there are no fluid leaks. Thoroughly test the operation of the brake before riding the motorcycle (see Chapter 1).

13 Wheels – inspection and repair

1 In order to carry out a proper inspection of the wheels, place the bike on its centrestand and support it so that the wheel being inspected is raised off the ground. Clean the wheels thoroughly to remove mud and dirt that may interfere with the inspection procedure or mask defects. Make a general check of the wheels (see Chapter 1) and tyres (see *Daily (pre-ride) checks*).
2 Attach a dial gauge to the fork slider or the swingarm and position its stem against the side of the rim. Spin the wheel slowly and check the axial (side-to-side) runout of the rim. In order to accurately check radial (out of round) runout with the dial gauge, the wheel would have to be removed from the machine, and the tyre from the wheel. With the axle clamped in a vice and the dial gauge positioned on the top of the rim, the wheel can be rotated to check the runout **(see illustration)**.

3 An easier, though slightly less accurate, method is to attach a stiff wire pointer to the fork slider or the swingarm and position the end a fraction of an inch from the wheel (where the wheel and tyre join). If the wheel is true, the distance from the pointer to the rim will be constant as the wheel is rotated. **Note:** *If wheel runout is excessive, check the wheel bearings very carefully before renewing the wheel.*
4 The wheels should also be visually inspected for cracks, flat spots on the rim and other damage. Look very closely for dents in the area where the tyre bead contacts the rim. Dents in this area may prevent complete sealing of the tyre against the rim, which leads to deflation of the tyre over a period of time. If damage is evident, or if runout in either direction is excessive, the wheel will have to be renewed. Never attempt to repair a damaged cast alloy wheel.

14 Wheels – alignment check

1 Misalignment of the wheels, which may be due to a bent frame, fork yokes or rear stub axle can cause strange and possibly serious handling problems. If the frame or yokes are at fault, repair by a frame specialist or renewal are the only alternatives.
2 To check the alignment you will need an assistant, a length of string or a perfectly straight piece of wood and a ruler. A plumb bob or other suitable weight will also be required.
3 In order to make a proper check of the wheels it is necessary to support the bike in an upright position, using an auxiliary stand. Measure the width of both tyres at their widest points. Subtract the smaller measurement from the larger measurement, then divide the difference by two. The result is the amount of offset that should exist between the front and rear tyres on both sides.
4 If a string is used, have your assistant hold one end of it about halfway between the floor and the rear axle, touching the rear sidewall of the tyre.

5 Run the other end of the string forward and pull it tight so that it is roughly parallel to the floor. Slowly bring the string into contact with the front sidewall of the rear tyre, then turn the front wheel until it is parallel with the string **(see illustration)**. Measure the distance from the front tyre sidewall to the string.
6 Repeat the procedure on the other side of the motorcycle. The distance from the front tyre sidewall to the string should be equal on both sides.
7 As was previously pointed out, a perfectly straight length of wood or metal bar may be substituted for the string **(see illustration)**. The procedure is the same.
8 If the distance between the string and tyre is greater on one side, or if the rear wheel appears to be cocked, the frame, swingarm or rear stub axle must be damaged.
9 If the front-to-back alignment is correct, the wheels still may be out of alignment vertically.
10 Using the plumb bob, or other suitable weight, and a length of string, check the rear wheel to make sure it is vertical. To do this, hold the string against the tyre upper sidewall and allow the weight to settle just off the floor.

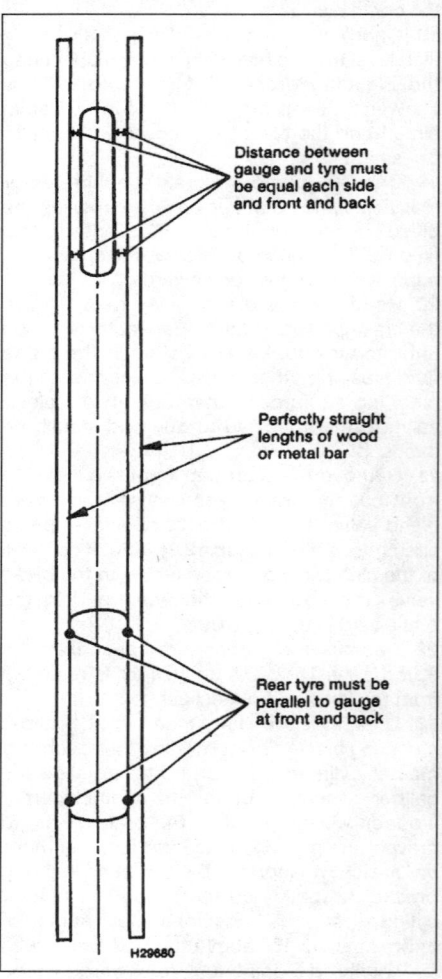

14.7 Wheel alignment checking using a straight-edge

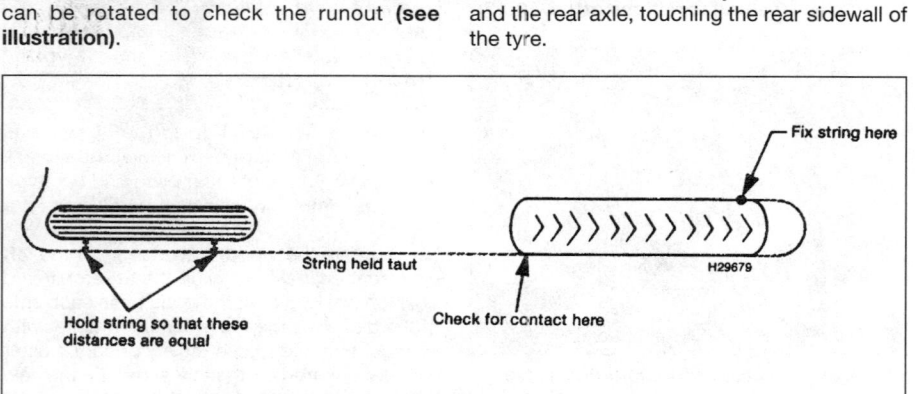

14.5 Wheel alignment checking using string

Brakes, wheels and tyres 7•19

When the string touches both the upper and lower tyre sidewalls and is perfectly straight, the wheel is vertical. If it is not, place thin spacers under one leg of the stand.
11 Once the rear wheel is vertical, check the front wheel in the same manner. If both wheels are not perfectly vertical, the frame and/or major suspension components are bent.

15 Front wheel – removal and installation

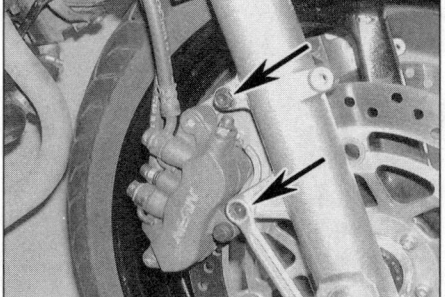

15.3a Slacken and remove the mounting bolts (arrows) . . .

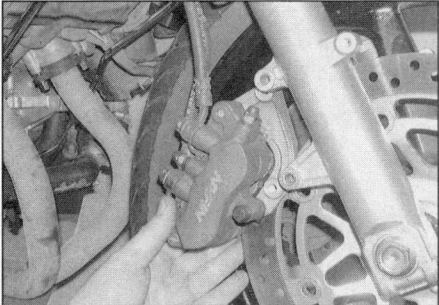

15.3b . . . then slide the right caliper assembly off from the brake disc

Removal

1 Position the motorcycle on its centrestand and tie down the rear of the motorcycle to raise the front wheel off the ground. Alternatively, remove the lower fairing panels (see Chapter 8) and support the bike under the engine to raise the front wheel.
2 Undo the mounting bolts and remove the front mudguard (see Chapter 8).
3 Remove the mounting bolts securing the right brake caliper to the fork slider. Slide the caliper assembly off from the disc and support the caliper with a piece of wire or a bungee cord so that no strain is placed on its hydraulic hose **(see illustrations)**. There is no need to disconnect the hoses from the caliper, or to displace the left caliper, unless required. **Note:** *Honda recommend that new brake caliper bracket bolts should be used on installation (the bolts are supplied with their threads pre-coated with locking compound). If* the old bolts are to be reused, remove all original locking compound from their threads and apply fresh compound before installing them.

Caution: *Do not operate the brake lever or pedal with the caliper removed.*

4 Slacken the axle clamp bolts on the bottom of the right fork slider then unscrew the axle bolt from the right end of the axle **(see illustrations)**.
5 Slacken the axle clamp bolts on the left fork slider **(see illustration)**. Support the wheel, then withdraw the axle from the left side and carefully lower the wheel. Use a screwdriver inserted through the holes in the end of the axle as a lever to aid removal **(see illustration)**.
6 Remove the long wheel spacer from the right side of the wheel and the short spacer from the left side for safekeeping **(see illustrations)**.

Caution: *Don't lay the wheel down and allow it to rest on a disc – the disc could become warped. Set the wheel on wood blocks so the disc doesn't support the weight of the wheel.*

7 Check the axle for straightness by rolling it on a flat surface such as a piece of plate glass (first wipe off all old grease and remove any corrosion using fine emery cloth). If the equipment is available, place the axle in V-blocks and measure the runout using a dial gauge. If the axle is bent or the runout exceeds the limit specified, renew it.
8 Refer to Section 16 if wheel bearing renewal is required.

Installation

9 Apply a smear of grease to the inside of the wheel spacers, and also to the outside where they fit into the dust seals. Fit the long spacer into the right side of the wheel and the short

15.4a Slacken the right fork slider axle clamp bolts (arrows) . . .

15.4b . . . then unscrew the axle bolt

15.5a Slacken the left fork slider axle clamp bolts . . .

15.5b . . . then withdraw the axle and lower the wheel out of position

15.6a Recover the long spacer from the right side of the wheel . . .

15.6b . . . and the short spacer from the left side

7•20 Brakes, wheels and tyres

15.11 Lift the front wheel into position ensuring the left disc passes correctly between the brake pads (arrow) and both spacers remain correctly fitted

15.12a Tighten the axle bolt to the specified torque . . .

15.12b . . . whilst retaining the axle with a screwdriver as shown

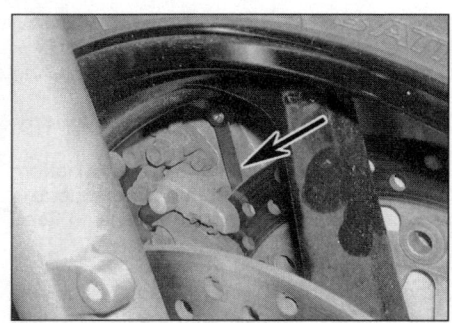

15.16a Use a feeler gauge (arrow) to ensure there is a clearance of at least 0.7 mm between each side of the left brake disc and the caliper mounting bracket . . .

15.16b . . . then tighten the left fork slider axle clamp bolts to the specified torque

spacer into the left side (see illustrations 15.6a and 15.6b). Each side of the wheel can be identified using the directional arrow cast into one of the spokes near the rim. The arrow denotes the normal direction of wheel rotation.

10 Manoeuvre the wheel into position, making sure the directional arrow is pointing in the normal direction of rotation. Apply a thin coat of grease to the axle.

11 Lift the wheel into place between the forks, making sure its disc passes between the left caliper brake pads (see illustration). Ensure both wheel spacers are still in position then slide the axle in from the left side.

12 Fit the axle bolt and tighten it to the specified torque setting. Use a screwdriver inserted through the holes in the end of the axle to counterhold it (see illustrations).

13 Tighten the axle clamp bolts on the right side fork slider to the specified torque setting.

14 Slide the right brake caliper into position, making sure the pads pass on each side of the disc, then fit the new mounting bolts and tighten them to the specified torque.

15 Apply the front brake a few times to bring the pads back into contact with the discs. Move the motorcycle off its stand, apply the front brake and pump the front forks a few times to settle all components in position.

16 Using feeler blades check that there is at least 0.7 mm clearance between each side of the left brake disc and caliper mounting bracket then tighten the axle clamp bolts on the left side fork slider to the specified torque (see illustrations).

17 Install the mudguard (see Chapter 8) and check the operation of the front brake before riding the motorcycle.

16 Front wheel bearings – renewal

Note: *Always renew the wheel bearings in pairs. Never renew the bearings individually. Avoid using a high pressure cleaner on the wheel bearing area.*

1 Remove the wheel (see Section 15).

2 Set the wheel on blocks so as not to allow the weight of the wheel to rest on either of the brake discs.

3 Prise out the dust seal on each side of the wheel using a flat-bladed screwdriver, taking care not to damage the rim of the hub (see illustration). Discard the seals as new ones should be used.

4 Using a metal rod (preferably a brass drift punch) inserted through the centre of the left side bearing, tap evenly around the inner race of the right bearing to drive it from the hub (see illustration). The bearing spacer will also come out.

5 Lay the wheel on its other side so that the left bearing faces down. Drive the bearing out of the wheel using the same technique as above.

6 If the bearings are of the unsealed type or are only sealed on one side, clean them with a high flash-point solvent (one which won't leave any residue) and blow them dry with compressed air (don't let the bearings spin as you dry them). Apply a few drops of oil to the bearing. **Note:** *If the bearing is sealed on both sides don't attempt to clean it.*

7 Hold the outer race of the bearing and rotate the inner race – if the bearing doesn't turn smoothly, has rough spots or is noisy, renew it.

 **HAYNES HINT** *Refer to Tools and Workshop Tips (Section 5) for more information about bearings.*

8 If the bearing is good and can be re-used and isn't sealed on both sides, wash it in

16.3 Prise out the dust seal with a large, flat-bladed screwdriver

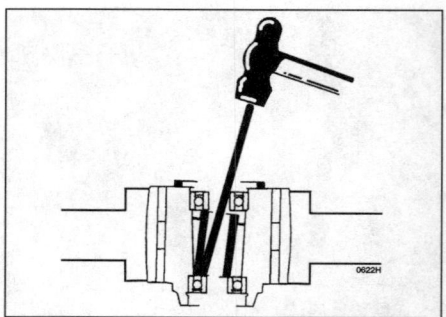

16.4 Locate the drift as shown when driving out a front wheel bearing

Brakes, wheels and tyres 7•21

16.9a Insert the first bearing . . .

16.9b . . . and tap it squarely into position using a tubular drift (such as a socket) which bears only on the bearing outer race

16.10 Turn the wheel over and insert the spacer into the hub before fitting the second bearing

solvent once again and dry it, then pack the bearing with grease.

9 Thoroughly clean the hub area of the wheel. First install the left side bearing into its recess in the hub, with the marked or sealed side facing outwards. Using the old bearing, a bearing driver or a socket large enough to contact the outer race of the bearing, drive it in until it's completely seated **(see illustrations)**.

10 Turn the wheel over and install the bearing spacer **(see illustration)**. Drive the right side bearing into place as described above.

11 Apply a smear of grease to the lips of the dust seals, then press them into the wheel. Gently drive them into place using a seal or bearing driver which contacts only the outer edge of the seal.

12 Clean off all grease from the brake discs using acetone or brake system cleaner then install the wheel (see Section 15).

17 Rear wheel – removal and installation

Removal

1 Place the motorcycle on its centrestand.
2 Slacken and remove the silencer mounting bolt nut then withdraw the bolt and washer **(see illustration)**. Slacken the silencer clamp bolts then carefully pivot the silencer downwards and out to gain the necessary clearance required for wheel removal **(see illustrations)**. Take care not to lose the collar from the rear of the silencer mounting.
3 Place the transmission in gear then have an assistant apply the rear brake hard whilst you slacken the four wheel nuts **(see illustration)**.
4 Remove the wheel nuts then free the wheel from the stub axle and remove it from the bike **(see illustration)**.

Installation

5 Fit the wheel to the stub axle and screw on the wheel nuts. Have an assistant apply the rear brake then tighten the wheel nuts evenly and progressively to the specified torque in a diagonal sequence.
6 Ensure the collar is in position in the silencer mounting then pivot the silencer back into position and insert the mounting bolt and washer. Fit the nut to the mounting bolt and tighten to the specified torque setting then securely tighten the silencer clamp bolts.

18 Rear wheel bearing holder – removal, overhaul and installation

Removal

1 Remove the centre cap from the end of the stub axle and release the staking of the stub axle nut using a hammer and a fine chisel or pin punch **(see illustrations)**. With the bike on its wheels, have an assistant sit on it and apply the rear brake whilst you slacken the stub axle nut and the four wheel nuts **(see illustration)**.
2 Remove the rear wheel as described in Section 17.

17.2a Unscrew the nut and remove the silencer mounting bolt and washer . . .

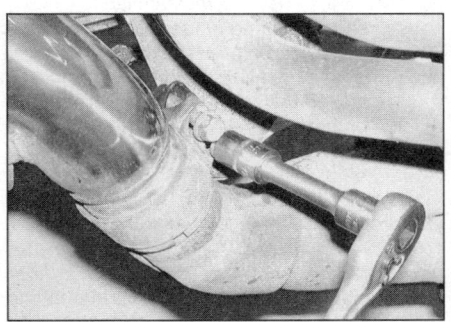

17.2b . . . then slacken the clamp bolts . . .

17.2c . . . and pivot the silencer clear of the wheel

17.3 Slacken the remove the four wheel bolts . . .

17.4 . . . then lift the rear wheel out of position

7

7•22 Brakes, wheels and tyres

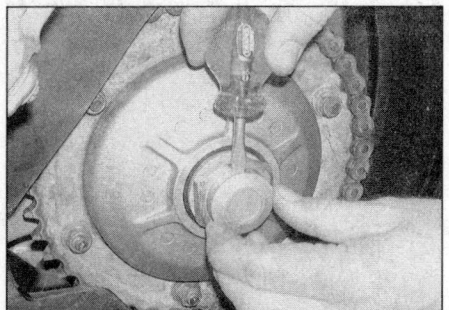

18.1a Prise out the centre cap . . .

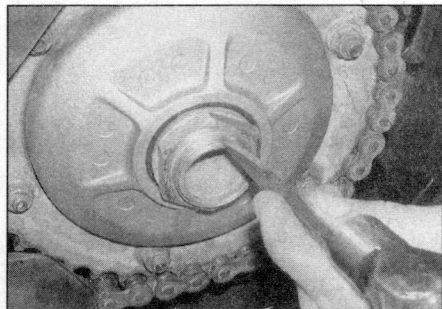

18.1b . . . then unstake the axle nut from the stub axle groove

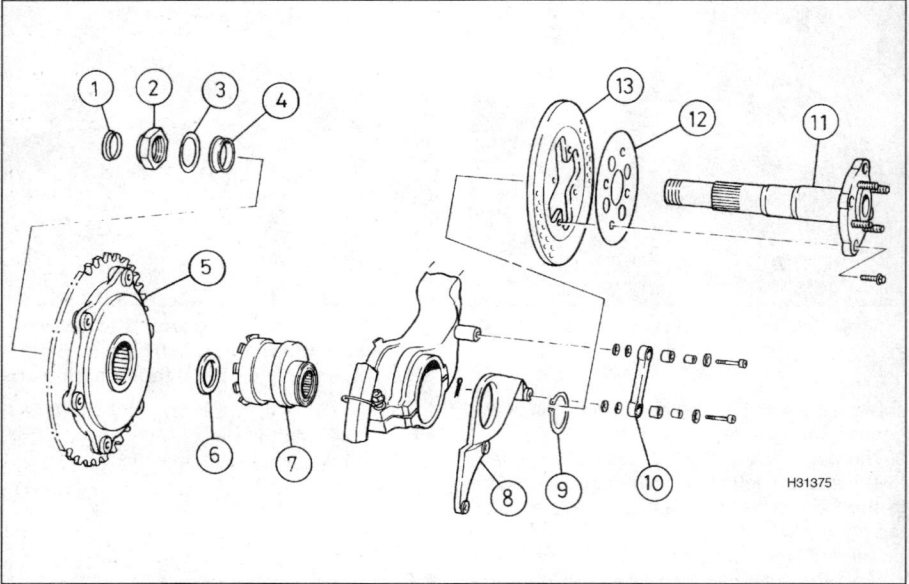

18.1c Rear wheel bearing holder and associated components

1 Centre cap	5 Sprocket coupling	9 Circlip
2 Axle nut	assembly	10 Torque rod
3 Conical washer	6 Spacer	11 Stub axle
4 Flanged	7 Bearing holder assembly	12 Spacer
spacer	8 Caliper mounting plate	13 Rear brake disc

3 Remove the chainguard retaining bolts and inner retaining clips and free the chainguard from the swingarm **(see illustration)**. Slacken and remove the retaining bolts and collars and remove the air guide from the base of the swingarm.
4 Unscrew the stub axle nut and remove the conical washer, noting which way around it is fitted **(see illustrations)**. Discard the nut; a new one will be needed for installation.
5 Slacken the bearing holder clamp bolt then release the drive chain tension (see Chapter 1) and unhook the chain from the rear sprocket.
6 Remove the flanged spacer from the outside of the sprocket coupling then slide the assembly off the stub axle **(see illustration)**.
7 Slacken and remove the caliper bracket mounting bolts then slide the caliper off the disc and tie it to the subframe to avoid any strain being placed on the hydraulic hoses. **Note:** *Honda recommend that new brake caliper bracket bolts should be used on installation (the bolts are supplied with their threads pre-coated with locking compound). If the old bolts are to be reused, remove all original locking compound from their threads and apply fresh compound before installing them.*
8 Withdraw the stub axle and brake disc assembly from the bearing holder, taking care not to lose the spacer from the left (sprocket) side of the bearing holder.
9 Remove the split pin from the bolt securing the torque rod to the brake caliper mounting plate. Unscrew both mounting bolts and remove the torque rod, taking care not to lose the collars which are fitted between the rod and plate/swingarm.
10 Using a pair of circlip pliers, extract the large circlip and remove the caliper mounting plate from the bearing holder, noting which way around it is fitted.
11 Fully slacken the clamp bolt and withdraw the bearing holder assembly from the swingarm.

18.3 Remove the retaining bolts and clips and free the chainguard from the swingarm

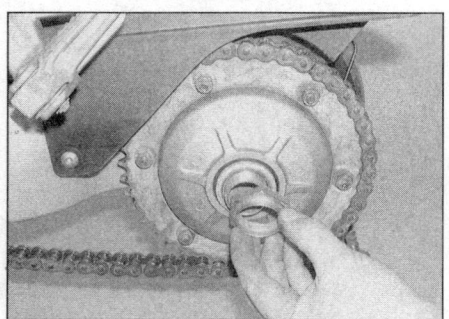

18.4a Unscrew the axle nut . . .

18.4b . . . and remove the conical washer and flanged spacer

18.6 Removing the sprocket coupling assembly from the stub axle

Overhaul

Note: *Do not remove the bearings from the holder unless they are to be renewed.*

12 Remove the spacer from the left (sprocket) side of the bearing holder and inspect all components closely, looking for obvious signs of wear such as heavy scoring, or for damage such as cracks or distortion **(see illustration)**. The left (sprocket) side bearing inner races should rotate freely and easily without any sign of freeplay or roughness. Renew worn components as necessary. Refer to Chapter 6 for details of the rear sprocket coupling assembly.

13 If the stub axle is damaged it must be renewed. Transfer the brake disc and spacer plate to the new axle as described in Section 8.

14 If the dust seals are damaged, the old seals can be levered out of position using a flat-bladed screwdriver **(see illustration)**. Ensure the sealing lip of each new seal is facing inwards then press the seal squarely into position until it is flush with the holder **(see illustration)**.

15 If the right (wheel) side needle roller bearing is damaged, remove the right side dust seal then prise out the bearing circlip from the holder. Support the bearing holder then press/drift the bearing out of position. The new bearing should be pressed or drawn into its bore rather than driven into position to prevent possible damage. In the absence of a press, a suitable drawbolt arrangement can be made up as described in *Tools and Workshop Tips* in the Reference section, ensuring the bearing is fitted with its marked surface outwards and integral seal innermost. Secure the bearing in position with the circlip, ensuring it is correctly located in the holder groove, then press a **new** dust seal into position.

16 To renew the left (sprocket) side ball bearing, prise out the left dust seal then extract the bearing circlip from inside the holder **(see illustration)**. Support the bearing holder and drift the bearing squarely out of position, using a hammer and suitable drift inserted through from the right side of the holder **(see illustration)**. Turn the bearing holder over and locate the new bearing in its bore, ensuring its marked surface is facing outwards. Tap it squarely into position using a tubular spacer (such as a socket) which bears only on the bearing outer race. Ensure the bearing is firmly against its stop then secure it in position with the circlip. Check that the circlip is correctly located in its groove then press a **new** dust seal into position.

17 Lubricate the dust seal lips with multi-purpose grease then fit the spacer to the left (sprocket) side dust seal.

Installation

18 Remove all traces of corrosion from the outside of the bearing holder assembly and the inside of the swingarm bore. Remove the spacer from the left (sprocket) side of the bearing holder. Apply a smear of multi-purpose grease to the needle roller bearing and both dust seal lips then install the spacer again.

19 Insert the bearing holder into the swingarm and locate the caliper mounting plate on the holder, making sure it is the correct way around. Secure the plate in position with the large circlip, ensuring it is correctly located in the holder groove.

20 Apply a drop of thread-locking compound to the threads of the torque rod retaining bolts and fit the collars to the torque rod. Install the rod, ensuring both collars are correctly positioned between the rod and swingarm/plate then fit both mounting bolts. Tighten the bolts to the specified torque then secure the rod to mounting plate bolt in position with a **new** split pin.

21 Apply a thin film of multi-purpose grease to the stub axle shaft then carefully slide the axle assembly into the bearing holder. Take

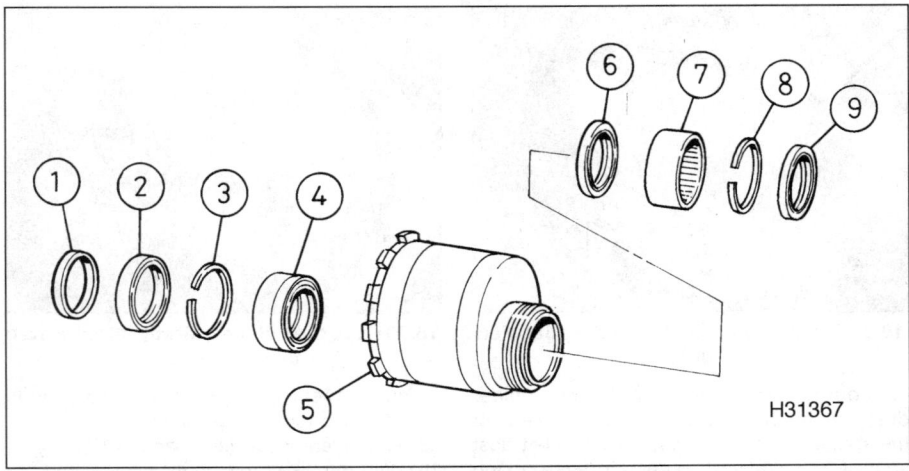

18.12 Exploded view of the rear wheel bearing holder assembly

1 Spacer	4 Ball bearing	7 Needle roller bearing
2 Dust seal	5 Bearing holder	8 Circlip
3 Circlip	6 Oil seal	9 Dust seal

18.14a Lever out the old dust seal with a large, flat-bladed screwdriver . . .

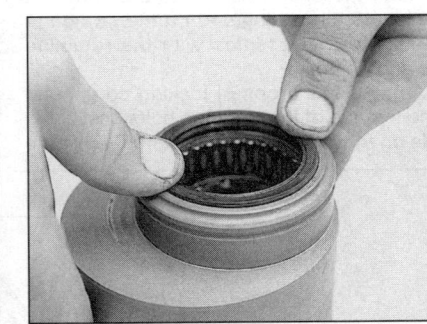

18.14b . . . and press the new seal into position ensuring its sealing lip is facing inwards

18.16a Prise out the circlip from the left side of the holder . . .

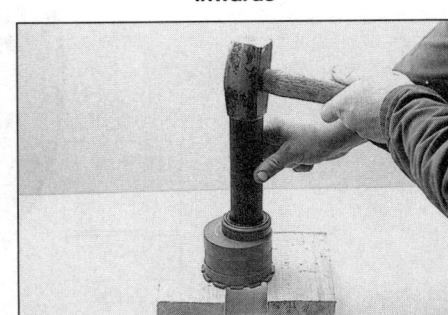

18.16b . . . then turn the holder over and drift the left bearing out from the other side

7•24 Brakes, wheels and tyres

18.29 Tighten the axle nut to the specified torque . . .

18.30a . . . then stake it firmly into the stub axle groove . . .

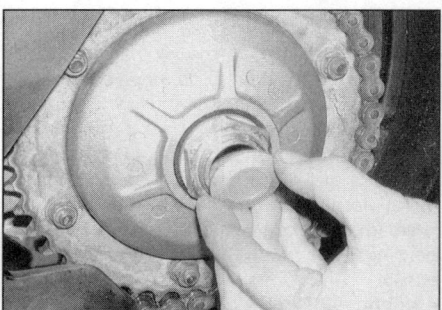

18.30b . . . before fitting the centre cap

care not to damage the right bearing holder dust seal as the axle is fitted and make sure the spacer remains in position in the left dust seal. Clean off all excess grease using acetone or brake system cleaner.

22 Slide the rear brake caliper into position, making sure the pads pass either side of the disc. Fit the **new** caliper bracket bolts and tighten to the specified torque.

23 Slide the sprocket coupling assembly onto the stub axle and engage it with the axle splines.

24 Lubricate the lip of the sprocket coupling dust seal with multi-purpose grease then carefully slide the flanged spacer into position.

25 Fitted the conical washer with its marked (convex) surface facing outwards. Lubricate the threads and face of the **new** axle nut with engine oil and thread the nut onto the axle.

26 Hook the drive chain onto the sprocket then locate the chainguard on the swingarm and secure it in position with the retaining clips and bolts.

27 Apply a few drops of locking compound to the threads of the air guide bolts. Install the guide on the base of the swingarm then fit the collars and retaining bolts, tightening them to the specified torque (see Chapter 6).

28 Fit the rear wheel (see Section 17).

29 With the bike back on its wheels, have an assistant sit on it and apply the rear brake hard whilst you tighten the stub axle nut to the specified torque **(see illustration)**.

30 Secure the axle nut in position by staking it firmly into the stub axle groove, using a hammer and punch **(see illustration)**. Fit the centre cap to the end of the stub axle.

31 Adjust the drive chain slack as described in Chapter 1 and check the operation of the rear brake before taking the machine on the road.

19 Tyres – general information and fitting

General information

1 The wheels fitted to all models are designed to take tubeless tyres only. Tyre sizes are given in the Specifications at the beginning of this Chapter.

2 Refer to *Daily (pre-ride) checks* at the beginning of this manual for tyre maintenance.

Fitting new tyres

3 When selecting new tyres, ensure that front and rear tyre types are compatible, the correct size and correct speed rating; if necessary seek advice from a tyre fitting specialist **(see illustration)**.

4 It is recommended that tyres are fitted by a motorcycle tyre specialist rather than attempted in the home workshop. This is particularly relevant in the case of tubeless tyres because the force required to break the seal between the wheel rim and tyre bead is substantial, and is usually beyond the capabilities of an individual working with normal tyre levers. Additionally, the specialist will be able to balance the wheels after tyre fitting.

5 Note that punctured tubeless tyres can in some cases be repaired, but such repairs must be carried out by a tyre fitting specialist. Note that Honda advise that a repaired tyre must not be used at speeds above 50 mph (80 kmh) for the first 24 hours, and not above 80 mph (130 kmh) thereafter.

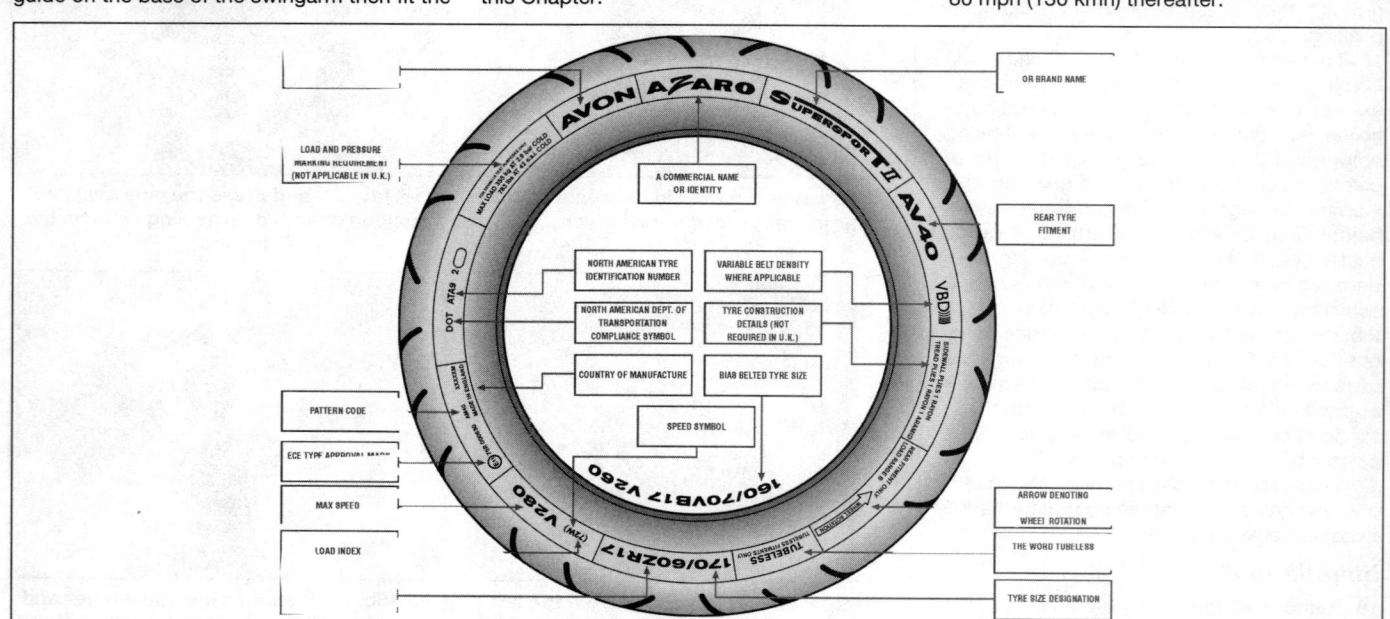

19.3 Common tyre sidewall markings

Chapter 8
Bodywork

Contents

Fairing panels – removal and installation 5
Front mudguard – removal and installation 6
General information 1
Rear view mirrors – removal and installation 4
Seat – removal and installation 2
Seat cowling – removal and installation 3

Degrees of difficulty

| Easy, suitable for novice with little experience | Fairly easy, suitable for beginner with some experience | Fairly difficult, suitable for competent DIY mechanic 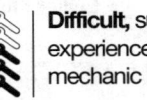 | Difficult, suitable for experienced DIY mechanic 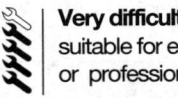 | Very difficult, suitable for expert DIY or professional  |

1 General information

This Chapter covers the procedures necessary to remove and install the body parts. Since many service and repair operations on these motorcycles require the removal of the body parts, the procedures are grouped here and referred to from other Chapters.

In the case of damage to the body parts, it is usually necessary to remove the broken component and replace it with a new or used one. The material that the body panels are composed of doesn't lend itself to conventional repair techniques. Note that there are however some companies that specialise in 'plastic welding' and there are a number of bodywork repair kits now available for motorcycles.

When attempting to remove any body panel, first study it closely, noting any fasteners and associated fittings, to be sure of returning everything to its correct place on installation. In some cases the aid of an assistant will be required when removing panels, to help avoid the risk of damage to paintwork. Once the evident fasteners have been removed, try to withdraw the panel as described but DO NOT FORCE IT – if it will not release, check that all fasteners have been removed and try again. Where a panel engages another by means of tabs, be careful not to break the tab or its mating slot or to damage the paintwork. Remember that a few moments of patience at this stage will save you a lot of money in replacing broken fairing panels!

When installing a body panel, first study it closely, noting any fasteners and associated fittings removed with it, to be sure of returning everything to its correct place. Check that all fasteners are in good condition, including all trim nuts or clips and damping/rubber mounts; any of these must be renewed if faulty before the panel is reassembled. Check also that all mounting brackets are straight and repair or renew them if necessary before attempting to install the panel. Where assistance was required to remove a panel, make sure your assistant is on hand to install it.

Tighten the fasteners securely, but be careful not to overtighten any of them or the panel may break (not always immediately) due to the uneven stress. Where quick-release fasteners are fitted, turn them 90° anti-clockwise to release them, and 90° clockwise to secure them.

2 Seat – removal and installation

Removal

1 If a pillion seat cover is fitted, unscrew the seat cowling bolt and washer which secure each side of the cover in position then free the cover and remove it from the motorcycle.
2 Insert the ignition key into the seat lock/helmet holder, and turn it clockwise to release the seat lock lever. Pull down on the lock lever then slide the seat assembly backwards and lift it away from the motorcycle.

Installation

3 Installation is the reverse of removal. Make sure all the tabs on the seat are correctly located in the brackets on the fuel tank and subframe then push down on the rear of the seat to secure it in position. Ensure the seat is securely locked then remove the ignition key. If necessary, clip the pillion seat cover correctly in position then securely refit the seat cowling bolts and washers.

8•2 Bodywork

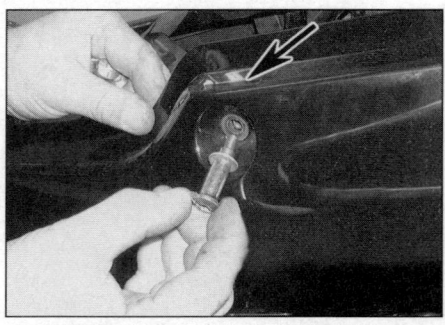

3.4 Unscrew the bolt and washer from each side of the seat cowling and recover the spacer (where fitted) from the cowling slot (arrow)

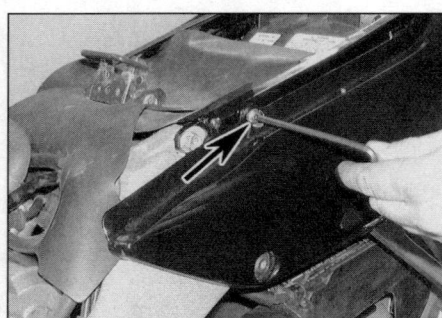

3.5a Unscrew the front mounting bolts (arrows) from each side . . .

3.5b . . . then manoeuvre the cowling assembly away from the bike

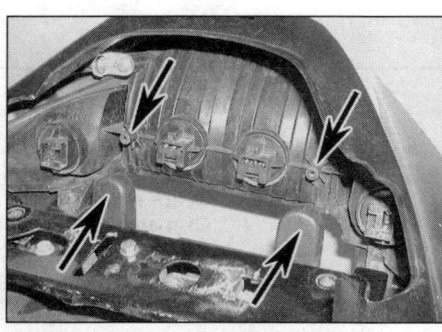

3.6 On installation ensure the pegs on the taillight locate correctly in the mudguard grommets (arrows)

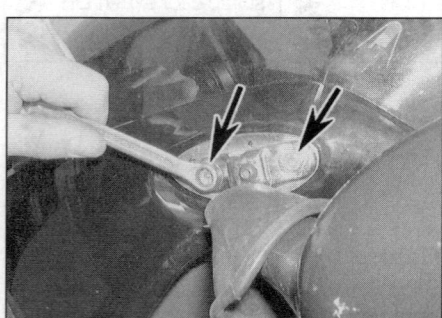

4.1a Unscrew the mounting bolts (arrows) and remove the mirror . . .

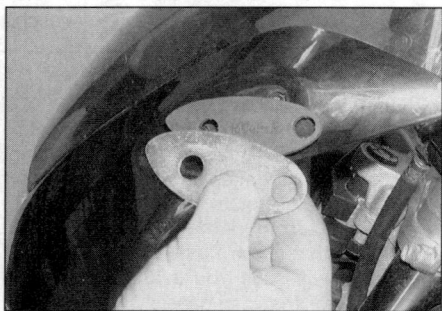

4.1b . . . along with its mounting plate and rubber spacer

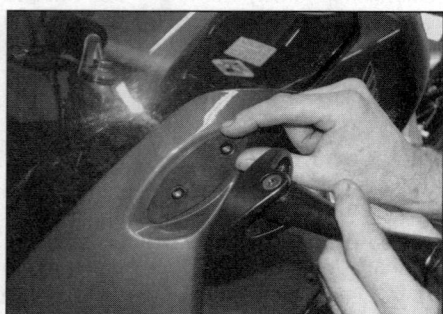

4.2 VFR800FI-Y and 1 mirror and rubber spacer

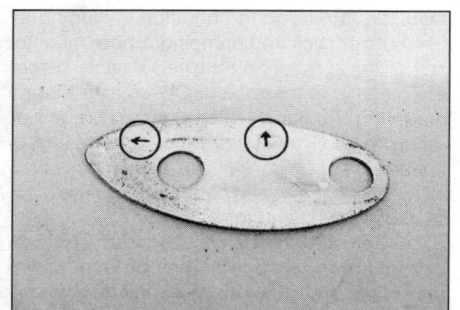

4.3 Ensure the mounting plate is fitted with its arrow markings facing forwards and upwards

3 Seat cowling – removal and installation

Removal

1 Remove the seat (see Section 2).
2 Remove the mounting bolts and washers and remove the pillion grab handles.
3 Disconnect the wiring connectors from the tail light assembly.
4 If a pillion seat cover was not installed, unscrew the bolt and washer from each side of seat cowling and lift out the spacers from the cowling slots which the front of the pillion seat cover (if fitted) would locate in (see illustration).
5 Slacken and remove the two bolts securing each side of the front of the seat cowling to the frame then move the cowling assembly backwards and away from the motorcycle (see illustrations).

Installation

6 Manoeuvre the cowling into position making sure the pegs on the tail light are correctly located in the grommets on the rear mudguard (see illustration).
7 Fit the bolts securing the front of the cowling in position and securely tighten them.
8 Where a pillion seat cover is not fitted, insert the spacers into the cowling slots then install the bolts and washers and tighten securely.
9 Securely reconnect all the wiring connectors to the tail light assembly
10 Install the pillion grab handles and fit the retaining bolts and washers, tightening them to 34 Nm.
11 Check the operation of the tail light/turn signals then install the seat (see Section 2).

4 Rear view mirrors – removal and installation

Removal

1 On VFR800FI-W and X models, peel back the rubber cover from the mirror base and unscrew the mounting bolts, then remove the mirror, mounting plate and rubber spacer (see illustrations).
2 On VFR800FI-Y and 1 models, unscrew the mounting bolts and remove the mirror and rubber spacer (see illustration).

Installation

3 On VFR800FI-W and X models, fit the rubber spacer with its identification marking facing out, and the mounting plate with its arrow markings facing forwards and upwards (see illustration). Install the mirror and tighten the mounting bolts securely, then locate the rubber cover over the mounting plate.

Bodywork 8•3

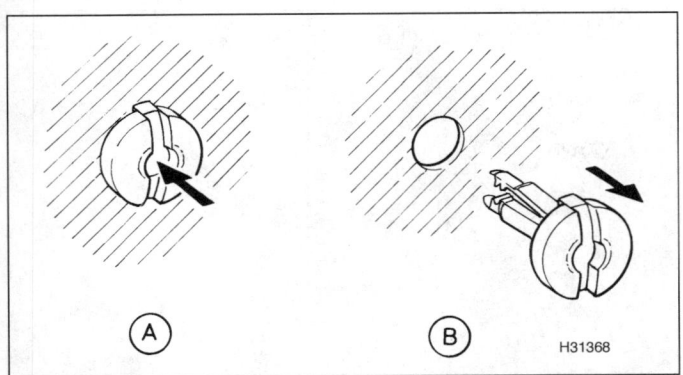

5.2a On VFR800FI-W and X models, press the trim clip centre section fully in (A) the pull out the complete clip (B)

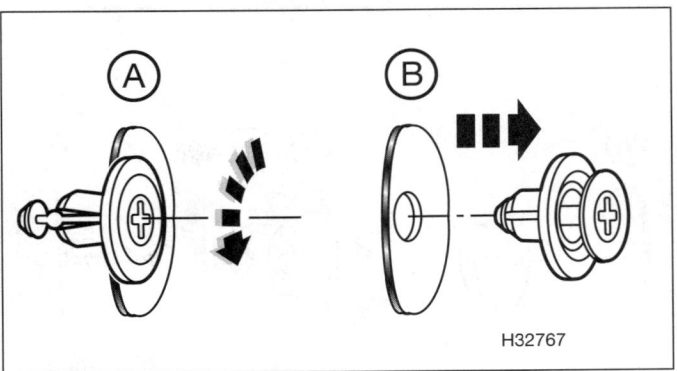

5.2b On VFR800FI-Y and 1 models, unscrew the centre of the clip (A) then pull out the complete clip (B)

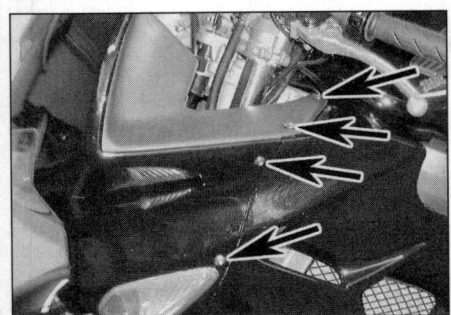

5.4a Undo the bolts securing the lower fairing panel to the upper fairing (four shown) . . .

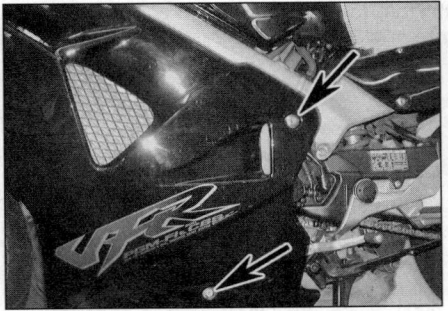

5.4b . . . and mountings (arrows) . . .

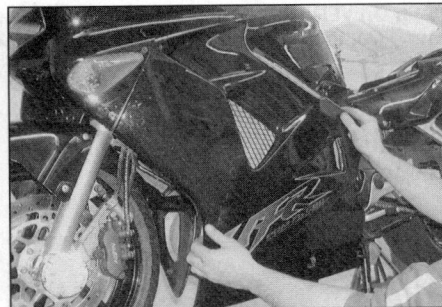

5.4c . . . then remove the panel from the bike

4 On VFR800FI-Y and 1 models, fit the rubber spacer then install the mirror and tighten the mounting bolts securely.

5 Fairing panels – removal and installation

Lower panels

Removal

1 Position the motorcycle on the centrestand. Each fairing lower panel can be removed as follows.
2 Remove the trim clips securing the base of the left and right side lower panels together (see illustrations).
3 Remove the trim clip and screws securing the inner fairing panel to the left and right side lower panels.
4 Remove the screws securing the lower panel to the upper fairing and mountings then carefully manoeuvre the panel away from the motorcycle (see illustrations).

Installation

5 Installation is the reverse of removal. Ensure the lower panels are correctly located with both the upper fairing and inner fairing panel, then install the trim clips (see illustrations).

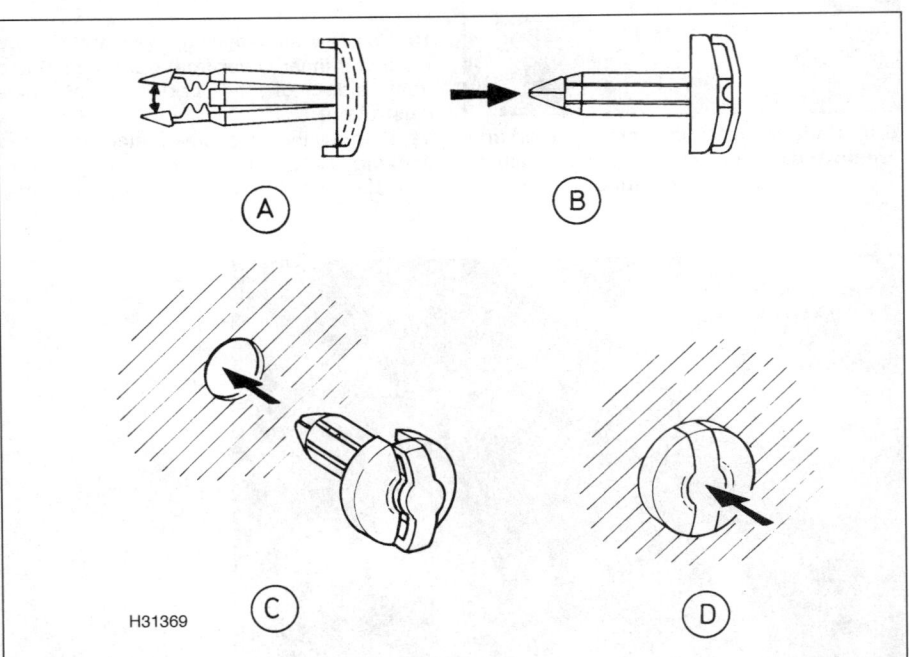

5.5a On VFR800FI-W and X models, get the trim clip ready for fitting by gently expanding and pushing its retaining tabs (A) into the clip until the centre section is fully extended (B). Fit the clip (C) and secure it in position by pressing the centre section in until it is flush with the clip body (D)

8

8•4 Bodywork

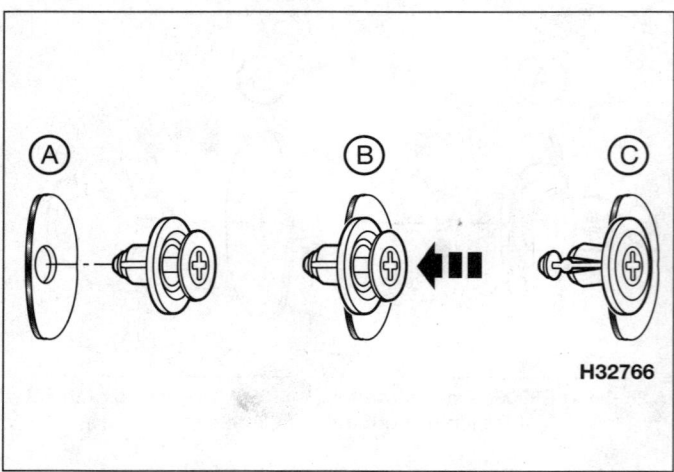

5.5b On VFR800FI-Y and 1 models, unscrew the centre of the clip (A), then fit the clip (B) and push the centre in to lock it

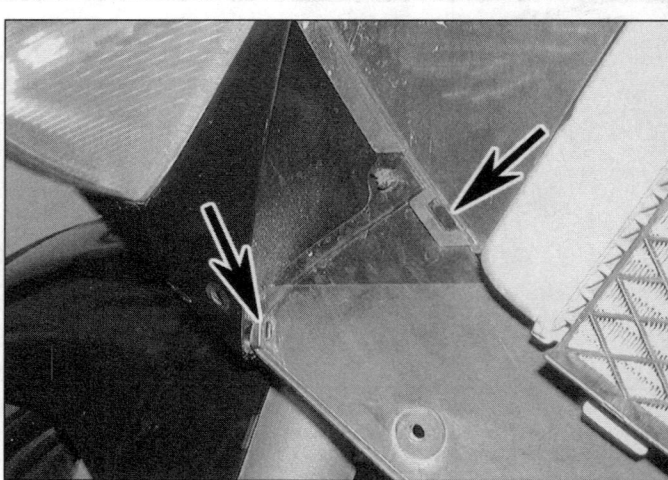

5.8 Ensure the inner panel is correctly engaged with the upper fairing before fitting its clips

Inner panel

Removal

6 Remove both the left and right side lower panels as described in Paragraphs 1 to 4.
7 Remove the trim clips securing the inner fairing panel to the upper fairing then carefully remove the panel from the motorcycle.

Installation

8 Installation is the reverse of removal, making sure the inner panel tabs are correctly located in the upper fairing (see illustration).

Windscreen

Removal

9 Remove both rear view mirrors (see Section 4).
10 Unscrew the retaining screw and remove the upper inner cover from both the left and right sides of the upper fairing (see illustration).
11 Remove the windscreen retaining screws (located between the rear view mirror mounting bolt holes on the left and right) then move the windscreen upwards and away from the upper fairing (see illustrations).

Installation

12 Installation is the reverse of removal making sure the windscreen tabs are correctly located in the upper fairing slots.

Upper fairing

Removal

13 Remove both rear view mirrors (see Section 4).
14 Remove the left and right side lower panels as described in Paragraphs 1 to 4 then remove the inner panel as described in Paragraph 5.
15 Remove the windscreen as described in Paragraphs 10 and 11.
16 Remove the two trim clips securing the base of the upper fairing in position (see

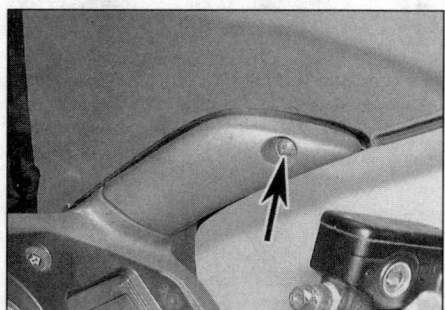

5.10 Undo the retaining screw (arrow) and remove the upper inner cover from each side of the fairing

5.11a Undo the windscreen retaining screws (left screw shown) . . .

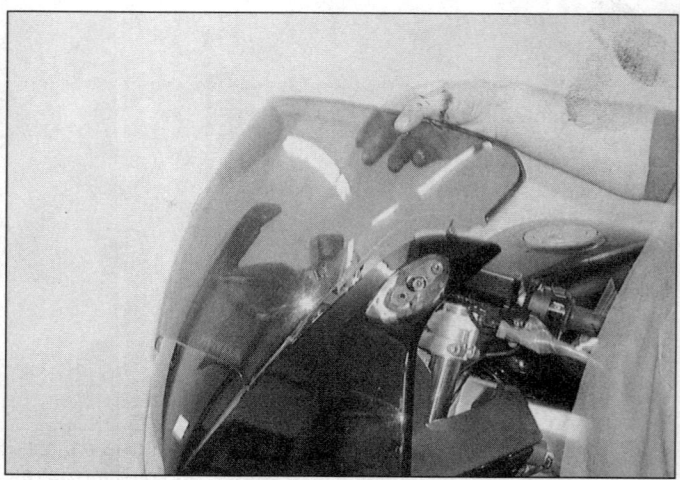

5.11b . . . then manoeuvre the windscreen out of position

Bodywork 8•5

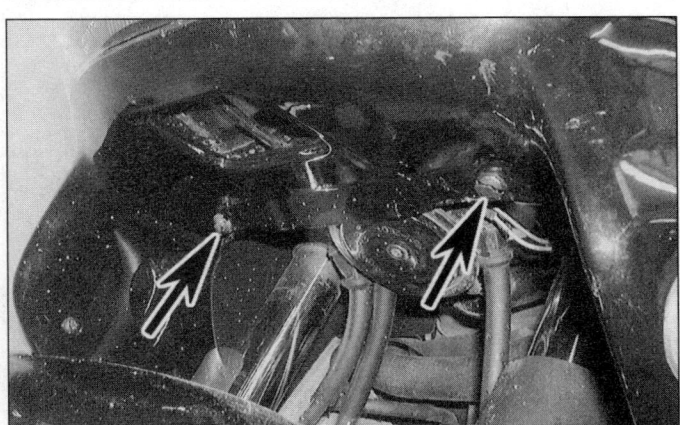

5.16 Remove the trim clips (arrows) from the base of the upper fairing

5.17 Undo the screws (arrows) securing the instrument cluster surround to the fairing

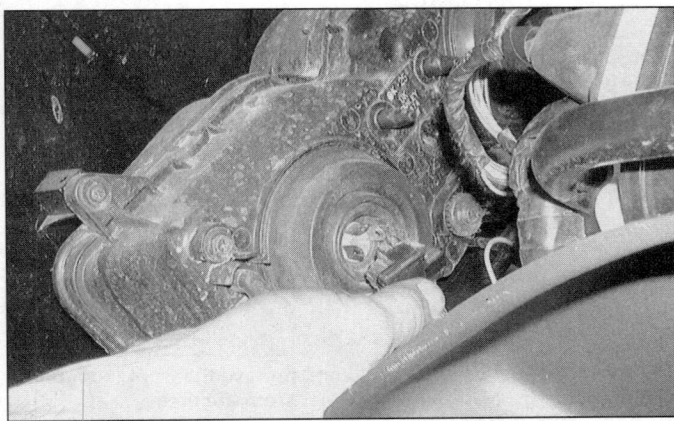

5.18a Manoeuvre the upper fairing away from the bike, disconnecting the headlight wiring connectors as they become accessible

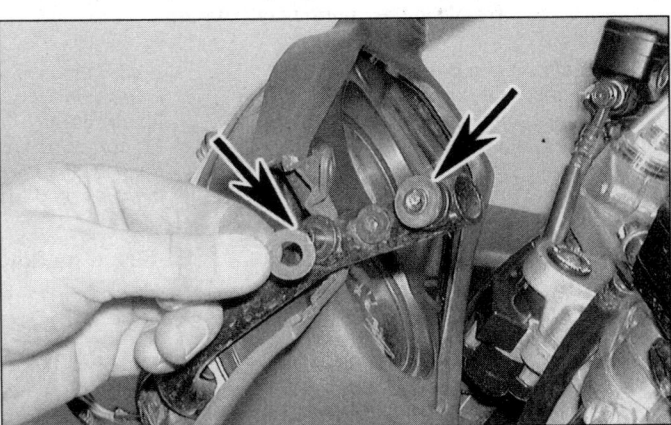

5.18b Recover the rubber spacers from the fairing stay and store them with the fairing

illustration).
17 Unscrew the two upper screws securing the instrument cluster surround panel to the upper fairing **(see illustration)**.
18 Free the upper fairing from the stay and move the assembly forwards. Disconnect the wiring connectors from the headlight and turn signals then remove the upper fairing assembly from the motorcycle **(see illustration)**. Recover the four rubber spacers which are fitted between the upper fairing and stay, one on each mirror mounting bolt point, and store them with the fairing **(see illustration)**.
19 If necessary, remove the two trim clips and lift the instrument cluster surround panel out of position **(see illustrations)**. The left and right side inner panels can also be removed

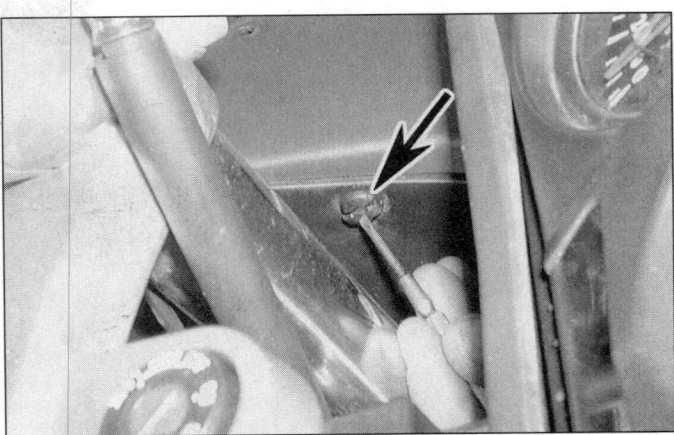

5.19a Remove the two trim clips (left clip shown) . . .

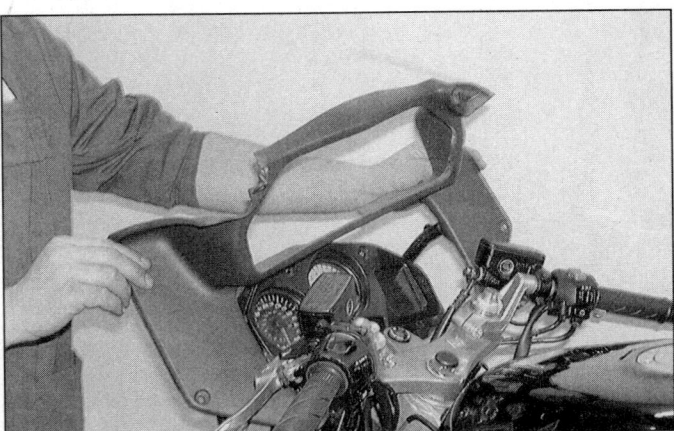

5.19b . . . and lift off the instrument cluster surround panel

8

8•6 Bodywork

5.20 On installation ensure the headlight pegs (arrows) locate correctly in the fairing stay grommets

6.1a Undo the retaining screws (left screws shown) . . .

once the clips have been removed and the wiring harness has been unclipped (the air temperature sensor is mounted on the right side inner panel and will need to be unclipped).

Installation

20 Installation is the reverse of removal, noting the following points.
 a) Make sure the rubber spacers and headlight locating peg grommets are in position on the fairing stay.
 b) Ensure all the wiring connectors are securely reconnected and the wiring harness correctly routed.
 c) As the upper fairing is installed make sure its tabs engage with the slots in the instrument cluster surround panel and the pegs on the rear of the headlight unit locate correctly in the stay grommets **(see illustration)**.
 d) On completion, check the operation of the headlight/turn signals before using the machine on the road.

6 Front mudguard – removal and installation

Removal

1 Unscrew the two bolts securing the mudguard to each fork slider and remove the mudguard in a forwards direction **(see illustrations)**.

6.1b . . . and remove the front mudguard from the bike

Installation

2 Installation is the reverse of removal.

Chapter 9
Electrical system

Contents

Alternator – removal, inspection and installation	32
Battery – charging	4
Battery – removal, installation, inspection and maintenance	3
Brake light switches – check and replacement	14
Brake/tail light bulbs – renewal	9
Charging system – leakage and output test	31
Charging system testing – general information and precautions	30
Clutch switch – check and replacement	24
Diode – check and replacement	25
Electrical system – fault finding	2
Fuel pump – check, removal and installation	see Chapter 4
Fuses – check and renewal	5
General information	1
Handlebar switches – check	20
Handlebar switches – removal and installation	21
Headlight aim – check and adjustment	see Chapter 1
Headlight assembly – removal and installation	8
Headlight and sidelight bulbs – renewal	7
Horn – check and replacement	26
Ignition switch – check, removal and installation	19
Ignition system components	see Chapter 5
Instrument and warning light bulbs – renewal	17
Instrument cluster – removal and installation	15
Instruments and sensors – check and replacement	16
Lighting system – check	6
Neutral switch – check, removal and installation	22
Oil pressure switch – check and replacement	18
Regulator/rectifier – check and replacement	33
Sidestand switch – check and replacement	23
Starter motor – disassembly, inspection and reassembly	29
Starter motor – removal and installation	28
Starter relay – check and replacement	27
Tail light assembly – removal and installation	10
Turn signal assemblies – removal and installation	13
Turn signal bulbs – renewal	12
Turn signal circuit – check	11

Degrees of difficulty

Easy, suitable for novice with little experience	Fairly easy, suitable for beginner with some experience	Fairly difficult, suitable for competent DIY mechanic	Difficult, suitable for experienced DIY mechanic	Very difficult, suitable for expert DIY or professional 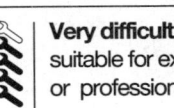

Specifications

Battery

Capacity	12V, 10Ah
Voltage	
Fully charged	13.0 to 13.2V
Discharged	below 12.3V
Charging rate	
Normal	1.2A for 5 to 10 hrs
Quick	5.0A for 1 hr

Charging system

Current leakage	1.2mA (max)
Regulated voltage output	14.0 to 14.8V @ 5000 rpm
Alternator stator coil resistance	0.1 to 1.0 ohms @ 20°C/68°F

Starter motor

Brush length	
Standard	12.0 to 13.0 mm
Service limit (min)	4.5 mm

Fuses

Main (A and B)	30A
Clock	10A
Fuel injection/ignition system	20A
Turn signal, brake light and horn	10A
Starter	10A
Fan	10A
Headlight	20A
Tail light and instruments	10A

Fuel gauge sender unit resistance
Full position . . . . 1 to 5 ohms
Empty position . . . . 92 to 96 ohms

Air temperature sensor resistance
At 25°C/77°F . . . . 4.8 to 5.2 K-ohm

Bulbs
European models
    Headlights . . . . 60/55W H4 halogen
    Sidelight 5W
    Brake/tail lights 21/5W
    Turn signal lights 21W
    Instrument and warning lights
        Turn signal indicator bulb 3.4W
        All other bulbs 1.7W
US models
    Headlights . . . . 45/45W halogen
    Brake/tail lights 21/5W
    Turn signal lights
        Front (with running light) 21/5W
        Rear 21W
    License plate light 8W
    Instrument and warning lights
        Turn signal indicator bulb 3.4W
        All other bulbs 1.7W

Torque settings
Alternator cover bolts . . . . 12 Nm
Alternator stator and wiring clip bolts . . . . 12 Nm
Alternator rotor bolt . . . . 103 Nm
Ignition (main) switch bolt . . . . 25 Nm
Neutral switch . . . . 12 Nm
Oil pressure switch . . . . 12 Nm
Sidestand switch bolt . . . . 10 Nm

1 General information

All models have a 12-volt electrical system charged by a three-phase alternator with a separate regulator/rectifier.

The regulator maintains the charging system output within the specified range to prevent overcharging, and the rectifier converts the ac (alternating current) output of the alternator to dc (direct current) to power the lights and other components and to charge the battery. The alternator rotor is mounted on the left end of the crankshaft.

The starter motor is on the front of the crankcase. The starting system includes the motor, the battery, the relay and the various wires and switches. If the engine kill switch in the RUN position and the ignition (main) switch is ON, the starter relay allows the starter motor to operate only if the transmission is in neutral (neutral switch on). If, however, the transmission is in gear, the clutch lever must also be pulled into the handlebar and the sidestand must be up.

Note: *Keep in mind that electrical parts, once purchased, cannot be returned. To avoid unnecessary expense, make very sure the faulty component has been positively identified before buying a new part.*

2 Electrical system – fault finding

Warning: To prevent the risk of short circuits, the ignition (main) switch must always be OFF and the battery negative (–) terminal should be disconnected before any of the bike's other electrical components are disturbed. Don't forget to reconnect the terminal securely once work is finished or if battery power is needed for circuit testing.

1 A typical electrical circuit consists of an electrical component, the switches, relays, etc. related to that component and the wiring and connectors that hook the component to both the battery and the frame. To aid in locating a problem in any electrical circuit, refer to the wiring diagrams at the end of this Chapter.

2 Before tackling any troublesome electrical circuit, first study the wiring diagram (see end of Chapter) thoroughly to get a complete picture of what makes up that individual circuit. Trouble spots, for instance, can often be narrowed down by noting if other components related to that circuit are operating properly or not. If several components or circuits fail at one time, chances are the fault lies in the fuse or earth (ground) connection, as several circuits often are routed through the same fuse and earth (ground) connections.

3 Electrical problems often stem from simple causes, such as loose or corroded connections or a blown fuse. Prior to any electrical fault finding, always visually check the condition of the fuse, wires and connections in the problem circuit. Intermittent failures can be especially frustrating, since you can't always duplicate the failure when it's convenient to test. In such situations, a good practice is to clean all connections in the affected circuit, whether or not they appear to be good. All of the connections and wires should also be wiggled to check for looseness which can cause intermittent failure.

4 If testing instruments are going to be utilised, use the wiring diagram to plan where you will make the necessary connections in order to accurately pinpoint the trouble spot.

5 The basic tools needed for electrical fault finding include a battery and bulb test circuit, a continuity tester, a test light, and a jumper wire. A multimeter capable of reading volts, ohms and amps is also very useful as an alternative to the above, and is necessary for performing more extensive tests and checks.

 *Refer to Fault Finding Equipment in the Reference section for details of how to use electrical test equipment.*

Electrical system 9•3

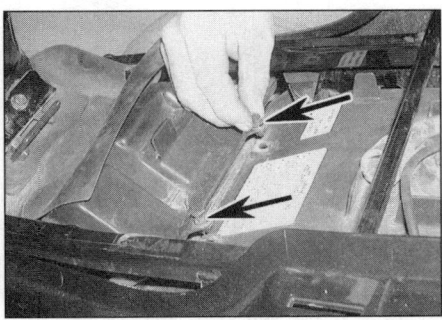

3.2 Remove the fasteners (arrows) and open up the battery cover

3.3a Unscrew the bolt and disconnect the negative (-) lead from the battery ...

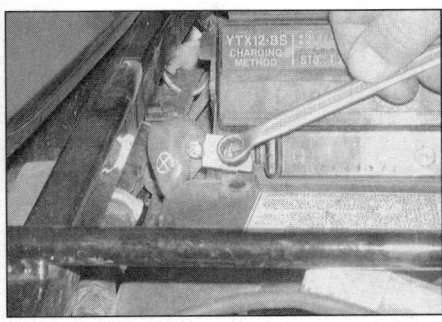

3.3b ... then lift the insulating cover and disconnect the positive (+) lead ...

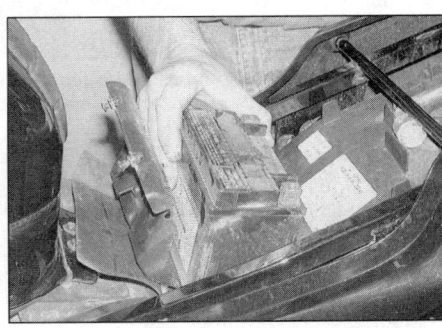

3.3c ... before lifting the battery out

3 Battery – removal, installation, inspection and maintenance

Caution: *Be extremely careful when handling or working around the battery. The electrolyte is very caustic and an explosive gas (hydrogen) is given off when the battery is charging.*

Removal and installation

1 Remove the seat (see Chapter 8).
2 Remove the fasteners (lift the centre pin then remove the complete fastener) and open up the battery cover **(see illustration)**.
3 Unscrew the negative (–) terminal bolt first and disconnect the lead from the battery **(see illustration)**. Lift up the red insulating cover to access the positive (+) terminal, then unscrew the bolt and disconnect the lead **(see illustration)**. Lift the battery from the bike **(see illustration)**.
4 On installation, clean the battery terminals and lead ends with a wire brush or knife and emery paper. Reconnect the leads, connecting the positive (+) terminal first.

 **Battery corrosion can be kept to a minimum by applying a layer of petroleum jelly to the terminals after the cables have been connected.**

5 Ensure the insulating cover is correctly fitted then secure the battery cover in position with the fasteners and install the seat (see Chapter 8).

Inspection and maintenance

6 The battery fitted to the models covered in this manual is of the maintenance-free (sealed) type, it therefore does not require topping up. However, the following checks should still be regularly performed.
7 Check the battery terminals and leads for tightness and corrosion. If corrosion is evident, unscrew the terminal screws and disconnect the leads from the battery, disconnecting the negative (–) terminal first, and clean the terminals and lead ends with a wire brush or knife and emery paper.

Reconnect the leads, connecting the negative (–) terminal last, and apply a thin coat of petroleum jelly to the connections to slow further corrosion.
8 The battery case should be kept clean to prevent current leakage, which can discharge the battery over a period of time (especially when it sits unused). Wash the outside of the case with a solution of baking soda and water. Rinse the battery thoroughly, then dry it.
9 Look for cracks in the case and renew the battery if any are found. If acid has been spilled on the frame or battery box, neutralise it with a baking soda and water solution, dry it thoroughly, then touch up any damaged paint.
10 If the motorcycle sits unused for long periods of time, disconnect the leads from the battery terminals, negative (–) first. Refer to Section 4 and charge the battery once every month to six weeks.
11 The condition of the battery can be assessed by measuring the voltage present at the battery terminals. Connect the voltmeter positive (+) probe to the battery positive (+) terminal, and the negative (–) probe to the battery negative (–) terminal. When fully charged there should be at least 13 volts present. If the voltage falls to 12.3 volts the battery must be removed, disconnecting the negative (–) terminal first, and recharged as described below in Section 4.

4 Battery – charging

Caution: *Be extremely careful when handling or working around the battery. The electrolyte is very caustic and an explosive gas (hydrogen) is given off when the battery is charging.*

1 Remove the battery (see Section 3). Connect the charger to the battery, making sure that the positive (+) lead on the charger is connected to the positive (+) terminal on the battery, and the negative (–) lead is connected to the negative (–) terminal.
2 Honda recommend that the battery is charged at a maximum rate of 1.2 amps for 5 to 10 hours. Exceeding this figure can cause the battery to overheat, buckling the plates and rendering it useless. Few owners will have access to an expensive current controlled charger, so if a normal domestic charger is used check that after a possible initial peak, the charge rate falls to a safe level **(see illustration)**. If the battery becomes hot during charging **stop**. Further charging will cause damage. **Note:** *In emergencies the battery can be charged at a higher rate of around 5.0 amps for a period of 1 hour. However, this is not recommended and the low amp charge is by far the safer method of charging the battery.*
3 If the recharged battery discharges rapidly when left disconnected it is likely that an internal short caused by physical damage or sulphation has occurred. A new battery will be required. A sound item will tend to lose its charge at about 1% per day.

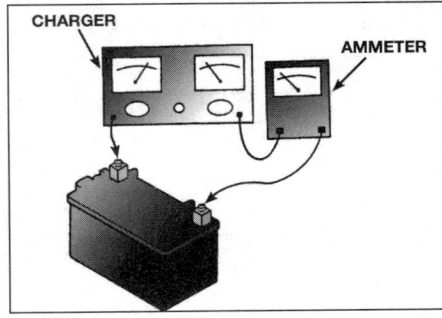

4.2 If the charger doesn't have an ammeter built-in, connect one in series as shown. DO NOT connect the ammeter between the battery terminals or it will be ruined

9•4 Electrical system

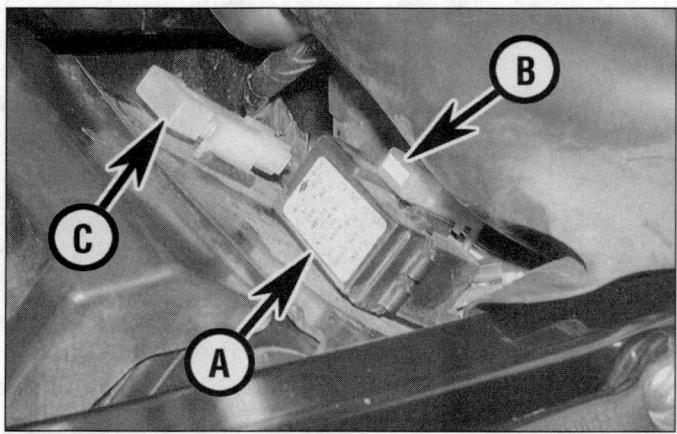

5.1 Most fuses are located in the fusebox (A). The clock fuse (B) and main fuse B (C) are located in separate housings next to the fusebox

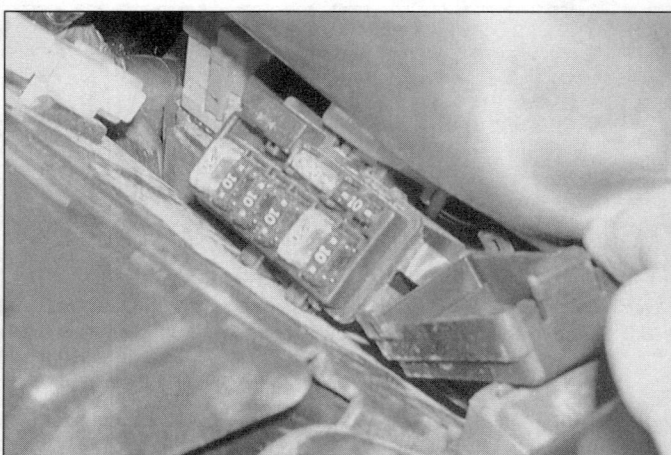

5.2a Unclip the fusebox cover to gain access to the fuses

4 Install the battery (see Section 3).
5 If the motorcycle sits unused for long periods of time, charge the battery once every month to six weeks and leave it disconnected.

5 Fuses – check and renewal

1 The electrical system is protected by fuses of different ratings which are all located under the seat. Most fuses, except the main fuses

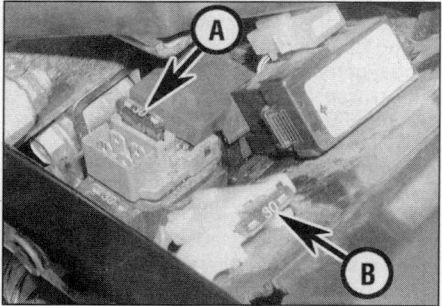

5.2b Disconnect the wiring connector from the starter relay to gain access to main fuse A (A). Main fuse B (B) is located in the holder behind the relay

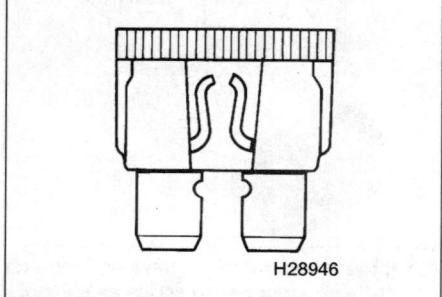

5.3 A blown fuse can be identified by a break in its element

and the clock fuse, are housed in the fusebox which is located in front of the battery (see illustration). The main fuse A is integral with the starter relay, which is just in front of and to the left of the fusebox, and main fuse B and the clock fuse are fitted to their own holders located to the left of and in front of the fusebox. Note: *The fusebox also houses the diode (see Section 25).*
2 To access the fuses, remove the seat (see Chapter 8) then unclip the relevant cover (see illustration). To gain access to the main fuse A, ensure the ignition is switched off then disconnect the starter relay wiring connector (see illustration).
3 To remove a fuse, first switch off the ignition, then pull the fuse out of its terminals. If you can't pull the fuse out with your fingertips, use a pair of suitable pliers. A blown fuse is easily identified by a break in the element (see illustration). Each fuse is clearly marked with its rating and must only be replaced by a fuse of the correct rating; the fuses are also colour-coded for easy recognition. A spare fuse of each rating is housed in the fusebox, and a spare main fuse is fitted to the starter relay. If a spare fuse is used, always renew it so that a spare of each rating is carried on the bike at all times.

⚠ Warning: *Never put in a fuse of a higher rating or bridge the terminals with any other substitute, however temporary it may be. Serious damage may be done to the circuit, or a fire may start.*

4 If a new fuse blows immediately, find the cause before renewing it again; a short to earth (ground) as a result of faulty insulation is most likely. Look for bare wires and chafed, melted or burned insulation.
5 Occasionally a fuse will blow or cause an open-circuit for no obvious reason. Corrosion of the fuse ends and fusebox terminals may occur and cause poor fuse contact. If this happens, remove the corrosion with a wire brush or emery paper, then spray the fuse end and terminals with electrical contact cleaner.

6 Lighting system – check

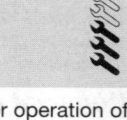

1 The battery provides power for operation of the headlight, tail light, brake light and instrument cluster lights. If none of the lights operate, always check battery voltage before proceeding. Low battery voltage indicates either a faulty battery or a defective charging system. Refer to Section 3 for battery checks and Sections 30 and 31 for charging system tests. Also, check the condition of the fuses.

Headlight

2 If the headlight (HI or LO beam) fails to work, first check the fuse with the key ON (see Section 5), and then the bulbs (see Section 7). If they are both good, the problem lies in the wiring, the relays (there are separate relays for the LO and HI beams), or one of the switches in the circuit.
3 If either relay is suspected of being faulty, the best way to determine this is to substitute it with another relay. Remove the upper fairing to access the relays – they are mounted behind the instrument cluster (see illustration). Both relays are identical so if only one is faulty the relays can be easily

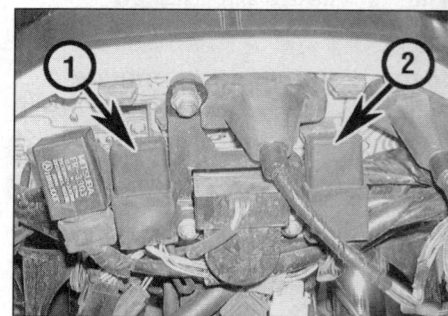

6.3 The headlight LO beam (1) and HI beam (2) relays are located behind the upper fairing

Electrical system 9•5

checked by swapping them over; if this transfers the problem to the other circuit then the relay is confirmed to be faulty. If both relays are faulty, check them as follows.

4 Remove the relay and connect an ohmmeter across the blue/black (HI beam relay) or the white/black (LO beam relay) and black/red terminals of the relay. Using a 12V battery and auxiliary wires, connect the battery positive (+) terminal to the blue (HI beam relay) or white (LO beam relay) terminal of the relay and the negative (–) terminal to the green terminal of the relay and note the meter reading obtained. If the relay is operating correctly there should be continuity (zero resistance) whilst the battery is connected and no continuity (infinite resistance) whilst the battery is disconnected. If this is not the case, renew the relay.

5 If the relays are proved good, check the wiring and switches testing (Sections 2 and 20) using the wiring diagrams at the end of this Chapter.

Tail light

6 If the tail light fails to work, check the bulbs and the bulb terminals first, then the fuse, then check for battery voltage at the brown/white (European models) or brown/blue (US models) terminal on the supply side of the tail light wiring connector. If voltage is present, check the earth (ground) circuit for an open or poor connection.

7 If no voltage is indicated, check the wiring between the tail light and the lighting switch.

Brake light

8 If the brake light fails to work, check the bulbs and the bulb terminals first, then the fuse. Check for battery voltage at the green/yellow terminal on the supply side of the tail light wiring connector, with the brake lever pulled in or the pedal depressed; if voltage is present, check the earth (ground) circuit for an open or poor connection.

9 If no voltage is indicated, check the brake light switches, then the wiring between the tail light and the switches.

10 See Section 14 for brake switch checks and Section 9 for tail/brake light bulb renewal.

Instrument and warning lights

11 See Section 17 for instrument and warning light bulb renewal.

Turn signal lights

10 See Section 11 for the turn signal circuit check.

7 Headlight and sidelight bulbs – renewal

Headlight

Note: *The headlight bulbs are of the quartz-halogen type. Do not touch the bulb glass as skin acids will shorten the bulb's service life. If the bulb is accidentally touched, it should be wiped carefully when cold with a rag soaked in methylated spirit and dried before fitting.*

⚠ **Warning: Allow the bulbs time to cool before removing them if the headlight has just been on.**

1 Disconnect the relevant wiring connector from the back of the headlight assembly and remove the rubber dust cover, noting how it fits **(see illustration)**.

2 Unhook the bulb retaining clip then remove the bulb.

3 Fit the new bulb, bearing in mind the information in the **Note** above. Make sure the tabs on the bulb fit correctly in the slots in the bulb housing, and secure it in position with the retaining clip.

4 Install the dust cover, making sure it is correctly seated and with the 'TOP' mark at the top, and connect the wiring connector.

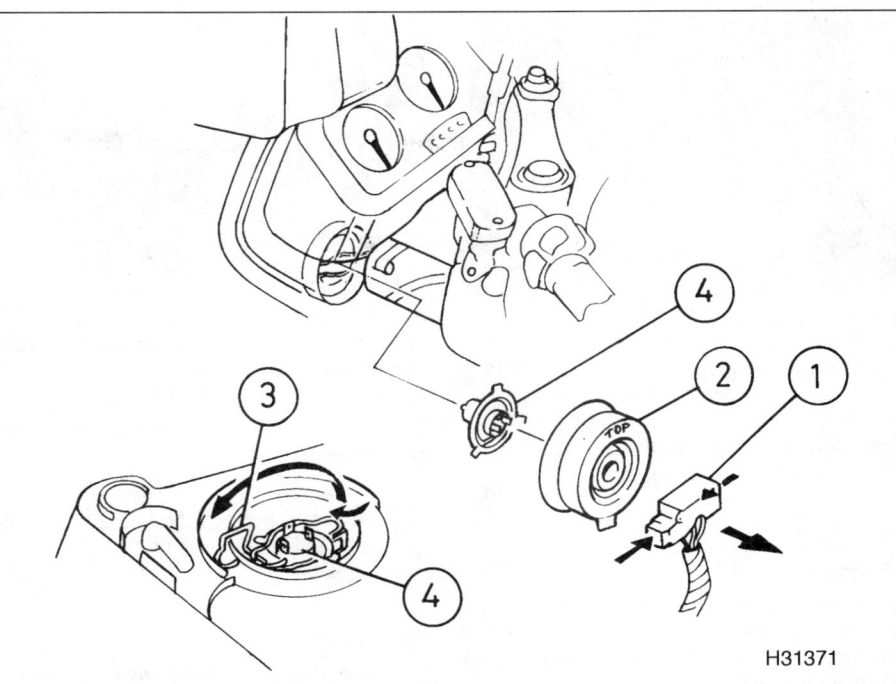

7.1 Disconnect the wiring connector (1) and remove the rubber cover (2) then unhook the spring clip (3) and withdraw the bulb (4) from the headlight unit

5 Check the operation of the headlight.

HAYNES HINT: *Always use a paper towel or dry cloth when handling new bulbs to prevent injury if the bulb should break and to increase bulb life.*

Sidelight – European models only

6 Remove the access cover from directly below the headlight unit.

7 Pull the bulbholder out of its socket in the base of the headlight, then carefully pull the bulb out of the holder **(see illustrations)**.

8 Install the new bulb in the bulbholder, then install the bulbholder by pressing it in. Make sure the bulbholder is correctly seated then securely fit the access cover to the fairing.

9 Check the operation of the sidelight.

7.7a Ease out the sidelight bulbholder from the base of the headlight unit . . .

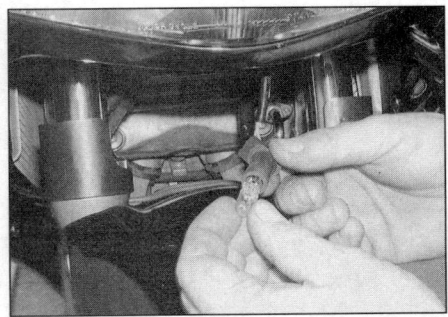

7.7b . . . then pull the bulb out from the holder

9•6 Electrical system

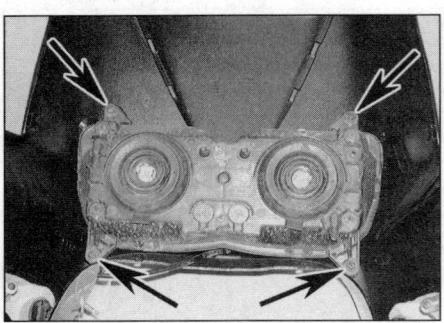

8.2 Undo the retaining screws (arrows) and remove the headlight unit from the upper fairing

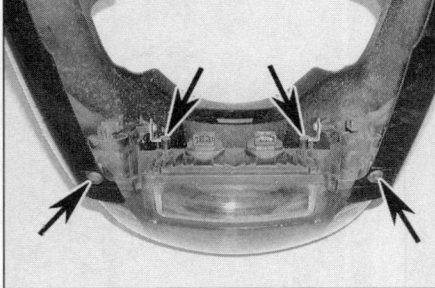

10.2 Undo the retaining screws (arrows) and remove the taillight assembly from the seat cowling

11.3 The turn signal relay (arrow) is located behind the instruments

8 Headlight assembly – removal and installation

Removal

1 Remove the upper fairing (see Chapter 8).
2 Undo the retaining screws and remove the headlight unit from the upper fairing **(see illustration)**.

Installation

3 Fit the headlight unit to the upper fairing, tightening its retaining screws securely, then install the fairing (see Chapter 8). Check the operation of the headlight unit and the headlight aim (see Chapter 1).

9 Brake/tail light bulbs – renewal

1 Remove the seat (see Chapter 8).
2 Disconnect the wiring connector from the relevant bulbholder then turn the bulbholder anti-clockwise and withdraw it from the tail light.
3 Pull the bulb out of the holder to remove it. Check the socket terminals for corrosion and clean them if necessary.
4 Fit the new bulb securely to the holder then install the bulbholder into the tail light and turn it clockwise to secure it.

5 Reconnect the wiring connector then install the seat (see Chapter 8).

10 Tail light assembly – removal and installation

Removal

1 Remove the seat cowling (see Chapter 8).
2 Undo the retaining screws and remove the tail light assembly from the seat cowling **(see illustration)**.

Installation

3 Fit the tail light assembly to the seat cowling, tightening its retaining screws securely, then install the seat cowling. Check the operation of the lights before using the motorcycle on the road.

11 Turn signal circuit – check

1 The battery provides power for operation of the turn signal lights, so if they do not operate, always check the battery voltage first. Low battery voltage indicates either a faulty battery or a defective charging system. Refer to Section 3 for battery checks and Sections 30 and 31 for charging system tests. Also, check the fuse (see Section 5) and the switch (see Section 20).

2 Most turn signal problems are the result of a burned out bulb or corroded socket. This is especially true when the turn signals function properly in one direction, but fail to flash in the other direction. Check the bulbs and the sockets (see Section 12).
3 If the bulbs and sockets are good, remove the upper fairing (see Chapter 8) for access to the relay, which is mounted behind the instrument cluster **(see illustration)**. Disconnect the wiring connector and check for power at the turn signal relay black/brown wire with the ignition ON. Turn the ignition OFF when the check is complete.
4 If no power was present at the relay, check the wiring from the relay to the ignition (main) switch for continuity.
5 If power was present at the relay, using the appropriate wiring diagram at the end of this Chapter, check the wiring between the relay, turn signal switch and turn signal lights for continuity. If the wiring and switch are sound, renew the relay.

12 Turn signal bulbs – renewal

Front

1 Undo the retaining screw and free the turn signal light unit from the upper fairing **(see illustrations)**.
2 Turn the bulbholder anti-clockwise and withdraw it from the light unit **(see illustration)**.

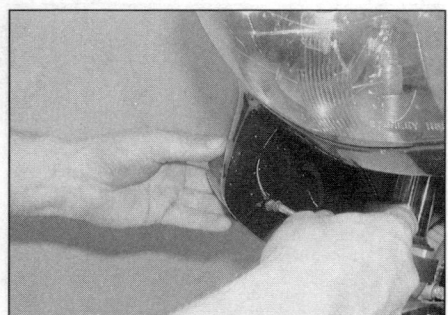

12.1a Undo the retaining screw . . .

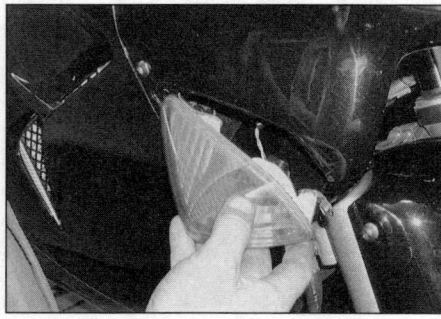

12.1b . . . and free the turn signal light from the fairing

12.2 Remove the bulbholder from the rear of the light unit . . .

Electrical system 9•7

3 Pull the bulb out of the holder to remove it **(see illustration)**. Check the socket terminals for corrosion and clean them if necessary.

4 Fit the new bulb to the holder then fit the bulbholder back into the light unit and turn it clockwise, making sure it is securely held. Secure the turn signal in the fairing with the screw.

Rear

5 Remove the seat (see Chapter 8).
6 Disconnect the wiring connector then turn the bulbholder anti-clockwise and withdraw it from the tail light assembly.
7 Pull the bulb out of the holder to remove it. Check the socket terminals for corrosion and clean them if necessary.
8 Fit the new bulb then install the bulbholder into the tail light and turn it clockwise to secure it.
9 Reconnect the wiring connector then install the seat (see Chapter 8).

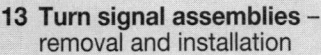

13 Turn signal assemblies – removal and installation

Front

Removal

1 Undo the retaining screw and free the turn signal light unit from the upper fairing **(see illustrations 12.1a and 12.1b)**.
2 Disconnect the wiring connector and remove the light unit from the motorcycle.

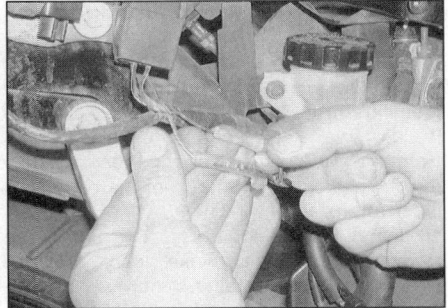

14.8 Disconnect the rear brake light switch wiring connectors . . .

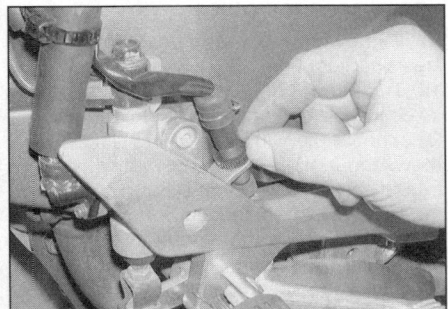

14.9 . . . then unscrew the knurled ring and remove the switch from the footrest bracket

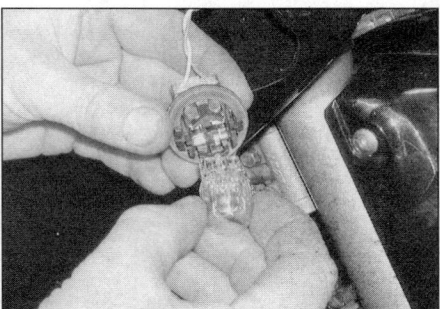

12.3 . . . then gently ease the bulb out of its holder

Installation

3 Installation is the reverse of removal. Check the operation of the turn signals.

Rear

4 The turn signals are an integral part of the tail light assembly (see Section 10).

14 Brake light switches – check and replacement

Circuit check

1 Before checking any electrical circuit, check the bulb (see Section 9) and fuse (see Section 5). If the front brake switch is being checked carry out the check at the switch terminals, if the rear brake light switch is being checked remove the seat cowling (see Chapter 8) to gain access to the switch wiring connectors.

2 Using a multimeter or test light connected to a good earth (ground), with the ignition ON check for voltage at the terminals of the brake light switch wiring connector(s) – there should be battery voltage on one terminal (the terminal connected to the wiring harness black/brown wire) and zero on the other with the lever/pedal at rest. If there's no voltage present at either, check the wiring between the switch and the fuse (see the wiring diagrams at the end of this Chapter).

3 If voltage is available, touch the probe of the test light to the other terminal of the switch, then pull the brake lever in or depress the brake pedal. If no reading is obtained or the test light doesn't light up, renew the switch.

4 If a reading is obtained or the test light does light up, yet the bulbs still do not come on, check the wiring between the switch and the brake light bulbs (see the wiring diagrams at the end of this Chapter).

Switch replacement

Front brake lever switch

5 The switch is mounted on the underside of the brake master cylinder. Disconnect the wiring connectors from the switch.
6 Remove the single screw securing the switch to the bottom of the master cylinder and remove the switch.
7 Installation is the reverse of removal. The switch is not adjustable.

Rear brake pedal switch

8 The switch is mounted on the inside of the right footrest bracket. Remove the seat cowling for access to the connectors (see Chapter 8). Trace the wiring from the switch and disconnect it at the connectors **(see illustration)**.
9 Detach the lower end of the switch spring from the brake pedal, then unscrew and remove the switch **(see illustration)**.
10 Installation is the reverse of removal. Make sure the brake light is activated just before the rear brake pedal takes effect. If adjustment is necessary, hold the switch and turn the adjusting ring on the switch body until the brake light is activated when required.

15 Instrument cluster – removal and installation

Removal

1 Remove the upper fairing and the instrument cluster surround panel (see Chapter 8).
2 Free the air temperature sensor wiring from its clips on the right side fairing inner panel and unclip the sensor so it is free to be removed with the instrument cluster **(see illustrations)**.

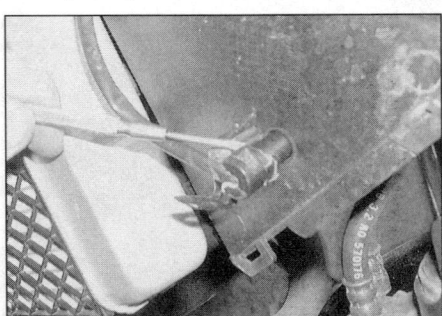

15.2a Unclip the air temperature sensor from the right side inner panel . . .

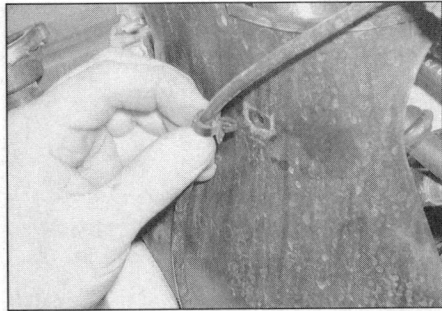

15.2b . . . and unclip the sensor wiring

9•8 Electrical system

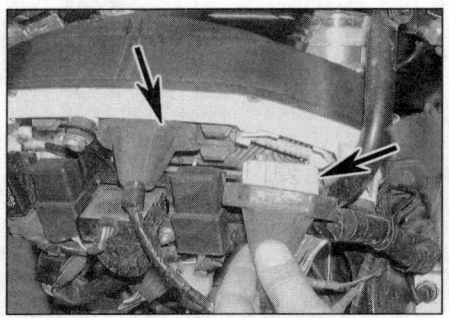

15.3 Disconnect the wiring connectors (arrows) . . .

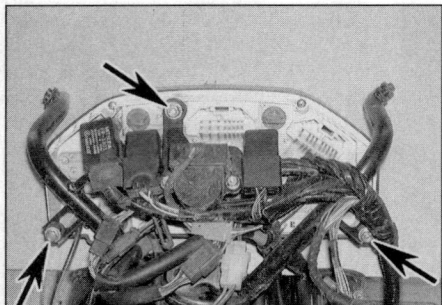

15.4a . . . then undo the retaining nuts (arrows) . . .

15.4b . . . and remove the instrument cluster

3 Disconnect the wiring connectors from the cluster **(see illustration)**.
4 Slacken and remove the three retaining nuts, noting the correct fitted position of the wiring clamp(s), and washers and remove the instrument cluster from its bracket **(see illustrations)**. Inspect the rubber mountings and renew them if they are damaged.

Installation

5 Ensure the rubber mountings are in position then install the instrument cluster. Fit the washers and retaining nuts, ensuring the wiring clamp(s) are correctly positioned, and tighten them securely. Reconnect the wiring connectors.
6 Ensure the air temperature sensor wiring is correctly routed then clip the sensor into position on the inner panel.
7 Install the surround panel and fit the fairing (see Chapter 8).

16 Instruments and sensors – check and replacement

Speedometer and speed sensor
Check

1 If the system malfunctions, first check that the battery is fully charged and that the terminals are clean, and that the wiring and fuses are in good condition.
2 Remove the fairing left side lower panel (see Chapter 8).

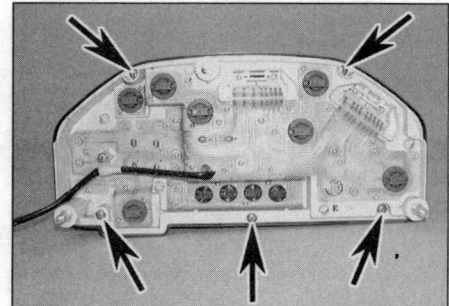

16.8 Undo the screws (arrows) and lift off the front cover from the instrument cluster

3 Trace the wiring back from the speed sensor (on the front sprocket cover) to its connector which is hidden behind the frame. Disconnect the connector and connect the positive (+) lead of a voltmeter to the brown/blue terminal of the wiring harness and the negative (–) lead to the green/black terminal of the connector. Switch on the ignition and check that battery voltage (around 12 volts) is present at the connector. If no reading is obtained, check the power supply to the sensor and the sensor earth (ground) (see *wiring diagrams* at the end of this Chapter).
4 If battery voltage is present, reconnect the speed sensor connector then remove the upper fairing (see Chapter 8). Make sure the instrument cluster connectors are free from corrosion and are securely connected then peel back the rubber covers to gain access to the rear of each connector. Connect the positive (+) lead of a voltmeter to the brown/blue terminal of one connector and the negative (–) lead to the green/black terminal of the other connector. Switch on the ignition and check that battery voltage (around 12 volts) is present at the connector. If no reading is obtained, check the wiring between the instrument cluster, fusebox and ignition switch.
5 If battery voltage is again present, switch off the ignition and use an ohmmeter to check for continuity between the pink terminals of the speed sensor wiring connector and the instrument cluster connector. If continuity (zero resistance) is not present, repair/replace the wiring harness.

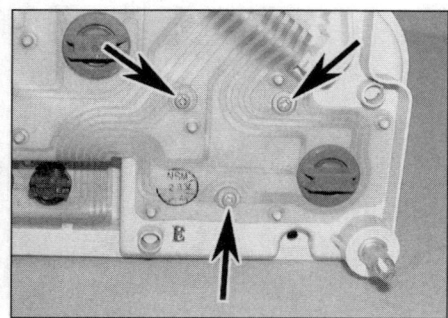

16.9 Speedometer retaining screws (arrows)

6 If the wiring and voltage supplies are proved good, place the bike on its centrestand so that the rear wheel is off the ground. Connect the positive (+) lead of a voltmeter to the pink terminal on the rear of the instrument cluster and the negative (–) lead to the green/black terminal of the cluster. Ensure all wiring connectors are securely connected then switch on the ignition and, with the bike in neutral, rotate the rear wheel by hand. If the speed sensor is operating correctly, a fluctuating voltage reading between 0 to 12 volts should be measured, the rate of fluctuation depending on the speed at which the rear wheel is being rotated. If the reading obtained is correct, the speedometer or the instrument cluster printed circuit board must be faulty. If no reading is obtained, it is the speed sensor which is at fault.

Speedometer replacement

7 Remove the instrument cluster (see Section 15).
8 Unscrew the five retaining screws and lift off the front cover from the instrument cluster **(see illustration)**.
9 Unscrew the speedometer retaining screws from the rear of the cluster then carefully remove the speedometer from the case **(see illustration)**.
10 Install the speedometer by reversing the removal sequence. Take care not to overtighten any retaining screws as the components are fragile and can easily be damaged.

Speed sensor replacement

11 Remove the fairing left side lower panel (see Chapter 8).
12 Trace the wiring back from the sensor to its connector behind the left side of the frame. Disconnect the wiring connector and release the wiring, noting its correct routing, so its is free to be removed with the sensor.
13 Unscrew the mounting bolts and remove the sensor from the sprocket cover **(see illustrations)**.
14 Install the speed sensor by reversing the removal sequence, making sure the sensor drive gear is correctly engaged with the front sprocket bolt.

Electrical system 9•9

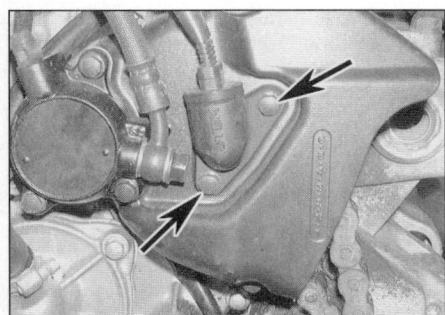

16.13a Unscrew the mounting bolts (arrows) . . .

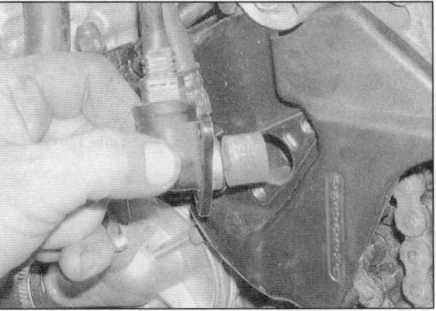

16.13b . . . and remove the speed sensor from the sprocket cover

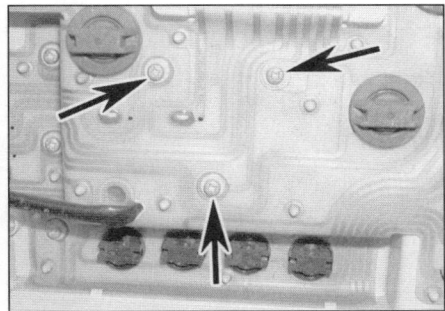

16.21 Tachometer retaining screws (arrows)

Tachometer

Check

15 If the system malfunctions, first check that the battery is fully charged and that the terminals are clean, and that the wiring and fuses are in good condition.
16 Remove the upper fairing and the seat (see Chapter 8). Remove the four trim clips and lift off the engine control module (ECM) cover from the rear of the mudguard.
17 Ensure the ignition is switched off then disconnect the wiring connectors from the instrument cluster and the ECM. Use an ohmmeter to check for continuity between the yellow/green terminals of the ECM wiring connector and the instrument cluster connector. If continuity (zero resistance) is not present, repair/replace the wiring harness.
18 If continuity exists, securely reconnect all connectors then peel back the rubber cover to gain access to the rear of the instrument cluster yellow/green terminal. Connect the positive (+) lead of a voltmeter and peak voltage adaptor arrangement* to the instrument cluster yellow/green terminal and the negative (–) lead to earth (ground). Start the engine and measure the tachometer input peak voltage; this should be at least 10.5 volts. If the peak voltage exceeds this amount then the tachometer or instrument cluster printed circuit board is faulty. If there is no reading then the ECM is at fault (see Chapter 4). *Note: Honda specify their own Imrie diagnostic tester (model 625), or the peak voltage adaptor (Pt. No. 07HGJ-0020100) with an aftermarket digital multimeter having an impedance of 10 M-ohm/DCV minimum, for this test.

Replacement

19 Remove the instrument cluster (see Section 15).
20 Unscrew the five retaining screws and lift off the front cover from the instrument cluster (see illustration 16.8).
21 Unscrew the tachometer retaining screws from the rear of the cluster then carefully remove the tachometer from the case (see illustration).
22 Install the tachometer by reversing the removal sequence. Take care not to overtighten any retaining screws as the components are fragile and can easily be damaged.

LCD (Liquid Crystal Display) unit and sensors

Coolant temperature gauge check

23 The coolant temperature gauge check is given in Chapter 3.

Fuel gauge check

24 Remove the fuel gauge sender unit (see below). Connect an ohmmeter across the sender unit terminals then check the resistance reading whilst moving the float arm slowly from the full to empty position and back again. Compare the readings obtained to those given in the Specifications, not only should the full and empty readings be as specified but the value should change evenly and progressively as the float arm is moved. If not the sender unit is faulty and should be renewed.
25 If the sender unit functions correctly, connect it to the wiring connector then switch on the ignition. Move the float arm and check the operation of the gauge. If the gauge does not function correctly, switch off the ignition and remove the upper fairing (see Chapter 8).
26 Disconnect the wiring connectors from the instrument cluster and fuel gauge sender unit. Use an ohmmeter to check for continuity between the grey/black wire terminals of the cluster and sender unit wiring connectors and then between the green/black wire terminal of the sender unit connector and earth (ground). If continuity (zero resistance) is not present, repair/replace the wiring harness. If continuity exists, then the LCD unit or instrument cluster printed circuit board is faulty.

16.31 Disconnect the air temperature sensor wiring connector (arrow) from the LCD unit

Air temperature gauge check

27 Remove the sensor (see below). Connect an ohmmeter across the sensor terminals and measure its resistance. Compare the reading obtained to that given in the Specifications noting that the specified value is only valid at 25°C (77°F); the sensor resistance will increase at lower temperatures and decrease at higher temperatures. If the resistance reading differs greatly from that specified, the sensor is probably faulty.
28 If the sensor appears to be functioning correctly then the LCD unit or instrument cluster printed circuit board is faulty.

LCD unit replacement

29 Remove the instrument cluster (see Section 15).
30 Unscrew the five retaining screws and lift off the front cover from the instrument cluster (see illustration 16.8).
31 Disconnect the air temperature sensor wiring connector from the front of the unit (see illustration).
32 Unscrew the LCD unit retaining screws from the rear of the cluster then carefully ease the unit out of position (see illustration).
33 Install the LCD unit by reversing the removal sequence. Do not overtighten any retaining screws as the components are fragile and can easily be damaged.

Coolant temperature gauge sensor replacement

34 The gauge is operated by the fuel injection system coolant temperature sensor (see Chapter 4).

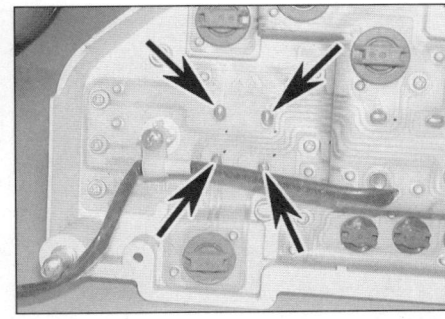

16.32 LCD unit retaining screws (arrows)

9•10 Electrical system

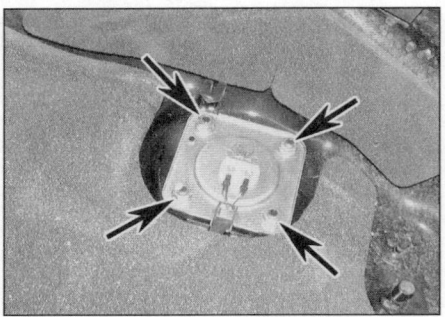

16.36 Unscrew the retaining nuts (arrows - note wiring clamp position) and manoeuvre the fuel gauge sender unit out from the tank

Fuel gauge sender unit replacement

35 Remove the fuel tank (see Chapter 4).
36 Unscrew the sender unit retaining nuts and remove the wiring clamp **(see illustration)**.
37 Free the wiring harness from behind the tank insulation then manoeuvre the sender unit out of position, taking care not to bend the float arm, and remove the O-ring.
38 Installation is the reverse of the removal sequence using a new O-ring.

Air temperature sensor replacement

39 Remove the instrument cluster (see Section 15).
40 Unscrew the five retaining screws and lift off the front cover from the instrument cluster **(see illustration 16.8)**.
41 Disconnect the sensor wiring connector from the LCD **(see illustration 16.31)** unit then undo the wiring clamp screw and remove the sensor from the instrument cluster **(see illustration)**.
42 Install the sensor by reversing the removal sequence.

17 Instrument and warning light bulbs – renewal

1 Remove the upper fairing (see Chapter 8).
2 The bulbs are accessible with the instrument cluster in place, but access to some is quite restricted. If it is too restricted

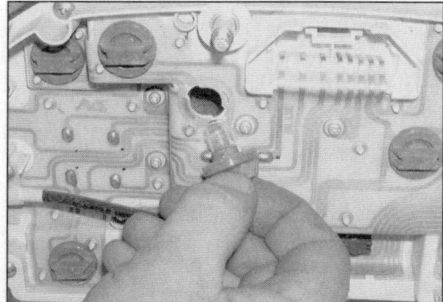

17.3a Twist the bulbholder anti-clockwise to free it from the rear of the cluster ...

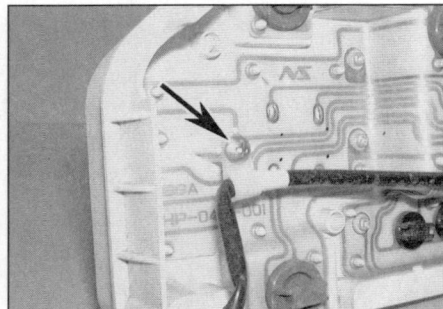

16.41 Undo the screw (arrow) and free the air temperature sensor wiring from the instrument cluster

for the bulb you are changing, displace the instrument cluster (see Section 15) to improve access.
3 Rotate the bulbholder anti-clockwise and remove it from the instrument case, then pull the bulb out of the bulbholder **(see illustrations)**.
4 If the socket contacts are dirty or corroded, scrape them clean and spray with electrical contact cleaner before a new bulb is installed.
5 Carefully push the new bulb into the holder, then securely fit the holder into the case.
6 Install the fairing (see Chapter 8).

18 Oil pressure switch – check, removal and installation

Check

1 The oil pressure warning light should come on when the ignition (main) switch is turned ON and extinguish a few seconds after the engine is started. If the oil pressure warning light comes on whilst the engine is running, stop the engine immediately and carry out an oil level check (see *Daily (pre-ride) checks*). If the level is correct, carry out an oil pressure check (see Chapter 1).
2 If the oil pressure warning light does not come on when the ignition is turned on, check the bulb (see Section 17) and fuse (see Section 5).
3 The oil pressure switch is screwed into the

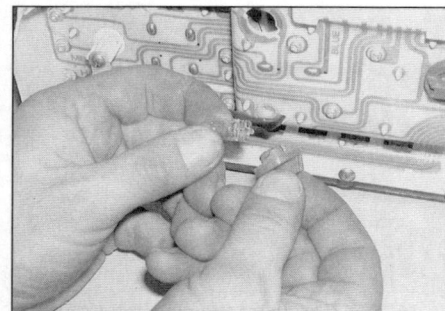

17.3b ... then pull the bulb out from the holder

top of the crankcase between the cylinders and is accessed by removing the throttle body assembly (see Chapter 4). Pull the rubber cover off the switch and remove the screw securing the wiring connector. With the ignition switched ON, earth (ground) the wire on the crankcase and check that the warning light comes on. If the light comes on, the switch is defective and must be renewed.
4 If the light still does not come on, check for voltage at the wire terminal. If there is no voltage present, check the wire between the switch, the instrument cluster and fuse for continuity (see the *wiring diagrams* at the end of this Chapter).
5 If the warning light comes on whilst the engine is running, yet the oil pressure is satisfactory, remove the wire from the oil pressure switch. With the wire detached and the ignition switched ON the light should be out. If it is illuminated, the wire between the switch and instrument cluster must be earthed (grounded) at some point. If the wiring is good, the switch must be assumed faulty and renewed.

Removal

6 Remove the throttle body assembly (see Chapter 4).
7 Pull the rubber cover off the switch and remove the screw securing the wiring connector.
8 Unscrew the oil pressure switch and withdraw it from the crankcase.

Installation

9 Apply a suitable sealant to the upper portion of the switch threads near the switch body, leaving the bottom 3 to 4 mm of thread clean. Install the switch in the crankcase and tighten it to the torque setting specified at the beginning of the Chapter.
10 Attach the wiring connector and secure it with the screw, then fit the rubber cover.
11 Install the throttle body assembly (see Chapter 4). Run the engine and check that the switch operates correctly.

19 Ignition switch – check, removal and installation

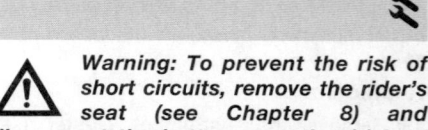

Warning: To prevent the risk of short circuits, remove the rider's seat (see Chapter 8) and disconnect the battery negative (–) lead before making any ignition (main) switch checks.

Check

1 Remove the upper fairing (see Chapter 8). Trace the ignition (main) switch wiring back from the base of the switch and disconnect it at the connector located beneath the bank of relays.
2 Using an ohmmeter or a continuity tester, check the continuity of the connector terminal

Electrical system 9•11

19.5 Disconnect the wiring connector . . .

19.6 . . . then unscrew the bolts (arrows) and remove the ignition (main) switch from the top yoke

pairs (see the *wiring diagrams* at the end of this Chapter). Insert the key in the switch. Continuity should exist between the terminals connected by a solid line on the diagram when the switch is in the indicated position.
3 If the switch fails any of the tests, renew it.

Removal

4 Remove the instrument cluster (see Section 15). On VFR800FI-Y and 1 models, remove the immobiliser receiver from the ignition (main) switch (see Chapter 5, Section 5).
5 Trace the ignition (main) switch wiring back from the base of the switch, freeing it from any retaining clips and ties, and disconnect it at the connector **(see illustration)**.
6 Unscrew the retaining bolts and withdraw the switch from the top yoke **(see illustration)**.

Installation

7 Remove all traces of locking compound from the threads of the lock bolts and top

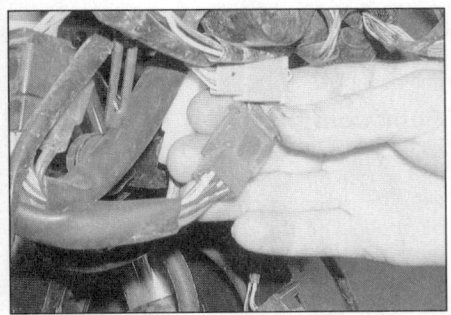

21.2 Disconnect the right handlebar switch wiring connector

21.4a Undo the switch retaining screws . . .

yoke and apply a few drops of fresh locking compound to each bolt.
8 Fit the lock to the top yoke and tighten its retaining bolts to the specified torque.
9 Ensure the lock wiring is correctly routed and retained by the necessary clips and ties then securely reconnect the wiring connector.
10 Refit the instrument cluster (see Section 15).

20 Handlebar switches – check

1 Generally speaking, the switches are reliable and trouble-free. Most troubles, when they do occur, are caused by dirty or corroded contacts, but wear and breakage of internal parts is a possibility that should not be overlooked. If breakage does occur, the entire switch and related wiring harness will have to be renewed as individual parts are not available.
2 The switches can be checked for continuity using an ohmmeter or a continuity test light. Always disconnect the battery negative (–) lead, which will prevent the possibility of a short circuit, before making the checks.
3 Remove the upper fairing (see Chapter 8). Trace the wiring harness of the switch in question back to its connector(s) and disconnect it/them.
4 Check for continuity between the terminals of the switch harness with the switch in the various positions (ie switch off – no continuity, switch on – continuity) – see the *wiring diagrams* at the end of this Chapter.

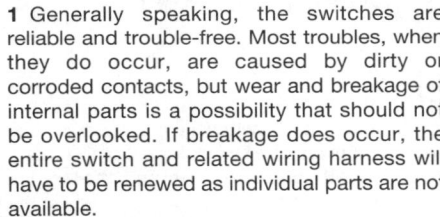

21.4b . . . then free the switch from the handlebar and disconnect the throttle cables (arrow) from the grip

5 If the continuity check indicates a problem exists, refer to Section 21, remove the switch and spray the switch contacts with electrical contact cleaner. If they are accessible, the contacts can be scraped clean with a knife or polished lightly with fine abrasive paper. If switch components are damaged or broken, it will be obvious when the switch is disassembled.

21 Handlebar switches – removal and installation

Removal

Right-hand switch

1 Remove the upper fairing (see Chapter 8).
2 Trace the wiring harness of the switch back to its connector and disconnect it **(see illustration)**. Work back along the harness, freeing it from all the relevant clips and ties, noting its correct routing.
3 Disconnect the wires from the front brake light switch.
4 Unscrew the two handlebar switch screws and free the switch upper half from the handlebar. Detach the throttle cables from the throttle grip then free the lower half of the switch from the handlebar **(see illustrations)**.
5 Unscrew the nut and detach the closing throttle cable from the switch then slacken the nut and unscrew the opening cable end fitting from the switch. The switch assembly can then be removed.

Left-hand switch

6 Remove the upper fairing (see Chapter 8).
7 Trace the wiring harness of the switch back and disconnect its connectors **(see illustration)**. Work back along the harness, freeing it from all the relevant clips and ties, noting its correct routing.
8 Disconnect the wires from the clutch switch and the horn.
9 Unscrew the handlebar switch screws and free the upper half of the switch from the handlebar **(see illustrations)**.
10 On VFR800FI-W and X models, detach the choke cable from the lever and free the lower half of the switch from the handlebar,

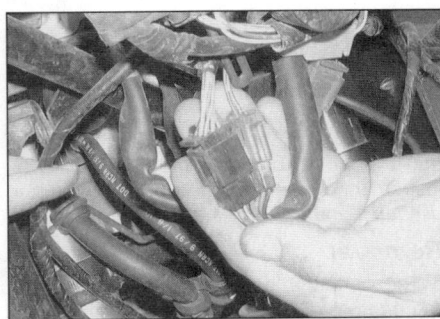

21.7 Disconnect the left handlebar switch wiring connectors

9•12 Electrical system

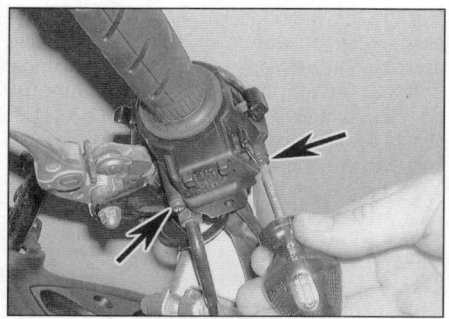

21.9a Undo the retaining screws (arrows) ...

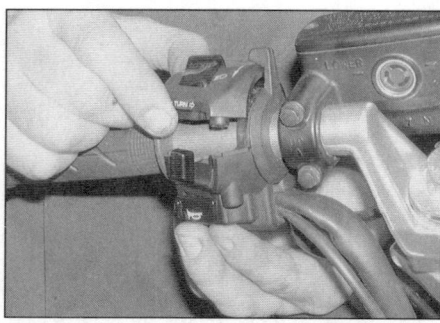

21.9b ... then free the switch assembly from the left handlebar

21.10a On VFR800FI-W and X models, detach the cable from the choke lever ...

then slacken the nut and unscrew the choke cable end fitting from the switch (see illustrations). Remove the switch assembly.

Installation

11 Installation is the reverse of removal. Make sure the locating pin in the lower half of the switch locates in the hole in the underside of the handlebar then assemble the switch and refit the retaining screws. Tighten the front screw first, followed by the rear screw.

22 Neutral switch – check, removal and installation

Check

1 Before checking the electrical circuit, check the bulb (see Section 17) and fuse (see Section 5).
2 The switch is located on the right side of the engine, directly below the oil filler cap. Make sure the transmission is in neutral then disconnect the wiring connector from the switch.
3 With the connector disconnected and the ignition switched ON, the neutral light should be out. If not, the wire between the connector and instrument cluster must be earthed (grounded) at some point.
4 With the ignition off, check for continuity between the switch terminal and the crankcase. With the transmission in neutral, there should be continuity. With the transmission in gear, there should be no continuity. If the tests prove otherwise, then the switch is faulty.
5 If the continuity tests prove the switch is good, check for voltage (ignition on) at the wire terminal using a test light. If there's no voltage present, check the diode (Section 25) and the wiring between the switch, diode, instrument cluster and fusebox (see the wiring diagrams at the end of this Chapter).

Removal

6 The switch is located on the right side of the engine, directly below the oil filler cap. Support the bike on its sidestand to prevent oil spillage when the switch is removed.

7 Detach the wiring connector then unscrew the switch and withdraw it from the crankcase.

Installation

8 Apply a smear of sealant to the threads of the switch, taking care not to cover the contact point.
9 Install the switch, tightening it securely, and reconnect the wiring connector. Check the operation of the neutral light.

23 Sidestand switch – check and replacement

Check

1 The sidestand switch is mounted on the sidestand pivot. The switch is part of the safety circuit which prevents or stops the engine running if the transmission is in gear whilst the sidestand is down, and prevents the engine from starting if the transmission is in gear unless the sidestand is up, and unless the clutch is pulled in. Before checking the electrical circuit, check the fuse (see Section 5).
2 Place the bike on its centrestand then remove the left side lower fairing panel (see Chapter 8).
3 Trace the wiring back from the switch to its connector, located behind the frame, and disconnect it.
4 Connect an ohmmeter across the switch terminals and check the operation of the switch. With the sidestand up there should be

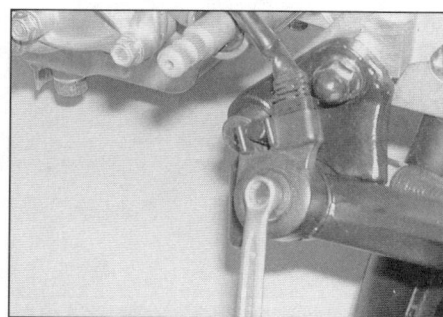

23.8 Unscrew the bolt and remove the sidestand switch from the bike

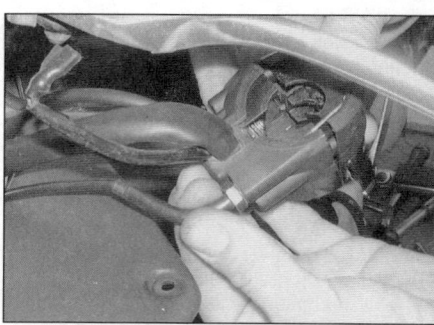

21.10b ... then slacken the nut and unscrew the choke cable end fitting from the switch

continuity (zero resistance) between the terminals, and with the stand down there should be no continuity (infinite resistance).
5 If the switch does not perform as expected, it is defective and must be renewed. If the switch is good, check the wiring between the various components in the starter safety circuit (see the wiring diagrams at the end of this book).

Replacement

Note: Honda recommend that a new retaining bolt is used on installation (the bolt is supplied with its threads pre-coated with locking compound). If the old bolt is to be reused, remove all original locking compound from its threads and apply fresh compound before installing it

6 Remove the left side lower fairing panel (see Chapter 8).
7 Trace the wiring back from the switch to its connector, located behind the frame, and disconnect it. Work back along the switch wiring, freeing it from any relevant retaining clips and ties, noting its correct routing.
8 Unscrew the switch bolt and remove the switch from the stand, noting how it fits (see illustration).
9 Fit the new switch onto the sidestand, making sure the pin locates in the hole in the sidestand, and the lug on the stand bracket locates into the cut-out in the switch body (see illustration). Fit the new retaining bolt and tighten it to the torque setting specified.
10 Make sure the wiring is correctly routed up to the connector and retained by all the

Electrical system 9•13

necessary clips and ties then reconnect the connector and check the operation of the sidestand switch.
11 Install the fairing lower panel (see Chapter 8).

24 Clutch switch – check and replacement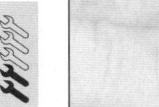

Check

1 The clutch switch is mounted on the base of the clutch master cylinder. The switch prevents the engine from being started whilst in gear unless the clutch lever is pulled in and the sidestand is up.
2 To check the switch, disconnect the wiring connectors from its terminals. Connect the probes of an ohmmeter or a continuity test light to the two switch terminals. With the clutch lever pulled in, continuity should be indicated. With the clutch lever out, no continuity (infinite resistance) should be indicated. Reconnect the wiring connectors.
3 If the switch is good, check for voltage at one of the terminals on the clutch switch wiring connectors with the ignition ON – there will be battery voltage on one terminal and zero on the other. If voltage is indicated, check the other components in the starter circuit as described in the relevant sections of this Chapter. If no voltage is indicated, or if all components are good, check the wiring between the various components (see the *wiring diagrams* at the end of this book).

Replacement

4 Disconnect the wiring connectors from the clutch switch.
5 Undo the retaining screw and remove the switch from the master cylinder.
6 Installation is the reverse of removal.

25 Diode – check and replacement

Check

1 The diode is located in the fusebox (see

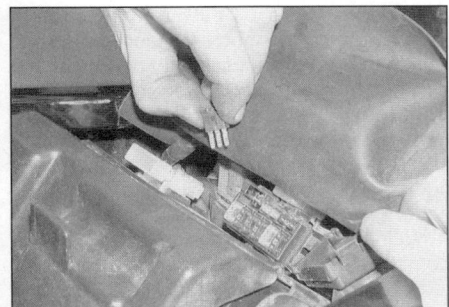

25.5 Removing the diode from the fusebox

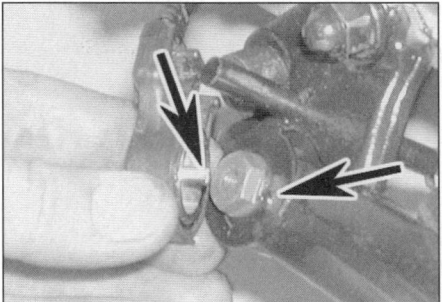

23.9 Fit the new switch ensuring the pin locates in the stand hole (arrows) and the switch cutout is aligned with the bracket lug

Section 5). The diode is part of the safety circuit which prevents or stops the engine running if the transmission is in gear whilst the sidestand is down, and prevents the engine from starting if the transmission is in gear unless the sidestand is up and the clutch lever is pulled in.
2 To check the diode, remove it from the fusebox (see below). Using an ohmmeter or continuity tester, connect one probe to one of the outer terminals of the diode and the other probe to the middle terminal of the diode (the terminal which connects to the light green wire in the wire harness). Now reverse the probes. The diode should only show continuity in one direction (as indicated by the symbol on the body of the diode). If the diode shows continuity or no continuity in both directions it should be renewed. Repeat the tests between the other outer terminal and the middle terminal. The same results should be achieved. If it doesn't behave as stated, renew the diode.
3 If the diode is good, check the other components in the starter circuit as described in the relevant sections of this Chapter. If all components are good, check the wiring between the various components (see the *wiring diagrams* at the end of this book).

Replacement

4 Remove the seat (see Chapter 8) then unclip the fusebox cover.
5 The diode is a small block that plugs into a connector in the fusebox. Ensure the ignition is switched off then pull out the diode **(see illustration)**.
6 Fit the new diode and check the operation of the neutral light and starter safety circuit. Fit the fusebox cover then install the seat (see Chapter 8).

26 Horn – check and replacement

Check

1 The horn is mounted on the underside of the bottom yoke.
2 Disconnect the wiring connectors from the horn. Using two jumper wires, apply battery voltage directly to the terminals on the horn. If the horn sounds, check the switch (see Section 20) and the wiring between the switch and the horn (see the *wiring diagrams* at the end of this Chapter).
3 If the horn doesn't sound, renew it.

Replacement

4 The horn is mounted on the underside of the bottom yoke.
5 Disconnect the wiring connectors from the horn, then unscrew the bolt securing the horn and remove it from the bike **(see illustration)**.
6 Install the horn and securely tighten the bolt. Connect the wiring connectors to the horn.

27 Starter relay – check and replacement

Check

1 If the starter circuit is faulty, first check the fuses (see Section 5).
2 Remove the seat (see Chapter 8). The starter relay is located underneath the rear of the fuel tank on the left side of the fusebox.
3 With the ignition switch ON, the engine kill switch in the RUN position, the transmission in neutral, the sidestand up and the clutch pulled in, press the starter switch. The relay should be heard to click.
4 If the relay doesn't click, switch off the ignition and disconnect the wiring connector from the starter relay.
5 Using an ohmmeter, check for continuity between the green/red terminal of the connector and a good earth (ground) point. Make the first check with the transmission in neutral; there should be continuity although a little resistance will be measured due to the diode in the circuit. Place the transmission in gear then check the resistance with the clutch lever pulled in and the sidestand up; there should be continuity again. If the first check shows an open circuit (infinite resistance), check the neutral switch circuit (Section 22) and if the second check shows an open circuit check the clutch and sidestand switches (Sections 23 and 24).

26.5 Disconnect the wiring connectors (arrows) from the horn (shown with upper fairing removed)

9•14 Electrical system

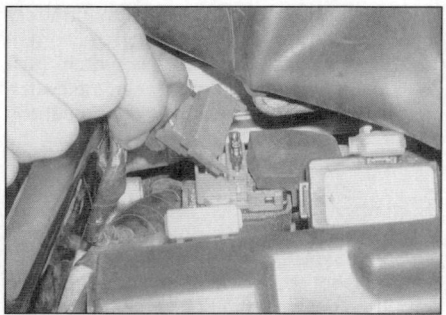

27.10a Disconnect the wiring connector from the starter relay . . .

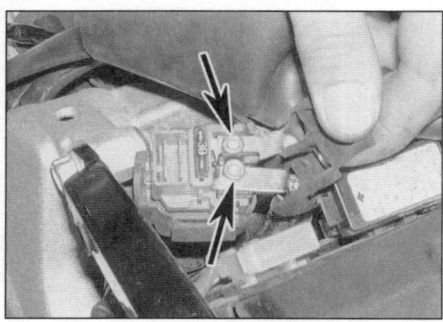

27.10b . . . then unscrew the bolts (arrows) and disconnect the starter and battery leads

6 Reconnect the wiring connector to the starter relay and peel back the rubber cover from the connector to gain access to the rear of its terminals. Connect the positive (+) lead of voltmeter to the rear of the starter relay yellow/red terminal and connect the negative (–) lead to a good earth (ground) point. Switch on the ignition and press the starter button whilst noting the reading obtained on the meter; battery voltage should be present only when the starter button is depressed. If not check, the handlebar switch (Section 20) and the wiring between the relay, switch and fusebox (see *wiring diagrams* at the end of this Chapter). If voltage is present, remove the relay as described below; test it as follows.

7 Set a multimeter to the ohms x 1 scale and connect it across the relay's starter motor and battery lead terminals. Using a fully-charged 12 volt battery and two insulated jumper wires, connect the positive (+) terminal of the battery to the yellow/red wire terminal of the relay, and the negative (–) terminal to the green/red wire terminal of the relay. If the relay is functioning correctly, it should be heard to click as the battery is connected and continuity (zero resistance) should be present between the battery and starter motor lead terminals. Disconnect the battery and check that the relay clicks again and an open circuit (infinite resistance) is present between the battery and starter motor terminals. If the starter relay does not perform as expected, it is faulty and must be renewed.

Replacement

8 The starter relay is located underneath the rear of the fuel tank on the left side of the fusebox. To gain access to the relay, remove the seat (see Chapter 8) and, if necessary, the seat cowling.
9 Disconnect the battery negative (–) terminal.
10 Disconnect the wiring connector from the starter relay, then unscrew the two bolts securing the starter motor and battery leads to the relay and detach the leads **(see illustrations)**.
11 Lift the relay out of position, to free it from its mounting lugs, and remove it from the bike.
12 Installation is the reverse of removal making sure the battery and starter motor leads are fitted the correct way around and their bolts are securely tightened. Ensure the rubber insulation cover is correctly fitted to the relay then reconnect the battery negative (–) lead.

28 Starter motor – removal and installation

Removal

1 Remove the seat (see Chapter 8). Disconnect the battery negative (–) lead. The starter motor is mounted on the front of the crankcase, to the right of the oil filter.
2 Remove the fairing left and right side lower panels and the inner panel (see Chapter 8).

28.3 Peel back the rubber cover then unscrew the nut and detach the starter lead from the motor

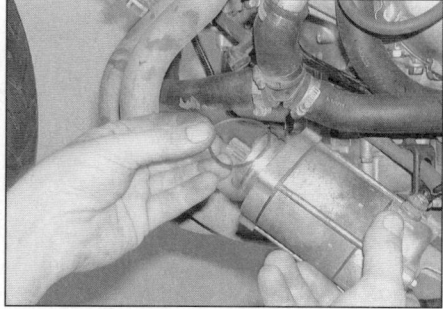

28.6 Always fit a new O-ring to the starter motor

3 Peel back the rubber terminal cover, then remove the nut securing the starter lead to the motor and detach the lead **(see illustration)**.
4 Unscrew the two bolts securing the starter motor to the crankcase then slide the starter motor out of position and remove it from the machine.
5 Remove the O-ring on the end of the starter motor and discard it as a new one must be used.

Installation

6 Install a new O-ring on the end of the starter motor and ensure it is seated in its groove **(see illustration)**. Apply a smear of engine oil to the O-ring to aid installation.
7 Manoeuvre the motor into position and slide it into the crankcase. Ensure that the starter motor teeth mesh correctly with those of the starter idle/reduction gear then fit the mounting bolts and tighten them securely.
8 Connect the starter lead to the motor and secure it with the nut. Make sure the rubber cover is correctly seated over the terminal.
9 Connect the battery negative (–) lead and check the operation of the starter motor before installing the fairing panels (see Chapter 8).

29 Starter motor – disassembly, inspection and reassembly

Disassembly

1 Remove the starter motor (see Section 28).
2 Note the alignment marks between the main housing and the front and rear covers, or make your own if they aren't clear.
3 Unscrew the three long bolts and withdraw them from the starter motor **(see illustration)**.
4 Wrap some insulating tape around the teeth on the end of the starter motor shaft – this will protect the oil seal from damage as the front cover is removed. Remove the front cover from the motor. Remove the cover seal from the main housing and discard it as a new one must be used. Remove the shims from the front end of the armature shaft or the inside of the front cover, noting their correct fitted locations. Also remove the toothed washer from the front cover.
5 Remove the rear cover and brushplate assembly from the motor. Remove the cover seal from the main housing and discard it as a new one must be used. Remove the shims from the rear end of the armature shaft or from inside the rear cover after the brushplate assembly has been removed.
6 Withdraw the armature from the main housing.
7 Noting the correct fitted location of each component, unscrew the terminal nut and remove it along with its washer, the insulating washers and O-ring. Withdraw the terminal bolt and brushplate assembly from the rear cover.

Electrical system 9•15

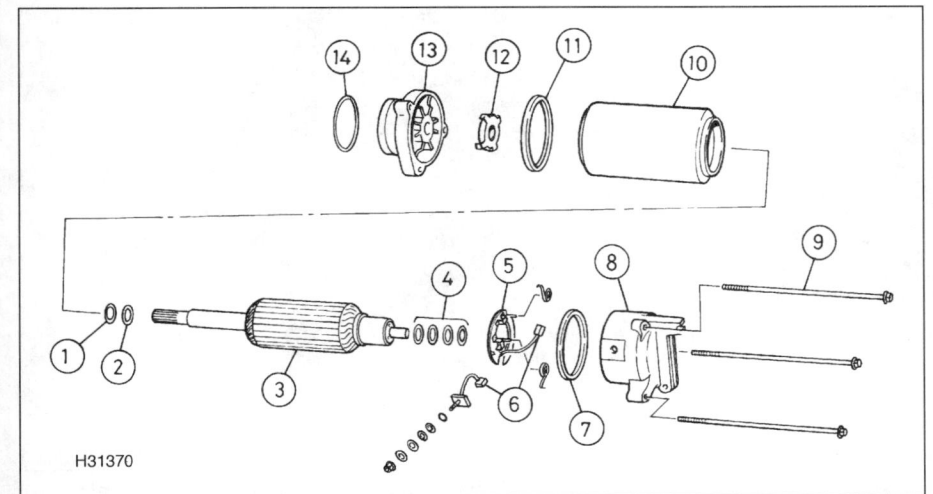

29.3 Exploded view of the starter motor

1 Insulated washer	4 Shims	8 Rear cover	12 Toothed washer
2 Shim	5 Brushplate	9 Bolts	13 Front cover
3 Armature	6 Brushes	10 Main housing	14 O-ring
	7 Seal	11 Seal	

8 Lift the brush springs and slide the brushes out from their holders.

Inspection

9 The parts of the starter motor that are most likely to require attention are the brushes. Measure the length of the brushes and compare the results to the brush length listed in this Chapter's Specifications (see illustration). If any of the brushes are worn beyond the service limit, renew the brush assembly. If the brushes are neither worn excessively, nor cracked, chipped, or otherwise damaged, they may be re-used.

10 Inspect the commutator bars on the armature for scoring, scratches and discoloration. The commutator can be cleaned and polished with crocus cloth, but do not use sandpaper or emery paper. After cleaning, wipe away any residue with a cloth soaked in electrical system cleaner or denatured alcohol.

11 Using an ohmmeter or a continuity test light, check for continuity between the commutator bars (see illustration). Continuity should exist between each bar and all of the others. Also, check for continuity between the commutator bars and the armature shaft (see illustration). There should be no continuity (infinite resistance) between the commutator and the shaft. If the checks indicate otherwise, the armature is defective.

12 Check for continuity between each brush and the terminal bolt. There should be continuity (zero resistance). Check for continuity between the terminal bolt and the housing (when assembled). There should be no continuity (infinite resistance).

13 Check the front end of the armature shaft for worn, cracked, chipped and broken teeth. If the shaft is damaged or worn, renew the armature.

14 Inspect the end covers for signs of cracks or wear. Inspect the magnets in the main housing and the housing itself for cracks.

15 Inspect the insulating washers and front cover oil seal for signs of damage and renew them if necessary.

Reassembly

16 Slide the brushes back into position in their holders and place the brush spring ends onto the brushes.

17 Ensure that the inner rubber insulator is in place on the terminal bolt, then insert the bolt through the rear cover and fit the brushplate assembly in the rear cover, making sure its tab is correctly located in the slot in the cover. Fit the O-ring and the insulating washers over the terminal, then fit the standard washer and the nut.

18 Slide the shims onto the rear end of the armature shaft, then lubricate the shaft with a drop of oil (see illustration). Insert the armature into the rear cover, locating the brushes on the commutator bars as you do so, taking care not to damage them (see illustration). Check that each brush is securely pressed against the commutator by its spring and is free to move easily in its holder.

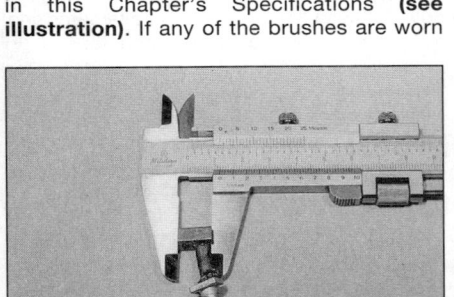

29.9 Measuring a brush length

29.11a Use an ohmmeter to check for continuity between commutator bars . . .

29.11b . . . and between the commutator bars and armature shaft for a short circuit

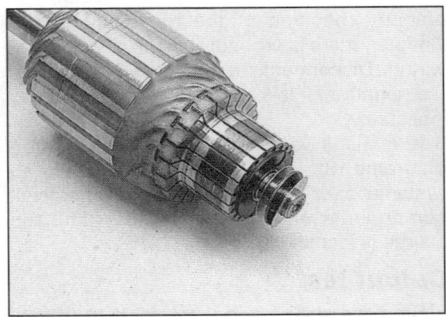

29.18a Fit the shims to the rear of the armature . . .

29.18b . . . then insert the armature into the rear cover, ensuring the brushes sit correctly on the commutator

9•16 Electrical system

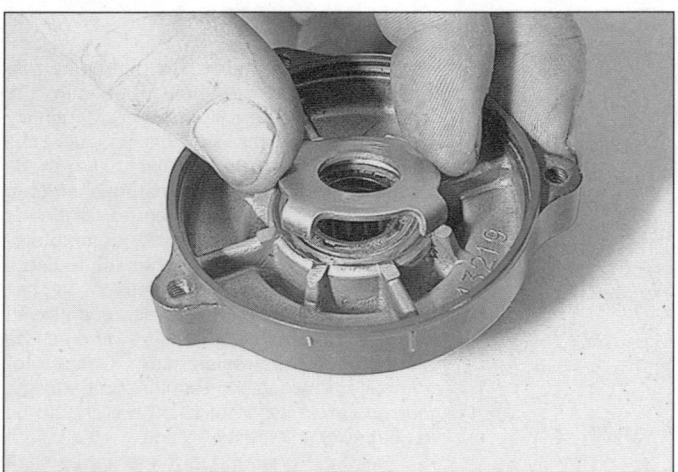

29.20a Fit the toothed washer to the front cover so that its teeth engage with the cover ribs

29.20b Fit the shims and insulated washer to the front of the armature ensuring the washer is outermost

19 Fit a new seal onto the main housing, then fit the housing over the armature and onto the rear cover, aligning the marks made on removal.
20 Apply a smear of grease to the lips of the front cover oil seal and fit a new seal onto the front of the main housing. Fit the toothed washer into the cover, making sure the teeth locate correctly **(see illustration)**. Slide the shims and insulated washer onto the front of the armature shaft, then install the cover, aligning the marks made on removal **(see illustration)**. Remove the protective tape from the shaft end.
21 Check that the marks made on removal are correctly aligned, then install the long bolts and tighten them securely.
22 Install the starter motor (see Section 28).

30 Charging system testing – general information and precautions

1 If the performance of the charging system is suspect, the system as a whole should be checked first, followed by testing of the individual components. **Note:** *Before beginning the checks, make sure the battery is fully charged and that all system connections are clean and tight.*
2 Checking the output of the charging system and the performance of the various components within the charging system requires the use of a multimeter (with voltage, current and resistance checking facilities).
3 When making the checks, follow the procedures carefully to prevent incorrect connections or short circuits, as irreparable damage to electrical system components may result if short circuits occur.
4 If a multimeter is not available, the job of checking the charging system should be left to a Honda dealer or automotive electrician.

31 Charging system – leakage and output test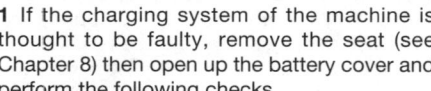

1 If the charging system of the machine is thought to be faulty, remove the seat (see Chapter 8) then open up the battery cover and perform the following checks.

Leakage test

Caution: Always connect an ammeter in series, never in parallel with the battery, otherwise it will be damaged. Do not turn the ignition ON or operate the starter motor when the ammeter is connected – a sudden surge in current will blow the meter's fuse.

2 Turn the ignition switch OFF and disconnect the lead from the battery negative (–) terminal.
3 Set the multimeter to the Amps function and connect its negative (–) probe to the battery negative (–) terminal, and positive (+) probe to the disconnected negative (–) lead. Always set the meter to a high amps range initially and then bring it down to the mA (milli Amps) range; if there is a high current flow in the circuit it may blow the meter's fuse.
4 If the current leakage indicated exceeds the amount specified at the beginning of the Chapter, there is probably a short circuit in the wiring. Disconnect the meter and reconnect the negative (–) lead to the battery, tightening it securely.
5 If leakage is indicated, use the wiring diagrams at the end of this book to systematically disconnect individual electrical components and repeat the test until the source is identified.

Output test

6 Start the engine and warm it up to normal operating temperature, then stop the engine. Remove the seat (see Chapter 8) and open up the battery cover.
7 Start the engine and allow it to idle. Connect a multimeter set to the 0 – 20 volts DC scale across the terminals of the battery (positive (+) lead to battery positive (+) terminal, negative (–) lead to battery negative (–) terminal). Slowly increase the engine speed to 5000 rpm and note the reading obtained, then stop the engine and turn the ignition OFF; do not allow the engine to overheat. The regulated voltage should be as specified at the beginning of the Chapter. If the voltage is outside these limits, check the alternator and the regulator (see Sections 32 and 33).

> **HAYNES HINT**: *Clues to a faulty regulator are constantly blowing bulbs, with brightness varying considerably with engine speed, and battery overheating.*

32 Alternator – check, removal and installation

Check

1 Remove the seat cowling (see Chapter 8).
2 Trace the wiring back from the regulator/rectifier unit on the left side of the subframe to its wiring connectors inside the protective cover and disconnect the alternator connector containing the three yellow wires **(see illustration)**.
3 Using a multimeter set to the ohms x 1 (ohmmeter) scale measure the resistance between each of the yellow wires on the alternator side of the connector, taking a total of three readings, then check for continuity between each terminal and ground (earth). If the stator coil windings are in good condition the three readings should be within the range shown in the Specifications at the start of this

Electrical system 9•17

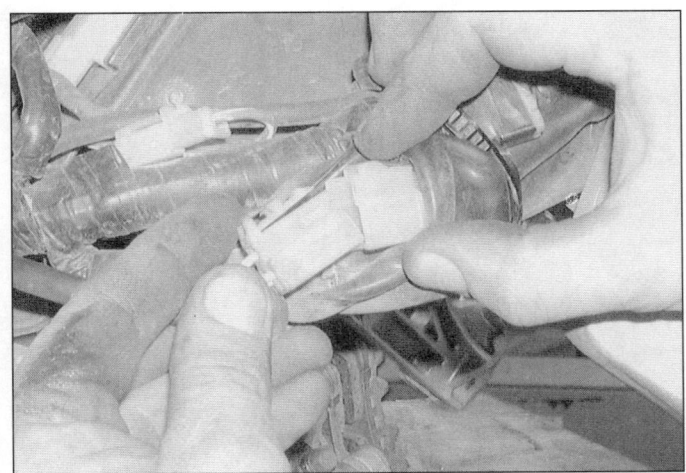

32.2 Disconnecting the alternator wiring connector from the regulator/rectifier

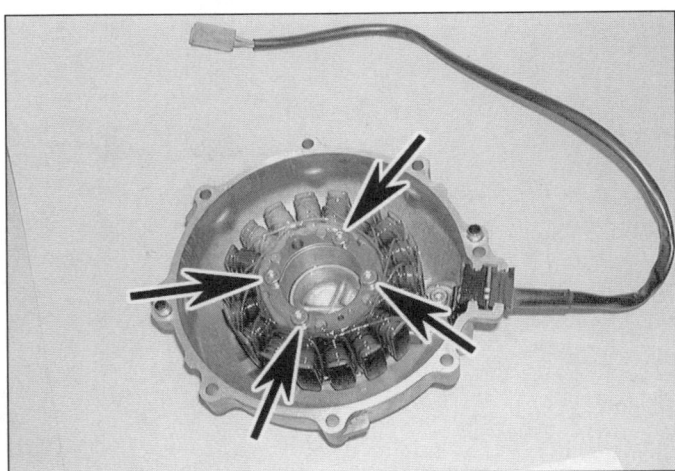

32.11 Undo the retaining bolts (arrows) then free the grommet and remove the alternator stator from the cover

Chapter and there should be no continuity (infinite resistance) between all of the terminals and earth. If not, the alternator stator coil assembly is at fault and should be renewed. **Note:** *Before condemning the stator coils, check the fault is not due to damaged wiring between the connector and coils.*

Removal

Alternator cover and stator

4 Remove the left side fairing lower panel and the seat cowling (see Chapter 8).
5 Remove the coolant reservoir (see Chapter 3).
6 Trace the alternator wiring back from the alternator cover to its connector inside the regulator/rectifier wiring protective cover (see illustration 32.2). Disconnect the wiring connector then work back along the wiring, freeing it from any clips or guides, and release it from the frame so it is free to be removed with the cover.
7 Drain the engine oil (see Chapter 1). Once the oil has drained, fit a new sealing washer to the drain plug then install the plug and tighten it to the specified torque.
8 Unscrew the alternator cover retaining bolts and withdraw the cover squarely from the engine unit. **Note:** *Due to the magnetic pull of the rotor, the cover may prove difficult to remove. Do not pry the cover away with a screwdriver as the mating surfaces will be damaged.*
10 Remove the cover locating dowels from the crankcase and discard the gasket.
11 To remove the stator, unscrew the stator retaining bolts and the bolt securing the wiring clamp, then remove the assembly, noting how the rubber wiring grommet fits (see illustration).

Alternator rotor

Note: *To remove the alternator rotor the special Honda rotor puller (Pt. No. 07733-0020001 or 07933-3950000) or a pattern equivalent will be required. Do not attempt to remove the rotor using any other method.*

12 Remove the alternator cover as described in Paragraphs 4 to 10.
13 Slacken and remove the rotor retaining bolt and washer whilst holding the rotor to prevent it turning. The rotor can be held using a large strap wrench (see illustration). Alternatively, if the engine is still in the frame, the engine can be locked through the transmission by selecting top gear and applying the rear brake hard.
14 Screw the rotor puller tool (see Note) into the centre of the rotor and tighten it securely (see illustration). Tap sharply on the end of puller tool to release the rotor's grip on the tapered crankshaft end. Remove the rotor.

Installation

Alternator cover and stator

15 Where necessary, remove all traces of sealant from the stator wiring grommet and the cover. Apply fresh sealant to the grommet then install the stator into the cover, ensuring the grommet is correctly located in the cover

32.13 Using a strap wrench to retain the rotor whilst the bolt is slackened

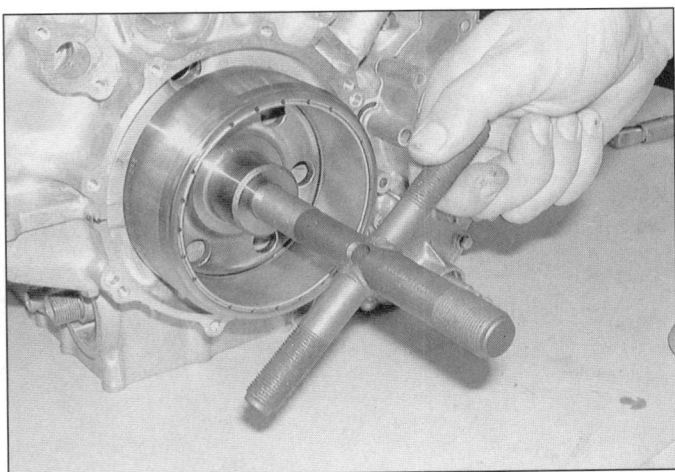

32.14 Screw the puller fully into the rotor to release it from the crankshaft end

9•18 Electrical system

32.16 Ensure the locating dowels (arrows) are in position then fit a new gasket to the engine

32.23 Ensure the crankshaft and rotor mating surfaces are clean before installing the rotor

groove. Apply a suitable non-permanent thread-locking compound to the stator and wiring clip bolt threads, then install the bolts and tighten them to the specified torque setting.

16 Ensure the mating surfaces are clean and dry then apply a smear of sealant (5 to 15 mm wide) to the area of the crankcase mating surface on either side of the crankcase half joints. Fit both locating dowels to the crankcase and locate the new gasket on the dowels **(see illustration)**.

17 Fit the alternator cover assembly to the engine and locate it on the dowels.

Caution: Take care not to trap your fingers between the cover and crankcase; the rotor will magnetically attract the stator as the cover is fitted.

18 Install the cover bolts and tighten them evenly to the specified torque setting.

19 Route the alternator wiring correctly through the frame, securing it in position with the retaining clips and guides, and securely reconnect it to the regulator/rectifier unit.

20 Fit the coolant reservoir and top-up the coolant (see Chapter 3).

21 Refill the engine with oil (see Chapter 1).

22 Install the lower fairing and seat cowling (see Chapter 8).

Alternator rotor

23 Ensure the tapered end of the crankshaft and the corresponding mating surface on the inside of the rotor are clean and free from debris. Make sure that no metal objects have attached themselves to the magnet on the inside of the rotor, then install the rotor onto the shaft **(see illustration)**.

24 Apply engine oil to the threads and washer of the rotor bolt then install the bolt and tighten it to the specified torque, using the method employed on removal to prevent the rotor from turning **(see illustrations)**.

25 Fit the alternator cover as described in Paragraphs 15 to 22.

33 Regulator/rectifier – check and replacement

Check

1 Remove the seat cowling (see Chapter 8). The regulator/rectifier is mounted on the left side of the rear subframe. Ensure the ignition is switched off then disconnect both wiring connectors from the regulator/rectifier.

2 Set the multimeter to the voltage setting and connect its positive (+) probe to the red/white terminal and the negative (–) probe to the green terminal on the wiring harness side of the connector. Turn the ignition switch on and check that full battery voltage is present between the terminals.

3 Turn the ignition switch OFF then set the multimeter to the resistance (ohms) scale. Check for continuity between the green

32.24a Lubricate the rotor bolt threads and washer with oil . . .

32.24b . . . then tighten the bolt to the specified torque whilst preventing rotation with a strap wrench

terminal of the wiring connector and earth (ground) on the frame. There should be continuity.
4 Check the alternator as described in Paragraph 3 of Section 32.
5 If the above checks do not provide the expected results check the wiring between the battery, regulator/rectifier and alternator (see the *wiring diagrams* at the end of this book).
6 If the wiring and alternator are proved good, the regulator/rectifier unit is probably faulty. The only way to test the unit is by substitution; Honda do not provide further test data.

Replacement

7 Remove the seat cowling (see Chapter 8). The regulator/rectifier is mounted onto the left side of the subframe.
8 Ensure the ignition is switched off then disconnect both the regulator/rectifier unit wiring connectors.
9 Slacken and remove the mounting bolts and nuts and remove the regulator/rectifier from the bike **(see illustration)**.
10 Installation is the reverse of removal ensuring the mounting bolts are securely tightened.

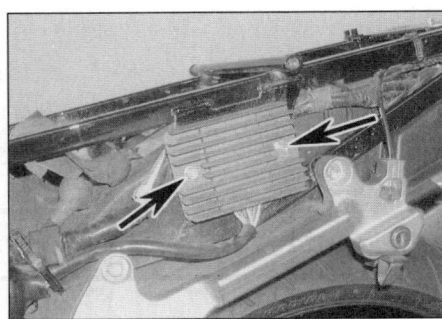

33.9 Regulator/rectifier unit retaining bolts (arrows)

9•20 Wiring diagrams

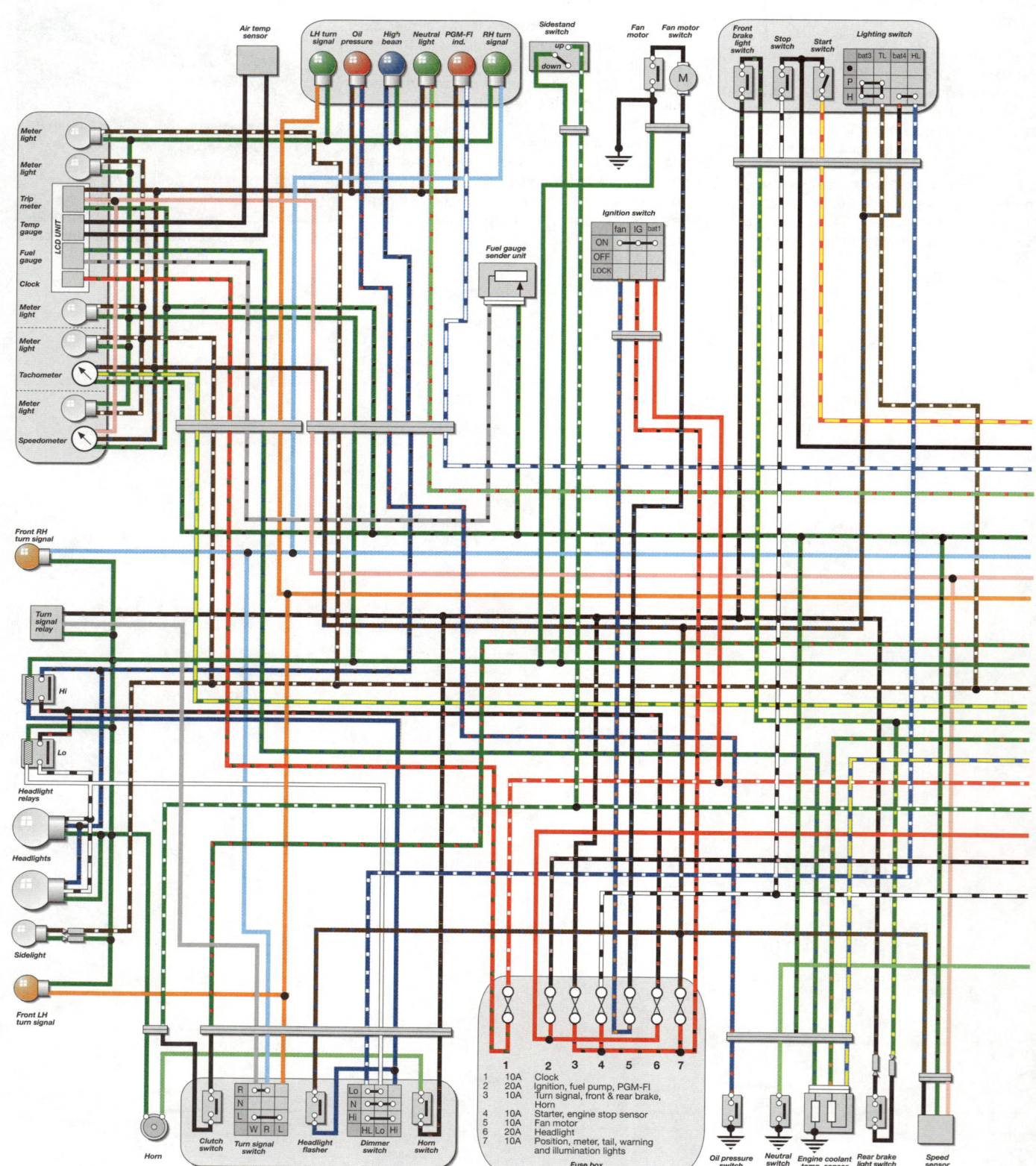

Honda VFR800FI-W and X European models - diagram 1a

Wiring diagrams 9•21

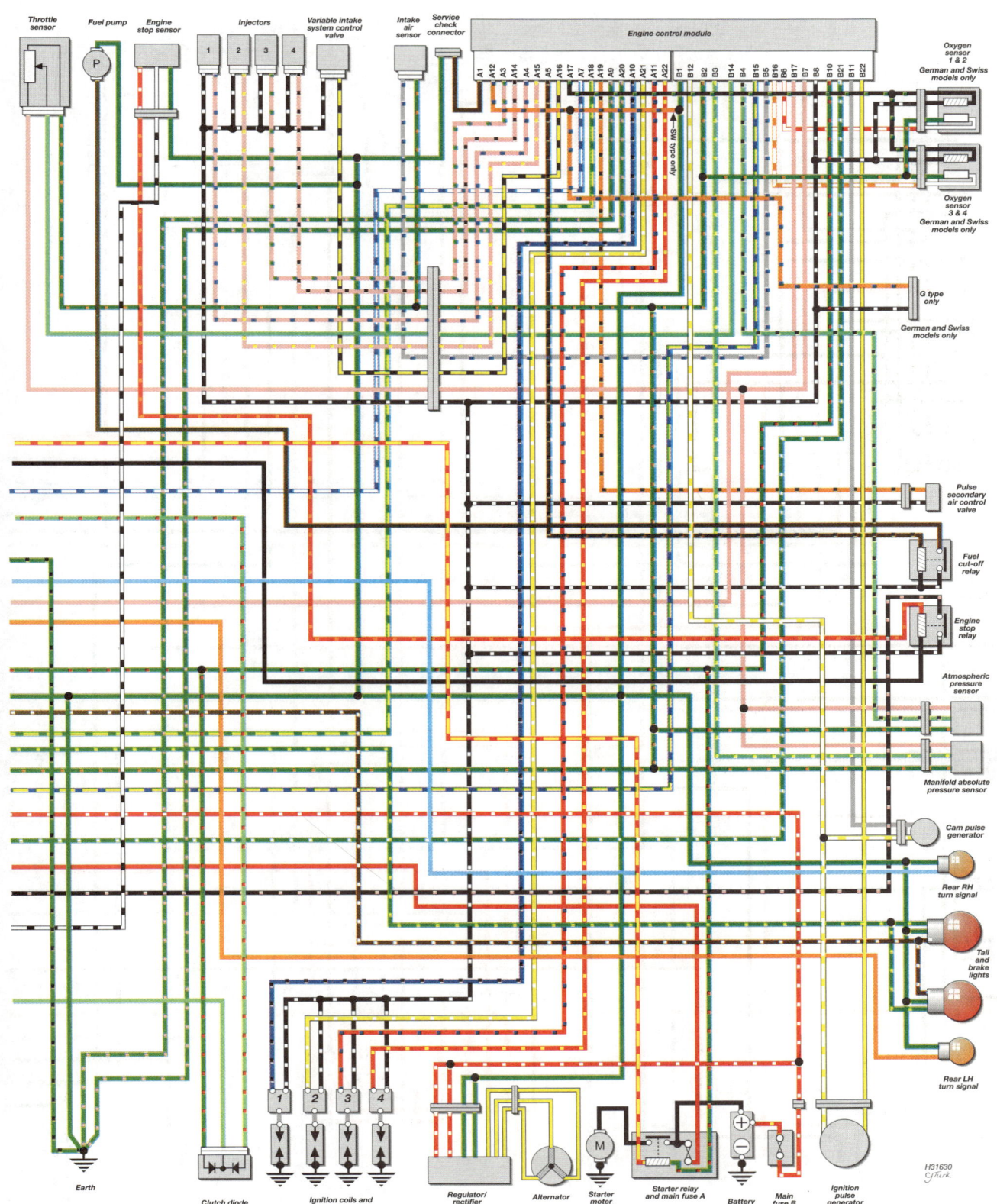

Honda VFR800FI-W and X European models - diagram 1b

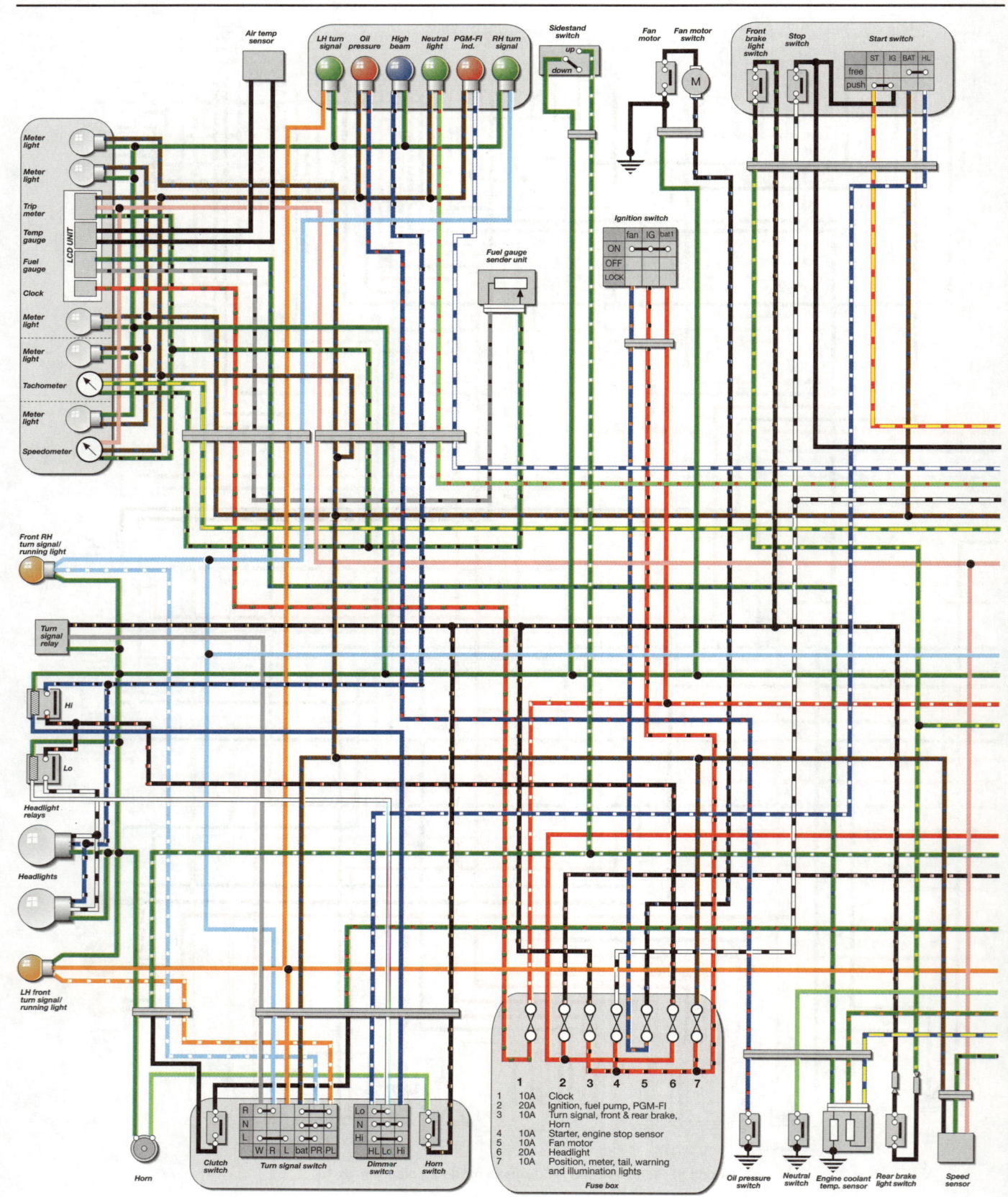

Honda VFR800FI-W and X US models - diagram 2a

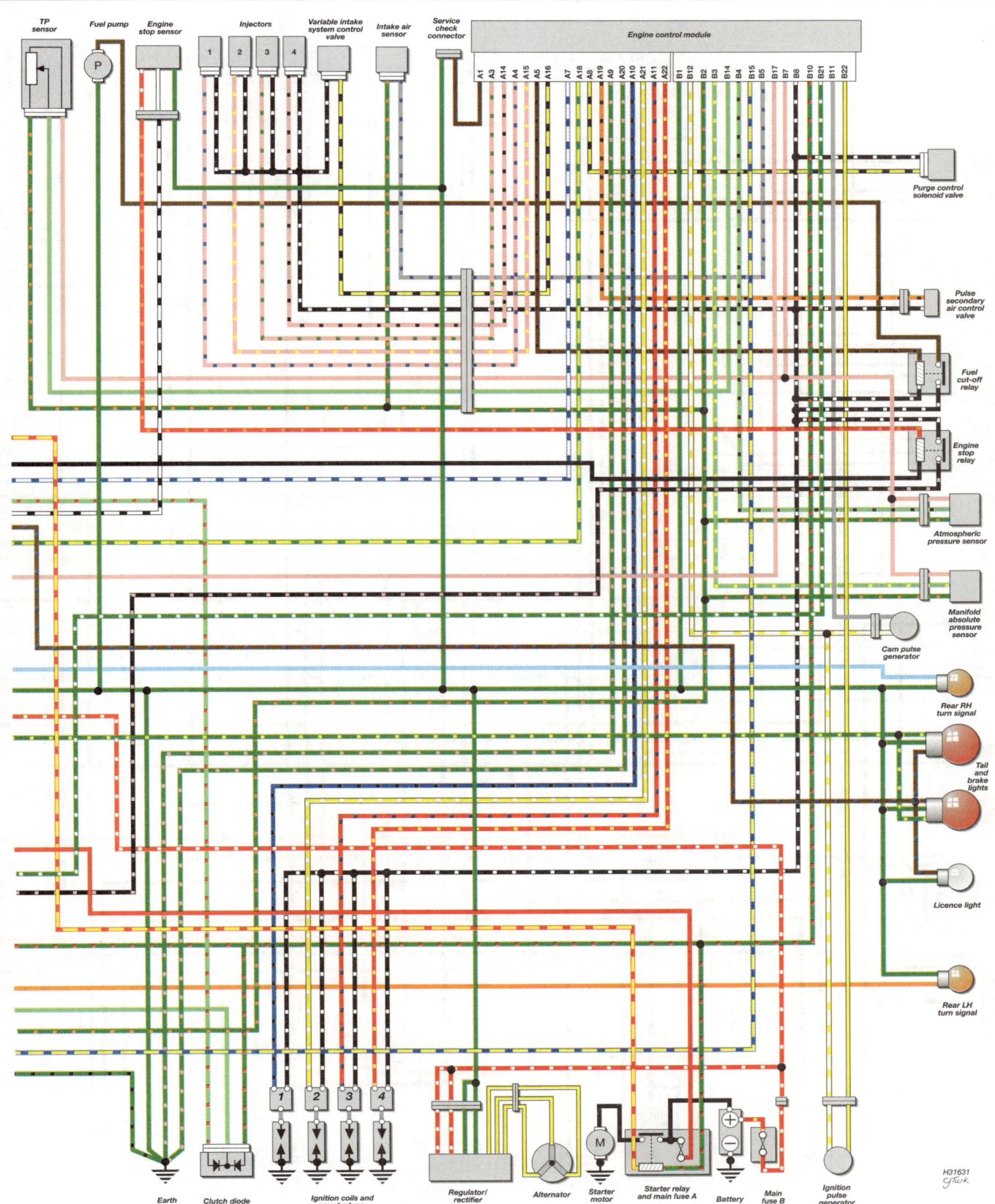
Honda VFR800FI-W and X US models - diagram 2b

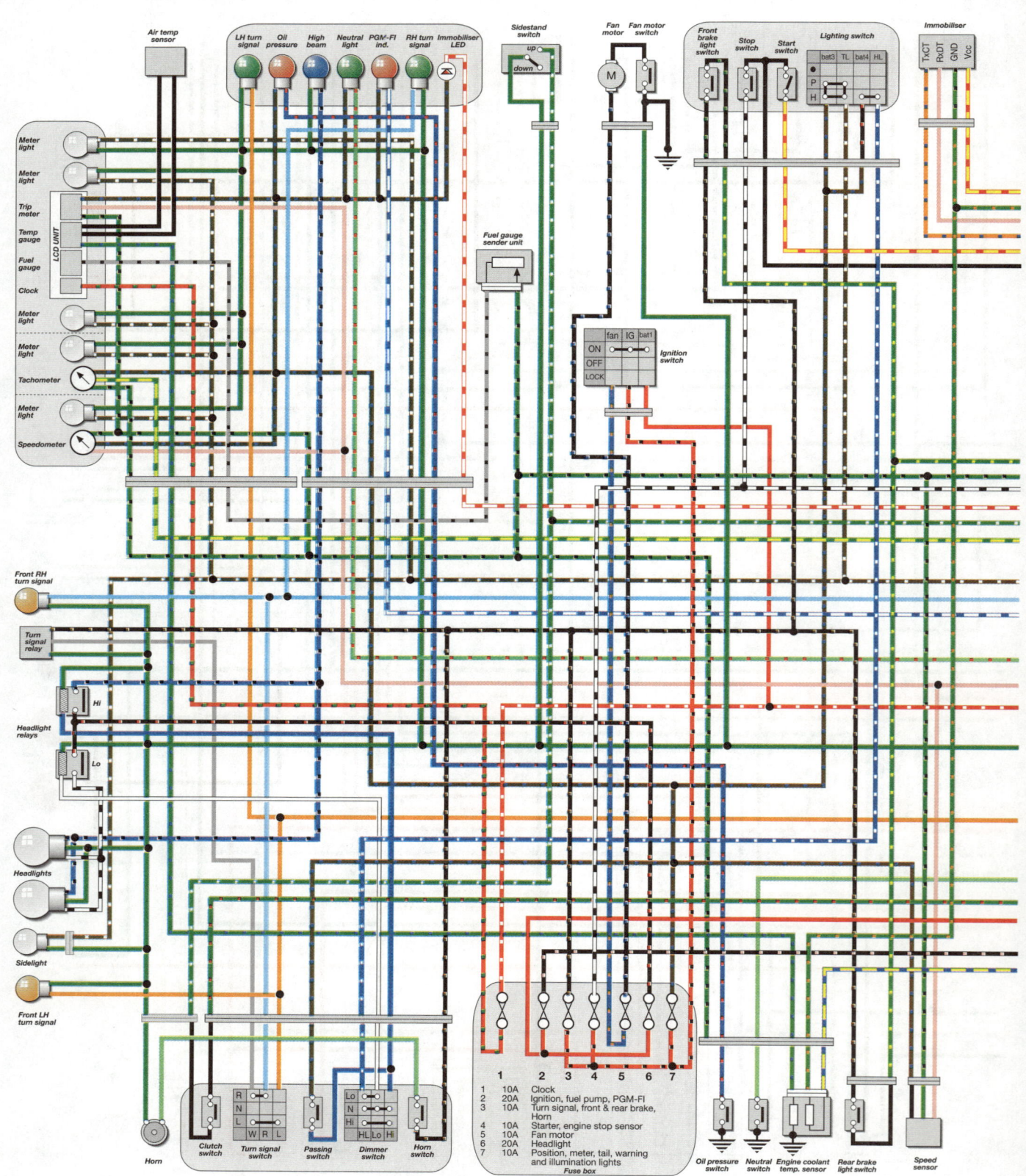

Honda VFR800FI-Y and 1 European models - diagram 3a

Wiring diagrams 9•25

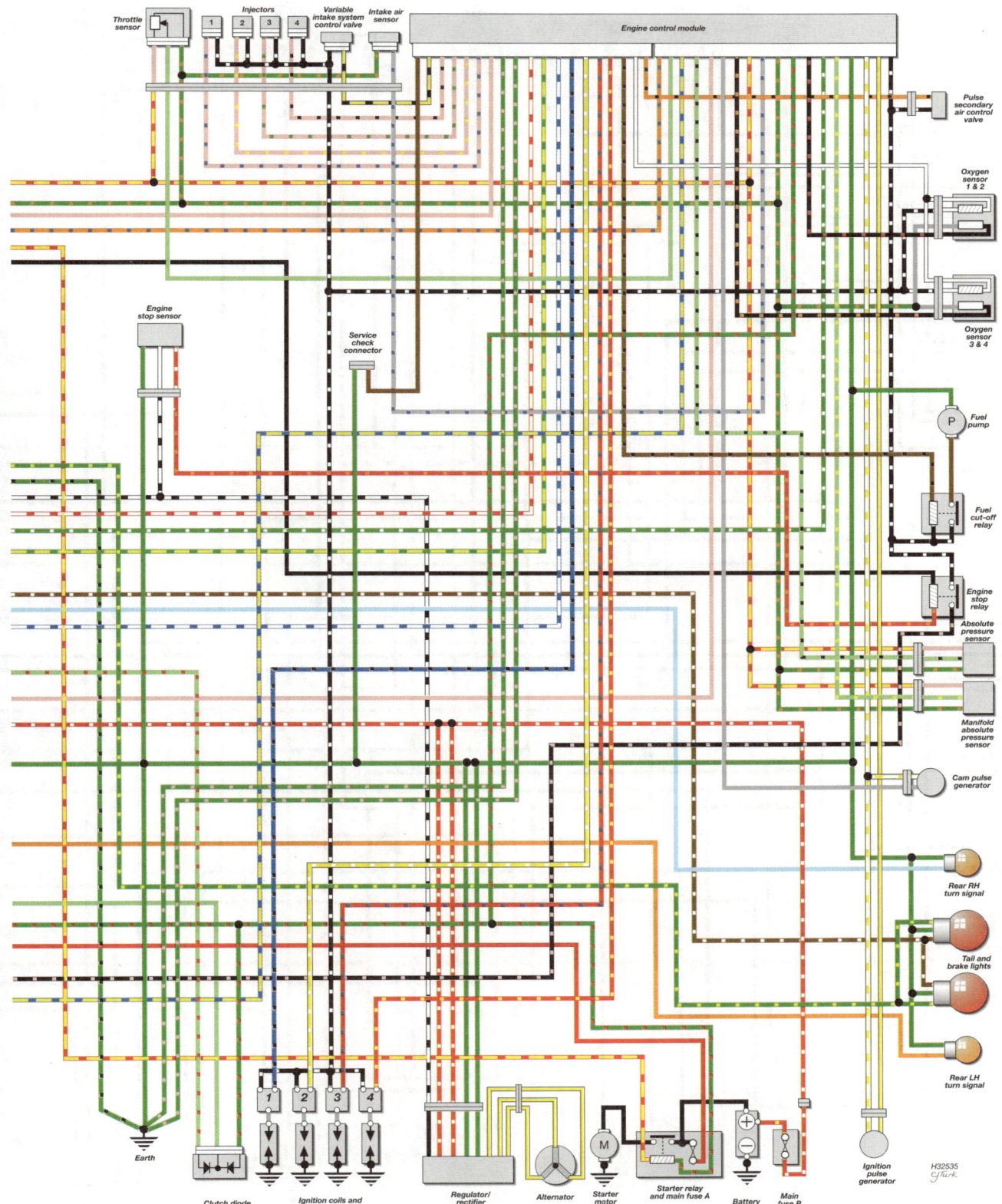

Honda VFR800FI-Y and 1 European models - diagram 3b

9•26 Wiring diagrams

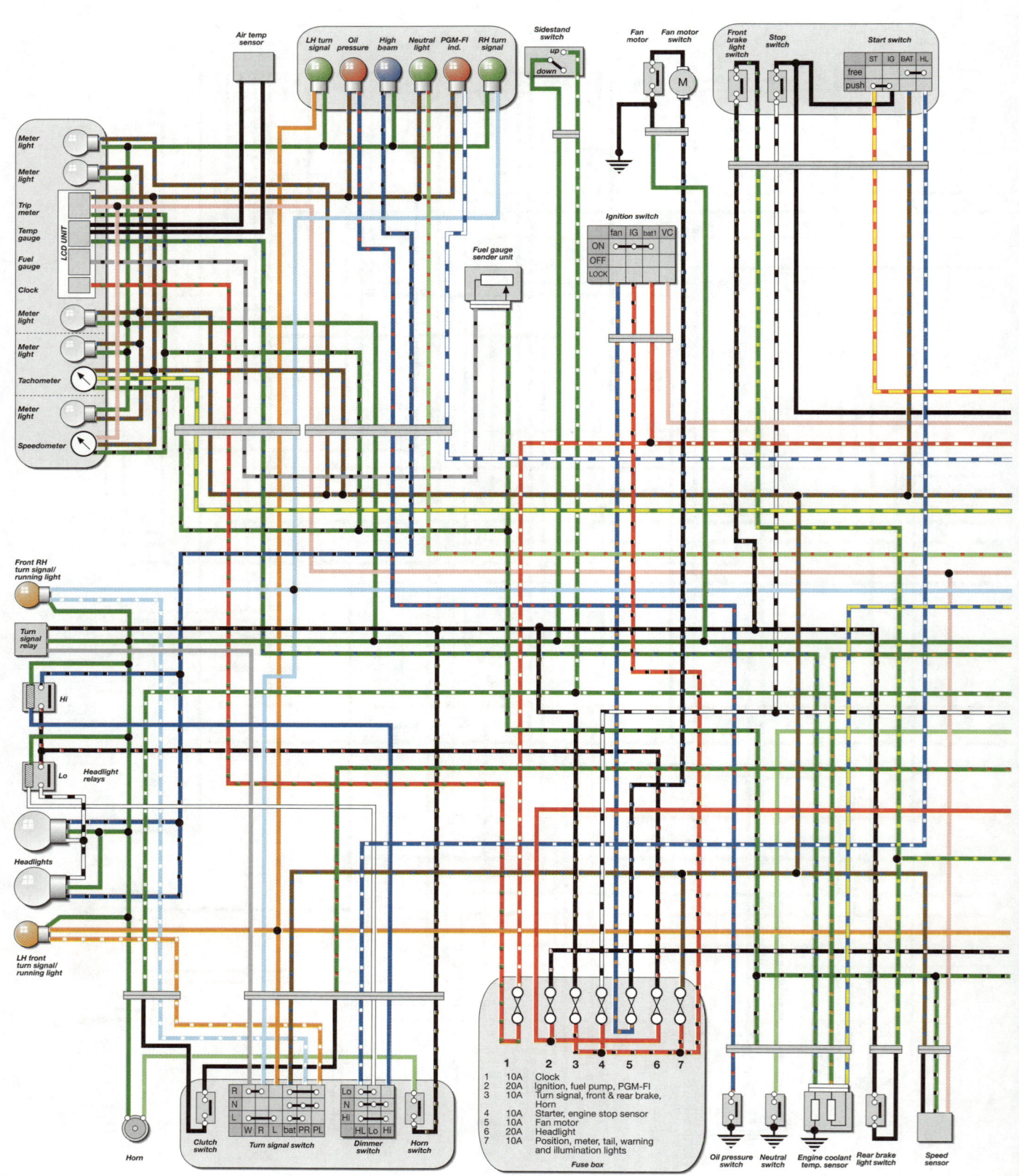

Honda VFR800FI-Y and 1 US models - diagram 4a

Wiring diagrams 9•27

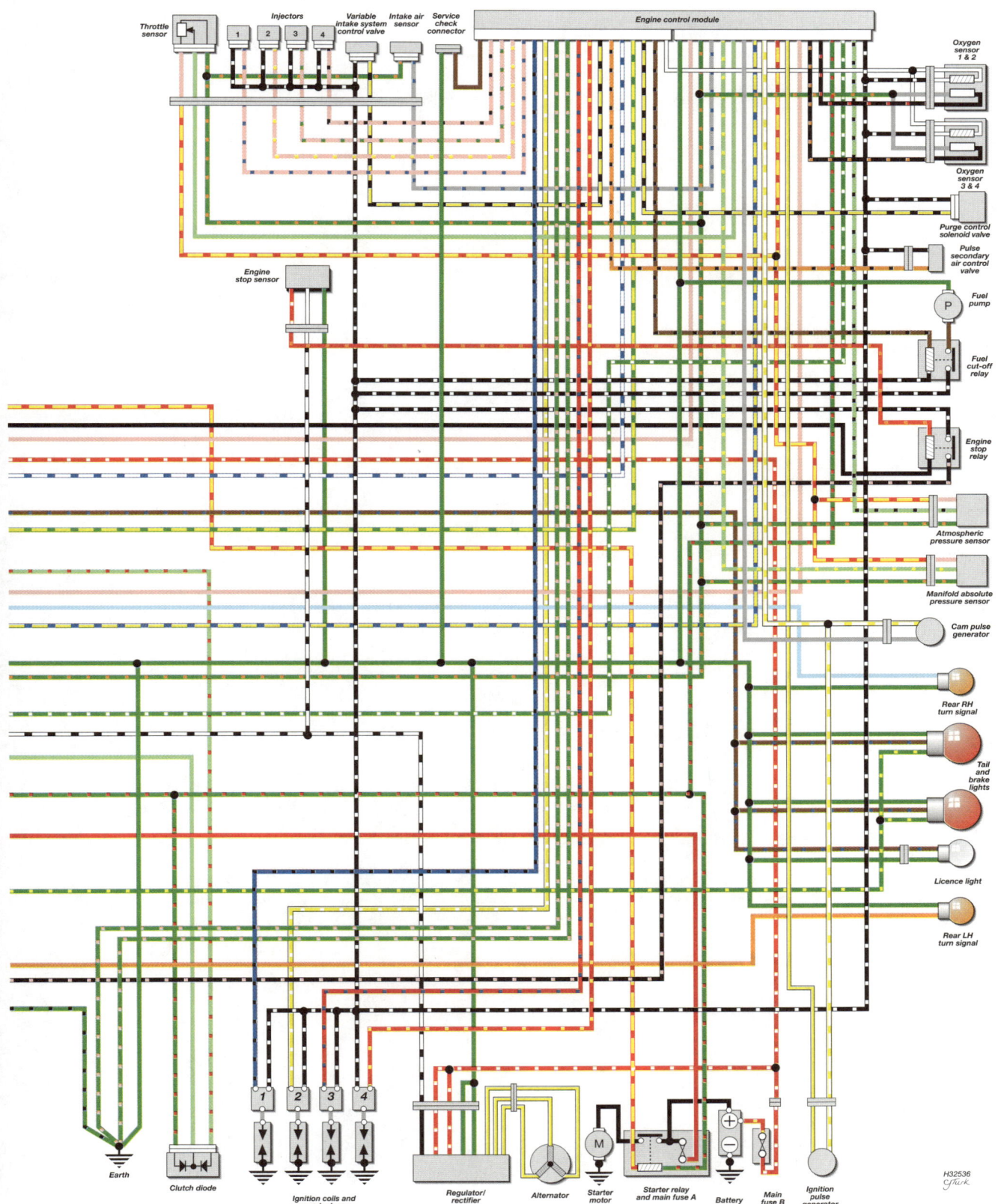

Honda VFR800FI-Y and 1 US models - diagram 4b

Notes

REF•1

Reference

Tools and Workshop Tips  REF•2

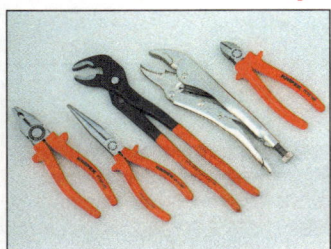

- Building up a tool kit and equipping your workshop ● Using tools
- Understanding bearing, seal, fastener and chain sizes and markings
- Repair techniques

Security  REF•20

- Locks and chains
- U-locks ● Disc locks
- Alarms and immobilisers
- Security marking systems ● Tips on how to prevent bike theft

Lubricants and fluids  REF•23

- Engine oils
- Transmission (gear) oils
- Coolant/anti-freeze
- Fork oils and suspension fluids ● Brake/clutch fluids
- Spray lubes, degreasers and solvents

Conversion Factors  REF•26

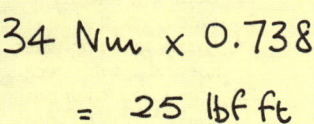

- Formulae for conversion of the metric (SI) units used throughout the manual into Imperial measures

MOT Test Checks  REF•27

- A guide to the UK MOT test ● Which items are tested ● How to prepare your motorcycle for the test and perform a pre-test check

Storage  REF•32

- How to prepare your motorcycle for going into storage and protect essential systems ● How to get the motorcycle back on the road

Fault Finding  REF•35

- Common faults and their likely causes ● How to check engine cylinder compression ● How to make electrical tests and use test meters

Technical Terms Explained 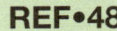 REF•48

- Component names, technical terms and common abbreviations explained

Index 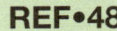 REF•52

REF•2 Tools and Workshop Tips

Buying tools

A toolkit is a fundamental requirement for servicing and repairing a motorcycle. Although there will be an initial expense in building up enough tools for servicing, this will soon be offset by the savings made by doing the job yourself. As experience and confidence grow, additional tools can be added to enable the repair and overhaul of the motorcycle. Many of the specialist tools are expensive and not often used so it may be preferable to hire them, or for a group of friends or motorcycle club to join in the purchase.

As a rule, it is better to buy more expensive, good quality tools. Cheaper tools are likely to wear out faster and need to be renewed more often, nullifying the original saving.

> **Warning:** To avoid the risk of a poor quality tool breaking in use, causing injury or damage to the component being worked on, always aim to purchase tools which meet the relevant national safety standards.

The following lists of tools do not represent the manufacturer's service tools, but serve as a guide to help the owner decide which tools are needed for this level of work. In addition, items such as an electric drill, hacksaw, files, soldering iron and a workbench equipped with a vice, may be needed. Although not classed as tools, a selection of bolts, screws, nuts, washers and pieces of tubing always come in useful.

For more information about tools, refer to the Haynes *Motorcycle Workshop Practice TechBook* (Bk. No. 3470).

Manufacturer's service tools

Inevitably certain tasks require the use of a service tool. Where possible an alternative tool or method of approach is recommended, but sometimes there is no option if personal injury or damage to the component is to be avoided. Where required, service tools are referred to in the relevant procedure.

Service tools can usually only be purchased from a motorcycle dealer and are identified by a part number. Some of the commonly-used tools, such as rotor pullers, are available in aftermarket form from mail-order motorcycle tool and accessory suppliers.

Maintenance and minor repair tools

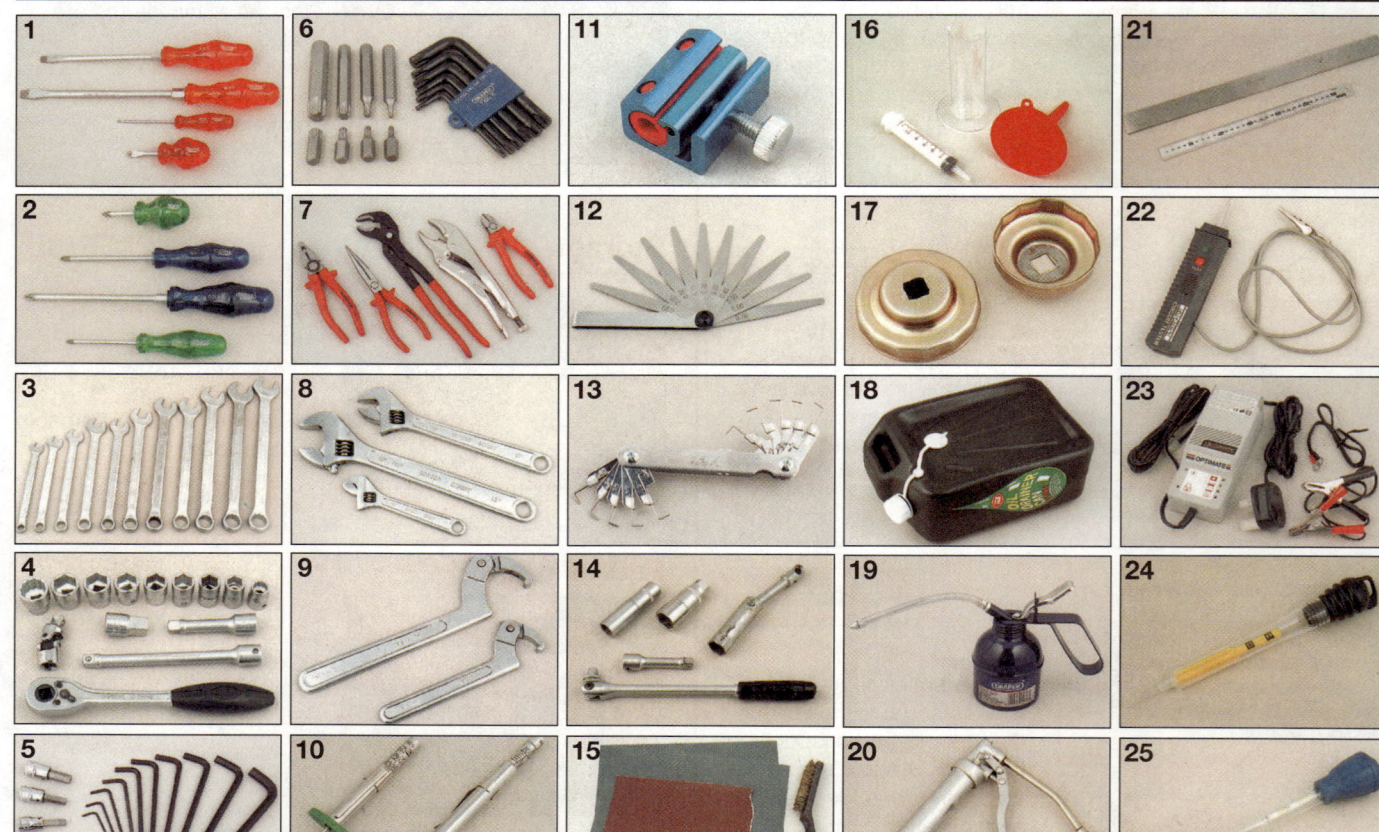

1 Set of flat-bladed screwdrivers
2 Set of Phillips head screwdrivers
3 Combination open-end and ring spanners
4 Socket set (3/8 inch or 1/2 inch drive)
5 Set of Allen keys or bits
6 Set of Torx keys or bits
7 Pliers, cutters and self-locking grips (Mole grips)
8 Adjustable spanners
9 C-spanners
10 Tread depth gauge and tyre pressure gauge
11 Cable oiler clamp
12 Feeler gauges
13 Spark plug gap measuring tool
14 Spark plug spanner or deep plug sockets
15 Wire brush and emery paper
16 Calibrated syringe, measuring vessel and funnel
17 Oil filter adapters
18 Oil drainer can or tray
19 Pump type oil can
20 Grease gun
21 Straight-edge and steel rule
22 Continuity tester
23 Battery charger
24 Hydrometer (for battery specific gravity check)
25 Anti-freeze tester (for liquid-cooled engines)

Tools and Workshop Tips REF•3

Repair and overhaul tools

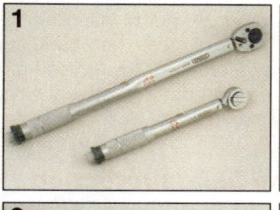
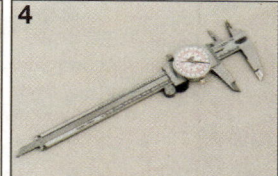
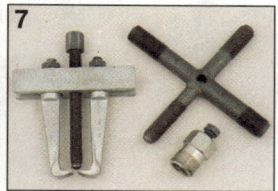
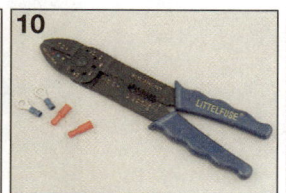

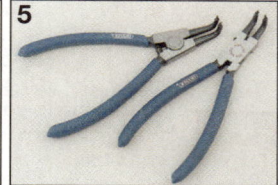
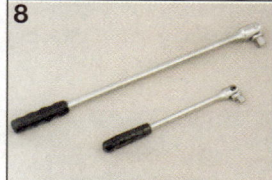
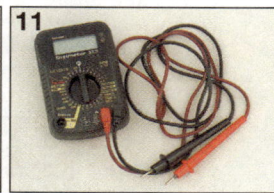
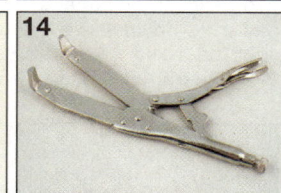

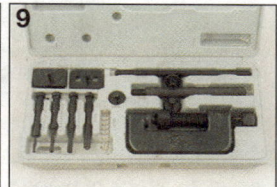

1 Torque wrench (small and mid-ranges)
2 Conventional, plastic or soft-faced hammers
3 Impact driver set
4 Vernier gauge
5 Circlip pliers (internal and external, or combination)
6 Set of cold chisels and punches
7 Selection of pullers
8 Breaker bars
9 Chain breaking/riveting tool set
10 Wire stripper and crimper tool
11 Multimeter (measures amps, volts and ohms)
12 Stroboscope (for dynamic timing checks)
13 Hose clamp (wingnut type shown)
14 Clutch holding tool
15 One-man brake/clutch bleeder kit

Specialist tools

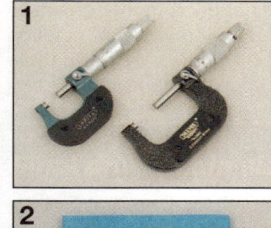
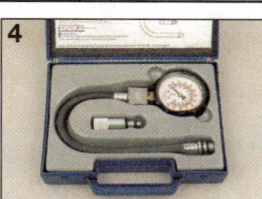
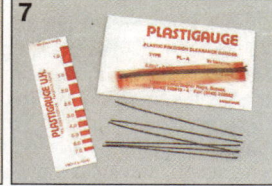
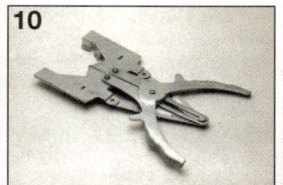

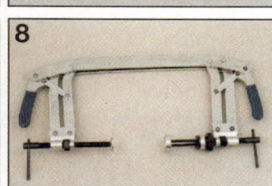
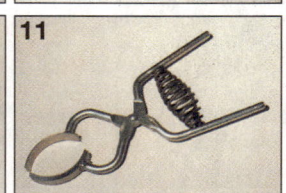
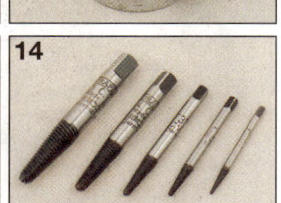

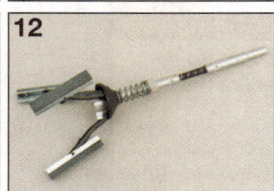
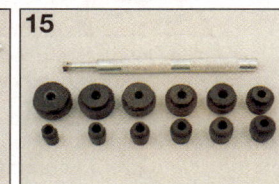

1 Micrometers (external type)
2 Telescoping gauges
3 Dial gauge
4 Cylinder compression gauge
5 Vacuum gauges (left) or manometer (right)
6 Oil pressure gauge
7 Plastigauge kit
8 Valve spring compressor (4-stroke engines)
9 Piston pin drawbolt tool
10 Piston ring removal and installation tool
11 Piston ring clamp
12 Cylinder bore hone (stone type shown)
13 Stud extractor
14 Screw extractor set
15 Bearing driver set

REF•4 Tools and Workshop Tips

1 Workshop equipment and facilities

The workbench

● Work is made much easier by raising the bike up on a ramp - components are much more accessible if raised to waist level. The hydraulic or pneumatic types seen in the dealer's workshop are a sound investment if you undertake a lot of repairs or overhauls **(see illustration 1.1)**.

1.1 Hydraulic motorcycle ramp

● If raised off ground level, the bike must be supported on the ramp to avoid it falling. Most ramps incorporate a front wheel locating clamp which can be adjusted to suit different diameter wheels. When tightening the clamp, take care not to mark the wheel rim or damage the tyre - use wood blocks on each side to prevent this.
● Secure the bike to the ramp using tie-downs **(see illustration 1.2)**. If the bike has only a sidestand, and hence leans at a dangerous angle when raised, support the bike on an auxiliary stand.

1.2 Tie-downs are used around the passenger footrests to secure the bike

● Auxiliary (paddock) stands are widely available from mail order companies or motorcycle dealers and attach either to the wheel axle or swingarm pivot **(see illustration 1.3)**. If the motorcycle has a centrestand, you can support it under the crankcase to prevent it toppling whilst either wheel is removed **(see illustration 1.4)**.

1.3 This auxiliary stand attaches to the swingarm pivot

1.4 Always use a block of wood between the engine and jack head when supporting the engine in this way

Fumes and fire

● Refer to the Safety first! page at the beginning of the manual for full details. Make sure your workshop is equipped with a fire extinguisher suitable for fuel-related fires (Class B fire - flammable liquids) - it is not sufficient to have a water-filled extinguisher.
● Always ensure adequate ventilation is available. Unless an exhaust gas extraction system is available for use, ensure that the engine is run outside of the workshop.
● If working on the fuel system, make sure the workshop is ventilated to avoid a build-up of fumes. This applies equally to fume build-up when charging a battery. Do not smoke or allow anyone else to smoke in the workshop.

Fluids

● If you need to drain fuel from the tank, store it in an approved container marked as suitable for the storage of petrol (gasoline) **(see illustration 1.5)**. Do not store fuel in glass jars or bottles.

1.5 Use an approved can only for storing petrol (gasoline)

● Use proprietary engine degreasers or solvents which have a high flash-point, such as paraffin (kerosene), for cleaning off oil, grease and dirt - never use petrol (gasoline) for cleaning. Wear rubber gloves when handling solvent and engine degreaser. The fumes from certain solvents can be dangerous - always work in a well-ventilated area.

Dust, eye and hand protection

● Protect your lungs from inhalation of dust particles by wearing a filtering mask over the nose and mouth. Many frictional materials still contain asbestos which is dangerous to your health. Protect your eyes from spouts of liquid and sprung components by wearing a pair of protective goggles **(see illustration 1.6)**.

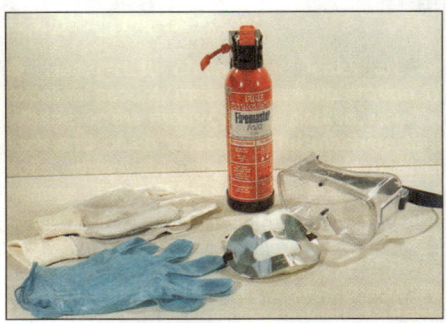

1.6 A fire extinguisher, goggles, mask and protective gloves should be at hand in the workshop

● Protect your hands from contact with solvents, fuel and oils by wearing rubber gloves. Alternatively apply a barrier cream to your hands before starting work. If handling hot components or fluids, wear suitable gloves to protect your hands from scalding and burns.

What to do with old fluids

● Old cleaning solvent, fuel, coolant and oils should not be poured down domestic drains or onto the ground. Package the fluid up in old oil containers, label it accordingly, and take it to a garage or disposal facility. Contact your local authority for location of such sites or ring the oil care hotline.

Note: It is antisocial and illegal to dump oil down the drain. To find the location of your local oil recycling bank, call this number free.

In the USA, note that any oil supplier must accept used oil for recycling.

Tools and Workshop Tips REF•5

2 Fasteners -
screws, bolts and nuts

Fastener types and applications

Bolts and screws

● Fastener head types are either of hexagonal, Torx or splined design, with internal and external versions of each type **(see illustrations 2.1 and 2.2)**; splined head fasteners are not in common use on motorcycles. The conventional slotted or Phillips head design is used for certain screws. Bolt or screw length is always measured from the underside of the head to the end of the item **(see illustration 2.11)**.

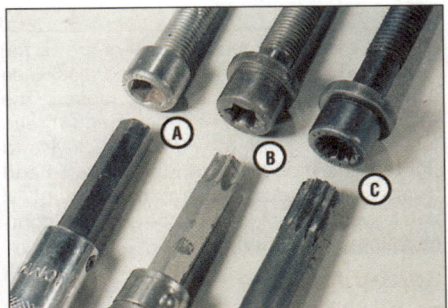

2.1 Internal hexagon/Allen (A), Torx (B) and splined (C) fasteners, with corresponding bits

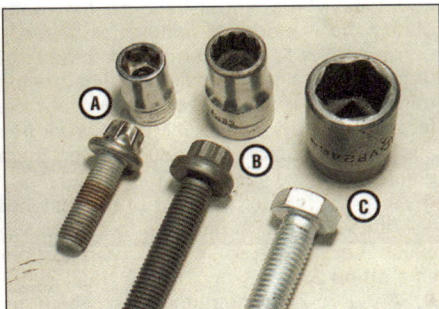

2.2 External Torx (A), splined (B) and hexagon (C) fasteners, with corresponding sockets

● Certain fasteners on the motorcycle have a tensile marking on their heads, the higher the marking the stronger the fastener. High tensile fasteners generally carry a 10 or higher marking. Never replace a high tensile fastener with one of a lower tensile strength.

Washers (see illustration 2.3)

● Plain washers are used between a fastener head and a component to prevent damage to the component or to spread the load when torque is applied. Plain washers can also be used as spacers or shims in certain assemblies. Copper or aluminium plain washers are often used as sealing washers on drain plugs.

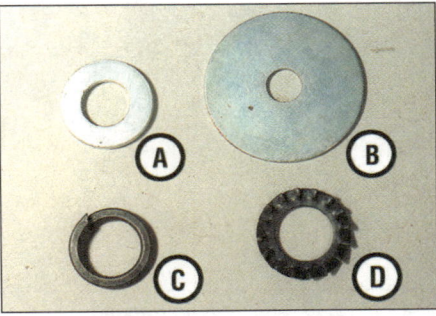

2.3 Plain washer (A), penny washer (B), spring washer (C) and serrated washer (D)

● The split-ring spring washer works by applying axial tension between the fastener head and component. If flattened, it is fatigued and must be renewed. If a plain (flat) washer is used on the fastener, position the spring washer between the fastener and the plain washer.
● Serrated star type washers dig into the fastener and component faces, preventing loosening. They are often used on electrical earth (ground) connections to the frame.
● Cone type washers (sometimes called Belleville) are conical and when tightened apply axial tension between the fastener head and component. They must be installed with the dished side against the component and often carry an OUTSIDE marking on their outer face. If flattened, they are fatigued and must be renewed.
● Tab washers are used to lock plain nuts or bolts on a shaft. A portion of the tab washer is bent up hard against one flat of the nut or bolt to prevent it loosening. Due to the tab washer being deformed in use, a new tab washer should be used every time it is disturbed.
● Wave washers are used to take up endfloat on a shaft. They provide light springing and prevent excessive side-to-side play of a component. Can be found on rocker arm shafts.

Nuts and split pins

● Conventional plain nuts are usually six-sided **(see illustration 2.4)**. They are sized by thread diameter and pitch. High tensile nuts carry a number on one end to denote their tensile strength.

2.4 Plain nut (A), shouldered locknut (B), nylon insert nut (C) and castellated nut (D)

● Self-locking nuts either have a nylon insert, or two spring metal tabs, or a shoulder which is staked into a groove in the shaft - their advantage over conventional plain nuts is a resistance to loosening due to vibration. The nylon insert type can be used a number of times, but must be renewed when the friction of the nylon insert is reduced, ie when the nut spins freely on the shaft. The spring tab type can be reused unless the tabs are damaged. The shouldered type must be renewed every time it is disturbed.
● Split pins (cotter pins) are used to lock a castellated nut to a shaft or to prevent slackening of a plain nut. Common applications are wheel axles and brake torque arms. Because the split pin arms are deformed to lock around the nut a new split pin must always be used on installation - always fit the correct size split pin which will fit snugly in the shaft hole. Make sure the split pin arms are correctly located around the nut **(see illustrations 2.5 and 2.6)**.

2.5 Bend split pin (cotter pin) arms as shown (arrows) to secure a castellated nut

2.6 Bend split pin (cotter pin) arms as shown to secure a plain nut

Caution: *If the castellated nut slots do not align with the shaft hole after tightening to the torque setting, tighten the nut until the next slot aligns with the hole - never slacken the nut to align its slot.*

● R-pins (shaped like the letter R), or slip pins as they are sometimes called, are sprung and can be reused if they are otherwise in good condition. Always install R-pins with their closed end facing forwards **(see illustration 2.7)**.

REF•6 Tools and Workshop Tips

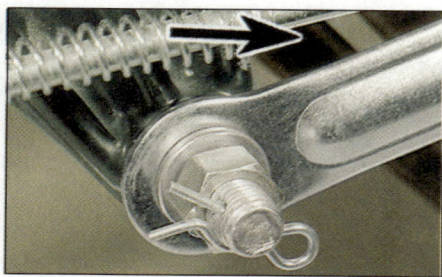

2.7 Correct fitting of R-pin. Arrow indicates forward direction

Circlips (see illustration 2.8)

● Circlips (sometimes called snap-rings) are used to retain components on a shaft or in a housing and have corresponding external or internal ears to permit removal. Parallel-sided (machined) circlips can be installed either way round in their groove, whereas stamped circlips (which have a chamfered edge on one face) must be installed with the chamfer facing away from the direction of thrust load **(see illustration 2.9)**.

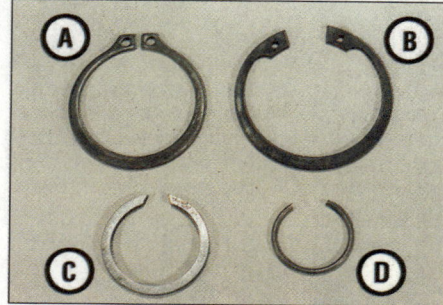

2.8 External stamped circlip (A), internal stamped circlip (B), machined circlip (C) and wire circlip (D)

● Always use circlip pliers to remove and install circlips; expand or compress them just enough to remove them. After installation, rotate the circlip in its groove to ensure it is securely seated. If installing a circlip on a splined shaft, always align its opening with a shaft channel to ensure the circlip ends are well supported and unlikely to catch **(see illustration 2.10)**.

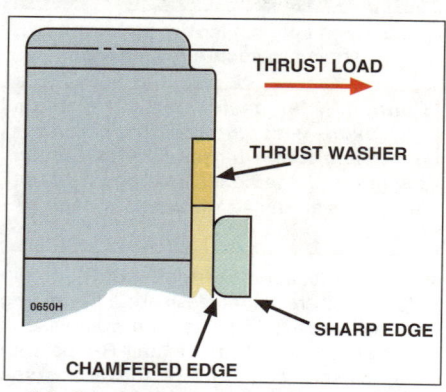

2.9 Correct fitting of a stamped circlip

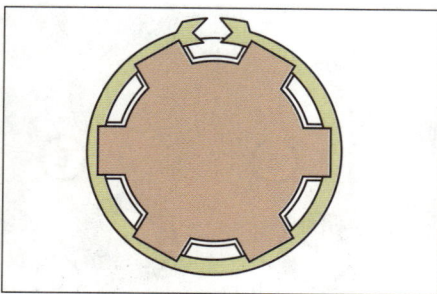

2.10 Align circlip opening with shaft channel

● Circlips can wear due to the thrust of components and become loose in their grooves, with the subsequent danger of becoming dislodged in operation. For this reason, renewal is advised every time a circlip is disturbed.

● Wire circlips are commonly used as piston pin retaining clips. If a removal tang is provided, long-nosed pliers can be used to dislodge them, otherwise careful use of a small flat-bladed screwdriver is necessary. Wire circlips should be renewed every time they are disturbed.

Thread diameter and pitch

● Diameter of a male thread (screw, bolt or stud) is the outside diameter of the threaded portion **(see illustration 2.11)**. Most motorcycle manufacturers use the ISO (International Standards Organisation) metric system expressed in millimetres, eg M6 refers to a 6 mm diameter thread. Sizing is the same for nuts, except that the thread diameter is measured across the valleys of the nut.

● Pitch is the distance between the peaks of the thread **(see illustration 2.11)**. It is expressed in millimetres, thus a common bolt size may be expressed as 6.0 x 1.0 mm (6 mm thread diameter and 1 mm pitch). Generally pitch increases in proportion to thread diameter, although there are always exceptions.

● Thread diameter and pitch are related for conventional fastener applications and the accompanying table can be used as a guide. Additionally, the AF (Across Flats), spanner or socket size dimension of the bolt or nut **(see illustration 2.11)** is linked to thread and pitch specification. Thread pitch can be measured with a thread gauge **(see illustration 2.12)**.

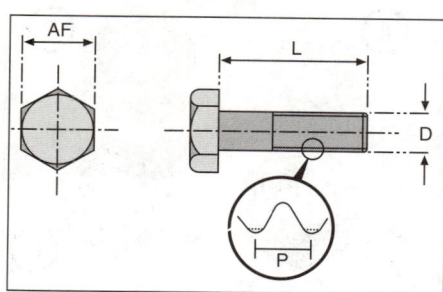

2.11 Fastener length (L), thread diameter (D), thread pitch (P) and head size (AF)

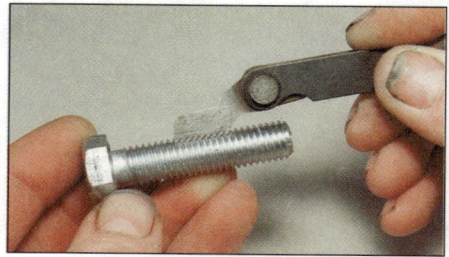

2.12 Using a thread gauge to measure pitch

AF size	Thread diameter x pitch (mm)
8 mm	M5 x 0.8
8 mm	M6 x 1.0
10 mm	M6 x 1.0
12 mm	M8 x 1.25
14 mm	M10 x 1.25
17 mm	M12 x 1.25

● The threads of most fasteners are of the right-hand type, ie they are turned clockwise to tighten and anti-clockwise to loosen. The reverse situation applies to left-hand thread fasteners, which are turned anti-clockwise to tighten and clockwise to loosen. Left-hand threads are used where rotation of a component might loosen a conventional right-hand thread fastener.

Seized fasteners

● Corrosion of external fasteners due to water or reaction between two dissimilar metals can occur over a period of time. It will build up sooner in wet conditions or in countries where salt is used on the roads during the winter. If a fastener is severely corroded it is likely that normal methods of removal will fail and result in its head being ruined. When you attempt removal, the fastener thread should be heard to crack free and unscrew easily - if it doesn't, stop there before damaging something.

● A smart tap on the head of the fastener will often succeed in breaking free corrosion which has occurred in the threads **(see illustration 2.13)**.

● An aerosol penetrating fluid (such as WD-40) applied the night beforehand may work its way down into the thread and ease removal. Depending on the location, you may be able to make up a Plasticine well around the fastener head and fill it with penetrating fluid.

2.13 A sharp tap on the head of a fastener will often break free a corroded thread

Tools and Workshop Tips   REF•7

• If you are working on an engine internal component, corrosion will most likely not be a problem due to the well lubricated environment. However, components can be very tight and an impact driver is a useful tool in freeing them **(see illustration 2.14)**.

2.14 Using an impact driver to free a fastener

• Where corrosion has occurred between dissimilar metals (eg steel and aluminium alloy), the application of heat to the fastener head will create a disproportionate expansion rate between the two metals and break the seizure caused by the corrosion. Whether heat can be applied depends on the location of the fastener - any surrounding components likely to be damaged must first be removed **(see illustration 2.15)**. Heat can be applied using a paint stripper heat gun or clothes iron, or by immersing the component in boiling water - wear protective gloves to prevent scalding or burns to the hands.

2.15 Using heat to free a seized fastener

• As a last resort, it is possible to use a hammer and cold chisel to work the fastener head unscrewed **(see illustration 2.16)**. This will damage the fastener, but more importantly extreme care must be taken not to damage the surrounding component.

Caution: Remember that the component being secured is generally of more value than the bolt, nut or screw - when the fastener is freed, do not unscrew it with force, instead work the fastener back and forth when resistance is felt to prevent thread damage.

2.16 Using a hammer and chisel to free a seized fastener

Broken fasteners and damaged heads

• If the shank of a broken bolt or screw is accessible you can grip it with self-locking grips. The knurled wheel type stud extractor tool or self-gripping stud puller tool is particularly useful for removing the long studs which screw into the cylinder mouth surface of the crankcase or bolts and screws from which the head has broken off **(see illustration 2.17)**. Studs can also be removed by locking two nuts together on the threaded end of the stud and using a spanner on the lower nut **(see illustration 2.18)**.

2.17 Using a stud extractor tool to remove a broken crankcase stud

2.18 Two nuts can be locked together to unscrew a stud from a component

• A bolt or screw which has broken off below or level with the casing must be extracted using a screw extractor set. Centre punch the fastener to centralise the drill bit, then drill a hole in the fastener **(see illustration 2.19)**. Select a drill bit which is approximately half to three-quarters the

2.19 When using a screw extractor, first drill a hole in the fastener . . .

diameter of the fastener and drill to a depth which will accommodate the extractor. Use the largest size extractor possible, but avoid leaving too small a wall thickness otherwise the extractor will merely force the fastener walls outwards wedging it in the casing thread.

• If a spiral type extractor is used, thread it anti-clockwise into the fastener. As it is screwed in, it will grip the fastener and unscrew it from the casing **(see illustration 2.20)**.

2.20 . . . then thread the extractor anti-clockwise into the fastener

• If a taper type extractor is used, tap it into the fastener so that it is firmly wedged in place. Unscrew the extractor (anti-clockwise) to draw the fastener out.

 **Warning: Stud extractors are very hard and may break off in the fastener if care is not taken - ask an engineer about spark erosion if this happens.**

• Alternatively, the broken bolt/screw can be drilled out and the hole retapped for an oversize bolt/screw or a diamond-section thread insert. It is essential that the drilling is carried out squarely and to the correct depth, otherwise the casing may be ruined - if in doubt, entrust the work to an engineer.

• Bolts and nuts with rounded corners cause the correct size spanner or socket to slip when force is applied. Of the types of spanner/socket available always use a six-point type rather than an eight or twelve-point type - better grip

REF•8 Tools and Workshop Tips

2.21 Comparison of surface drive ring spanner (left) with 12-point type (right)

is obtained. Surface drive spanners grip the middle of the hex flats, rather than the corners, and are thus good in cases of damaged heads **(see illustration 2.21)**.

● Slotted-head or Phillips-head screws are often damaged by the use of the wrong size screwdriver. Allen-head and Torx-head screws are much less likely to sustain damage. If enough of the screw head is exposed you can use a hacksaw to cut a slot in its head and then use a conventional flat-bladed screwdriver to remove it. Alternatively use a hammer and cold chisel to tap the head of the fastener around to slacken it. Always replace damaged fasteners with new ones, preferably Torx or Allen-head type.

HAYNES HiNT

A dab of valve grinding compound between the screw head and screwdriver tip will often give a good grip.

Thread repair

● Threads (particularly those in aluminium alloy components) can be damaged by overtightening, being assembled with dirt in the threads, or from a component working loose and vibrating. Eventually the thread will fail completely, and it will be impossible to tighten the fastener.

● If a thread is damaged or clogged with old locking compound it can be renovated with a thread repair tool (thread chaser) **(see illustrations 2.22 and 2.23)**; special thread

2.22 A thread repair tool being used to correct an internal thread

2.23 A thread repair tool being used to correct an external thread

chasers are available for spark plug hole threads. The tool will not cut a new thread, but clean and true the original thread. Make sure that you use the correct diameter and pitch tool. Similarly, external threads can be cleaned up with a die or a thread restorer file **(see illustration 2.24)**.

2.24 Using a thread restorer file

● It is possible to drill out the old thread and retap the component to the next thread size. This will work where there is enough surrounding material and a new bolt or screw can be obtained. Sometimes, however, this is not possible - such as where the bolt/screw passes through another component which must also be suitably modified, also in cases where a spark plug or oil drain plug cannot be obtained in a larger diameter thread size.

● The diamond-section thread insert (often known by its popular trade name of Heli-Coil) is a simple and effective method of renewing the thread and retaining the original size. A kit can be purchased which contains the tap, insert and installing tool **(see illustration 2.25)**. Drill out the damaged thread with the size drill specified **(see illustration 2.26)**. Carefully retap the thread **(see illustration 2.27)**. Install the

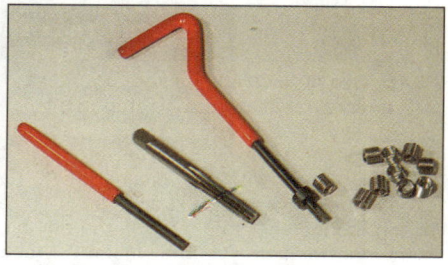

2.25 Obtain a thread insert kit to suit the thread diameter and pitch required

2.26 To install a thread insert, first drill out the original thread . . .

2.27 . . . tap a new thread . . .

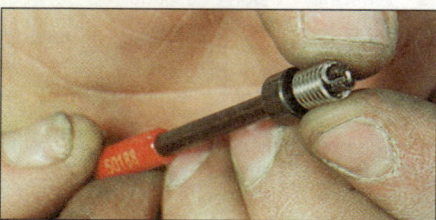

2.28 . . . fit insert on the installing tool . . .

2.29 . . . and thread into the component . . .

2.30 . . . break off the tang when complete

insert on the installing tool and thread it slowly into place using a light downward pressure **(see illustrations 2.28 and 2.29)**. When positioned between a 1/4 and 1/2 turn below the surface withdraw the installing tool and use the break-off tool to press down on the tang, breaking it off **(see illustration 2.30)**.

● There are epoxy thread repair kits on the market which can rebuild stripped internal threads, although this repair should not be used on high load-bearing components.

Tools and Workshop Tips   REF•9

Thread locking and sealing compounds

● Locking compounds are used in locations where the fastener is prone to loosening due to vibration or on important safety-related items which might cause loss of control of the motorcycle if they fail. It is also used where important fasteners cannot be secured by other means such as lockwashers or split pins.

● Before applying locking compound, make sure that the threads (internal and external) are clean and dry with all old compound removed. Select a compound to suit the component being secured - a non-permanent general locking and sealing type is suitable for most applications, but a high strength type is needed for permanent fixing of studs in castings. Apply a drop or two of the compound to the first few threads of the fastener, then thread it into place and tighten to the specified torque. Do not apply excessive thread locking compound otherwise the thread may be damaged on subsequent removal.

● Certain fasteners are impregnated with a dry film type coating of locking compound on their threads. Always renew this type of fastener if disturbed.

● Anti-seize compounds, such as copper-based greases, can be applied to protect threads from seizure due to extreme heat and corrosion. A common instance is spark plug threads and exhaust system fasteners.

3 Measuring tools and gauges

Feeler gauges

● Feeler gauges (or blades) are used for measuring small gaps and clearances **(see illustration 3.1)**. They can also be used to measure endfloat (sideplay) of a component on a shaft where access is not possible with a dial gauge.

● Feeler gauge sets should be treated with care and not bent or damaged. They are etched with their size on one face. Keep them clean and very lightly oiled to prevent corrosion build-up.

● When measuring a clearance, select a gauge which is a light sliding fit between the two components. You may need to use two gauges together to measure the clearance accurately.

Micrometers

● A micrometer is a precision tool capable of measuring to 0.01 or 0.001 of a millimetre. It should always be stored in its case and not in the general toolbox. It must be kept clean and never dropped, otherwise its frame or measuring anvils could be distorted resulting in inaccurate readings.

● External micrometers are used for measuring outside diameters of components and have many more applications than internal micrometers. Micrometers are available in different size ranges, eg 0 to 25 mm, 25 to 50 mm, and upwards in 25 mm steps; some large micrometers have interchangeable anvils to allow a range of measurements to be taken. Generally the largest precision measurement you are likely to take on a motorcycle is the piston diameter.

● Internal micrometers (or bore micrometers) are used for measuring inside diameters, such as valve guides and cylinder bores. Telescoping gauges and small hole gauges are used in conjunction with an external micrometer, whereas the more expensive internal micrometers have their own measuring device.

External micrometer

Note: *The conventional analogue type instrument is described. Although much easier to read, digital micrometers are considerably more expensive.*

● Always check the calibration of the micrometer before use. With the anvils closed (0 to 25 mm type) or set over a test gauge (for

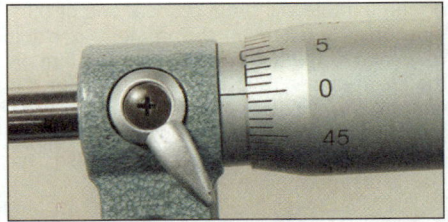

3.2 Check micrometer calibration before use

the larger types) the scale should read zero **(see illustration 3.2)**; make sure that the anvils (and test piece) are clean first. Any discrepancy can be adjusted by referring to the instructions supplied with the tool. Remember that the micrometer is a precision measuring tool - don't force the anvils closed, use the ratchet (4) on the end of the micrometer to close it. In this way, a measured force is always applied.

● To use, first make sure that the item being measured is clean. Place the anvil of the micrometer (1) against the item and use the thimble (2) to bring the spindle (3) lightly into contact with the other side of the item **(see illustration 3.3)**. Don't tighten the thimble down because this will damage the micrometer - instead use the ratchet (4) on the end of the micrometer. The ratchet mechanism applies a measured force preventing damage to the instrument.

● The micrometer is read by referring to the linear scale on the sleeve and the annular scale on the thimble. Read off the sleeve first to obtain the base measurement, then add the fine measurement from the thimble to obtain the overall reading. The linear scale on the sleeve represents the measuring range of the micrometer (eg 0 to 25 mm). The annular scale

3.1 Feeler gauges are used for measuring small gaps and clearances - thickness is marked on one face of gauge

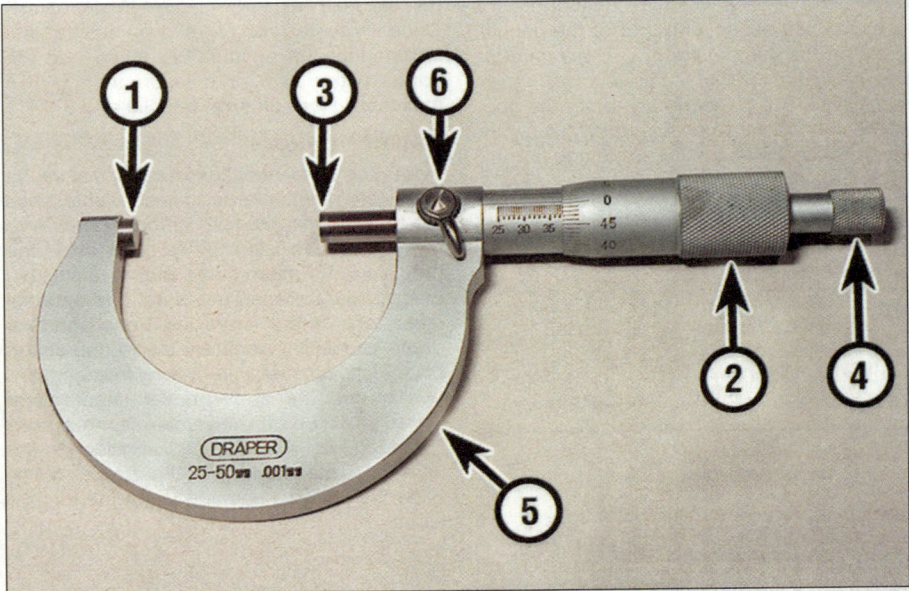

3.3 Micrometer component parts

| 1 | Anvil | 3 | Spindle | 5 | Frame |
| 2 | Thimble | 4 | Ratchet | 6 | Locking lever |

REF•10 Tools and Workshop Tips

on the thimble will be in graduations of 0.01 mm (or as marked on the frame) - one full revolution of the thimble will move 0.5 mm on the linear scale. Take the reading where the datum line on the sleeve intersects the thimble's scale. Always position the eye directly above the scale otherwise an inaccurate reading will result.

In the example shown the item measures 2.95 mm **(see illustration 3.4)**:

Linear scale	2.00 mm
Linear scale	0.50 mm
Annular scale	0.45 mm
Total figure	**2.95 mm**

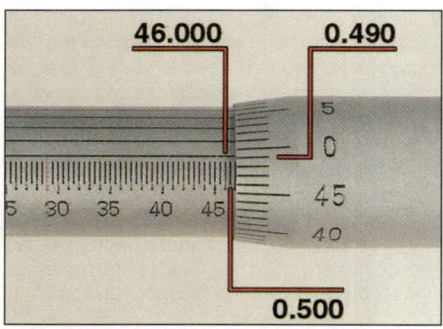

3.5 Micrometer reading of 46.99 mm on linear and annular scales . . .

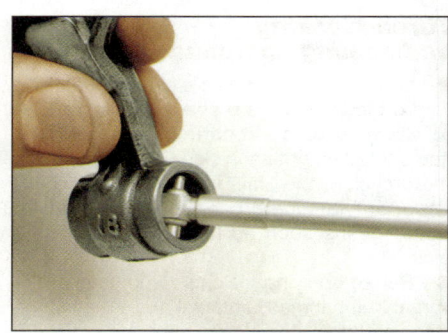

3.7 Expand the telescoping gauge in the bore, lock its position . . .

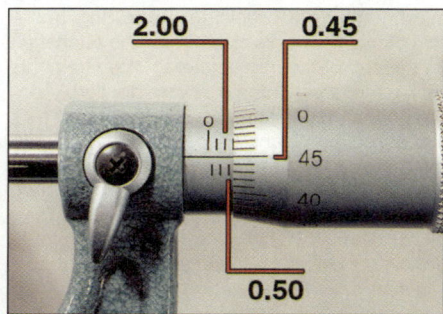

3.4 Micrometer reading of 2.95 mm

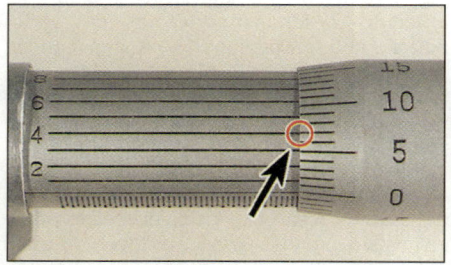

3.6 . . . and 0.004 mm on vernier scale

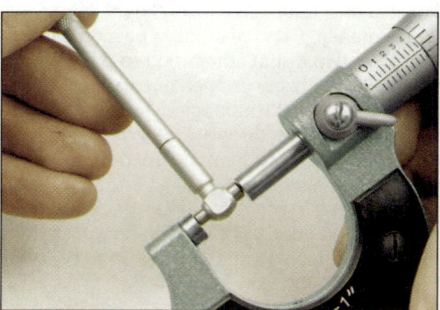

3.8 . . . then measure the gauge with a micrometer

Most micrometers have a locking lever (6) on the frame to hold the setting in place, allowing the item to be removed from the micrometer.
● Some micrometers have a vernier scale on their sleeve, providing an even finer measurement to be taken, in 0.001 increments of a millimetre. Take the sleeve and thimble measurement as described above, then check which graduation on the vernier scale aligns with that of the annular scale on the thimble **Note:** *The eye must be perpendicular to the scale when taking the vernier reading - if necessary rotate the body of the micrometer to ensure this.* Multiply the vernier scale figure by 0.001 and add it to the base and fine measurement figures.

In the example shown the item measures 46.994 mm **(see illustrations 3.5 and 3.6)**:

Linear scale (base)	46.000 mm
Linear scale (base)	00.500 mm
Annular scale (fine)	00.490 mm
Vernier scale	00.004 mm
Total figure	**46.994 mm**

Internal micrometer

● Internal micrometers are available for measuring bore diameters, but are expensive and unlikely to be available for home use. It is suggested that a set of telescoping gauges and small hole gauges, both of which must be used with an external micrometer, will suffice for taking internal measurements on a motorcycle.
● Telescoping gauges can be used to measure internal diameters of components. Select a gauge with the correct size range, make sure its ends are clean and insert it into the bore. Expand the gauge, then lock its position and withdraw it from the bore **(see illustration 3.7)**. Measure across the gauge ends with a micrometer **(see illustration 3.8)**.
● Very small diameter bores (such as valve guides) are measured with a small hole gauge. Once adjusted to a slip-fit inside the component, its position is locked and the gauge withdrawn for measurement with a micrometer **(see illustrations 3.9 and 3.10)**.

Vernier caliper

Note: *The conventional linear and dial gauge type instruments are described. Digital types are easier to read, but are far more expensive.*
● The vernier caliper does not provide the precision of a micrometer, but is versatile in being able to measure internal and external diameters. Some types also incorporate a depth gauge. It is ideal for measuring clutch plate friction material and spring free lengths.
● To use the conventional linear scale vernier, slacken off the vernier clamp screws (1) and set its jaws over (2), or inside (3), the item to be measured **(see illustration 3.11)**. Slide the jaw into contact, using the thumbwheel (4) for fine movement of the sliding scale (5) then tighten the clamp screws (1). Read off the main scale (6) where the zero on the sliding scale (5) intersects it, taking the whole number to the left of the zero; this provides the base measurement. View along the sliding scale and select the division which

3.9 Expand the small hole gauge in the bore, lock its position . . .

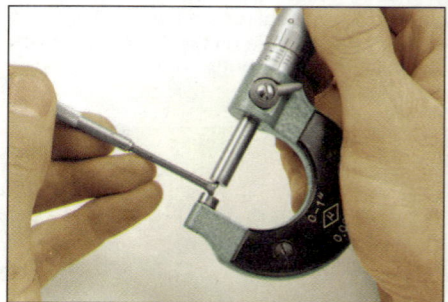

3.10 . . . then measure the gauge with a micrometer

lines up exactly with any of the divisions on the main scale, noting that the divisions usually represents 0.02 of a millimetre. Add this fine measurement to the base measurement to obtain the total reading.

Tools and Workshop Tips REF•11

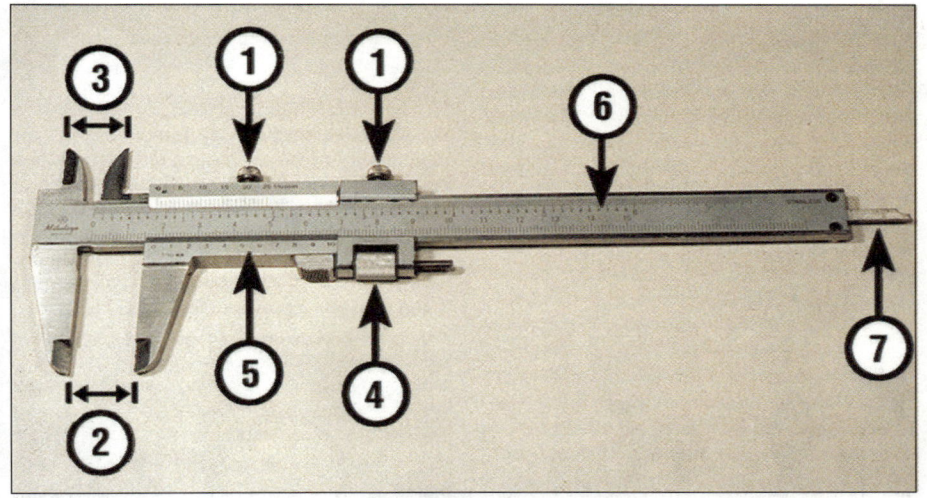

3.11 Vernier component parts (linear gauge)

| 1 Clamp screws | 3 Internal jaws | 5 Sliding scale | 7 Depth gauge |
| 2 External jaws | 4 Thumbwheel | 6 Main scale | |

In the example shown the item measures 55.92 mm **(see illustration 3.12)**:

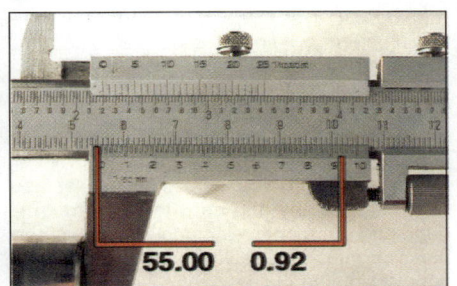

3.12 Vernier gauge reading of 55.92 mm

Base measurement	55.00 mm
Fine measurement	00.92 mm
Total figure	**55.92 mm**

● Some vernier calipers are equipped with a dial gauge for fine measurement. Before use, check that the jaws are clean, then close them fully and check that the dial gauge reads zero. If necessary adjust the gauge ring accordingly. Slacken the vernier clamp screw (1) and set its jaws over (2), or inside (3), the item to be measured **(see illustration 3.13)**. Slide the jaws into contact, using the thumbwheel (4) for fine movement. Read off the main scale (5) where the edge of the sliding scale (6) intersects it, taking the whole number to the left of the zero; this provides the base measurement. Read off the needle position on the dial gauge (7) scale to provide the fine measurement; each division represents 0.05 of a millimetre. Add this fine measurement to the base measurement to obtain the total reading.

In the example shown the item measures 55.95 mm **(see illustration 3.14)**:

Base measurement	55.00 mm
Fine measurement	00.95 mm
Total figure	**55.95 mm**

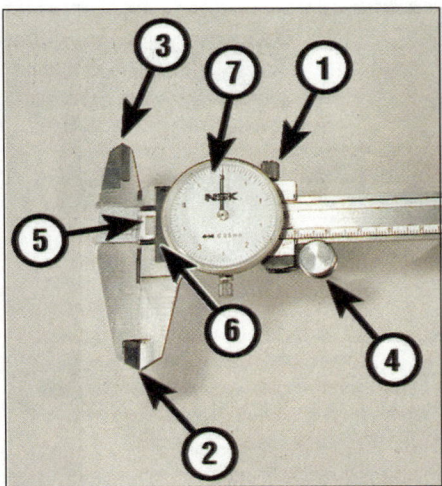

3.13 Vernier component parts (dial gauge)

1 Clamp screw	5 Main scale
2 External jaws	6 Sliding scale
3 Internal jaws	7 Dial gauge
4 Thumbwheel	

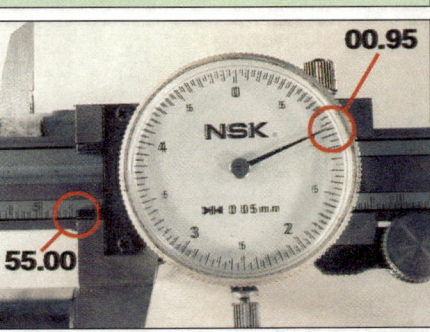

3.14 Vernier gauge reading of 55.95 mm

Plastigauge

● Plastigauge is a plastic material which can be compressed between two surfaces to measure the oil clearance between them. The width of the compressed Plastigauge is measured against a calibrated scale to determine the clearance.

● Common uses of Plastigauge are for measuring the clearance between crankshaft journal and main bearing inserts, between crankshaft journal and big-end bearing inserts, and between camshaft and bearing surfaces. The following example describes big-end oil clearance measurement.

● Handle the Plastigauge material carefully to prevent distortion. Using a sharp knife, cut a length which corresponds with the width of the bearing being measured and place it carefully across the journal so that it is parallel with the shaft **(see illustration 3.15)**. Carefully install both bearing shells and the connecting rod. Without rotating the rod on the journal tighten its bolts or nuts (as applicable) to the specified torque. The connecting rod and bearings are then disassembled and the crushed Plastigauge examined.

3.15 Plastigauge placed across shaft journal

● Using the scale provided in the Plastigauge kit, measure the width of the material to determine the oil clearance **(see illustration 3.16)**. Always remove all traces of Plastigauge after use using your fingernails.

Caution: Arriving at the correct clearance demands that the assembly is torqued correctly, according to the settings and sequence (where applicable) provided by the motorcycle manufacturer.

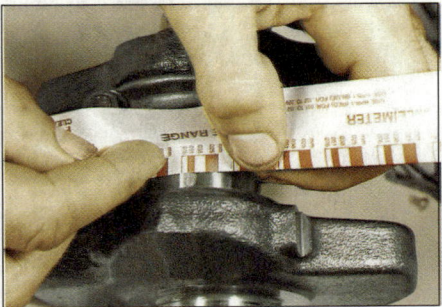

3.16 Measuring the width of the crushed Plastigauge

REF•12 Tools and Workshop Tips

Dial gauge or DTI (Dial Test Indicator)

● A dial gauge can be used to accurately measure small amounts of movement. Typical uses are measuring shaft runout or shaft endfloat (sideplay) and setting piston position for ignition timing on two-strokes. A dial gauge set usually comes with a range of different probes and adapters and mounting equipment.

● The gauge needle must point to zero when at rest. Rotate the ring around its periphery to zero the gauge.

● Check that the gauge is capable of reading the extent of movement in the work. Most gauges have a small dial set in the face which records whole millimetres of movement as well as the fine scale around the face periphery which is calibrated in 0.01 mm divisions. Read off the small dial first to obtain the base measurement, then add the measurement from the fine scale to obtain the total reading.

In the example shown the gauge reads 1.48 mm **(see illustration 3.17)**:

Base measurement	1.00 mm
Fine measurement	0.48 mm
Total figure	**1.48 mm**

3.17 Dial gauge reading of 1.48 mm

● If measuring shaft runout, the shaft must be supported in vee-blocks and the gauge mounted on a stand perpendicular to the shaft. Rest the tip of the gauge against the centre of the shaft and rotate the shaft slowly whilst watching the gauge reading **(see illustration 3.18)**. Take several measurements along the length of the shaft and record the maximum gauge reading as the amount of runout in the shaft. **Note:** *The reading obtained will be total runout at that point - some manufacturers specify that the runout figure is halved to compare with their specified runout limit.*

● Endfloat (sideplay) measurement requires that the gauge is mounted securely to the surrounding component with its probe touching the end of the shaft. Using hand pressure, push and pull on the shaft noting the maximum endfloat recorded on the gauge **(see illustration 3.19)**.

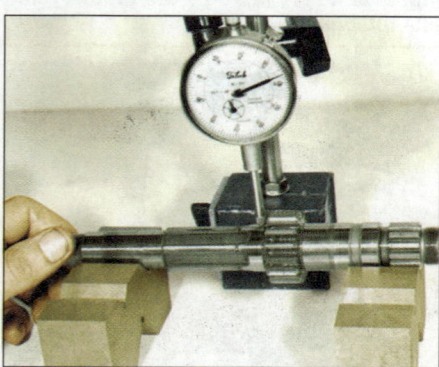

3.18 Using a dial gauge to measure shaft runout

3.19 Using a dial gauge to measure shaft endfloat

● A dial gauge with suitable adapters can be used to determine piston position BTDC on two-stroke engines for the purposes of ignition timing. The gauge, adapter and suitable length probe are installed in the place of the spark plug and the gauge zeroed at TDC. If the piston position is specified as 1.14 mm BTDC, rotate the engine back to 2.00 mm BTDC, then slowly forwards to 1.14 mm BTDC.

Cylinder compression gauges

● A compression gauge is used for measuring cylinder compression. Either the rubber-cone type or the threaded adapter type can be used. The latter is preferred to ensure a perfect seal against the cylinder head. A 0 to 300 psi (0 to 20 Bar) type gauge (for petrol/gasoline engines) will be suitable for motorcycles.

● The spark plug is removed and the gauge either held hard against the cylinder head (cone type) or the gauge adapter screwed into the cylinder head (threaded type) **(see illustration 3.20)**. Cylinder compression is measured with the engine turning over, but not running - carry out the compression test as described in *Fault Finding Equipment*. The gauge will hold the reading until manually released.

Oil pressure gauge

● An oil pressure gauge is used for measuring engine oil pressure. Most gauges come with a set of adapters to fit the thread of the take-off point **(see illustration 3.21)**. If the take-off point specified by the motorcycle manufacturer is an external oil pipe union, make sure that the specified replacement union is used to prevent oil starvation.

3.21 Oil pressure gauge and take-off point adapter (arrow)

● Oil pressure is measured with the engine running (at a specific rpm) and often the manufacturer will specify pressure limits for a cold and hot engine.

Straight-edge and surface plate

● If checking the gasket face of a component for warpage, place a steel rule or precision straight-edge across the gasket face and measure any gap between the straight-edge and component with feeler gauges **(see illustration 3.22)**. Check diagonally across the component and between mounting holes **(see illustration 3.23)**.

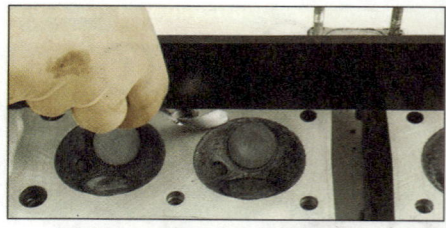

3.22 Use a straight-edge and feeler gauges to check for warpage

3.20 Using a rubber-cone type cylinder compression gauge

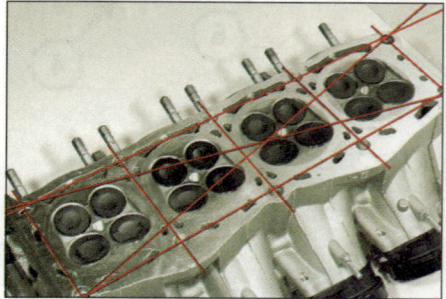

3.23 Check for warpage in these directions

Tools and Workshop Tips   REF•13

● Checking individual components for warpage, such as clutch plain (metal) plates, requires a perfectly flat plate or piece or plate glass and feeler gauges.

4 Torque and leverage

What is torque?

● Torque describes the twisting force about a shaft. The amount of torque applied is determined by the distance from the centre of the shaft to the end of the lever and the amount of force being applied to the end of the lever; distance multiplied by force equals torque.

● The manufacturer applies a measured torque to a bolt or nut to ensure that it will not slacken in use and to hold two components securely together without movement in the joint. The actual torque setting depends on the thread size, bolt or nut material and the composition of the components being held.

● Too little torque may cause the fastener to loosen due to vibration, whereas too much torque will distort the joint faces of the component or cause the fastener to shear off. Always stick to the specified torque setting.

Using a torque wrench

● Check the calibration of the torque wrench and make sure it has a suitable range for the job. Torque wrenches are available in Nm (Newton-metres), kgf m (kilograms-force metre), lbf ft (pounds-feet), lbf in (inch-pounds). Do not confuse lbf ft with lbf in.

● Adjust the tool to the desired torque on the scale (see illustration 4.1). If your torque wrench is not calibrated in the units specified, carefully convert the figure (see *Conversion Factors*). A manufacturer sometimes gives a torque setting as a range (8 to 10 Nm) rather than a single figure - in this case set the tool midway between the two settings. The same torque may be expressed as 9 Nm ± 1 Nm. Some torque wrenches have a method of locking the setting so that it isn't inadvertently altered during use.

● Install the bolts/nuts in their correct location and secure them lightly. Their threads must be clean and free of any old locking compound. Unless specified the threads and flange should be dry - oiled threads are necessary in certain circumstances and the manufacturer will take this into account in the specified torque figure. Similarly, the manufacturer may also specify the application of thread-locking compound.

● Tighten the fasteners in the specified sequence until the torque wrench clicks, indicating that the torque setting has been reached. Apply the torque again to double-check the setting. Where different thread diameter fasteners secure the component, as a rule tighten the larger diameter ones first.

● When the torque wrench has been finished with, release the lock (where applicable) and fully back off its setting to zero - do not leave the torque wrench tensioned. Also, do not use a torque wrench for slackening a fastener.

Angle-tightening

● Manufacturers often specify a figure in degrees for final tightening of a fastener. This usually follows tightening to a specific torque setting.

● A degree disc can be set and attached to the socket (see illustration 4.2) or a protractor can be used to mark the angle of movement on the bolt/nut head and the surrounding casting (see illustration 4.3).

4.2 Angle tightening can be accomplished with a torque-angle gauge . . .

4.3 . . . or by marking the angle on the surrounding component

Loosening sequences

● Where more than one bolt/nut secures a component, loosen each fastener evenly a little at a time. In this way, not all the stress of the joint is held by one fastener and the components are not likely to distort.

● If a tightening sequence is provided, work in the REVERSE of this, but if not, work from the outside in, in a criss-cross sequence (see illustration 4.4).

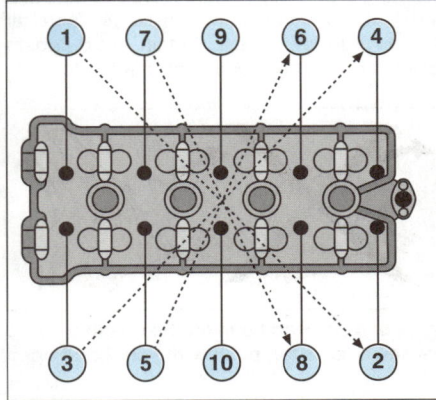

4.4 When slackening, work from the outside inwards

Tightening sequences

● If a component is held by more than one fastener it is important that the retaining bolts/nuts are tightened evenly to prevent uneven stress build-up and distortion of sealing faces. This is especially important on high-compression joints such as the cylinder head.

● A sequence is usually provided by the manufacturer, either in a diagram or actually marked in the casting. If not, always start in the centre and work outwards in a criss-cross pattern (see illustration 4.5). Start off by securing all bolts/nuts finger-tight, then set the torque wrench and tighten each fastener by a small amount in sequence until the final torque is reached. By following this practice,

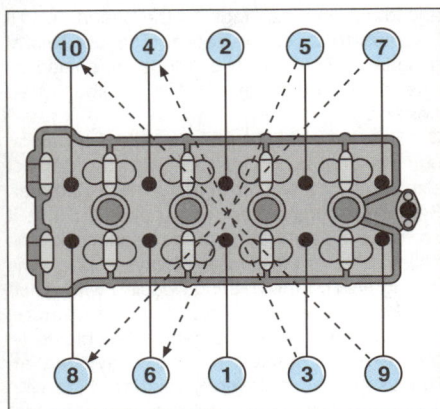

4.5 When tightening, work from the inside outwards

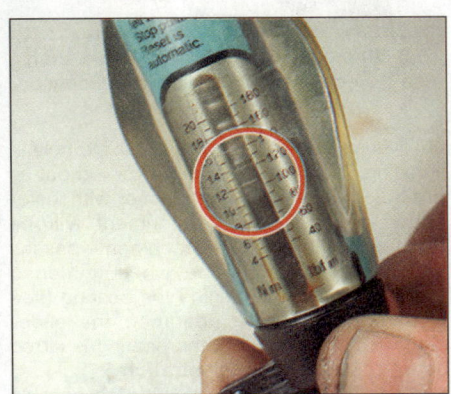

4.1 Set the torque wrench index mark to the setting required, in this case 12 Nm

REF•14 Tools and Workshop Tips

the joint will be held evenly and will not be distorted. Important joints, such as the cylinder head and big-end fasteners often have two- or three-stage torque settings.

Applying leverage

● Use tools at the correct angle. Position a socket wrench or spanner on the bolt/nut so that you pull it towards you when loosening. If this can't be done, push the spanner without curling your fingers around it **(see illustration 4.6)** - the spanner may slip or the fastener loosen suddenly, resulting in your fingers being crushed against a component.

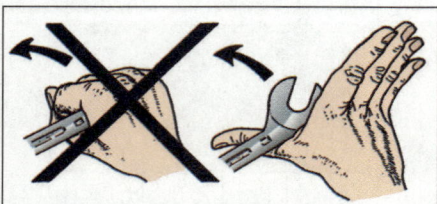

4.6 If you can't pull on the spanner to loosen a fastener, push with your hand open

● Additional leverage is gained by extending the length of the lever. The best way to do this is to use a breaker bar instead of the regular length tool, or to slip a length of tubing over the end of the spanner or socket wrench.
● If additional leverage will not work, the fastener head is either damaged or firmly corroded in place (see *Fasteners*).

5 Bearings

Bearing removal and installation
Drivers and sockets

● Before removing a bearing, always inspect the casing to see which way it must be driven out - some casings will have retaining plates or a cast step. Also check for any identifying markings on the bearing and if installed to a certain depth, measure this at this stage. Some roller bearings are sealed on one side - take note of the original fitted position.
● Bearings can be driven out of a casing using a bearing driver tool (with the correct size head) or a socket of the correct diameter. Select the driver head or socket so that it contacts the outer race of the bearing, not the balls/rollers or inner race. Always support the casing around the bearing housing with wood blocks, otherwise there is a risk of fracture. The bearing is driven out with a few blows on the driver or socket from a heavy mallet. Unless access is severely restricted (as with wheel bearings), a pin-punch is not recommended unless it is moved around the bearing to keep it square in its housing.

● The same equipment can be used to install bearings. Make sure the bearing housing is supported on wood blocks and line up the bearing in its housing. Fit the bearing as noted on removal - generally they are installed with their marked side facing outwards. Tap the bearing squarely into its housing using a driver or socket which bears only on the bearing's outer race - contact with the bearing balls/rollers or inner race will destroy it **(see illustrations 5.1 and 5.2)**.
● Check that the bearing inner race and balls/rollers rotate freely.

5.1 Using a bearing driver against the bearing's outer race

5.2 Using a large socket against the bearing's outer race

Pullers and slide-hammers

● Where a bearing is pressed on a shaft a puller will be required to extract it **(see illustration 5.3)**. Make sure that the puller clamp or legs fit securely behind the bearing and are unlikely to slip out. If pulling a bearing

5.3 This bearing puller clamps behind the bearing and pressure is applied to the shaft end to draw the bearing off

off a gear shaft for example, you may have to locate the puller behind a gear pinion if there is no access to the race and draw the gear pinion off the shaft as well **(see illustration 5.4)**.

Caution: Ensure that the puller's centre bolt locates securely against the end of the shaft and will not slip when pressure is applied. Also ensure that puller does not damage the shaft end.

5.4 Where no access is available to the rear of the bearing, it is sometimes possible to draw off the adjacent component

● Operate the puller so that its centre bolt exerts pressure on the shaft end and draws the bearing off the shaft.
● When installing the bearing on the shaft, tap only on the bearing's inner race - contact with the balls/rollers or outer race with destroy the bearing. Use a socket or length of tubing as a drift which fits over the shaft end **(see illustration 5.5)**.

5.5 When installing a bearing on a shaft use a piece of tubing which bears only on the bearing's inner race

● Where a bearing locates in a blind hole in a casing, it cannot be driven or pulled out as described above. A slide-hammer with knife-edged bearing puller attachment will be required. The puller attachment passes through the bearing and when tightened expands to fit firmly behind the bearing **(see illustration 5.6)**. By operating the slide-hammer part of the tool the bearing is jarred out of its housing **(see illustration 5.7)**.
● It is possible, if the bearing is of reasonable weight, for it to drop out of its housing if the casing is heated as described opposite. If this

Tools and Workshop Tips REF•15

5.6 Expand the bearing puller so that it locks behind the bearing . . .

5.7 . . . attach the slide hammer to the bearing puller

method is attempted, first prepare a work surface which will enable the casing to be tapped face down to help dislodge the bearing - a wood surface is ideal since it will not damage the casing's gasket surface. Wearing protective gloves, tap the heated casing several times against the work surface to dislodge the bearing under its own weight **(see illustration 5.8)**.

5.8 Tapping a casing face down on wood blocks can often dislodge a bearing

● Bearings can be installed in blind holes using the driver or socket method described above.

Drawbolts

● Where a bearing or bush is set in the eye of a component, such as a suspension linkage arm or connecting rod small-end, removal by drift may damage the component. Furthermore, a rubber bushing in a shock absorber eye cannot successfully be driven out of position. If access is available to a engineering press, the task is straightforward. If not, a drawbolt can be fabricated to extract the bearing or bush.

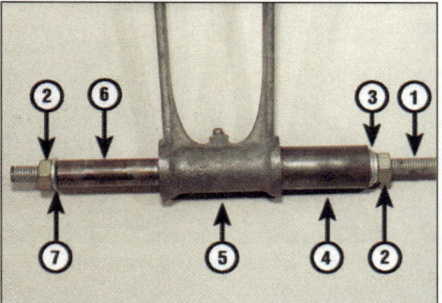

5.9 Drawbolt component parts assembled on a suspension arm

1 Bolt or length of threaded bar
2 Nuts
3 Washer (external diameter greater than tubing internal diameter)
4 Tubing (internal diameter sufficient to accommodate bearing)
5 Suspension arm with bearing
6 Tubing (external diameter slightly smaller than bearing)
7 Washer (external diameter slightly smaller than bearing)

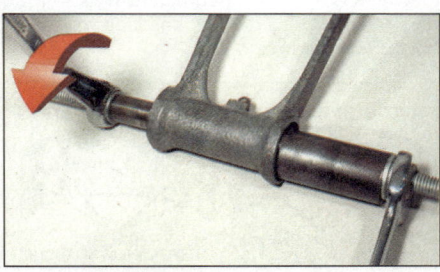

5.10 Drawing the bearing out of the suspension arm

● To extract the bearing/bush you will need a long bolt with nut (or piece of threaded bar with two nuts), a piece of tubing which has an internal diameter larger than the bearing/bush, another piece of tubing which has an external diameter slightly smaller than the bearing/bush, and a selection of washers **(see illustrations 5.9 and 5.10)**. Note that the pieces of tubing must be of the same length, or longer, than the bearing/bush.

● The same kit (without the pieces of tubing) can be used to draw the new bearing/bush back into place **(see illustration 5.11)**.

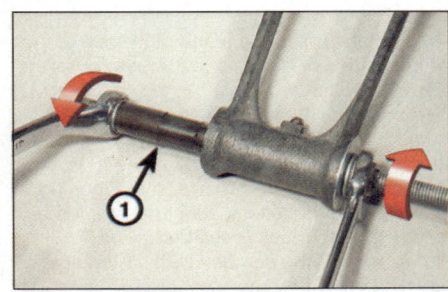

5.11 Installing a new bearing (1) in the suspension arm

Temperature change

● If the bearing's outer race is a tight fit in the casing, the aluminium casing can be heated to release its grip on the bearing. Aluminium will expand at a greater rate than the steel bearing outer race. There are several ways to do this, but avoid any localised extreme heat (such as a blow torch) - aluminium alloy has a low melting point.

● Approved methods of heating a casing are using a domestic oven (heated to 100°C) or immersing the casing in boiling water **(see illustration 5.12)**. Low temperature range localised heat sources such as a paint stripper heat gun or clothes iron can also be used **(see illustration 5.13)**. Alternatively, soak a rag in boiling water, wring it out and wrap it around the bearing housing.

> ⚠️ **Warning: All of these methods require care in use to prevent scalding and burns to the hands. Wear protective gloves when handling hot components.**

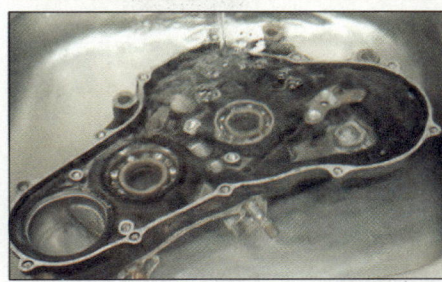

5.12 A casing can be immersed in a sink of boiling water to aid bearing removal

5.13 Using a localised heat source to aid bearing removal

● If heating the whole casing note that plastic components, such as the neutral switch, may suffer - remove them beforehand.

● After heating, remove the bearing as described above. You may find that the expansion is sufficient for the bearing to fall out of the casing under its own weight or with a light tap on the driver or socket.

● If necessary, the casing can be heated to aid bearing installation, and this is sometimes the recommended procedure if the motorcycle manufacturer has designed the housing and bearing fit with this intention.

REF•16 Tools and Workshop Tips

● Installation of bearings can be eased by placing them in a freezer the night before installation. The steel bearing will contract slightly, allowing easy insertion in its housing. This is often useful when installing steering head outer races in the frame.

Bearing types and markings

● Plain shell bearings, ball bearings, needle roller bearings and tapered roller bearings will all be found on motorcycles (see illustrations 5.14 and 5.15). The ball and roller types are usually caged between an inner and outer race, but uncaged variations may be found.

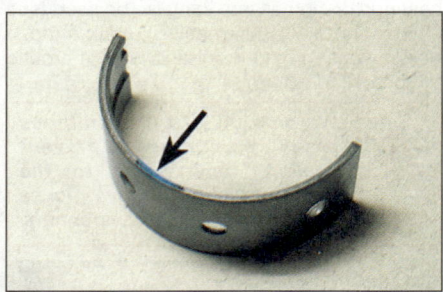

5.14 Shell bearings are either plain or grooved. They are usually identified by colour code (arrow)

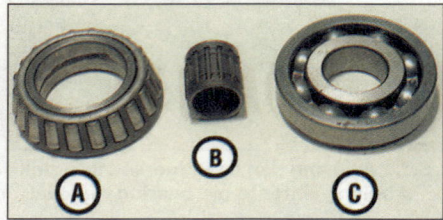

5.15 Tapered roller bearing (A), needle roller bearing (B) and ball journal bearing (C)

● Shell bearings (often called inserts) are usually found at the crankshaft main and connecting rod big-end where they are good at coping with high loads. They are made of a phosphor-bronze material and are impregnated with self-lubricating properties.

● Ball bearings and needle roller bearings consist of a steel inner and outer race with the balls or rollers between the races. They require constant lubrication by oil or grease and are good at coping with axial loads. Taper roller bearings consist of rollers set in a tapered cage set on the inner race; the outer race is separate. They are good at coping with axial loads and prevent movement along the shaft - a typical application is in the steering head.

● Bearing manufacturers produce bearings to ISO size standards and stamp one face of the bearing to indicate its internal and external diameter, load capacity and type (see illustration 5.16).

● Metal bushes are usually of phosphor-bronze material. Rubber bushes are used in suspension mounting eyes. Fibre bushes have also been used in suspension pivots.

5.16 Typical bearing marking

Bearing fault finding

● If a bearing outer race has spun in its housing, the housing material will be damaged. You can use a bearing locking compound to bond the outer race in place if damage is not too severe.

● Shell bearings will fail due to damage of their working surface, as a result of lack of lubrication, corrosion or abrasive particles in the oil (see illustration 5.17). Small particles of dirt in the oil may embed in the bearing material whereas larger particles will score the bearing and shaft journal. If a number of short journeys are made, insufficient heat will be generated to drive off condensation which has built up on the bearings.

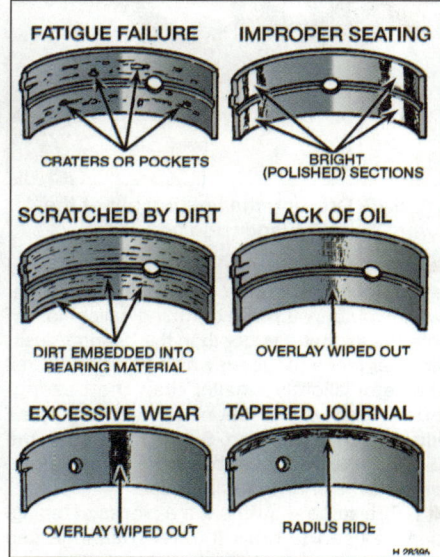

5.17 Typical bearing failures

● Ball and roller bearings will fail due to lack of lubrication or damage to the balls or rollers. Tapered-roller bearings can be damaged by overloading them. Unless the bearing is sealed on both sides, wash it in paraffin (kerosene) to remove all old grease then allow it to dry. Make a visual inspection looking to dented balls or rollers, damaged cages and worn or pitted races (see illustration 5.18).

● A ball bearing can be checked for wear by listening to it when spun. Apply a film of light oil to the bearing and hold it close to the ear - hold the outer race with one hand and spin the inner

5.18 Example of ball journal bearing with damaged balls and cages

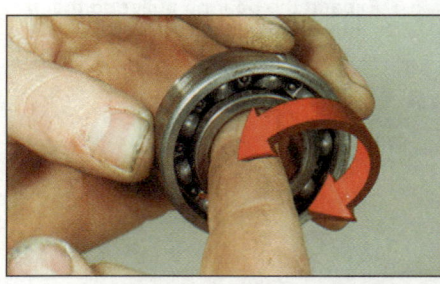

5.19 Hold outer race and listen to inner race when spun

race with the other hand (see illustration 5.19). The bearing should be almost silent when spun; if it grates or rattles it is worn.

6 Oil seals

Oil seal removal and installation

● Oil seals should be renewed every time a component is dismantled. This is because the seal lips will become set to the sealing surface and will not necessarily reseal.

● Oil seals can be prised out of position using a large flat-bladed screwdriver (see illustration 6.1). In the case of crankcase seals, check first that the seal is not lipped on the inside, preventing its removal with the crankcases joined.

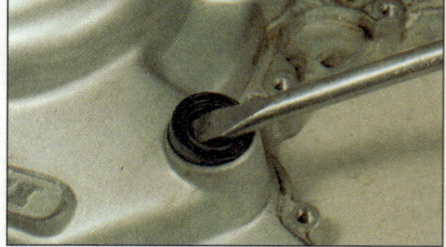

6.1 Prise out oil seals with a large flat-bladed screwdriver

● New seals are usually installed with their marked face (containing the seal reference code) outwards and the spring side towards the fluid being retained. In certain cases, such as a two-stroke engine crankshaft seal, a double lipped seal may be used due to there being fluid or gas on each side of the joint.

Tools and Workshop Tips   REF•17

● Use a bearing driver or socket which bears only on the outer hard edge of the seal to install it in the casing - tapping on the inner edge will damage the sealing lip.

Oil seal types and markings

● Oil seals are usually of the single-lipped type. Double-lipped seals are found where a liquid or gas is on both sides of the joint.
● Oil seals can harden and lose their sealing ability if the motorcycle has been in storage for a long period - renewal is the only solution.
● Oil seal manufacturers also conform to the ISO markings for seal size - these are moulded into the outer face of the seal (see illustration 6.2).

6.2 These oil seal markings indicate inside diameter, outside diameter and seal thickness

7 Gaskets and sealants

Types of gasket and sealant

● Gaskets are used to seal the mating surfaces between components and keep lubricants, fluids, vacuum or pressure contained within the assembly. Aluminium gaskets are sometimes found at the cylinder joints, but most gaskets are paper-based. If the mating surfaces of the components being joined are undamaged the gasket can be installed dry, although a dab of sealant or grease will be useful to hold it in place during assembly.
● RTV (Room Temperature Vulcanising) silicone rubber sealants cure when exposed to moisture in the atmosphere. These sealants are good at filling pits or irregular gasket faces, but will tend to be forced out of the joint under very high torque. They can be used to replace a paper gasket, but first make sure that the width of the paper gasket is not essential to the shimming of internal components. RTV sealants should not be used on components containing petrol (gasoline).
● Non-hardening, semi-hardening and hard setting liquid gasket compounds can be used with a gasket or between a metal-to-metal joint. Select the sealant to suit the application: universal non-hardening sealant can be used on virtually all joints; semi-hardening on joint faces which are rough or damaged; hard setting sealant on joints which require a permanent bond and are subjected to high temperature and pressure. **Note:** *Check first if the paper gasket has a bead of sealant impregnated in its surface before applying additional sealant.*
● When choosing a sealant, make sure it is suitable for the application, particularly if being applied in a high-temperature area or in the vicinity of fuel. Certain manufacturers produce sealants in either clear, silver or black colours to match the finish of the engine. This has a particular application on motorcycles where much of the engine is exposed.
● Do not over-apply sealant. That which is squeezed out on the outside of the joint can be wiped off, whereas an excess of sealant on the inside can break off and clog oilways.

Breaking a sealed joint

● Age, heat, pressure and the use of hard setting sealant can cause two components to stick together so tightly that they are difficult to separate using finger pressure alone. Do not resort to using levers unless there is a pry point provided for this purpose (see illustration 7.1) or else the gasket surfaces will be damaged.
● Use a soft-faced hammer (see illustration 7.2) or a wood block and conventional hammer to strike the component near the mating surface. Avoid hammering against cast extremities since they may break off. If this method fails, try using a wood wedge between the two components.

Caution: If the joint will not separate, double-check that you have removed all the fasteners.

7.1 If a pry point is provided, apply gently pressure with a flat-bladed screwdriver

7.2 Tap around the joint with a soft-faced mallet if necessary - don't strike cooling fins

Removal of old gasket and sealant

● Paper gaskets will most likely come away complete, leaving only a few traces stuck on

Most components have one or two hollow locating dowels between the two gasket faces. If a dowel cannot be removed, do not resort to gripping it with pliers - it will almost certainly be distorted. Install a close-fitting socket or Phillips screwdriver into the dowel and then grip the outer edge of the dowel to free it.

the sealing faces of the components. It is imperative that all traces are removed to ensure correct sealing of the new gasket.
● Very carefully scrape all traces of gasket away making sure that the sealing surfaces are not gouged or scored by the scraper (see illustrations 7.3, 7.4 and 7.5). Stubborn deposits can be removed by spraying with an aerosol gasket remover. Final preparation of

7.3 Paper gaskets can be scraped off with a gasket scraper tool . . .

7.4 . . . a knife blade . . .

7.5 . . . or a household scraper

REF•18 Tools and Workshop Tips

7.6 Fine abrasive paper is wrapped around a flat file to clean up the gasket face

7.7 A kitchen scourer can be used on stubborn deposits

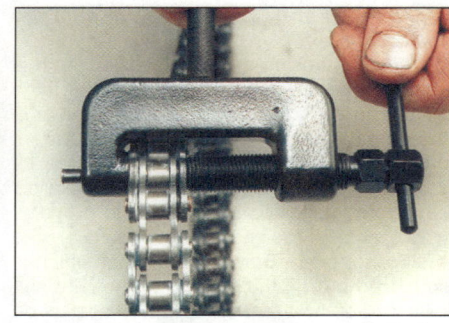

8.1 Tighten the chain breaker to push the pin out of the link . . .

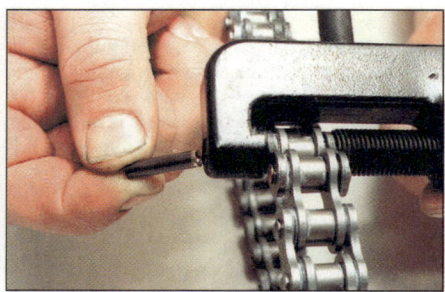

8.2 . . . withdraw the pin, remove the tool . . .

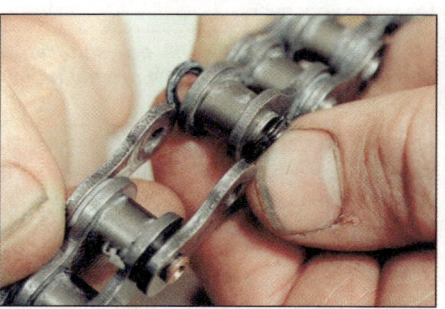

8.3 . . . and separate the chain link

8.4 Insert the new soft link, with O-rings, through the chain ends . . .

8.5 . . . install the O-rings over the pin ends . . .

8.6 . . . followed by the sideplate

8.7 Push the sideplate into position using a clamp

the gasket surface can be made with very fine abrasive paper or a plastic kitchen scourer **(see illustrations 7.6 and 7.7)**.
● Old sealant can be scraped or peeled off components, depending on the type originally used. Note that gasket removal compounds are available to avoid scraping the components clean; make sure the gasket remover suits the type of sealant used.

8 Chains

Breaking and joining final drive chains

● Drive chains for all but small bikes are continuous and do not have a clip-type connecting link. The chain must be broken using a chain breaker tool and the new chain securely riveted together using a new soft rivet-type link. Never use a clip-type connecting link instead of a rivet-type link, except in an emergency. Various chain breaking and riveting tools are available, either as separate tools or combined as illustrated in the accompanying photographs - read the instructions supplied with the tool carefully.

> **Warning:** *The need to rivet the new link pins correctly cannot be overstressed - loss of control of the motorcycle is very likely to result if the chain breaks in use.*

● Rotate the chain and look for the soft link. The soft link pins look like they have been deeply centre-punched instead of peened over like all the other pins **(see illustration 8.9)** and its sideplate may be a different colour. Position the soft link midway between the sprockets and assemble the chain breaker tool over one of the soft link pins **(see illustration 8.1)**. Operate the tool to push the pin out through the chain **(see illustration 8.2)**. On an O-ring chain, remove the O-rings **(see illustration 8.3)**. Carry out the same procedure on the other soft link pin.

> *Caution: Certain soft link pins (particularly on the larger chains) may require their ends to be filed or ground off before they can be pressed out using the tool.*

● Check that you have the correct size and strength (standard or heavy duty) new soft link - do not reuse the old link. Look for the size marking on the chain sideplates **(see illustration 8.10)**.
● Position the chain ends so that they are engaged over the rear sprocket. On an O-ring chain, install a new O-ring over each pin of the link and insert the link through the two chain ends **(see illustration 8.4)**. Install a new O-ring over the end of each pin, followed by the sideplate (with the chain manufacturer's marking facing outwards) **(see illustrations 8.5 and 8.6)**. On an unsealed chain, insert the link through the two chain ends, then install the sideplate with the chain manufacturer's marking facing outwards.
● Note that it may not be possible to install the sideplate using finger pressure alone. If using a joining tool, assemble it so that the plates of the tool clamp the link and press the sideplate over the pins **(see illustration 8.7)**. Otherwise, use two small sockets placed over

Tools and Workshop Tips REF•19

8.8 Assemble the chain riveting tool over one pin at a time and tighten it fully

8.9 Pin end correctly riveted (A), pin end unriveted (B)

the rivet ends and two pieces of the wood between a G-clamp. Operate the clamp to press the sideplate over the pins.
● Assemble the joining tool over one pin (following the maker's instructions) and tighten the tool down to spread the pin end securely **(see illustrations 8.8 and 8.9)**. Do the same on the other pin.

 **Warning: Check that the pin ends are secure and that there is no danger of the sideplate coming loose. If the pin ends are cracked the soft link must be renewed.**

Final drive chain sizing

● Chains are sized using a three digit number, followed by a suffix to denote the chain type **(see illustration 8.10)**. Chain type is either standard or heavy duty (thicker sideplates), and also unsealed or O-ring/X-ring type.
● The first digit of the number relates to the pitch of the chain, ie the distance from the centre of one pin to the centre of the next pin **(see illustration 8.11)**. Pitch is expressed in eighths of an inch, as follows:

8.10 Typical chain size and type marking

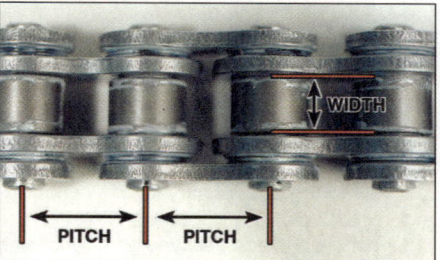

8.11 Chain dimensions

Sizes commencing with a 4 (eg 428) have a pitch of 1/2 inch (12.7 mm)

Sizes commencing with a 5 (eg 520) have a pitch of 5/8 inch (15.9 mm)

Sizes commencing with a 6 (eg 630) have a pitch of 3/4 inch (19.1 mm)

● The second and third digits of the chain size relate to the width of the rollers, again in imperial units, eg the 525 shown has 5/16 inch (7.94 mm) rollers **(see illustration 8.11)**.

9 Hoses

Clamping to prevent flow

● Small-bore flexible hoses can be clamped to prevent fluid flow whilst a component is worked on. Whichever method is used, ensure that the hose material is not permanently distorted or damaged by the clamp.
a) A brake hose clamp available from auto accessory shops **(see illustration 9.1)**.
b) A wingnut type hose clamp **(see illustration 9.2)**.

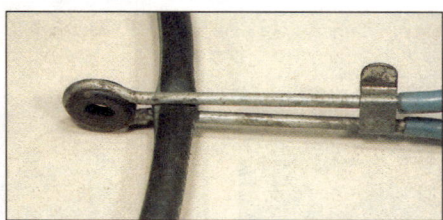

9.1 Hoses can be clamped with an automotive brake hose clamp . . .

9.2 . . . a wingnut type hose clamp . . .

c) Two sockets placed each side of the hose and held with straight-jawed self-locking grips **(see illustration 9.3)**.
d) Thick card each side of the hose held between straight-jawed self-locking grips **(see illustration 9.4)**.

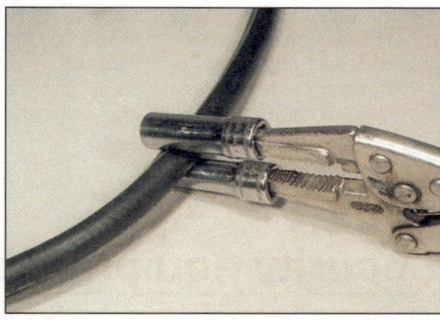

9.3 . . . two sockets and a pair of self-locking grips . . .

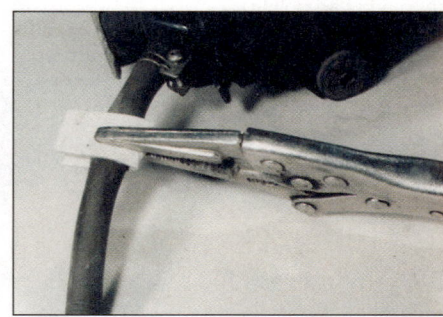

9.4 . . . or thick card and self-locking grips

Freeing and fitting hoses

● Always make sure the hose clamp is moved well clear of the hose end. Grip the hose with your hand and rotate it whilst pulling it off the union. If the hose has hardened due to age and will not move, slit it with a sharp knife and peel its ends off the union **(see illustration 9.5)**.
● Resist the temptation to use grease or soap on the unions to aid installation; although it helps the hose slip over the union it will equally aid the escape of fluid from the joint. It is preferable to soften the hose ends in hot water and wet the inside surface of the hose with water or a fluid which will evaporate.

9.5 Cutting a coolant hose free with a sharp knife

Security

Introduction

In less time than it takes to read this introduction, a thief could steal your motorcycle. Returning only to find your bike has gone is one of the worst feelings in the world. Even if the motorcycle is insured against theft, once you've got over the initial shock, you will have the inconvenience of dealing with the police and your insurance company.

The motorcycle is an easy target for the professional thief and the joyrider alike and the official figures on motorcycle theft make for depressing reading; on average a motorcycle is stolen every 16 minutes in the UK!

Motorcycle thefts fall into two categories, those stolen 'to order' and those taken by opportunists. The thief stealing to order will be on the look out for a specific make and model and will go to extraordinary lengths to obtain that motorcycle. The opportunist thief on the other hand will look for easy targets which can be stolen with the minimum of effort and risk.

Whilst it is never going to be possible to make your machine 100% secure, it is estimated that around half of all stolen motorcycles are taken by opportunist thieves. Remember that the opportunist thief is always on the look out for the easy option: if there are two similar motorcycles parked side-by-side, they will target the one with the lowest level of security. By taking a few precautions, you can reduce the chances of your motorcycle being stolen.

Security equipment

There are many specialised motorcycle security devices available and the following text summarises their applications and their good and bad points.

Once you have decided on the type of security equipment which best suits your needs, we recommended that you read one of the many equipment tests regularly carried out by the motorcycle press. These tests compare the products from all the major manufacturers and give impartial ratings on their effectiveness, value-for-money and ease of use.

No one item of security equipment can provide complete protection. It is highly recommended that two or more of the items described below are combined to increase the security of your motorcycle (a lock and chain plus an alarm system is just about ideal). The more security measures fitted to the bike, the less likely it is to be stolen.

Lock and chain

Pros: *Very flexible to use; can be used to secure the motorcycle to almost any immovable object. On some locks and chains, the lock can be used on its own as a disc lock (see below).*

Cons: *Can be very heavy and awkward to carry on the motorcycle, although some types will be supplied with a carry bag which can be strapped to the pillion seat.*

● Heavy-duty chains and locks are an excellent security measure **(see illustration 1)**. Whenever the motorcycle is parked, use the lock and chain to secure the machine to a solid, immovable object such as a post or railings. This will prevent the machine from being ridden away or being lifted into the back of a van.

● When fitting the chain, always ensure the chain is routed around the motorcycle frame or swingarm **(see illustrations 2 and 3)**. Never merely pass the chain around one of the wheel rims; a thief may unbolt the wheel and lift the rest of the machine into a van, leaving you with just the wheel! Try to avoid having excess chain free, thus making it difficult to use cutting tools, and keep the chain and lock off the ground to prevent thieves attacking it with a cold chisel. Position the lock so that its lock barrel is facing downwards; this will make it harder for the thief to attack the lock mechanism.

Ensure the lock and chain you buy is of good quality and long enough to shackle your bike to a solid object

Pass the chain through the bike's frame, rather than just through a wheel . . .

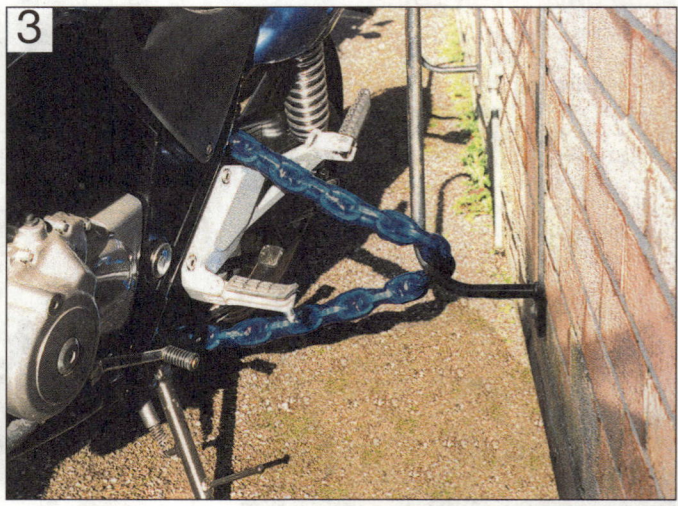

. . . and loop it around a solid object

Security REF•21

U-locks

Pros: *Highly effective deterrent which can be used to secure the bike to a post or railings. Most U-locks come with a carrier which allows the lock to be easily carried on the bike.*

Cons: *Not as flexible to use as a lock and chain.*

● These are solid locks which are similar in use to a lock and chain. U-locks are lighter than a lock and chain but not so flexible to use. The length and shape of the lock shackle limit the objects to which the bike can be secured **(see illustration 4)**.

Disc locks

Pros: *Small, light and very easy to carry; most can be stored underneath the seat.*

Cons: *Does not prevent the motorcycle being lifted into a van. Can be very embarrassing if you forget to remove the lock before attempting to ride off!*

● Disc locks are designed to be attached to the front brake disc. The lock passes through one of the holes in the disc and prevents the wheel rotating by jamming against the fork/brake caliper **(see illustration 5)**. Some are equipped with an alarm siren which sounds if the disc lock is moved; this not only acts as a theft deterrent but also as a handy reminder if you try to move the bike with the lock still fitted.

● Combining the disc lock with a length of cable which can be looped around a post or railings provides an additional measure of security **(see illustration 6)**.

Alarms and immobilisers

Pros: *Once installed it is completely hassle-free to use. If the system is 'Thatcham' or 'Sold Secure-approved', insurance companies may give you a discount.*

Cons: *Can be expensive to buy and complex to install. No system will prevent the motorcycle from being lifted into a van and taken away.*

● Electronic alarms and immobilisers are available to suit a variety of budgets. There are three different types of system available: pure alarms, pure immobilisers, and the more expensive systems which are combined alarm/immobilisers **(see illustration 7)**.

● An alarm system is designed to emit an audible warning if the motorcycle is being tampered with.

● An immobiliser prevents the motorcycle being started and ridden away by disabling its electrical systems.

● When purchasing an alarm/immobiliser system, check the cost of installing the system unless you are able to do it yourself. If the motorcycle is not used regularly, another consideration is the current drain of the system. All alarm/immobiliser systems are powered by the motorcycle's battery; purchasing a system with a very low current drain could prevent the battery losing its charge whilst the motorcycle is not being used.

U-locks can be used to secure the bike to a solid object – ensure you purchase one which is long enough

A typical disc lock attached through one of the holes in the disc

A disc lock combined with a security cable provides additional protection

A typical alarm/immobiliser system

REF•22 Security

Indelible markings can be applied to most areas of the bike – always apply the manufacturer's sticker to warn off thieves

Chemically-etched code numbers can be applied to main body panels . . .

. . . again, always ensure that the kit manufacturer's sticker is applied in a prominent position

Security marking kits

Pros: Very cheap and effective deterrent. Many insurance companies will give you a discount on your insurance premium if a recognised security marking kit is used on your motorcycle.

Cons: Does not prevent the motorcycle being stolen by joyriders.

● There are many different types of security marking kits available. The idea is to mark as many parts of the motorcycle as possible with a unique security number **(see illustrations 8, 9 and 10)**. A form will be included with the kit to register your personal details and those of the motorcycle with the kit manufacturer. This register is made available to the police to help them trace the rightful owner of any motorcycle or components which they recover should all other forms of identification have been removed. Always apply the warning stickers provided with the kit to deter thieves.

Ground anchors, wheel clamps and security posts

Pros: An excellent form of security which will deter all but the most determined of thieves.

Cons: Awkward to install and can be expensive.

● Whilst the motorcycle is at home, it is a good idea to attach it securely to the floor or a solid wall, even if it is kept in a securely locked garage. Various types of ground anchors, security posts and wheel clamps are available for this purpose **(see illustration 11)**. These security devices are either bolted to a solid concrete or brick structure or can be cemented into the ground.

Permanent ground anchors provide an excellent level of security when the bike is at home

Security at home

A high percentage of motorcycle thefts are from the owner's home. Here are some things to consider whenever your motorcycle is at home:

✔ Where possible, always keep the motorcycle in a securely locked garage. Never rely solely on the standard lock on the garage door, these are usual hopelessly inadequate. Fit an additional locking mechanism to the door and consider having the garage alarmed. A security light, activated by a movement sensor, is also a good investment.

✔ Always secure the motorcycle to the ground or a wall, even if it is inside a securely locked garage.

✔ Do not regularly leave the motorcycle outside your home, try to keep it out of sight wherever possible. If a garage is not available, fit a motorcycle cover over the bike to disguise its true identity.

✔ It is not uncommon for thieves to follow a motorcyclist home to find out where the bike is kept. They will then return at a later date. Be aware of this whenever you are returning home on your motorcycle. If you suspect you are being followed, do not return home, instead ride to a garage or shop and stop as a precaution.

✔ When selling a motorcycle, do not provide your home address or the location where the bike is normally kept. Arrange to meet the buyer at a location away from your home. Thieves have been known to pose as potential buyers to find out where motorcycles are kept and then return later to steal them.

Security away from the home

As well as fitting security equipment to your motorcycle here are a few general rules to follow whenever you park your motorcycle.

✔ Park in a busy, public place.

✔ Use car parks which incorporate security features, such as CCTV.

✔ At night, park in a well-lit area, preferably directly underneath a street light.

✔ Engage the steering lock.

✔ Secure the motorcycle to a solid, immovable object such as a post or railings with an additional lock. If this is not possible, secure the bike to a friend's motorcycle. Some public parking places provide security loops for motorcycles.

✔ Never leave your helmet or luggage attached to the motorcycle. Take them with you at all times.

Lubricants and fluids

A wide range of lubricants, fluids and cleaning agents is available for motor-cycles. This is a guide as to what is available, its applications and properties.

Four-stroke engine oil

● Engine oil is without doubt the most important component of any four-stroke engine. Modern motorcycle engines place a lot of demands on their oil and choosing the right type is essential. Using an unsuitable oil will lead to an increased rate of engine wear and could result in serious engine damage. Before purchasing oil, always check the recommended oil specification given by the manufacturer. The manufacturer will state a recommended 'type or classification' and also a specific 'viscosity' range for engine oil.

● The oil 'type or classification' is identified by its API (American Petroleum Institute) rating. The API rating will be in the form of two letters, e.g. SG. The S identifies the oil as being suitable for use in a petrol (gasoline) engine (S stands for spark ignition) and the second letter, ranging from A to J, identifies the oil's performance rating. The later this letter, the higher the specification of the oil; for example API SG oil exceeds the requirements of API SF oil. **Note:** *On some oils there may also be a second rating consisting of another two letters, the first letter being C, e.g. API SF/CD. This rating indicates the oil is also suitable for use in a diesel engines (the C stands for compression ignition) and is thus of no relevance for motorcycle use.*

● The 'viscosity' of the oil is identified by its SAE (Society of Automotive Engineers) rating. All modern engines require multigrade oils and the SAE rating will consist of two numbers, the first followed by a W, e.g. 10W/40. The first number indicates the viscosity rating of the oil at low temperatures (W stands for winter – tested at –20°C) and the second number represents the viscosity of the oil at high temperatures (tested at 100°C). The lower the number, the thinner the oil. For example an oil with an SAE 10W/40 rating will give better cold starting and running than an SAE 15W/40 oil.

● As well as ensuring the 'type' and 'viscosity' of the oil match the recommendations, another consideration to make when buying engine oil is whether to purchase a standard mineral-based oil, a semi-synthetic oil (also known as a synthetic blend or synthetic-based oil) or a fully-synthetic oil. Although all oils will have a similar rating and viscosity, their cost will vary considerably; mineral-based oils are the cheapest, the fully-synthetic oils the most expensive with the semi-synthetic oils falling somewhere in-between. This decision is very much up to the owner, but it should be noted that modern synthetic oils have far better lubricating and cleaning qualities than traditional mineral-based oils and tend to retain these properties for far longer. Bearing in mind the operating conditions inside a modern, high-revving motorcycle engine it is highly recommended that a fully synthetic oil is used. The extra expense at each service could save you money in the long term by preventing premature engine wear.

● As a final note always ensure that the oil is specifically designed for use in motorcycle engines. Engine oils designed primarily for use in car engines sometimes contain additives or friction modifiers which could cause clutch slip on a motorcycle fitted with a wet-clutch.

Two-stroke engine oil

● Modern two-stroke engines, with their high power outputs, place high demands on their oil. If engine seizure is to be avoided it is essential that a high-quality oil is used. Two-stroke oils differ hugely from four-stroke oils. The oil lubricates only the crankshaft and piston(s) (the transmission has its own lubricating oil) and is used on a total-loss basis where it is burnt completely during the combustion process.

● The Japanese have recently introduced a classification system for two-stroke oils, the JASO rating. This rating is in the form of two letters, either FA, FB or FC – FA is the lowest classification and FC the highest. Ensure the oil being used meets or exceeds the recommended rating specified by the manufacturer.

● As well as ensuring the oil rating matches the recommendation, another consideration to make when buying engine oil is whether to purchase a standard mineral-based oil, a semi-synthetic oil (also known as a synthetic blend or synthetic-based oil) or a fully-synthetic oil. The cost of each type of oil varies considerably; mineral-based oils are the cheapest, the fully-synthetic oils the most expensive with the semi-synthetic oils falling somewhere in-between. This decision is very much up to the owner, but it should be noted that modern synthetic oils have far better lubricating properties and burn cleaner than traditional mineral-based oils. It is therefore recommended that a fully synthetic oil is used. The extra expense could save you money in the long term by preventing premature engine wear, engine performance will be improved, carbon deposits and exhaust smoke will be reduced.

REF•24 Lubricants and fluids

● Always ensure that the oil is specifically designed for use in an injector system. Many high quality two-stroke oils are designed for competition use and need to be pre-mixed with fuel. These oils are of a much higher viscosity and are not designed to flow through the injector pumps used on road-going two-stroke motorcycles.

Transmission (gear) oil

● On a two-stroke engine, the transmission and clutch are lubricated by their own separate oil bath which must be changed in accordance with the Maintenance Schedule.
● Although the engine and transmission units of most four-strokes use a common lubrication supply, there are some exceptions where the engine and gearbox have separate oil reservoirs and a dry clutch is used.
● Motorcycle manufacturers will either recommend a monograde transmission oil or a four-stroke multigrade engine oil to lubricate the transmission.
● Transmission oils, or gear oils as they are often called, are designed specifically for use in transmission systems. The viscosity of these oils is represented by an SAE number, but the scale of measurement applied is different to that used to grade engine oils. As a rough guide a SAE90 gear oil will be of the same viscosity as an SAE50 engine oil.

Shaft drive oil

● On models equipped with shaft final drive, the shaft drive gears are will have their own oil supply. The manufacturer will state a recommended 'type or classification' and also a specific 'viscosity' range in the same manner as for four-stroke engine oil.
● Gear oil classification is given by the number which follows the API GL (GL standing for gear lubricant) rating, the higher the number, the higher the specification of the oil, e.g. API GL5 oil is a higher specification than API GL4 oil. Ensure the oil meets or exceeds the classification specified and is of the correct viscosity. The viscosity of gear oils is also represented by an SAE number but the scale of measurement used is different to that used to grade engine oils. As a rough guide an SAE90 gear oil will be of the same viscosity as an SAE50 engine oil.
● If the use of an EP (Extreme Pressure) gear oil is specified, ensure the oil purchased is suitable.

Fork oil and suspension fluid

● Conventional telescopic front forks are hydraulic and require fork oil to work. To ensure the forks function correctly, the fork oil must be changed in accordance with the Maintenance Schedule.
● Fork oil is available in a variety of viscosities, identified by their SAE rating; fork oil ratings vary from light (SAE 5) to heavy (SAE 30). When purchasing fork oil, ensure the viscosity rating matches that specified by the manufacturer.
● Some lubricant manufacturers also produce a range of high-quality suspension fluids which are very similar to fork oil but are designed mainly for competition use. These fluids may have a different viscosity rating system which is not to be confused with the SAE rating of normal fork oil. Refer to the manufacturer's instructions if in any doubt.

Brake and clutch fluid

● All disc brake systems and some clutch systems are hydraulically operated. To ensure correct operation, the hydraulic fluid must be changed in accordance with the Maintenance Schedule.
● Brake and clutch fluid is classified by its DOT rating with most motorcycle manufacturers specifying DOT 3 or 4 fluid. Both fluid types are glycol-based and can be mixed together without adverse effect; DOT 4 fluid exceeds the requirements of DOT 3 fluid. Although it is safe to use DOT 4 fluid in a system designed for use with DOT 3 fluid, never use DOT 3 fluid in a system which specifies the use of DOT 4 as this will adversely affect the system's performance. The type required for the system will be marked on the fluid reservoir cap.
● Some manufacturers also produce a DOT 5 hydraulic fluid. DOT 5 hydraulic fluid is silicone-based and is not compatible with the glycol-based DOT 3 and 4 fluids. Never mix DOT 5 fluid with DOT 3 or 4 fluid as this will seriously affect the performance of the hydraulic system.

Coolant/antifreeze

● When purchasing coolant/antifreeze, always ensure it is suitable for use in an aluminium engine and contains corrosion inhibitors to prevent possible blockages of the internal coolant passages of the system. As a general rule, most coolants are designed to be used neat and should not be diluted whereas antifreeze can be mixed with distilled water to provide a coolant solution of the required strength. Refer to the manufacturer's instructions on the bottle.
● Ensure the coolant is changed in accordance with the Maintenance Schedule.

Chain lube

● Chain lube is an aerosol-type spray lubricant specifically designed for use on motorcycle final drive chains. Chain lube has two functions, to minimise friction between the final drive chain and sprockets and to prevent corrosion of the chain. Regular use of a good-quality chain lube will extend the life of the drive chain and sprockets and thus maximise the power being transmitted from the transmission to the rear wheel.
● When using chain lube, always allow some time for the solvents in the lube to evaporate before riding the motorcycle. This will minimise the amount of lube which will

Lubricants and fluids REF•25

'fling' off from the chain when the motorcycle is used. If the motorcycle is equipped with an 'O-ring' chain, ensure the chain lube is labelled as being suitable for use on 'O-ring' chains.

Degreasers and solvents

● There are many different types of solvents and degreasers available to remove the grime and grease which accumulate around the motorcycle during normal use. Degreasers and solvents are usually available as an aerosol-type spray or as a liquid which you apply with a brush. Always closely follow the manufacturer's instructions and wear eye protection during use. Be aware that many solvents are flammable and may give off noxious fumes; take adequate precautions when using them (see *Safety First!*).

● For general cleaning, use one of the many solvents or degreasers available from most motorcycle accessory shops. These solvents are usually applied then left for a certain time before being washed off with water.

Brake cleaner is a solvent specifically designed to remove all traces of oil, grease and dust from braking system components. Brake cleaner is designed to evaporate quickly and leaves behind no residue.

Carburettor cleaner is an aerosol-type solvent specifically designed to clear carburettor blockages and break down the hard deposits and gum often found inside carburettors during overhaul.

Contact cleaner is an aerosol-type solvent designed for cleaning electrical components. The cleaner will remove all traces of oil and dirt from components such as switch contacts or fouled spark plugs and then dry, leaving behind no residue.

Gasket remover is an aerosol-type solvent designed for removing stubborn gaskets from engine components during overhaul. Gasket remover will minimise the amount of scraping required to remove the gasket and therefore reduce the risk of damage to the mating surface.

Spray lubricants

● Aerosol-based spray lubricants are widely available and are excellent for lubricating lever pivots and exposed cables and switches. Try to use a lubricant which is of the dry-film type as the fluid evaporates, leaving behind a dry-film of lubricant. Lubricants which leave behind an oily residue will attract dust and dirt which will increase the rate of wear of the cable/lever.

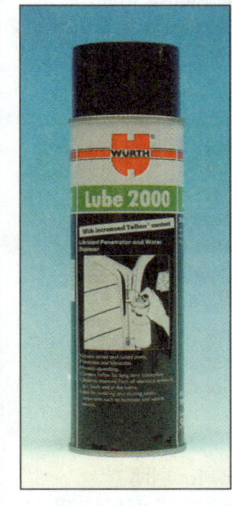

● Most lubricants also act as a moisture dispersant and a penetrating fluid. This means they can also be used to 'dry out' electrical components such as wiring connectors or switches as well as helping to free seized fasteners.

Greases

● Grease is used to lubricate many of the pivot-points. A good-quality multi-purpose grease is suitable for most applications but some manufacturers will specify the use of specialist greases for use on components such as swingarm and suspension linkage bushes. These specialist greases can be purchased from most motorcycle (or car) accessory shops; commonly specified types include molybdenum disulphide grease, lithium-based grease, graphite-based grease, silicone-based grease and high-temperature copper-based grease.

Gasket sealing compounds

● Gasket sealing compounds can be used in conjunction with gaskets, to improve their sealing capabilities, or on their own to seal metal-to-metal joints. Depending on their type, sealing compounds either set hard or stay relatively soft and pliable.

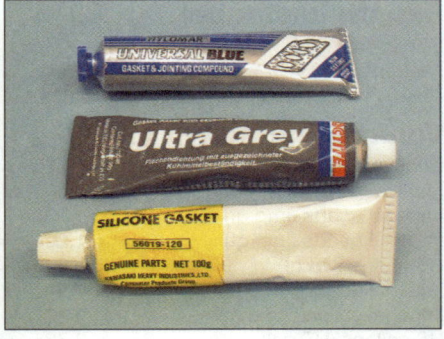

● When purchasing a gasket sealing compound, ensure that it is designed specifically for use on an internal combustion engine. General multi-purpose sealants available from DIY stores may appear visibly similar but they are not designed to withstand the extreme heat or contact with fuel and oil encountered when used on an engine (see *'Tools and Workshop Tips'* for further information).

Thread locking compound

● Thread locking compounds are used to secure certain threaded fasteners in position to prevent them from loosening due to vibration. Thread locking compounds can be purchased from most motorcycle (and car) accessory shops. Ensure the threads of the both components are completely clean and dry before sparingly applying the locking compound (see *'Tools and Workshop Tips'* for further information).

Fuel additives

● Fuel additives which protect and clean the fuel system components are widely available. These additives are designed to remove all traces of deposits that build up on the carburettors/injectors and prevent wear, helping the fuel system to operate more efficiently. If a fuel additive is being used, check that it is suitable for use with your motorcycle, especially if your motorcycle is equipped with a catalytic converter.

● Octane boosters are also available. These additives are designed to improve the performance of highly-tuned engines being run on normal pump-fuel and are of no real use on standard motorcycles.

Conversion Factors

Length (distance)
Inches (in)	x 25.4	= Millimetres (mm)	x 0.0394	= Inches (in)
Feet (ft)	x 0.305	= Metres (m)	x 3.281	= Feet (ft)
Miles	x 1.609	= Kilometres (km)	x 0.621	= Miles

Volume (capacity)
Cubic inches (cu in; in^3)	x 16.387	= Cubic centimetres (cc; cm^3)	x 0.061	= Cubic inches (cu in; in^3)
Imperial pints (Imp pt)	x 0.568	= Litres (l)	x 1.76	= Imperial pints (Imp pt)
Imperial quarts (Imp qt)	x 1.137	= Litres (l)	x 0.88	= Imperial quarts (Imp qt)
Imperial quarts (Imp qt)	x 1.201	= US quarts (US qt)	x 0.833	= Imperial quarts (Imp qt)
US quarts (US qt)	x 0.946	= Litres (l)	x 1.057	= US quarts (US qt)
Imperial gallons (Imp gal)	x 4.546	= Litres (l)	x 0.22	= Imperial gallons (Imp gal)
Imperial gallons (Imp gal)	x 1.201	= US gallons (US gal)	x 0.833	= Imperial gallons (Imp gal)
US gallons (US gal)	x 3.785	= Litres (l)	x 0.264	= US gallons (US gal)

Mass (weight)
Ounces (oz)	x 28.35	= Grams (g)	x 0.035	= Ounces (oz)
Pounds (lb)	x 0.454	= Kilograms (kg)	x 2.205	= Pounds (lb)

Force
Ounces-force (ozf; oz)	x 0.278	= Newtons (N)	x 3.6	= Ounces-force (ozf; oz)
Pounds-force (lbf; lb)	x 4.448	= Newtons (N)	x 0.225	= Pounds-force (lbf; lb)
Newtons (N)	x 0.1	= Kilograms-force (kgf; kg)	x 9.81	= Newtons (N)

Pressure
Pounds-force per square inch (psi; lbf/in^2; lb/in^2)	x 0.070	= Kilograms-force per square centimetre (kgf/cm^2; kg/cm^2)	x 14.223	= Pounds-force per square inch (psi; lbf/in^2; lb/in^2)
Pounds-force per square inch (psi; lbf/in^2; lb/in^2)	x 0.068	= Atmospheres (atm)	x 14.696	= Pounds-force per square inch (psi; lbf/in^2; lb/in^2)
Pounds-force per square inch (psi; lbf/in^2; lb/in^2)	x 0.069	= Bars	x 14.5	= Pounds-force per square inch (psi; lbf/in^2; lb/in^2)
Pounds-force per square inch (psi; lbf/in^2; lb/in^2)	x 6.895	= Kilopascals (kPa)	x 0.145	= Pounds-force per square inch (psi; lbf/in^2; lb/in^2)
Kilopascals (kPa)	x 0.01	= Kilograms-force per square centimetre (kgf/cm^2; kg/cm^2)	x 98.1	= Kilopascals (kPa)
Millibar (mbar)	x 100	= Pascals (Pa)	x 0.01	= Millibar (mbar)
Millibar (mbar)	x 0.0145	= Pounds-force per square inch (psi; lbf/in^2; lb/in^2)	x 68.947	= Millibar (mbar)
Millibar (mbar)	x 0.75	= Millimetres of mercury (mmHg)	x 1.333	= Millibar (mbar)
Millibar (mbar)	x 0.401	= Inches of water (inH$_2$O)	x 2.491	= Millibar (mbar)
Millimetres of mercury (mmHg)	x 0.535	= Inches of water (inH$_2$O)	x 1.868	= Millimetres of mercury (mmHg)
Inches of water (inH$_2$O)	x 0.036	= Pounds-force per square inch (psi; lbf/in^2; lb/in^2)	x 27.68	= Inches of water (inH$_2$O)

Torque (moment of force)
Pounds-force inches (lbf in; lb in)	x 1.152	= Kilograms-force centimetre (kgf cm; kg cm)	x 0.868	= Pounds-force inches (lbf in; lb in)
Pounds-force inches (lbf in; lb in)	x 0.113	= Newton metres (Nm)	x 8.85	= Pounds-force inches (lbf in; lb in)
Pounds-force inches (lbf in; lb in)	x 0.083	= Pounds-force feet (lbf ft; lb ft)	x 12	= Pounds-force inches (lbf in; lb in)
Pounds-force feet (lbf ft; lb ft)	x 0.138	= Kilograms-force metres (kgf m; kg m)	x 7.233	= Pounds-force feet (lbf ft; lb ft)
Pounds-force feet (lbf ft; lb ft)	x 1.356	= Newton metres (Nm)	x 0.738	= Pounds-force feet (lbf ft; lb ft)
Newton metres (Nm)	x 0.102	= Kilograms-force metres (kgf m; kg m)	x 9.804	= Newton metres (Nm)

Power
Horsepower (hp)	x 745.7	= Watts (W)	x 0.0013	= Horsepower (hp)

Velocity (speed)
Miles per hour (miles/hr; mph)	x 1.609	= Kilometres per hour (km/hr; kph)	x 0.621	= Miles per hour (miles/hr; mph)

Fuel consumption*
Miles per gallon (mpg)	x 0.354	= Kilometres per litre (km/l)	x 2.825	= Miles per gallon (mpg)

Temperature

Degrees Fahrenheit = (°C x 1.8) + 32   Degrees Celsius (Degrees Centigrade; °C) = (°F - 32) x 0.56

It is common practice to convert from miles per gallon (mpg) to litres/100 kilometres (l/100km), where mpg x l/100 km = 282

MOT Test Checks REF•27

About the MOT Test

In the UK, all vehicles more than three years old are subject to an annual test to ensure that they meet minimum safety requirements. A current test certificate must be issued before a machine can be used on public roads, and is required before a road fund licence can be issued. Riding without a current test certificate will also invalidate your insurance.

For most owners, the MOT test is an annual cause for anxiety, and this is largely due to owners not being sure what needs to be checked prior to submitting the motorcycle for testing. The simple answer is that a fully roadworthy motorcycle will have no difficulty in passing the test.

This is a guide to getting your motorcycle through the MOT test. Obviously it will not be possible to examine the motorcycle to the same standard as the professional MOT tester, particularly in view of the equipment required for some of the checks. However, working through the following procedures will enable you to identify any problem areas before submitting the motorcycle for the test.

It has only been possible to summarise the test requirements here, based on the regulations in force at the time of printing. Test standards are becoming increasingly stringent, although there are some exemptions for older vehicles. More information about the MOT test can be obtained from the TSO publications, *How Safe is your Motorcycle* and *The MOT Inspection Manual for Motorcycle Testing*.

Many of the checks require that one of the wheels is raised off the ground. If the motorcycle doesn't have a centre stand, note that an auxiliary stand will be required. Additionally, the help of an assistant may prove useful.

Certain exceptions apply to machines under 50 cc, machines without a lighting system, and Classic bikes - if in doubt about any of the requirements listed below seek confirmation from an MOT tester prior to submitting the motorcycle for the test.

Check that the frame number is clearly visible.

> **HAYNES HiNT** *If a component is in borderline condition, the tester has discretion in deciding whether to pass or fail it. If the motorcycle presented is clean and evidently well cared for, the tester may be more inclined to pass a borderline component than if the motorcycle is scruffy and apparently neglected.*

Electrical System

Lights, turn signals, horn and reflector

✔ With the ignition on, check the operation of the following electrical components. **Note:** *The electrical components on certain small-capacity machines are powered by the generator, requiring that the engine is run for this check.*

a) Headlight and tail light. Check that both illuminate in the low and high beam switch positions.
b) Position lights. Check that the front position (or sidelight) and tail light illuminate in this switch position.
c) Turn signals. Check that all flash at the correct rate, and that the warning light(s) function correctly. Check that the turn signal switch works correctly.
d) Hazard warning system (where fitted). Check that all four turn signals flash in this switch position.
e) Brake stop light. Check that the light comes on when the front and rear brakes are independently applied. Models first used on or after 1st April 1986 must have a brake light switch on each brake.
f) Horn. Check that the sound is continuous and of reasonable volume.

✔ Check that there is a red reflector on the rear of the machine, either mounted separately or as part of the tail light lens.
✔ Check the condition of the headlight, tail light and turn signal lenses.

Headlight beam height

✔ The MOT tester will perform a headlight beam height check using specialised beam setting equipment **(see illustration 1)**. This equipment will not be available to the home mechanic, but if you suspect that the headlight is incorrectly set or may have been maladjusted in the past, you can perform a rough test as follows.
✔ Position the bike in a straight line facing a brick wall. The bike must be off its stand, upright and with a rider seated. Measure the height from the ground to the centre of the headlight and mark a horizontal line on the wall at this height. Position the motorcycle 3.8 metres from the wall and draw a vertical line up the wall central to the centreline of the motorcycle. Switch to dipped beam and check that the beam pattern falls slightly lower than the horizontal line and to the left of the vertical line **(see illustration 2)**.

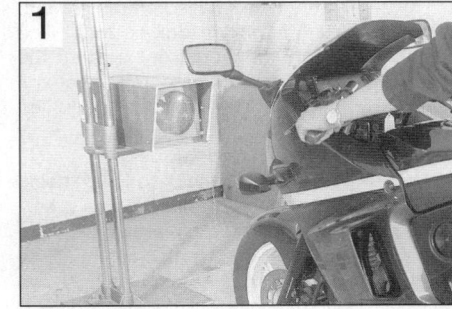

Headlight beam height checking equipment

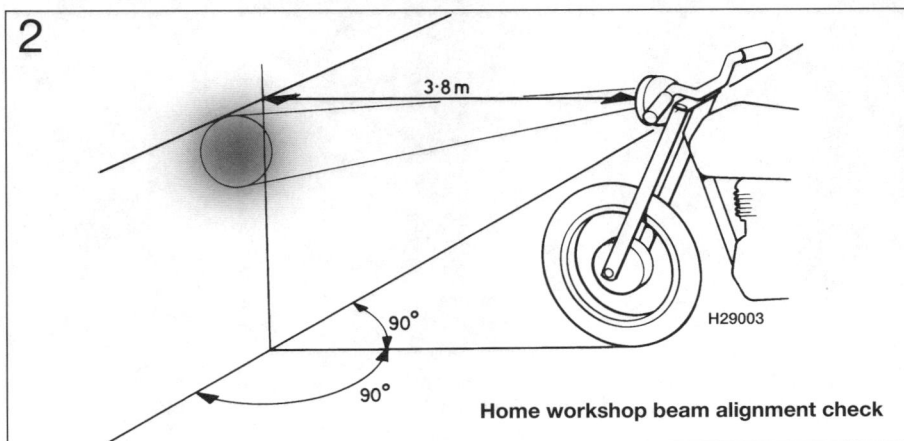

Home workshop beam alignment check

REF•28 MOT Test Checks

Exhaust System and Final Drive

Exhaust

✔ Check that the exhaust mountings are secure and that the system does not foul any of the rear suspension components.
✔ Start the motorcycle. When the revs are increased, check that the exhaust is neither holed nor leaking from any of its joints. On a linked system, check that the collector box is not leaking due to corrosion.
✔ Note that the exhaust decibel level ("loudness" of the exhaust) is assessed at the discretion of the tester. If the motorcycle was first used on or after 1st January 1985 the silencer must carry the BSAU 193 stamp, or a marking relating to its make and model, or be of OE (original equipment) manufacture. If the silencer is marked NOT FOR ROAD USE, RACING USE ONLY or similar, it will fail the MOT.

Final drive

✔ On chain or belt drive machines, check that the chain/belt is in good condition and does not have excessive slack. Also check that the sprocket is securely mounted on the rear wheel hub. Check that the chain/belt guard is in place.
✔ On shaft drive bikes, check for oil leaking from the drive unit and fouling the rear tyre.

Steering and Suspension

Steering

✔ With the front wheel raised off the ground, rotate the steering from lock to lock. The handlebar or switches must not contact the fuel tank or be close enough to trap the rider's hand. Problems can be caused by damaged lock stops on the lower yoke and frame, or by the fitting of non-standard handlebars.
✔ When performing the lock to lock check, also ensure that the steering moves freely without drag or notchiness. Steering movement can be impaired by poorly routed cables, or by overtight head bearings or worn bearings. The tester will perform a check of the steering head bearing lower race by mounting the front wheel on a surface plate, then performing a lock to lock check with the weight of the machine on the lower bearing (see illustration 3).
✔ Grasp the fork sliders (lower legs) and attempt to push and pull on the forks (see illustration 4). Any play in the steering head bearings will be felt. Note that in extreme cases, wear of the front fork bushes can be misinterpreted for head bearing play.
✔ Check that the handlebars are securely mounted.
✔ Check that the handlebar grip rubbers are secure. They should by bonded to the bar left end and to the throttle cable pulley on the right end.

Front wheel mounted on a surface plate for steering head bearing lower race check

Front suspension

✔ With the motorcycle off the stand, hold the front brake on and pump the front forks up and down (see illustration 5). Check that they are adequately damped.

Checking the steering head bearings for freeplay

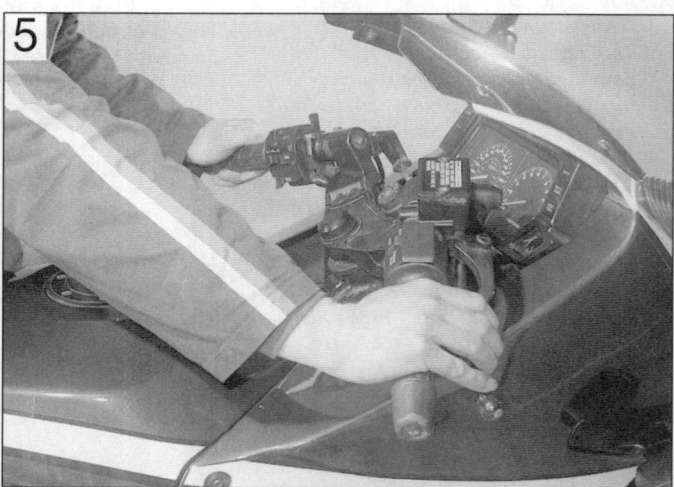

Hold the front brake on and pump the front forks up and down to check operation

MOT Test Checks REF•29

Inspect the area around the fork dust seal for oil leakage (arrow)

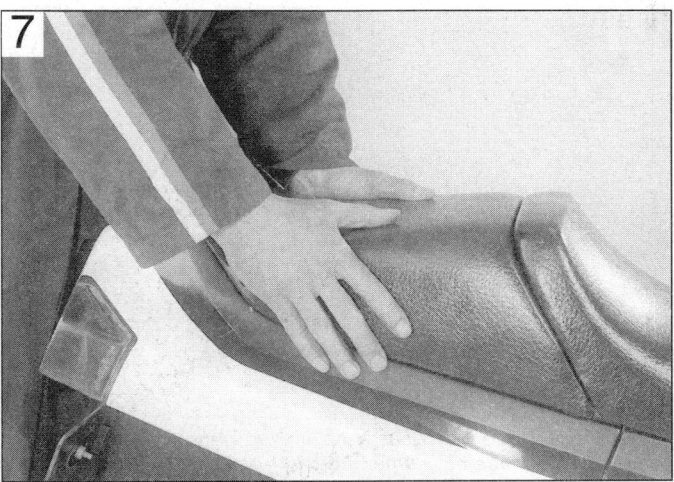

Bounce the rear of the motorcycle to check rear suspension operation

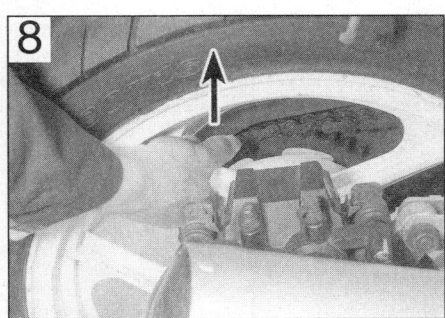

Checking for rear suspension linkage play

✔ Inspect the area above and around the front fork oil seals **(see illustration 6)**. There should be no sign of oil on the fork tube (stanchion) nor leaking down the slider (lower leg). On models so equipped, check that there is no oil leaking from the anti-dive units.

✔ On models with swingarm front suspension, check that there is no freeplay in the linkage when moved from side to side.

Rear suspension

✔ With the motorcycle off the stand and an assistant supporting the motorcycle by its handlebars, bounce the rear suspension **(see illustration 7)**. Check that the suspension components do not foul on any of the cycle parts and check that the shock absorber(s) provide adequate damping.

✔ Visually inspect the shock absorber(s) and check that there is no sign of oil leakage from its damper. This is somewhat restricted on certain single shock models due to the location of the shock absorber.

✔ With the rear wheel raised off the ground, grasp the wheel at the highest point and attempt to pull it up **(see illustration 8)**. Any play in the swingarm pivot or suspension linkage bearings will be felt as movement. **Note:** *Do not confuse play with actual suspension movement.* Failure to lubricate suspension linkage bearings can lead to bearing failure **(see illustration 9)**.

✔ With the rear wheel raised off the ground, grasp the swingarm ends and attempt to move the swingarm from side to side and forwards and backwards - any play indicates wear of the swingarm pivot bearings **(see illustration 10)**.

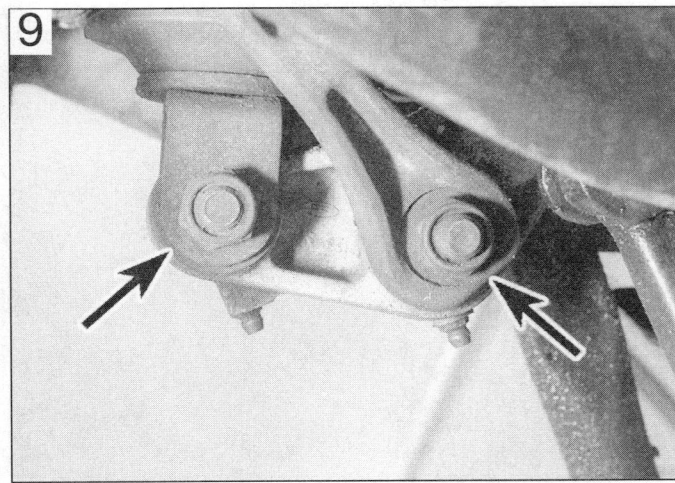

Worn suspension linkage pivots (arrows) are usually the cause of play in the rear suspension

Grasp the swingarm at the ends to check for play in its pivot bearings

REF•30 MOT Test Checks

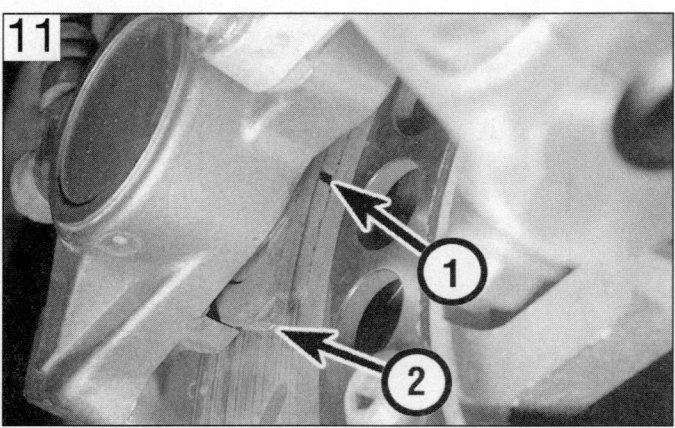

Brake pad wear can usually be viewed without removing the caliper. Most pads have wear indicator grooves (1) and some also have indicator tangs (2)

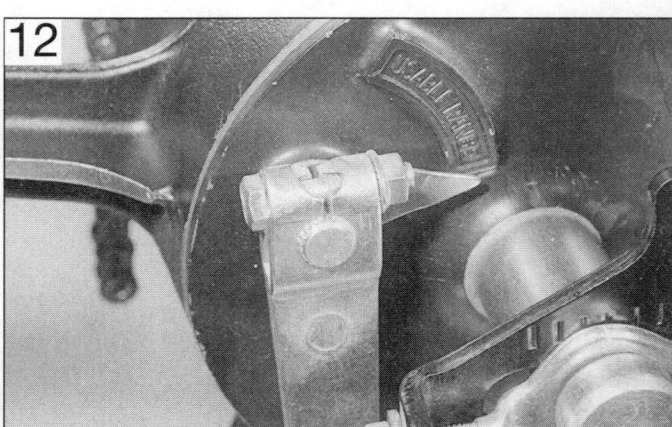

On drum brakes, check the angle of the operating lever with the brake fully applied. Most drum brakes have a wear indicator pointer and scale.

Brakes, Wheels and Tyres

Brakes

✔ With the wheel raised off the ground, apply the brake then free it off, and check that the wheel is about to revolve freely without brake drag.
✔ On disc brakes, examine the disc itself. Check that it is securely mounted and not cracked.
✔ On disc brakes, view the pad material through the caliper mouth and check that the pads are not worn down beyond the limit **(see illustration 11)**.
✔ On drum brakes, check that when the brake is applied the angle between the operating lever and cable or rod is not too great **(see illustration 12)**. Check also that the operating lever doesn't foul any other components.
✔ On disc brakes, examine the flexible hoses from top to bottom. Have an assistant hold the brake on so that the fluid in the hose is under pressure, and check that there is no sign of fluid leakage, bulges or cracking. If there are any metal brake pipes or unions, check that these are free from corrosion and damage. Where a brake-linked anti-dive system is fitted, check the hoses to the anti-dive in a similar manner.
✔ Check that the rear brake torque arm is secure and that its fasteners are secured by self-locking nuts or castellated nuts with split-pins or R-pins **(see illustration 13)**.
✔ On models with ABS, check that the self-check warning light in the instrument panel works.
✔ The MOT tester will perform a test of the motorcycle's braking efficiency based on a calculation of rider and motorcycle weight. Although this cannot be carried out at home, you can at least ensure that the braking systems are properly maintained. For hydraulic disc brakes, check the fluid level, lever/pedal feel (bleed of air if its spongy) and pad material. For drum brakes, check adjustment, cable or rod operation and shoe lining thickness.

Wheels and tyres

✔ Check the wheel condition. Cast wheels should be free from cracks and if of the built-up design, all fasteners should be secure. Spoked wheels should be checked for broken, corroded, loose or bent spokes.
✔ With the wheel raised off the ground, spin the wheel and visually check that the tyre and wheel run true. Check that the tyre does not foul the suspension or mudguards.
✔ With the wheel raised off the ground, grasp the wheel and attempt to move it about the axle (spindle) **(see illustration 14)**. Any play felt here indicates wheel bearing failure.

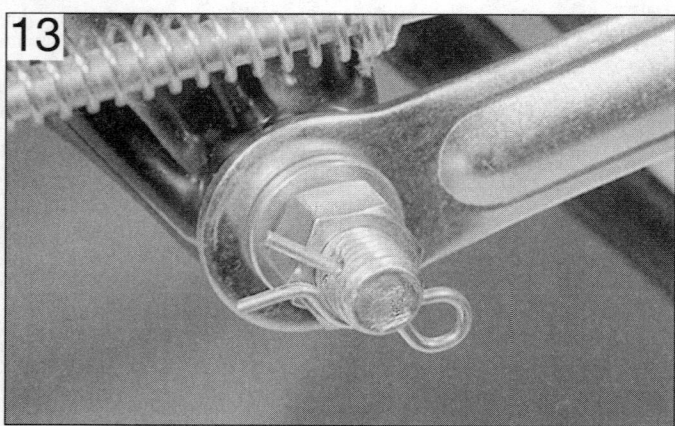

Brake torque arm must be properly secured at both ends

Check for wheel bearing play by trying to move the wheel about the axle (spindle)

MOT Test Checks REF•31

Checking the tyre tread depth

Tyre direction of rotation arrow can be found on tyre sidewall

Castellated type wheel axle (spindle) nut must be secured by a split pin or R-pin

Two straightedges are used to check wheel alignment

✔ Check the tyre tread depth, tread condition and sidewall condition **(see illustration 15)**.
✔ Check the tyre type. Front and rear tyre types must be compatible and be suitable for road use. Tyres marked NOT FOR ROAD USE, COMPETITION USE ONLY or similar, will fail the MOT.
✔ If the tyre sidewall carries a direction of rotation arrow, this must be pointing in the direction of normal wheel rotation **(see illustration 16)**.
✔ Check that the wheel axle (spindle) nuts (where applicable) are properly secured. A self-locking nut or castellated nut with a split-pin or R-pin can be used **(see illustration 17)**.
✔ Wheel alignment is checked with the motorcycle off the stand and a rider seated. With the front wheel pointing straight ahead, two perfectly straight lengths of metal or wood and placed against the sidewalls of both tyres **(see illustration 18)**. The gap each side of the front tyre must be equidistant on both sides. Incorrect wheel alignment may be due to a cocked rear wheel (often as the result of poor chain adjustment) or in extreme cases, a bent frame.

General checks and condition

✔ Check the security of all major fasteners, bodypanels, seat, fairings (where fitted) and mudguards.

✔ Check that the rider and pillion footrests, handlebar levers and brake pedal are securely mounted.

✔ Check for corrosion on the frame or any load-bearing components. If severe, this may affect the structure, particularly under stress.

Sidecars

A motorcycle fitted with a sidecar requires additional checks relating to the stability of the machine and security of attachment and swivel joints, plus specific wheel alignment (toe-in) requirements. Additionally, tyre and lighting requirements differ from conventional motorcycle use. Owners are advised to check MOT test requirements with an official test centre.

REF•32 Storage

Preparing for storage

Before you start

If repairs or an overhaul is needed, see that this is carried out now rather than left until you want to ride the bike again.

Give the bike a good wash and scrub all dirt from its underside. Make sure the bike dries completely before preparing for storage.

Engine

● Remove the spark plug(s) and lubricate the cylinder bores with approximately a teaspoon of motor oil using a spout-type oil can (see illustration 1). Reinstall the spark plug(s). Crank the engine over a couple of times to coat the piston rings and bores with oil. If the bike has a kickstart, use this to turn the engine over. If not, flick the kill switch to the OFF position and crank the engine over on the starter (see illustration 2). If the nature on the ignition system prevents the starter operating with the kill switch in the OFF position, remove the spark plugs and fit them back in their caps; ensure that the plugs are earthed (grounded) against the cylinder head when the starter is operated (see illustration 3).

 Warning: It is important that the plugs are earthed (grounded) away from the spark plug holes otherwise there is a risk of atomised fuel from the cylinders igniting.

 **HAYNES HINT** *On a single cylinder four-stroke engine, you can seal the combustion chamber completely by positioning the piston at TDC on the compression stroke.*

● Drain the carburettor(s) otherwise there is a risk of jets becoming blocked by gum deposits from the fuel (see illustration 4).

● If the bike is going into long-term storage, consider adding a fuel stabiliser to the fuel in the tank. If the tank is drained completely, corrosion of its internal surfaces may occur if left unprotected for a long period. The tank can be treated with a rust preventative especially for this purpose. Alternatively, remove the tank and pour half a litre of motor oil into it, install the filler cap and shake the tank to coat its internals with oil before draining off the excess. The same effect can also be achieved by spraying WD40 or a similar water-dispersant around the inside of the tank via its flexible nozzle.

● Make sure the cooling system contains the correct mix of antifreeze. Antifreeze also contains important corrosion inhibitors.

● The air intakes and exhaust can be sealed off by covering or plugging the openings. Ensure that you do not seal in any condensation; run the engine until it is hot,

Squirt a drop of motor oil into each cylinder

Flick the kill switch to OFF . . .

. . . and ensure that the metal bodies of the plugs (arrows) are earthed against the cylinder head

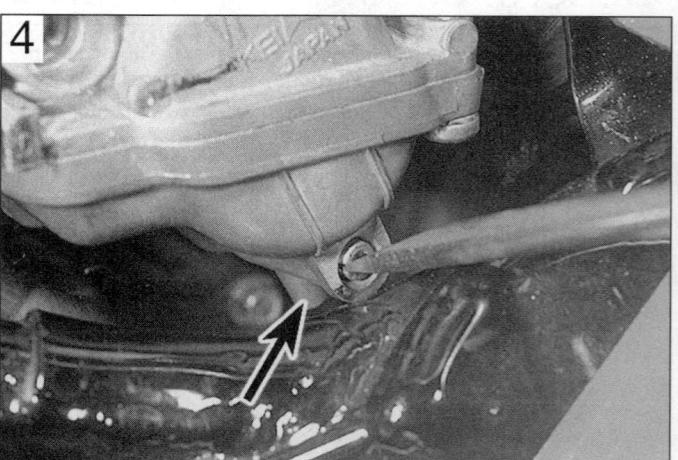
Connect a hose to the carburettor float chamber drain stub (arrow) and unscrew the drain screw

Storage REF•33

Exhausts can be sealed off with a plastic bag

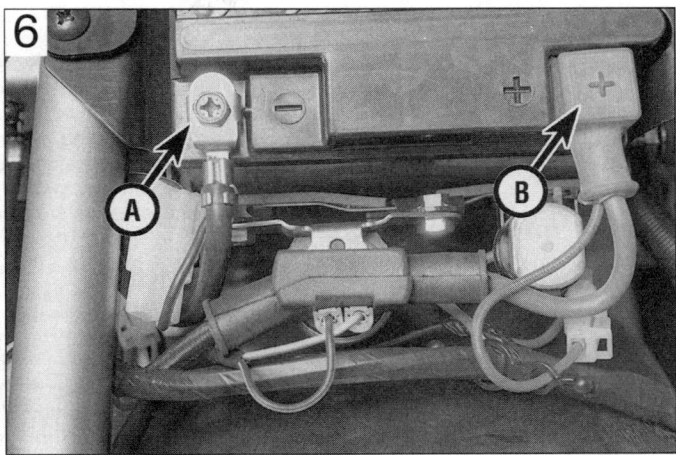
Disconnect the negative lead (A) first, followed by the positive lead (B)

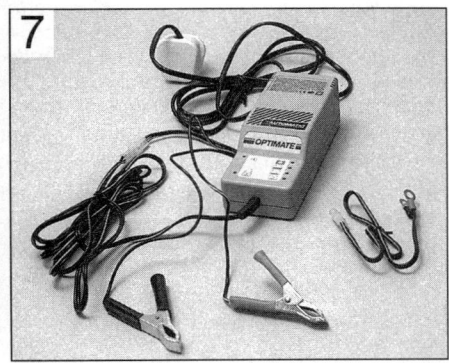
Use a suitable battery charger - this kit also assess battery condition

then switch off and allow to cool. Tape a piece of thick plastic over the silencer end(s) **(see illustration 5)**. Note that some advocate pouring a tablespoon of motor oil into the silencer(s) before sealing them off.

Battery

● Remove it from the bike - in extreme cases of cold the battery may freeze and crack its case **(see illustration 6)**.

● Check the electrolyte level and top up if necessary (conventional refillable batteries). Clean the terminals.
● Store the battery off the motorcycle and away from any sources of fire. Position a wooden block under the battery if it is to sit on the ground.
● Give the battery a trickle charge for a few hours every month **(see illustration 7)**.

Tyres

● Place the bike on its centrestand or an auxiliary stand which will support the motorcycle in an upright position. Position wood blocks under the tyres to keep them off the ground and to provide insulation from damp. If the bike is being put into long-term storage, ideally both tyres should be off the ground; not only will this protect the tyres, but will also ensure that no load is placed on the steering head or wheel bearings.
● Deflate each tyre by 5 to 10 psi, no more or the beads may unseat from the rim, making subsequent inflation difficult on tubeless tyres.

Pivots and controls

● Lubricate all lever, pedal, stand and footrest pivot points. If grease nipples are fitted to the rear suspension components, apply lubricant to the pivots.
● Lubricate all control cables.

Cycle components

● Apply a wax protectant to all painted and plastic components. Wipe off any excess, but don't polish to a shine. Where fitted, clean the screen with soap and water.
● Coat metal parts with Vaseline (petroleum jelly). When applying this to the fork tubes, do not compress the forks otherwise the seals will rot from contact with the Vaseline.
● Apply a vinyl cleaner to the seat.

Storage conditions

● Aim to store the bike in a shed or garage which does not leak and is free from damp.
● Drape an old blanket or bedspread over the bike to protect it from dust and direct contact with sunlight (which will fade paint). This also hides the bike from prying eyes. Beware of tight-fitting plastic covers which may allow condensation to form and settle on the bike.

Getting back on the road

Engine and transmission

● Change the oil and replace the oil filter. If this was done prior to storage, check that the oil hasn't emulsified - a thick whitish substance which occurs through condensation.
● Remove the spark plugs. Using a spout-type oil can, squirt a few drops of oil into the cylinder(s). This will provide initial lubrication as the piston rings and bores comes back into contact. Service the spark plugs, or fit new ones, and install them in the engine.

● Check that the clutch isn't stuck on. The plates can stick together if left standing for some time, preventing clutch operation. Engage a gear and try rocking the bike back and forth with the clutch lever held against the handlebar. If this doesn't work on cable-operated clutches, hold the clutch lever back against the handlebar with a strong elastic band or cable tie for a couple of hours **(see illustration 8)**.
● If the air intakes or silencer end(s) were blocked off, remove the bung or cover used.
● If the fuel tank was coated with a rust

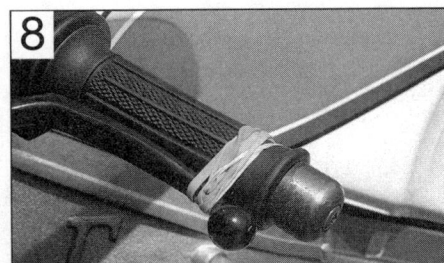
Hold clutch lever back against the handlebar with elastic bands or a cable tie

Storage

preventative, oil or a stabiliser added to the fuel, drain and flush the tank and dispose of the fuel sensibly. If no action was taken with the fuel tank prior to storage, it is advised that the old fuel is disposed of since it will go off over a period of time. Refill the fuel tank with fresh fuel.

Frame and running gear

- Oil all pivot points and cables.
- Check the tyre pressures. They will definitely need inflating if pressures were reduced for storage.
- Lubricate the final drive chain (where applicable).
- Remove any protective coating applied to the fork tubes (stanchions) since this may well destroy the fork seals. If the fork tubes weren't protected and have picked up rust spots, remove them with very fine abrasive paper and refinish with metal polish.
- Check that both brakes operate correctly. Apply each brake hard and check that it's not possible to move the motorcycle forwards, then check that the brake frees off again once released. Brake caliper pistons can stick due to corrosion around the piston head, or on the sliding caliper types, due to corrosion of the slider pins. If the brake doesn't free after repeated operation, take the caliper off for examination. Similarly drum brakes can stick due to a seized operating cam, cable or rod linkage.
- If the motorcycle has been in long-term storage, renew the brake fluid and clutch fluid (where applicable).
- Depending on where the bike has been stored, the wiring, cables and hoses may have been nibbled by rodents. Make a visual check and investigate disturbed wiring loom tape.

Battery

- If the battery has been previously removal and given top up charges it can simply be reconnected. Remember to connect the positive cable first and the negative cable last.
- On conventional refillable batteries, if the battery has not received any attention, remove it from the motorcycle and check its electrolyte level. Top up if necessary then charge the battery. If the battery fails to hold a charge and a visual checks show heavy white sulphation of the plates, the battery is probably defective and must be renewed. This is particularly likely if the battery is old. Confirm battery condition with a specific gravity check.
- On sealed (MF) batteries, if the battery has not received any attention, remove it from the motorcycle and charge it according to the information on the battery case - if the battery fails to hold a charge it must be renewed.

Starting procedure

- If a kickstart is fitted, turn the engine over a couple of times with the ignition OFF to distribute oil around the engine. If no kickstart is fitted, flick the engine kill switch OFF and the ignition ON and crank the engine over a couple of times to work oil around the upper cylinder components. If the nature of the ignition system is such that the starter won't work with the kill switch OFF, remove the spark plugs, fit them back into their caps and earth (ground) their bodies on the cylinder head. Reinstall the spark plugs afterwards.
- Switch the kill switch to RUN, operate the choke and start the engine. If the engine won't start don't continue cranking the engine - not only will this flatten the battery, but the starter motor will overheat. Switch the ignition off and try again later. If the engine refuses to start, go through the fault finding procedures in this manual. **Note:** *If the bike has been in storage for a long time, old fuel or a carburettor blockage may be the problem. Gum deposits in carburettors can block jets - if a carburettor cleaner doesn't prove successful the carburettors must be dismantled for cleaning.*
- Once the engine has started, check that the lights, turn signals and horn work properly.
- Treat the bike gently for the first ride and check all fluid levels on completion. Settie the bike back into the maintenance schedule.

Fault Finding REF•35

This Section provides an easy reference-guide to the more common faults that are likely to afflict your machine. Obviously, the opportunities are almost limitless for faults to occur as a result of obscure failures, and to try and cover all eventualities would require a book. Indeed, a number have been written on the subject.

Successful troubleshooting is not a mysterious 'black art' but the application of a bit of knowledge combined with a systematic and logical approach to the problem. Approach any troubleshooting by first accurately identifying the symptom and then checking through the list of possible causes, starting with the simplest or most obvious and progressing in stages to the most complex.

Take nothing for granted, but above all apply liberal quantities of common sense.

The main symptom of a fault is given in the text as a major heading below which are listed the various systems or areas which may contain the fault. Details of each possible cause for a fault and the remedial action to be taken are given, in brief, in the paragraphs below each heading. Further information should be sought in the relevant Chapter.

1 Engine doesn't start or is difficult to start
- [ ] Starter motor doesn't rotate
- [ ] Starter motor rotates but engine does not turn over
- [ ] Starter works but engine won't turn over (seized)
- [ ] No fuel flow
- [ ] Engine flooded
- [ ] No spark or weak spark
- [ ] Compression low
- [ ] Stalls after starting
- [ ] Rough idle

2 Poor running at low speed
- [ ] Spark weak
- [ ] Fuel/air mixture incorrect
- [ ] Compression low
- [ ] Poor acceleration

3 Poor running or no power at high speed
- [ ] Firing incorrect
- [ ] Fuel/air mixture incorrect
- [ ] Compression low
- [ ] Knocking or pinking
- [ ] Miscellaneous causes

4 Overheating
- [ ] Engine overheats
- [ ] Firing incorrect
- [ ] Fuel/air mixture incorrect
- [ ] Compression too high
- [ ] Engine load excessive
- [ ] Lubrication inadequate

5 Clutch problems
- [ ] Clutch slipping
- [ ] Clutch not disengaging completely

6 Gearchanging problems
- [ ] Doesn't go into gear, or lever doesn't return
- [ ] Jumps out of gear
- [ ] Overselects

7 Abnormal engine noise
- [ ] Knocking or pinking
- [ ] Piston slap or rattling
- [ ] Valve noise
- [ ] Other noise

8 Abnormal driveline noise
- [ ] Clutch noise
- [ ] Transmission noise
- [ ] Final drive noise

9 Abnormal frame and suspension noise
- [ ] Front end noise
- [ ] Shock absorber noise
- [ ] Brake noise

10 Oil level indicator light comes on
- [ ] Engine lubrication system
- [ ] Electrical system

11 Excessive exhaust smoke
- [ ] White smoke
- [ ] Black smoke

12 Poor handling or stability
- [ ] Handlebar hard to turn
- [ ] Handlebar shakes or vibrates excessively
- [ ] Handlebar pulls to one side
- [ ] Poor shock absorbing qualities

13 Braking problems
- [ ] Brakes are spongy, don't hold
- [ ] Brake lever or pedal pulsates
- [ ] Brakes drag

14 Electrical problems
- [ ] Battery dead or weak
- [ ] Battery overcharged

… Fault Finding

1 Engine doesn't start or is difficult to start

Starter motor doesn't rotate
- [ ] Engine kill switch OFF.
- [ ] Fuse blown. Check main fuse A and starter circuit fuse (Chapter 9).
- [ ] Battery voltage low. Check and recharge battery (Chapter 9).
- [ ] Starter motor defective. Make sure the wiring to the starter is secure. Make sure the starter relay clicks when the start button is pushed. If the relay clicks, then the fault is in the wiring or motor.
- [ ] Starter relay faulty. Check it according to the procedure in Chapter 9.
- [ ] Starter switch not contacting. The contacts could be wet, corroded or dirty. Disassemble and clean the switch (Chapter 9).
- [ ] Wiring open or shorted. Check all wiring connections and harnesses to make sure that they are dry, tight and not corroded. Also check for broken or frayed wires that can cause a short to ground (earth) (see wiring diagram, Chapter 9).
- [ ] Ignition switch defective. Check the switch according to the procedure in Chapter 9. Renew the switch if it is defective.
- [ ] Engine kill switch defective. Check for wet, dirty or corroded contacts. Clean or renew the switch as necessary (Chapter 9).
- [ ] Faulty neutral, sidestand or clutch switch. Check the wiring to each switch and the switch itself according to the procedures in Chapter 9.

Starter motor rotates but engine does not turn over
- [ ] Starter motor clutch defective. Inspect and repair or renew (Chapter 2).
- [ ] Damaged idler or starter gears. Inspect and renew the damaged parts (Chapter 2).

Starter works but engine won't turn over (seized)
- [ ] Seized engine caused by one or more internally damaged components. Failure due to wear, abuse or lack of lubrication. Damage can include seized valves, followers, camshafts, pistons, crankshaft, connecting rod bearings, or transmission gears or bearings. Refer to Chapter 2 for engine disassembly.

No fuel flow
- [ ] No fuel in tank.
- [ ] Fuel tank breather hose obstructed.
- [ ] Fuel pump faulty, or filter is blocked (see Chapter 4).

Engine flooded
- [ ] Starting technique incorrect. Under normal circumstances the machine should start with little or no throttle. When the engine is cold, the choke should be operated and the engine started without opening the throttle. When the engine is at operating temperature, only a very slight amount of throttle should be necessary.

No spark or weak spark
- [ ] Ignition switch OFF.
- [ ] Engine kill switch turned to the OFF position.
- [ ] Battery voltage low. Check and recharge the battery as necessary (Chapter 9).
- [ ] Spark plugs dirty, defective or worn out. Locate reason for fouled plugs using spark plug condition chart and follow the plug maintenance procedures (Chapter 1).
- [ ] Spark plug caps or secondary (HT) wiring faulty. Check condition. Renew either or both components if cracks or deterioration are evident (Chapter 5).
- [ ] Spark plug caps not making good contact. Make sure that the plug caps fit snugly over the plug ends.
- [ ] Engine control module (ECM) defective. Refer to Chapter 4 for details.
- [ ] Ignition pulse generator defective. Check the unit, referring to Chapter 4 for details.
- [ ] Ignition HT coils defective. Check the coils, referring to Chapter 5.
- [ ] Ignition or kill switch shorted. This is usually caused by water, corrosion, damage or excessive wear. The switches can be disassembled and cleaned with electrical contact cleaner. If cleaning does not help, renew the switches (Chapter 9).
- [ ] Wiring shorted or broken between:
    a) Ignition switch and engine kill switch (or blown fuse)
    b) Engine control module (ECM) and engine kill switch
    c) Engine control module (ECM) and ignition HT coils
    d) Ignition HT coils and spark plugs
    e) Engine control module (ECM) and ignition pulse generator
- [ ] Make sure that all wiring connections are clean, dry and tight. Look for chafed and broken wires (Chapters 5 and 9).

Compression low
- [ ] Spark plugs loose. Remove the plugs and inspect their threads. Reinstall and tighten to the specified torque (Chapter 1).
- [ ] Cylinder heads not sufficiently tightened down. If a cylinder head is suspected of being loose, then there's a chance that the gasket or head is damaged if the problem has persisted for any length of time. The head bolts should be tightened to the proper torque in the correct sequence (Chapter 2).
- [ ] Improper valve clearance. This means that the valve is not closing completely and compression pressure is leaking past the valve. Check and adjust the valve clearances (Chapter 1).
- [ ] Cylinder and/or piston worn. Excessive wear will cause compression pressure to leak past the rings. This is usually accompanied by worn rings as well. A top-end overhaul is necessary (Chapter 2).
- [ ] Piston rings worn, weak, broken, or sticking. Broken or sticking piston rings usually indicate a lubrication or fuelling problem that causes excess carbon deposits or seizures to form on the pistons and rings. Top-end overhaul is necessary (Chapter 2).
- [ ] Piston ring-to-groove clearance excessive. This is caused by excessive wear of the piston ring lands. Piston renewal is necessary (Chapter 2).
- [ ] Cylinder head gasket damaged. If a head is allowed to become loose, or if excessive carbon build-up on the piston crown and combustion chamber causes extremely high compression, the head gasket may leak. Retorquing the head is not always sufficient to restore the seal, so gasket renewal is necessary (Chapter 2).
- [ ] Cylinder head warped. This is caused by overheating or improperly tightened head bolts. Machine shop resurfacing or head renewal is necessary (Chapter 2).
- [ ] Valve spring broken or weak. Caused by component failure or wear; the springs must be renewed (Chapter 2).
- [ ] Valve not seating properly. This is caused by a bent valve (from over-revving or improper valve adjustment), burned valve or seat (improper fuelling) or an accumulation of carbon deposits on the seat (from fuelling or lubrication problems). The valves must be cleaned and/or renewed and the seats serviced if possible (Chapter 2).

Stalls after starting
- [ ] Improper choke action. On VFR800FI-W and VFR800FI-X models check the operation of the choke cable (Chapter 1). On VFR800FI-Y and VFR800FI-1 models check the operation of the fast idle wax unit (Chapter 4).
- [ ] Ignition malfunction. See Chapter 5.
- [ ] Fuel injection system malfunction. See Chapter 4.
- [ ] Fuel contaminated. The fuel can be contaminated with either dirt or water, or can change chemically if the machine is allowed to sit for several months or more. Drain the tank (Chapter 4).
- [ ] Intake air leak. Check for loose throttle body intake rubber retaining clips and damaged/disconnected vacuum hoses (Chapter 4).
- [ ] Engine idle speed incorrect. Turn idle adjusting screw until the engine idles at the specified rpm (Chapter 1).

Fault Finding REF•37

1 Engine doesn't start or is difficult to start (continued)

Rough idle
- [ ] Ignition malfunction. See Chapter 5.
- [ ] Idle speed incorrect. See Chapter 1.
- [ ] Throttle body assembly starter valves not synchronised. Adjust them with vacuum gauge or manometer set as described in Chapter 4.
- [ ] Fuel injection system malfunction. See Chapter 4.
- [ ] Fuel contaminated. The fuel can be contaminated with either dirt or water, or can change chemically if the machine is allowed to sit for several months or more. Drain the tank (Chapter 4).
- [ ] Intake air leak. Check for loose throttle body intake rubber retaining clips and damaged/disconnected vacuum hoses. Renew the intake rubbers if they are split or perished (Chapter 4).
- [ ] Air filter clogged. Renew the air filter element (Chapter 1).

2 Poor running at low speeds

Spark weak
- [ ] Battery voltage low. Check and recharge battery (Chapter 9).
- [ ] Spark plugs fouled, defective or worn out. Refer to Chapter 1 for spark plug maintenance.
- [ ] Spark plug cap or HT wiring defective. Refer to Chapters 1 and 5 for details on the ignition system.
- [ ] Spark plug caps not making contact.
- [ ] Incorrect spark plugs. Wrong type, heat range or cap configuration. Check and install correct plugs listed in Chapter 1.
- [ ] Engine control module (ECM) faulty. See Chapter 4.
- [ ] Ignition pulse generator defective. See Chapter 4.
- [ ] Ignition HT coils defective. See Chapter 5.

Fuel/air mixture incorrect
- [ ] Fuel injector clogged or fuel injection system malfunction (see Chapter 4).
- [ ] Air filter clogged, poorly sealed or missing (Chapter 1).
- [ ] Air filter housing poorly sealed. Look for cracks, holes or loose clamps and renew or repair defective parts.
- [ ] Fuel tank breather hose obstructed.
- [ ] Intake air leak. Check for loose throttle body intake rubber retaining clips and damaged/disconnected vacuum hoses. Renew the intake rubbers if they are split or perished (Chapter 4).

Compression low
- [ ] Spark plugs loose. Remove the plugs and inspect their threads. Reinstall and tighten to the specified torque (Chapter 1).
- [ ] Cylinder heads not sufficiently tightened down. If a cylinder head is suspected of being loose, then there's a chance that the gasket and head are damaged if the problem has persisted for any length of time. The head bolts should be tightened to the proper torque in the correct sequence (Chapter 2).
- [ ] Improper valve clearance. This means that the valve is not closing completely and compression pressure is leaking past the valve. Check and adjust the valve clearances (Chapter 1).
- [ ] Cylinder and/or piston worn. Excessive wear will cause compression pressure to leak past the rings. This is usually accompanied by worn rings as well. A top-end overhaul is necessary (Chapter 2).
- [ ] Piston rings worn, weak, broken, or sticking. Broken or sticking piston rings usually indicate a lubrication or fuelling problem that causes excess carbon deposits or seizures to form on the pistons and rings. Top-end overhaul is necessary (Chapter 2).
- [ ] Piston ring-to-groove clearance excessive. This is caused by excessive wear of the piston ring lands. Piston renewal is necessary (Chapter 2).
- [ ] Cylinder head gasket damaged. If a head is allowed to become loose, or if excessive carbon build-up on the piston crown and combustion chamber causes extremely high compression, the head gasket may leak. Retorquing the head is not always sufficient to restore the seal, so gasket renewal is necessary (Chapter 2).
- [ ] Cylinder head warped. This is caused by overheating or improperly tightened head bolts. Machine shop resurfacing or head renewal is necessary (Chapter 2).
- [ ] Valve spring broken or weak. Caused by component failure or wear; the springs must be renewed (Chapter 2).
- [ ] Valve not seating properly. This is caused by a bent valve (from over-revving or improper valve adjustment), burned valve or seat (improper fuelling) or an accumulation of carbon deposits on the seat (from fuelling, lubrication problems). The valves must be cleaned and/or renewed and the seats serviced if possible (Chapter 2).

Poor acceleration
- [ ] Timing not advancing. The ignition pulse generator or ECM may be defective. If so, they must be replaced with new ones as they can't be repaired. Check them (see Chapter 4).
- [ ] Throttle body assembly starter valves not synchronised. Adjust them with vacuum gauge or manometer set (see Chapter 4).
- [ ] Fuel injection system fault. See Chapter 4.
- [ ] Engine oil viscosity too high. Using a heavier oil than that recommended in Chapter 1 can damage the oil pump or lubrication system and cause drag on the engine.
- [ ] Brakes dragging. Usually caused by debris which has entered the brake piston seals, or from a warped disc or bent axle. Repair as necessary (Chapter 7).

REF•38 Fault Finding

3 Poor running or no power at high speed

Firing incorrect
- [ ] Air filter restricted. Clean or renew filter (Chapter 1).
- [ ] Spark plugs fouled, defective or worn out. See Chapter 1 for spark plug maintenance.
- [ ] Spark plug caps or HT wiring defective. See Chapters 1 and 5 for details of the ignition system.
- [ ] Spark plug caps not in good contact. See Chapter 5.
- [ ] Incorrect spark plugs. Wrong type, heat range or cap configuration. Check and install correct plugs listed in Chapter 1.
- [ ] Engine control module (ECM) defective. See Chapter 4.
- [ ] Ignition pulse generator defective. See Chapter 4.
- [ ] Ignition coils defective. See Chapter 5.

Fuel/air mixture incorrect
- [ ] Fuel injector clogged or fuel injection system malfunction (see Chapter 4).
- [ ] Air filter clogged, poorly sealed, or missing (Chapter 1).
- [ ] Air filter housing poorly sealed. Look for cracks, holes or loose clamps, and renew or repair defective parts.
- [ ] Fuel tank breather hose obstructed.
- [ ] Intake air leak. Check for loose throttle body intake rubber retaining clips and damaged/disconnected vacuum hoses. Renew the intake rubbers if they are split or perished (Chapter 4).

Compression low
- [ ] Spark plugs loose. Remove the plugs and inspect their threads. Reinstall and tighten to the specified torque (Chapter 1).
- [ ] Cylinder heads not sufficiently tightened down. If a cylinder head is suspected of being loose, then there's a chance that the gasket and head are damaged if the problem has persisted for any length of time. The head bolts should be tightened to the proper torque in the correct sequence (Chapter 2).
- [ ] Improper valve clearance. This means that the valve is not closing completely and compression pressure is leaking past the valve. Check and adjust the valve clearances (Chapter 1).
- [ ] Cylinder and/or piston worn. Excessive wear will cause compression pressure to leak past the rings. This is usually accompanied by worn rings as well. A top-end overhaul is necessary (Chapter 2).
- [ ] Piston rings worn, weak, broken, or sticking. Broken or sticking piston rings usually indicate a lubrication or fuelling problem that causes excess carbon deposits or seizures to form on the pistons and rings. Top-end overhaul is necessary (Chapter 2).
- [ ] Piston ring-to-groove clearance excessive. This is caused by excessive wear of the piston ring lands. Piston renewal is necessary (Chapter 2).
- [ ] Cylinder head gasket damaged. If a head is allowed to become loose, or if excessive carbon build-up on the piston crown and combustion chamber causes extremely high compression, the head gasket may leak. Retorquing the head is not always sufficient to restore the seal, so gasket renewal is necessary (Chapter 2).
- [ ] Cylinder head warped. This is caused by overheating or improperly tightened head bolts. Machine shop resurfacing or head renewal is necessary (Chapter 2).
- [ ] Valve spring broken or weak. Caused by component failure or wear; the springs must be renewed (Chapter 2).
- [ ] Valve not seating properly. This is caused by a bent valve (from over-revving or improper valve adjustment), burned valve or seat (improper fuelling) or an accumulation of carbon deposits on the seat (from fuelling or lubrication problems). The valves must be cleaned and/or renewed and the seats serviced if possible (Chapter 2).

Knocking or pinking
- [ ] Carbon build-up in combustion chamber. Use of a fuel additive that will dissolve the adhesive bonding the carbon particles to the crown and chamber is the easiest way to remove the build-up. Otherwise, the cylinder heads will have to be removed and decarbonised (Chapter 2).
- [ ] Incorrect or poor quality fuel. Old or improper grades of fuel can cause detonation. This causes the piston to rattle, thus the knocking or pinking sound. Drain old fuel and always use the recommended fuel grade.
- [ ] Spark plug heat range incorrect. Uncontrolled detonation indicates the plug heat range is too hot. The plug in effect becomes a glow plug, raising cylinder temperatures. Install the proper heat range plug (Chapter 1).
- [ ] Improper air/fuel mixture. This will cause the cylinders to run hot, which leads to detonation. An intake air leak can cause this imbalance. See Chapter 4.

Miscellaneous causes
- [ ] Throttle valve doesn't open fully. Adjust the throttle grip freeplay (Chapter 1).
- [ ] Clutch slipping. May be caused by loose or worn clutch components. Refer to Chapter 2 for clutch overhaul procedures.
- [ ] Engine oil viscosity too high. Using a heavier oil than the one recommended in Chapter 1 can damage the oil pump or lubrication system and cause drag on the engine.
- [ ] Brakes dragging. Usually caused by debris which has entered the brake piston seals, or from a warped disc or bent axle. Repair as necessary.

Fault Finding REF•39

4 Overheating

Engine overheats
- ☐ Coolant level low. Check and add coolant (Chapter 1).
- ☐ Leak in cooling system. Check cooling system hoses and radiators for leaks and other damage. Repair or renew parts as necessary (Chapter 3).
- ☐ Thermostat sticking open or closed. Check and renew as described in Chapter 3.
- ☐ Faulty pressure cap. Remove the cap and have it pressure tested (Chapter 3).
- ☐ Coolant passages clogged. Have the entire system drained and flushed, then refill with fresh coolant.
- ☐ Water pump defective. Remove the pump and check the components (Chapter 3).
- ☐ Clogged radiator fins. Clean them by blowing compressed air through the fins from the backside.
- ☐ Cooling fan or fan switch fault (Chapter 3).

Firing incorrect
- ☐ Spark plugs fouled, defective or worn out. See Chapter 1 for spark plug maintenance.
- ☐ Incorrect spark plugs.
- ☐ Engine control module (ECM) defective. See Chapter 4.
- ☐ Ignition pulse generator faulty. See Chapter 4.
- ☐ Faulty ignition HT coils. See Chapter 5.

Fuel/air mixture incorrect
- ☐ Fuel injector clogged or fuel injection system malfunction (see Chapter 4).
- ☐ Air filter clogged, poorly sealed, or missing (Chapter 1).
- ☐ Air filter housing poorly sealed. Look for cracks, holes or loose clamps, and renew or repair defective parts.
- ☐ Fuel tank breather hose obstructed.
- ☐ Intake air leak. Check for loose throttle body intake rubber retaining clips and damaged/disconnected vacuum hoses. Renew the intake rubbers if they are split or perished (Chapter 4).

Compression too high
- ☐ Carbon build-up in combustion chamber. Use of a fuel additive that will dissolve the adhesive bonding the carbon particles to the piston crown and chamber is the easiest way to remove the build-up. Otherwise, the cylinder heads will have to be removed and decarbonised (Chapter 2).
- ☐ Improperly machined head surface or installation of incorrect gasket during engine assembly.

Engine load excessive
- ☐ Clutch slipping. Can be caused by damaged, loose or worn clutch components. Refer to Chapter 2 for overhaul procedures.
- ☐ Engine oil level too high. The addition of too much oil will cause pressurisation of the crankcase and inefficient engine operation. Check Specifications and drain to proper level (Chapter 1).
- ☐ Engine oil viscosity too high. Using a heavier oil than the one recommended in Chapter 1 can damage the oil pump or lubrication system as well as cause drag on the engine.
- ☐ Brakes dragging. Usually caused by debris which has entered the brake piston seals, or from a warped disc or bent axle. Repair as necessary.

Lubrication inadequate
- ☐ Engine oil level too low. Friction caused by intermittent lack of lubrication or from oil that is overworked can cause overheating. The oil provides a definite cooling function in the engine. Check the oil level (see Daily (pre-ride) checks).
- ☐ Poor quality engine oil or incorrect viscosity or type. Oil is rated not only according to viscosity but also according to type. Some oils are not rated high enough for use in this engine. Check the Specifications section and change to the correct oil (Chapter 1).

5 Clutch problems

Clutch slipping
- ☐ Clutch fluid level too high. Check fluid level (see Daily (pre-ride) checks).
- ☐ Friction plates worn or warped. Overhaul the clutch assembly (Chapter 2).
- ☐ Plain plates warped (Chapter 2).
- ☐ Clutch springs broken or weak. Old or heat-damaged (from slipping clutch) springs should be renewed (Chapter 2).
- ☐ Clutch pushrod bent. Check and, if necessary, renew (Chapter 2).
- ☐ Clutch centre or housing unevenly worn. This causes improper engagement of the plates. Renew the damaged or worn parts (Chapter 2).

Clutch not disengaging completely
- ☐ Clutch fluid level too low. Top-up and bleed the hydraulic system (Chapter 2).
- ☐ Clutch plates warped or damaged. This will cause clutch drag, which in turn will cause the machine to creep. Overhaul the clutch assembly (Chapter 2).
- ☐ Clutch spring tension uneven. Usually caused by a sagged or broken spring. Check and renew the springs as a set (Chapter 2).
- ☐ Engine oil deteriorated. Old, thin, worn out oil will not provide proper lubrication for the plates, causing the clutch to drag. Renew the oil and filter (Chapter 1).
- ☐ Engine oil viscosity too high. Using a heavier oil than recommended in Chapter 1 can cause the plates to stick together, putting a drag on the engine. Change to the correct weight oil (Chapter 1).
- ☐ Clutch drum bearing seized. Lack of lubrication, severe wear or damage can cause the bearing to seize on the input shaft. Overhaul of the clutch, and perhaps transmission, may be necessary to repair the damage (Chapter 2).
- ☐ Clutch hydraulic release mechanism defective. Bleed the hydraulic system (Chapter 2). If this fails to cure the problem overhaul the clutch master and slave cylinders (Chapter 2).
- ☐ Loose clutch centre nut. Causes housing and centre misalignment putting a drag on the engine. Engagement adjustment continually varies. Overhaul the clutch assembly (Chapter 2).

6 Gearchanging problems

Doesn't go into gear or lever doesn't return
- ☐ Clutch not disengaging. See above.
- ☐ Selector fork(s) bent or seized. Often caused by lack of lubrication or gears sticking on shaft. Overhaul the transmission (Chapter 2).
- ☐ Gear(s) stuck on shaft. Most often caused by a lack of lubrication or excessive wear in transmission bearings and bushes. Overhaul the transmission (Chapter 2).
- ☐ Selector drum binding. Caused by lubrication failure or excessive wear. Renew the drum and bearing (Chapter 2).
- ☐ Gearchange lever centralising spring weak or broken (Chapter 2).
- ☐ Gearchange lever broken. Splines stripped out of lever or shaft, caused by a loose lever clamp bolt. Renew necessary parts (Chapter 2).
- ☐ Gearchange mechanism stopper arm broken or worn. Gear engagement becomes difficult and rotary movement of selector drum results. Renew the arm (Chapter 2).
- ☐ Stopper arm spring broken. Allows arm to float, causing sporadic gearchange operation. Renew spring (Chapter 2).

Jumps out of gear
- ☐ Selector fork(s) worn. Overhaul the transmission (Chapter 2).
- ☐ Gear groove(s) worn. Overhaul the transmission (Chapter 2).
- ☐ Gear dogs or dog slots worn or damaged. The gears should be inspected and renewed. No attempt should be made to repair the worn parts.

Overselects
- ☐ Stopper arm spring weak or broken (Chapter 2).
- ☐ Gearchange shaft centralising spring post broken or distorted (Chapter 2).

7 Abnormal engine noise

Knocking or pinking
- ☐ Carbon build-up in combustion chamber. Use of a fuel additive that will dissolve the adhesive bonding the carbon particles to the piston crown and chamber is the easiest way to remove the build-up. Otherwise, the cylinder head will have to be removed and decarbonised (Chapter 2).
- ☐ Incorrect or poor quality fuel. Old or improper fuel can cause detonation. This causes the pistons to rattle, thus the knocking or pinking sound. Drain the old fuel and always use the recommended grade fuel (Chapter 4).
- ☐ Spark plug heat range incorrect. Uncontrolled detonation indicates that the plug heat range is too hot. The plug in effect becomes a glow plug, raising cylinder temperatures. Install the proper heat range plug (Chapter 1).
- ☐ Improper air/fuel mixture. This will cause the cylinders to run hot and lead to detonation. Blocked injectors or an air leak can cause this imbalance (Chapter 4).

Piston slap or rattling
- ☐ Cylinder-to-piston clearance excessive. Caused by improper assembly. Inspect and overhaul top-end parts (Chapter 2).
- ☐ Connecting rod bent. Caused by over-revving, trying to start a badly flooded engine or from ingesting a foreign object into the combustion chamber. Renew the damaged parts (Chapter 2).
- ☐ Piston pin or piston pin bore worn or seized from wear or lack of lubrication. Renew damaged parts (Chapter 2).
- ☐ Piston ring(s) worn, broken or sticking. Overhaul the top-end (Chapter 2).
- ☐ Piston seizure damage. Usually from lack of lubrication or overheating. Renew the pistons and bore the cylinders, as necessary (Chapter 2).
- ☐ Connecting rod bearing clearance excessive. Caused by excessive wear or lack of lubrication. Renew worn parts.

Valve noise
- ☐ Incorrect valve clearances. Adjust the clearances by referring to Chapter 1.
- ☐ Valve spring broken or weak. Check and renew weak valve springs (Chapter 2).
- ☐ Camshaft or cylinder head worn or damaged. Oil starvation at high rpm is usually the cause of damage. Low oil level or failure to change the oil at the recommended intervals are the chief causes. Since there are no replaceable bearings in the head the head itself will have to be renewed if there is excessive wear or damage (Chapter 2).

Other noise
- ☐ Cylinder head gasket leaking.
- ☐ Exhaust pipe leaking at cylinder head connection. Caused by improper fit of pipe(s), loose exhaust nuts or damaged exhaust port gasket. All exhaust fasteners should be tightened evenly and carefully. Failure to do this will lead to a leak.
- ☐ Crankshaft runout excessive. Caused by a bent crankshaft (from over-revving) or damage from an upper cylinder component failure.
- ☐ Engine mounting bolts loose. Tighten all engine mount bolts (Chapter 2).
- ☐ Crankshaft bearings worn (Chapter 2).
- ☐ Camshaft drive gear assembly defective. Renew according to the procedure in Chapter 2.

Fault Finding REF•41

8 Abnormal driveline noise

Clutch noise
- [ ] Clutch outer drum/friction plate clearance excessive (Chapter 2).
- [ ] Loose or damaged clutch pressure plate and/or bolts (Chapter 2).

Transmission noise
- [ ] Bearings worn. Also includes the possibility that the shafts are worn. Overhaul the transmission (Chapter 2).
- [ ] Gears worn or chipped (Chapter 2).
- [ ] Metal chips jammed in gear teeth. Probably pieces from a broken clutch, gear or selector mechanism that were picked up by the gears. This will cause early bearing failure (Chapter 2).
- [ ] Engine oil level too low. Causes a howl from transmission. Also affects engine power and clutch operation (Chapter 1).

Final drive noise
- [ ] Chain not adjusted properly (Chapter 1).
- [ ] Front or rear sprocket loose. Tighten fasteners (Chapter 6).
- [ ] Sprockets worn. Renew sprockets and chain (Chapter 6).
- [ ] Rear sprocket warped. Renew sprockets and chain (Chapter 6).

9 Abnormal frame and suspension noise

Front end noise
- [ ] Low fluid level or improper viscosity oil in forks. This can sound like spurting and is usually accompanied by irregular fork action (Chapter 6).
- [ ] Spring weak or broken. Makes a clicking or scraping sound. Fork oil, when drained, will have a lot of metal particles in it (Chapter 6).
- [ ] Steering head bearings loose or damaged. Clicks when braking. Check and adjust or renew as necessary (Chapters 1 and 6).
- [ ] Fork yokes loose. Make sure all clamp bolts are tightened to the specified torque (Chapter 6).
- [ ] Fork tube bent. Good possibility if machine has been dropped. Renew tube (Chapter 6).
- [ ] Front axle bolt or axle clamp bolts loose. Tighten them to the specified torque (Chapter 7).
- [ ] Loose or worn wheel bearings. Check and renew as needed (Chapter 7).

Shock absorber noise
- [ ] Fluid level incorrect. Indicates a leak caused by defective seal. Shock will be covered with oil. Renew the shock. (Chapter 6).
- [ ] Defective shock absorber with internal damage. This is in the body of the shock and can't be remedied. The shock must be renewed (Chapter 6).
- [ ] Bent or damaged shock body. Renew the shock (Chapter 6).
- [ ] Loose or worn suspension linkage components. Check and renew as necessary (Chapter 6).

Brake noise
- [ ] Squeal caused by dust or dirt on brake pads. Usually found in combination with glazed pads. Clean using brake cleaning solvent (Chapter 7).
- [ ] Contamination of brake pads. Oil, grease or brake fluid causing brake to chatter or squeal. Pads must be renewed (Chapter 7).
- [ ] Pads glazed. Caused by excessive heat from prolonged use or from contamination. Do not use sandpaper/emery cloth or any other abrasive to roughen the pad surfaces as abrasives will stay in the pad material and damage the disc. A very fine flat file can be used, but pad renewal is recommended (Chapter 7).
- [ ] Disc warped. Can cause a chattering, clicking or intermittent squeal. Usually accompanied by a pulsating lever and uneven braking. Renew the disc (Chapter 7).
- [ ] Loose or worn wheel bearings. Check and renew as needed (Chapter 7).

10 Oil pressure warning light comes on

Engine lubrication system
- [ ] Engine oil pump defective, blocked oil strainer gauze or failed relief valve. Carry out oil pressure check (Chapter 1).
- [ ] Engine oil level low. Inspect for leak or other problem causing low oil level and add recommended oil (Chapter 1).
- [ ] Engine oil viscosity too low. Very old, thin oil or an improper weight of oil used in the engine. Change to correct oil (Chapter 1).
- [ ] Camshaft or journals worn. Excessive wear causing drop in oil pressure. Renew cam and/or cylinder head. Abnormal wear could be caused by oil starvation at high rpm from low oil level or improper weight or type of oil (Chapter 1).
- [ ] Crankshaft and/or bearings worn. Same problems as above. Check and renew crankshaft and/or bearings (Chapter 2).

Electrical system
- [ ] Oil pressure switch defective. Check the switch according to the procedure in Chapter 9. Renew it if it is defective.
- [ ] Oil pressure indicator light circuit defective. Check for pinched, shorted, disconnected or damaged wiring (Chapter 9).

REF•42 Fault Finding

11 Excessive exhaust smoke

White smoke
- [ ] Piston oil ring worn. The ring may be broken or damaged, causing oil from the crankcase to be pulled past the piston into the combustion chamber. Renew the rings (Chapter 2).
- [ ] Cylinders worn, scored or cracked. Caused by overheating or oil starvation. Depending on the extent of the damage, worn or scored cylinders can be rebored, otherwise install a new cylinder block/crankcase (Chapter 2).
- [ ] Valve oil seal damaged or worn. Renew oil seals (Chapter 2).
- [ ] Valve guide worn. Perform a complete valve job (Chapter 2).
- [ ] Engine oil level too high, which causes the oil to be forced past the rings. Drain oil to the proper level (Chapter 1).
- [ ] Head gasket broken between oil return and cylinder. Causes oil to be pulled into the combustion chamber. Renew the head gasket and check the head for warpage (Chapter 2).
- [ ] Abnormal crankcase pressurisation, which forces oil past the rings. Clogged breather is usually the cause.

Black smoke
- [ ] Air filter clogged. Clean or renew the element (Chapter 1).
- [ ] Fuel injection system malfunction (Chapter 4).

12 Poor handling or stability

Handlebar hard to turn
- [ ] Steering head bearing adjuster nut too tight. Check adjustment as described in Chapter 1.
- [ ] Bearings damaged. Roughness can be felt as the bars are turned from side-to-side. Renew bearings and races (Chapter 6).
- [ ] Races dented or worn. Denting results from wear in only one position (eg, straight ahead), from a collision or hitting a pothole or from dropping the machine. Renew races and bearings (Chapter 6).
- [ ] Steering stem lubrication inadequate. Causes are grease getting hard from age or being washed out by high pressure car washes. Disassemble steering head and repack bearings (Chapter 6).
- [ ] Steering stem bent. Caused by a collision, hitting a pothole or by dropping the machine. Renew damaged part. Don't try to straighten the steering stem (Chapter 6).
- [ ] Front tyre air pressure too low (Chapter 1).

Handlebar shakes or vibrates excessively
- [ ] Tyres worn or out of balance (Chapter 7).
- [ ] Swingarm bearings worn. Renew worn bearings (Chapter 6).
- [ ] Wheel rim(s) warped or damaged. Inspect wheels for runout (Chapter 7).
- [ ] Wheel bearings worn. Worn front or rear wheel bearings can cause poor tracking. Worn front bearings will cause wobble (Chapter 7).
- [ ] Handlebar clamp bolts loose (Chapter 6).
- [ ] Fork yoke bolts loose. Tighten them to the specified torque (Chapter 6).
- [ ] Engine mounting bolts loose. Will cause excessive vibration with increased engine rpm (Chapter 2).

Handlebar pulls to one side
- [ ] Frame bent. Definitely suspect this if the machine has been dropped. May or may not be accompanied by cracking near the bend. Renew the frame (Chapter 6).
- [ ] Wheels out of alignment. Caused by improper location of axle spacers or from bent steering stem or frame (Chapter 6).
- [ ] Swingarm bent or twisted. Caused by age (metal fatigue) or impact damage. Renew the arm (Chapter 6).
- [ ] Steering stem bent. Caused by impact damage or by dropping the motorcycle. Renew the steering stem (Chapter 6).
- [ ] Fork tube bent. Disassemble the forks and renew the damaged parts (Chapter 6).
- [ ] Fork oil level uneven. Check and add or drain as necessary (Chapter 6).

Poor shock absorbing qualities
- [ ] Too hard:
  - a) Fork oil level excessive (Chapter 6).
  - b) Fork oil viscosity too high. Use a lighter oil (see the Specifications in Chapter 6).
  - c) Fork tube bent. Causes a harsh, sticking action (Chapter 6).
  - d) Shock shaft or body bent or damaged (Chapter 6).
  - e) Fork internal damage (Chapter 6).
  - f) Shock internal damage.
  - g) Tyre pressure too high (Chapter 1).
- [ ] Too soft:
  - a) Fork or shock oil insufficient and/or leaking (Chapter 6).
  - b) Fork oil level too low (Chapter 6).
  - c) Fork oil viscosity too light (Chapter 6).
  - d) Fork springs weak or broken (Chapter 6).
  - e) Shock internal damage or leakage (Chapter 6).

Fault Finding REF•43

13 Braking problems

Brakes are spongy, don't hold
- [ ] Air in brake line. Caused by inattention to master cylinder fluid level or by leakage. Correct the problem and bleed brakes (Chapter 7).
- [ ] Pad or disc worn (Chapters 1 and 7).
- [ ] Brake fluid leak. See paragraph 1.
- [ ] Contaminated pads. Caused by contamination with oil, grease, brake fluid, etc. Pads must be renewed (Chapter 7).
- [ ] Brake fluid deteriorated. Fluid is old or contaminated. Drain system, replenish with new fluid and bleed the system (Chapter 7).
- [ ] Master cylinder internal parts worn or damaged causing fluid to bypass (Chapter 7).
- [ ] Master cylinder bore scratched by foreign material or broken spring. Repair or renew master cylinder (Chapter 7).
- [ ] Disc warped. Renew disc (Chapter 7).

Brake lever or pedal pulsates
- [ ] Disc warped. Renew disc (Chapter 7).
- [ ] Axle bent. Renew axle (Chapter 7).
- [ ] Brake caliper bolts loose (Chapter 7).
- [ ] Brake caliper slider pins damaged or sticking, causing caliper to bind. Lubricate the sliders or renew them if they are corroded or bent (Chapter 7).
- [ ] Wheel warped or otherwise damaged (Chapter 7).
- [ ] Wheel bearings damaged or worn (Chapter 7).

Brakes drag
- [ ] Master cylinder piston seized. Caused by wear or damage to piston or cylinder bore (Chapter 7).
- [ ] Lever binding. Check pivot and lubricate (Chapter 7).
- [ ] Brake caliper binds. Caused by inadequate lubrication or damage to caliper slider pins (Chapter 7).
- [ ] Brake caliper piston seized in bore. Caused by wear or ingestion of dirt past deteriorated seal (Chapter 7).
- [ ] Brake pad damaged. Pad material separated from backing plate. Usually caused by faulty manufacturing process or from contact with chemicals. Renew pads (Chapter 7).
- [ ] Pads improperly installed (Chapter 7).
- [ ] Fault in dual combined brake system (DCBS) – could be the delay valve or the proportional control valve (Chapter 7).

14 Electrical problems

Battery dead or weak
- [ ] Battery faulty. Caused by sulphated plates which are shorted through sedimentation. Also, broken battery terminal making only occasional contact (Chapter 9).
- [ ] Battery cables making poor contact (Chapter 9).
- [ ] Load excessive. Caused by addition of high wattage lights or other electrical accessories.
- [ ] Ignition switch defective. Switch either grounds (earths) internally or fails to shut off system. Renew the switch (Chapter 9).
- [ ] Regulator/rectifier defective (Chapter 9).
- [ ] Alternator stator coil open or shorted (Chapter 9).
- [ ] Wiring faulty. Wiring grounded (earthed) or connections loose in ignition, charging or lighting circuits (Chapter 9).

Battery overcharged
- [ ] Regulator/rectifier defective. Overcharging is noticed when battery gets excessively warm (Chapter 9).
- [ ] Battery defective. Renew battery (Chapter 9).
- [ ] Battery amperage too low, wrong type or size. Install manufacturer's specified amp-hour battery to handle charging load (Chapter 9).

REF•44 Fault Finding Equipment

Checking engine compression

● Low compression will result in exhaust smoke, heavy oil consumption, poor starting and poor performance. A compression test will provide useful information about an engine's condition and if performed regularly, can give warning of trouble before any other symptoms become apparent.

● A compression gauge will be required, along with an adapter to suit the spark plug hole thread size. Note that the screw-in type gauge/adapter set up is preferable to the rubber cone type.

● Before carrying out the test, first check the valve clearances as described in Chapter 1.

1 Run the engine until it reaches normal operating temperature, then stop it and remove the spark plug(s), taking care not to scald your hands on the hot components.

2 Install the gauge adapter and compression gauge in No. 1 cylinder spark plug hole (see illustration 1).

Screw the compression gauge adapter into the spark plug hole, then screw the gauge into the adapter

3 On kickstart-equipped motorcycles, make sure the ignition switch is OFF, then open the throttle fully and kick the engine over a couple of times until the gauge reading stabilises.

4 On motorcycles with electric start only, the procedure will differ depending on the nature of the ignition system. Flick the engine kill switch (engine stop switch) to OFF and turn the ignition switch ON; open the throttle fully and crank the engine over on the starter motor for a couple of revolutions until the gauge reading stabilises. If the starter will not operate with the kill switch OFF, turn the ignition switch OFF and refer to the next paragraph.

5 Install the spark plugs back into their suppressor caps and arrange the plug electrodes so that their metal bodies are earthed (grounded) against the cylinder head; this is essential to prevent damage to the ignition system as the engine is spun over (see illustration 2). Position the plugs well

All spark plugs must be earthed (grounded) against the cylinder head

away from the plug holes otherwise there is a risk of atomised fuel escaping from the combustion chambers and igniting. As a safety precaution, cover the cylinder head covers with rags and disconnect the fuel pump wiring connector (see Chapter 4). Now turn the ignition switch ON and kill switch ON, open the throttle fully and crank the engine over on the starter motor for a couple of revolutions until the gauge reading stabilises.

6 After one or two revolutions the pressure should build up to a maximum figure and then stabilise. Take a note of this reading and on multi-cylinder engines repeat the test on the remaining cylinders.

7 The correct pressures are given in Chapter 1 Specifications. If the results fall within the specified range and on multi-cylinder engines all are relatively equal, the engine is in good condition. If there is a marked difference between the readings, or if the readings are lower than specified, inspection of the top-end components will be required.

8 Low compression pressure may be due to worn cylinder bores, pistons or rings, failure of the cylinder head gasket, worn valve seals, or poor valve seating.

9 To distinguish between cylinder/piston wear and valve leakage, pour a small quantity of oil into the bore to temporarily seal the piston rings, then repeat the compression tests (see illustration 3). If the readings show

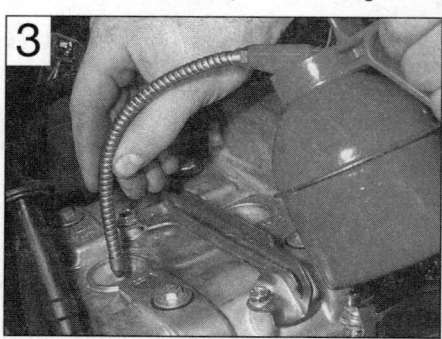

Bores can be temporarily sealed with a squirt of motor oil

a noticeable increase in pressure this confirms that the cylinder bore, piston, or rings are worn. If, however, no change is indicated, the cylinder head gasket or valves should be examined.

10 High compression pressure indicates excessive carbon build-up in the combustion chamber and on the piston crown. If this is the case the cylinder head should be removed and the deposits removed. Note that excessive carbon build-up is less likely with the used on modern fuels.

Checking battery open-circuit voltage

 Warning: The gases produced by the battery are explosive - never smoke or create any sparks in the vicinity of the battery. Never allow the electrolyte to contact your skin or clothing - if it does, wash it off and seek immediate medical attention.

Fault Finding Equipment REF•45

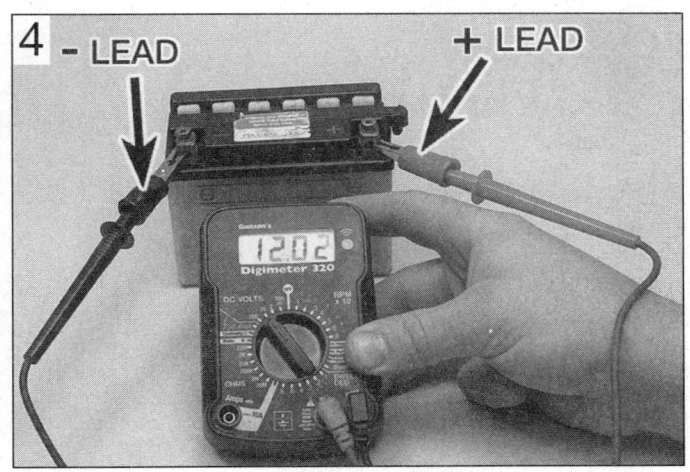

Measuring open-circuit battery voltage

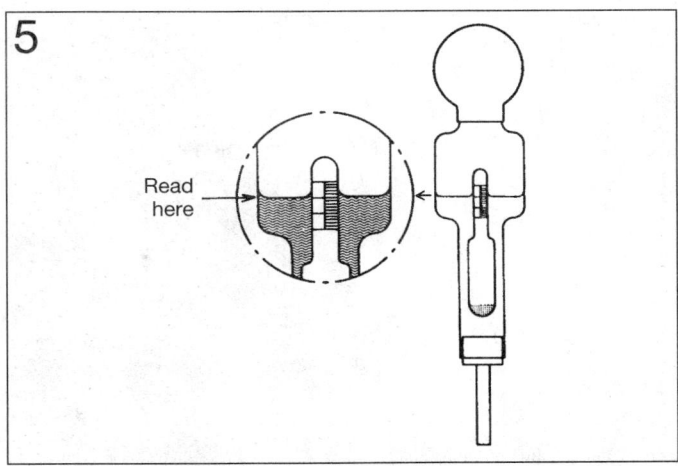

Float-type hydrometer for measuring battery specific gravity

● Before any electrical fault is investigated the battery should be checked.

● You'll need a dc voltmeter or multimeter to check battery voltage. Check that the leads are inserted in the correct terminals on the meter, red lead to positive (+ve), black lead to negative (-ve). Incorrect connections can damage the meter.

● A sound fully-charged 12 volt battery should produce between 12.3 and 12.6 volts across its terminals (12.8 volts for a maintenance-free battery). On machines with a 6 volt battery, voltage should be between 6.1 and 6.3 volts.

1 Set a multimeter to the 0 to 20 volts dc range and connect its probes across the battery terminals. Connect the meter's positive (+ve) probe, usually red, to the battery positive (+ve) terminal, followed by the meter's negative (-ve) probe, usually black, to the battery negative terminal (-ve) **(see illustration 4)**.

2 If battery voltage is low (below 10 volts on a 12 volt battery or below 4 volts on a six volt battery), charge the battery and test the voltage again. If the battery repeatedly goes flat, investigate the motorcycle's charging system.

Checking battery specific gravity (SG)

⚠ **Warning: The gases produced by the battery are explosive - never smoke or create any sparks in the vicinity of the battery. Never allow the electrolyte to contact your skin or clothing - if it does, wash it off and seek immediate medical attention.**

● The specific gravity check gives an indication of a battery's state of charge.

● A hydrometer is used for measuring specific gravity. Make sure you purchase one which has a small enough hose to insert in the aperture of a motorcycle battery.

● Specific gravity is simply a measure of the electrolyte's density compared with that of water. Water has an SG of 1.000 and fully-charged battery electrolyte is about 26% heavier, at 1.260.

● Specific gravity checks are not possible on maintenance-free batteries. Testing the open-circuit voltage is the only means of determining their state of charge.

1 To measure SG, remove the battery from the motorcycle and remove the first cell cap. Draw

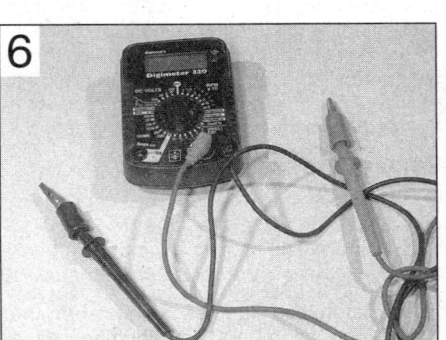

Digital multimeter can be used for all electrical tests

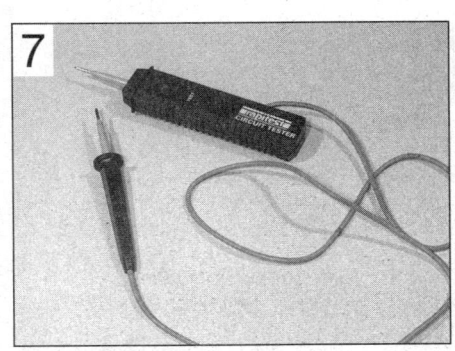

Battery-powered continuity tester

some electrolyte into the hydrometer and note the reading **(see illustration 5)**. Return the electrolyte to the cell and install the cap.

2 The reading should be in the region of 1.260 to 1.280. If SG is below 1.200 the battery needs charging. Note that SG will vary with temperature; it should be measured at 20°C (68°F). Add 0.007 to the reading for every 10°C above 20°C, and subtract 0.007 from the reading for every 10°C below 20°C. Add 0.004 to the reading for every 10°F above 68°F, and subtract 0.004 from the reading for every 10°F below 68°F.

3 When the check is complete, rinse the hydrometer thoroughly with clean water.

Checking for continuity

● The term continuity describes the uninterrupted flow of electricity through an electrical circuit. A continuity check will determine whether an **open-circuit** situation exists.

● Continuity can be checked with an ohmmeter, multimeter, continuity tester or battery and bulb test circuit **(see illustrations 6, 7 and 8)**.

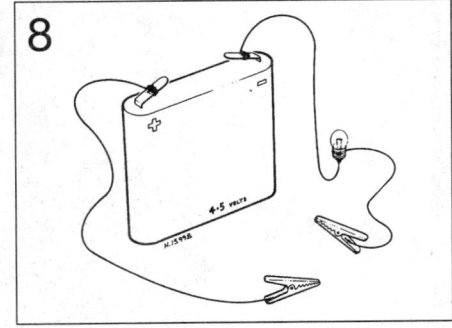

Battery and bulb test circuit

REF•46 Fault Finding Equipment

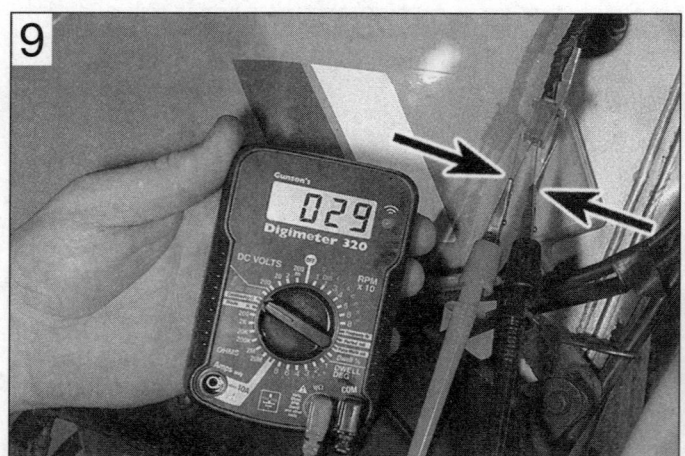

Continuity check of front brake light switch using a meter - note split pins used to access connector terminals

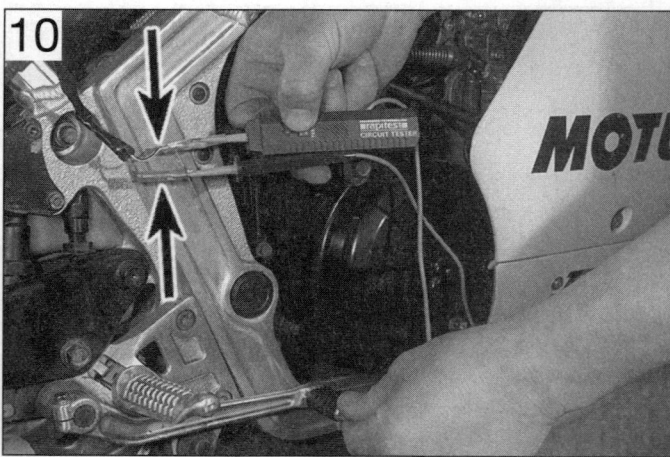

Continuity check of rear brake light switch using a continuity tester

- All of these instruments are self-powered by a battery, therefore the checks are made with the ignition OFF.
- As a safety precaution, always disconnect the battery negative (-ve) lead before making checks, particularly if ignition switch checks are being made.
- If using a meter, select the appropriate ohms scale and check that the meter reads infinity (∞). Touch the meter probes together and check that meter reads zero; where necessary adjust the meter so that it reads zero.
- After using a meter, always switch it OFF to conserve its battery.

Switch checks

1 If a switch is at fault, trace its wiring up to the wiring connectors. Separate the wire connectors and inspect them for security and condition. A build-up of dirt or corrosion here will most likely be the cause of the problem - clean up and apply a water dispersant such as WD40.

2 If using a test meter, set the meter to the ohms x 10 scale and connect its probes across the wires from the switch **(see illustration 9)**. Simple ON/OFF type switches, such as brake light switches, only have two wires whereas combination switches, like the ignition switch, have many internal links. Study the wiring diagram to ensure that you are connecting across the correct pair of wires. Continuity (low or no measurable resistance - 0 ohms) should be indicated with the switch ON and no continuity (high resistance) with it OFF.

3 Note that the polarity of the test probes doesn't matter for continuity checks, although care should be taken to follow specific test procedures if a diode or solid-state component is being checked.

4 A continuity tester or battery and bulb circuit can be used in the same way. Connect its probes as described above **(see illustration 10)**. The light should come on to indicate continuity in the ON switch position, but should extinguish in the OFF position.

Wiring checks

- Many electrical faults are caused by damaged wiring, often due to incorrect routing or chaffing on frame components.
- Loose, wet or corroded wire connectors can also be the cause of electrical problems, especially in exposed locations.

1 A continuity check can be made on a single length of wire by disconnecting it at each end and connecting a meter or continuity tester across both ends of the wire **(see illustration 11)**.

2 Continuity (low or no resistance - 0 ohms) should be indicated if the wire is good. If no continuity (high resistance) is shown, suspect a broken wire.

Checking for voltage

- A voltage check can determine whether current is reaching a component.
- Voltage can be checked with a dc voltmeter, multimeter set on the dc volts scale, test light or buzzer **(see illustrations 12 and 13)**. A meter has the advantage of being able to measure actual voltage.
- When using a meter, check that its leads are inserted in the correct terminals on the meter, red to positive (+ve), black to negative (-ve). Incorrect connections can damage the meter.
- A voltmeter (or multimeter set to the dc volts scale) should always be connected in parallel (across the load). Connecting it in series will destroy the meter.
- Voltage checks are made with the ignition ON.

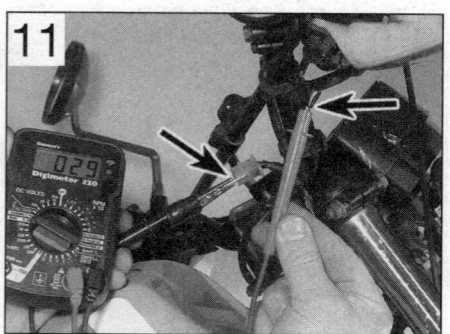

Continuity check of front brake light switch sub-harness

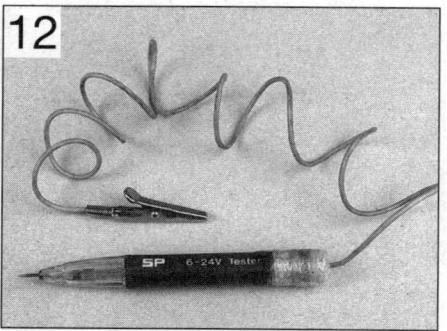

A simple test light can be used for voltage checks

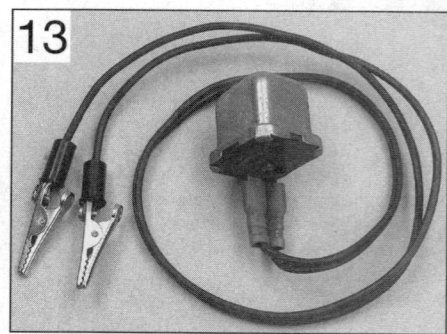

A buzzer is useful for voltage checks

Fault Finding Equipment   REF•47

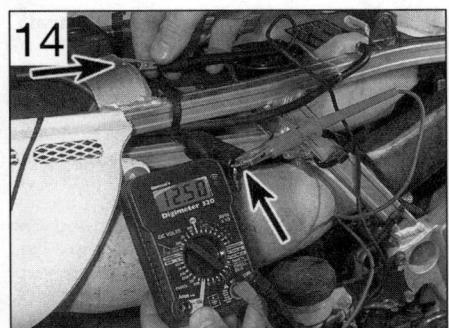

Checking for voltage at the rear brake light power supply wire using a meter . . .

1 First identify the relevant wiring circuit by referring to the wiring diagram at the end of this manual. If other electrical components share the same power supply (ie are fed from the same fuse), take note whether they are working correctly - this is useful information in deciding where to start checking the circuit.
2 If using a meter, check first that the meter leads are plugged into the correct terminals on the meter (see above). Set the meter to the dc volts function, at a range suitable for the battery voltage. Connect the meter red probe (+ve) to the power supply wire and the black probe to a good metal earth (ground) on the motorcycle's frame or directly to the battery negative (-ve) terminal **(see illustration 14)**. Battery voltage should be shown on the meter

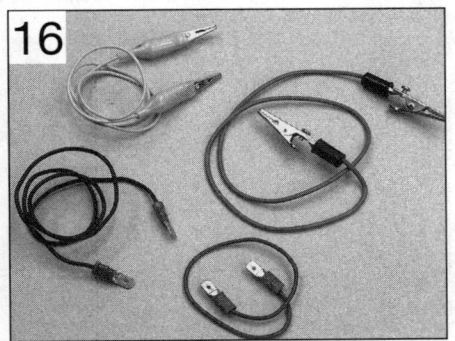

A selection of jumper wires for making earth (ground) checks

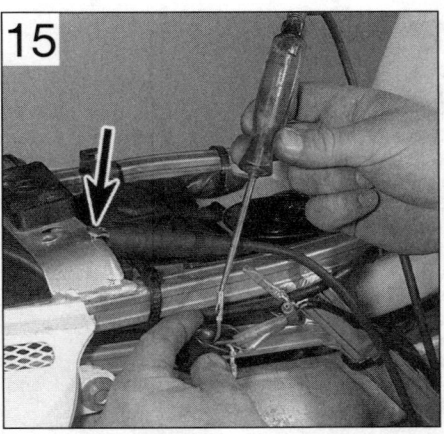

. . . or a test light - note the earth connection to the frame (arrow)

with the ignition switched ON.
3 If using a test light or buzzer, connect its positive (+ve) probe to the power supply terminal and its negative (-ve) probe to a good earth (ground) on the motorcycle's frame or directly to the battery negative (-ve) terminal **(see illustration 15)**. With the ignition ON, the test light should illuminate or the buzzer sound.
4 If no voltage is indicated, work back towards the fuse continuing to check for voltage. When you reach a point where there is voltage, you know the problem lies between that point and your last check point.

Checking the earth (ground)

● Earth connections are made either directly to the engine or frame (such as sensors, neutral switch etc. which only have a positive feed) or by a separate wire into the earth circuit of the wiring harness. Alternatively a short earth wire is sometimes run directly from the component to the motorcycle's frame.
● Corrosion is often the cause of a poor earth connection.
● If total failure is experienced, check the security of the main earth lead from the

negative (-ve) terminal of the battery and also the main earth (ground) point on the wiring harness. If corroded, dismantle the connection and clean all surfaces back to bare metal.
1 To check the earth on a component, use an insulated jumper wire to temporarily bypass its earth connection **(see illustration 16)**. Connect one end of the jumper wire between the earth terminal or metal body of the component and the other end to the motorcycle's frame.
2 If the circuit works with the jumper wire installed, the original earth circuit is faulty. Check the wiring for open-circuits or poor connections. Clean up direct earth connections, removing all traces of corrosion and remake the joint. Apply petroleum jelly to the joint to prevent future corrosion.

Tracing a short-circuit

● A short-circuit occurs where current shorts to earth (ground) bypassing the circuit components. This usually results in a blown fuse.

● A short-circuit is most likely to occur where the insulation has worn through due to wiring chafing on a component, allowing a direct path to earth (ground) on the frame.

1 Remove any bodypanels necessary to access the circuit wiring.
2 Check that all electrical switches in the circuit are OFF, then remove the circuit fuse and connect a test light, buzzer or voltmeter (set to the dc scale) across the fuse terminals. No voltage should be shown.
3 Move the wiring from side to side whilst observing the test light or meter. When the test light comes on, buzzer sounds or meter shows voltage, you have found the cause of the short. It will usually shown up as damaged or burned insulation.
4 Note that the same test can be performed on each component in the circuit, even the switch.

REF•48 Technical Terms Explained

A

ABS (Anti-lock braking system) A system, usually electronically controlled, that senses incipient wheel lockup during braking and relieves hydraulic pressure at wheel which is about to skid.
Aftermarket Components suitable for the motorcycle, but not produced by the motorcycle manufacturer.
Allen key A hexagonal wrench which fits into a recessed hexagonal hole.
Alternating current (ac) Current produced by an alternator. Requires converting to direct current by a rectifier for charging purposes.
Alternator Converts mechanical energy from the engine into electrical energy to charge the battery and power the electrical system.
Ampere (amp) A unit of measurement for the flow of electrical current. Current = Volts ÷ Ohms.
Ampere-hour (Ah) Measure of battery capacity.
Angle-tightening A torque expressed in degrees. Often follows a conventional tightening torque for cylinder head or main bearing fasteners **(see illustration)**.

Angle-tightening cylinder head bolts

Antifreeze A substance (usually ethylene glycol) mixed with water, and added to the cooling system, to prevent freezing of the coolant in winter. Antifreeze also contains chemicals to inhibit corrosion and the formation of rust and other deposits that would tend to clog the radiator and coolant passages and reduce cooling efficiency.
Anti-dive System attached to the fork lower leg (slider) to prevent fork dive when braking hard.
Anti-seize compound A coating that reduces the risk of seizing on fasteners that are subjected to high temperatures, such as exhaust clamp bolts and nuts.
API American Petroleum Institute. A quality standard for 4-stroke motor oils.
Asbestos A natural fibrous mineral with great heat resistance, commonly used in the composition of brake friction materials. Asbestos is a health hazard and the dust created by brake systems should never be inhaled or ingested.
ATF Automatic Transmission Fluid. Often used in front forks.
ATU Automatic Timing Unit. Mechanical device for advancing the ignition timing on early engines.
ATV All Terrain Vehicle. Often called a Quad.
Axial play Side-to-side movement.
Axle A shaft on which a wheel revolves. Also known as a spindle.

B

Backlash The amount of movement between meshed components when one component is held still. Usually applies to gear teeth.
Ball bearing A bearing consisting of a hardened inner and outer race with hardened steel balls between the two races.
Bearings Used between two working surfaces to prevent wear of the components and a build-up of heat. Four types of bearing are commonly used on motorcycles: plain shell bearings, ball bearings, tapered roller bearings and needle roller bearings.
Bevel gears Used to turn the drive through 90°. Typical applications are shaft final drive and camshaft drive **(see illustration)**.

Bevel gears are used to turn the drive through 90°

BHP Brake Horsepower. The British measurement for engine power output. Power output is now usually expressed in kilowatts (kW).
Bias-belted tyre Similar construction to radial tyre, but with outer belt running at an angle to the wheel rim.
Big-end bearing The bearing in the end of the connecting rod that's attached to the crankshaft.
Bleeding The process of removing air from an hydraulic system via a bleed nipple or bleed screw.
Bottom-end A description of an engine's crankcase components and all components contained there-in.
BTDC Before Top Dead Centre in terms of piston position. Ignition timing is often expressed in terms of degrees or millimetres BTDC.
Bush A cylindrical metal or rubber component used between two moving parts.
Burr Rough edge left on a component after machining or as a result of excessive wear.

C

Cam chain The chain which takes drive from the crankshaft to the camshaft(s).
Canister The main component in an evaporative emission control system (California market only); contains activated charcoal granules to trap vapours from the fuel system rather than allowing them to vent to the atmosphere.
Castellated Resembling the parapets along the top of a castle wall. For example, a castellated wheel axle or spindle nut.
Catalytic converter A device in the exhaust system of some machines which converts certain pollutants in the exhaust gases into less harmful substances.
Charging system Description of the components which charge the battery, ie the alternator, rectifier and regulator.
Circlip A ring-shaped clip used to prevent endwise movement of cylindrical parts and shafts. An internal circlip is installed in a groove in a housing; an external circlip fits into a groove on the outside of a cylindrical piece such as a shaft. Also known as a snap-ring.
Clearance The amount of space between two parts. For example, between a piston and a cylinder, between a bearing and a journal, etc.
Coil spring A spiral of elastic steel found in various sizes throughout a vehicle, for example as a springing medium in the suspension and in the valve train.
Compression Reduction in volume, and increase in pressure and temperature, of a gas, caused by squeezing it into a smaller space.
Compression damping Controls the speed the suspension compresses when hitting a bump.
Compression ratio The relationship between cylinder volume when the piston is at top dead centre and cylinder volume when the piston is at bottom dead centre.
Continuity The uninterrupted path in the flow of electricity. Little or no measurable resistance.
Continuity tester Self-powered bleeper or test light which indicates continuity.
Cp Candlepower. Bulb rating commonly found on US motorcycles.
Crossply tyre Tyre plies arranged in a criss-cross pattern. Usually four or six plies used, hence 4PR or 6PR in tyre size codes.
Cush drive Rubber damper segments fitted between the rear wheel and final drive sprocket to absorb transmission shocks **(see illustration)**.

Cush drive rubbers dampen out transmission shocks

D

Degree disc Calibrated disc for measuring piston position. Expressed in degrees.
Dial gauge Clock-type gauge with adapters for measuring runout and piston position. Expressed in mm or inches.
Diaphragm The rubber membrane in a master cylinder or carburettor which seals the upper chamber.
Diaphragm spring A single sprung plate often used in clutches.
Direct current (dc) Current produced by a dc generator.

Technical Terms Explained

Decarbonisation The process of removing carbon deposits - typically from the combustion chamber, valves and exhaust port/system.
Detonation Destructive and damaging explosion of fuel/air mixture in combustion chamber instead of controlled burning.
Diode An electrical valve which only allows current to flow in one direction. Commonly used in rectifiers and starter interlock systems.
Disc valve (or rotary valve) A induction system used on some two-stroke engines.
Double-overhead camshaft (DOHC) An engine that uses two overhead camshafts, one for the intake valves and one for the exhaust valves.
Drivebelt A toothed belt used to transmit drive to the rear wheel on some motorcycles. A drivebelt has also been used to drive the camshafts. Drivebelts are usually made of Kevlar.
Driveshaft Any shaft used to transmit motion. Commonly used when referring to the final driveshaft on shaft drive motorcycles.

E

Earth return The return path of an electrical circuit, utilising the motorcycle's frame.
ECU (Electronic Control Unit) A computer which controls (for instance) an ignition system, or an anti-lock braking system.
EGO Exhaust Gas Oxygen sensor. Sometimes called a Lambda sensor.
Electrolyte The fluid in a lead-acid battery.
EMS (Engine Management System) A computer controlled system which manages the fuel injection and the ignition systems in an integrated fashion.
Endfloat The amount of lengthways movement between two parts. As applied to a crankshaft, the distance that the crankshaft can move side-to-side in the crankcase.
Endless chain A chain having no joining link. Common use for cam chains and final drive chains.
EP (Extreme Pressure) Oil type used in locations where high loads are applied, such as between gear teeth.
Evaporative emission control system Describes a charcoal filled canister which stores fuel vapours from the tank rather than allowing them to vent to the atmosphere. Usually only fitted to California models and referred to as an EVAP system.
Expansion chamber Section of two-stroke engine exhaust system so designed to improve engine efficiency and boost power.

F

Feeler blade or gauge A thin strip or blade of hardened steel, ground to an exact thickness, used to check or measure clearances between parts.
Final drive Description of the drive from the transmission to the rear wheel. Usually by chain or shaft, but sometimes by belt.
Firing order The order in which the engine cylinders fire, or deliver their power strokes, beginning with the number one cylinder.
Flooding Term used to describe a high fuel level in the carburettor float chambers, leading to fuel overflow. Also refers to excess fuel in the combustion chamber due to incorrect starting technique.

Free length The no-load state of a component when measured. Clutch, valve and fork spring lengths are measured at rest, without any preload.
Freeplay The amount of travel before any action takes place. The looseness in a linkage, or an assembly of parts, between the initial application of force and actual movement. For example, the distance the rear brake pedal moves before the rear brake is actuated.
Fuel injection The fuel/air mixture is metered electronically and directed into the engine intake ports (indirect injection) or into the cylinders (direct injection). Sensors supply information on engine speed and conditions.
Fuel/air mixture The charge of fuel and air going into the engine. See **Stoichiometric ratio**.
Fuse An electrical device which protects a circuit against accidental overload. The typical fuse contains a soft piece of metal which is calibrated to melt at a predetermined current flow (expressed as amps) and break the circuit.

G

Gap The distance the spark must travel in jumping from the centre electrode to the side electrode in a spark plug. Also refers to the distance between the ignition rotor and the pickup coil in an electronic ignition system.
Gasket Any thin, soft material - usually cork, cardboard, asbestos or soft metal - installed between two metal surfaces to ensure a good seal. For instance, the cylinder head gasket seals the joint between the block and the cylinder head.
Gauge An instrument panel display used to monitor engine conditions. A gauge with a movable pointer on a dial or a fixed scale is an analogue gauge. A gauge with a numerical readout is called a digital gauge.
Gear ratios The drive ratio of a pair of gears in a gearbox, calculated on their number of teeth.
Glaze-busting see **Honing**
Grinding Process for renovating the valve face and valve seat contact area in the cylinder head.
Gudgeon pin The shaft which connects the connecting rod small-end with the piston. Often called a piston pin or wrist pin.

H

Helical gears Gear teeth are slightly curved and produce less gear noise that straight-cut gears. Often used for primary drives.

Installing a Helicoil thread insert in a cylinder head

Helicoil A thread insert repair system. Commonly used as a repair for stripped spark plug threads **(see illustration)**.
Honing A process used to break down the glaze on a cylinder bore (also called glaze-busting). Can also be carried out to roughen a rebored cylinder to aid ring bedding-in.
HT (High Tension) Description of the electrical circuit from the secondary winding of the ignition coil to the spark plug.
Hydraulic A liquid filled system used to transmit pressure from one component to another. Common uses on motorcycles are brakes and clutches.
Hydrometer An instrument for measuring the specific gravity of a lead-acid battery.
Hygroscopic Water absorbing. In motorcycle applications, braking efficiency will be reduced if DOT 3 or 4 hydraulic fluid absorbs water from the air - care must be taken to keep new brake fluid in tightly sealed containers.

I

lbf ft Pounds-force feet. An imperial unit of torque. Sometimes written as ft-lbs.
lbf in Pound-force inch. An imperial unit of torque, applied to components where a very low torque is required. Sometimes written as in-lbs.
IC Abbreviation for Integrated Circuit.
Ignition advance Means of increasing the timing of the spark at higher engine speeds. Done by mechanical means (ATU) on early engines or electronically by the ignition control unit on later engines.
Ignition timing The moment at which the spark plug fires, expressed in the number of crankshaft degrees before the piston reaches the top of its stroke, or in the number of millimetres before the piston reaches the top of its stroke.
Infinity (∞) Description of an open-circuit electrical state, where no continuity exists.
Inverted forks (upside down forks) The sliders or lower legs are held in the yokes and the fork tubes or stanchions are connected to the wheel axle (spindle). Less unsprung weight and stiffer construction than conventional forks.

J

JASO Quality standard for 2-stroke oils.
Joule The unit of electrical energy.
Journal The bearing surface of a shaft.

K

Kickstart Mechanical means of turning the engine over for starting purposes. Only usually fitted to mopeds, small capacity motorcycles and off-road motorcycles.
Kill switch Handebar-mounted switch for emergency ignition cut-out. Cuts the ignition circuit on all models, and additionally prevent starter motor operation on others.
km Symbol for kilometre.
kmh Abbreviation for kilometres per hour.

L

Lambda (λ) sensor A sensor fitted in the exhaust system to measure the exhaust gas oxygen content (excess air factor).

REF•50 Technical Terms Explained

Lapping see **Grinding**.
LCD Abbreviation for Liquid Crystal Display.
LED Abbreviation for Light Emitting Diode.
Liner A steel cylinder liner inserted in a aluminium alloy cylinder block.
Locknut A nut used to lock an adjustment nut, or other threaded component, in place.
Lockstops The lugs on the lower triple clamp (yoke) which abut those on the frame, preventing handlebar-to-fuel tank contact.
Lockwasher A form of washer designed to prevent an attaching nut from working loose.
LT Low Tension Description of the electrical circuit from the power supply to the primary winding of the ignition coil.

M

Main bearings The bearings between the crankshaft and crankcase.
Maintenance-free (MF) battery A sealed battery which cannot be topped up.
Manometer Mercury-filled calibrated tubes used to measure intake tract vacuum. Used to synchronise carburettors on multi-cylinder engines.
Micrometer A precision measuring instrument that measures component outside diameters **(see illustration)**.

Tappet shims are measured with a micrometer

MON (Motor Octane Number) A measure of a fuel's resistance to knock.
Monograde oil An oil with a single viscosity, eg SAE80W.
Monoshock A single suspension unit linking the swingarm or suspension linkage to the frame.
mph Abbreviation for miles per hour.
Multigrade oil Having a wide viscosity range (eg 10W40). The W stands for Winter, thus the viscosity ranges from SAE10 when cold to SAE40 when hot.
Multimeter An electrical test instrument with the capability to measure voltage, current and resistance. Some meters also incorporate a continuity tester and buzzer.

N

Needle roller bearing Inner race of caged needle rollers and hardened outer race. Examples of uncaged needle rollers can be found on some engines. Commonly used in rear suspension applications and in two-stroke engines.
Nm Newton metres.
NOx Oxides of Nitrogen. A common toxic pollutant emitted by petrol engines at higher temperatures.

O

Octane The measure of a fuel's resistance to knock.
OE (Original Equipment) Relates to components fitted to a motorcycle as standard or replacement parts supplied by the motorcycle manufacturer.
Ohm The unit of electrical resistance. Ohms = Volts ÷ Current.
Ohmmeter An instrument for measuring electrical resistance.
Oil cooler System for diverting engine oil outside of the engine to a radiator for cooling purposes.
Oil injection A system of two-stroke engine lubrication where oil is pump-fed to the engine in accordance with throttle position.
Open-circuit An electrical condition where there is a break in the flow of electricity - no continuity (high resistance).
O-ring A type of sealing ring made of a special rubber-like material; in use, the O-ring is compressed into a groove to provide the sealing action.
Oversize (OS) Term used for piston and ring size options fitted to a rebored cylinder.
Overhead cam (sohc) engine An engine with single camshaft located on top of the cylinder head.
Overhead valve (ohv) engine An engine with the valves located in the cylinder head, but with the camshaft located in the engine block or crankcase.
Oxygen sensor A device installed in the exhaust system which senses the oxygen content in the exhaust and converts this information into an electric current. Also called a Lambda sensor.

P

Plastigauge A thin strip of plastic thread, available in different sizes, used for measuring clearances. For example, a strip of Plastigauge is laid across a bearing journal. The parts are assembled and dismantled; the width of the crushed strip indicates the clearance between journal and bearing.
Polarity Either negative or positive earth (ground), determined by which battery lead is connected to the frame (earth return). Modern motorcycles are usually negative earth.
Pre-ignition A situation where the fuel/air mixture ignites before the spark plug fires. Often due to a hot spot in the combustion chamber caused by carbon build-up. Engine has a tendency to 'run-on'.
Pre-load (suspension) The amount a spring is compressed when in the unloaded state. Preload can be applied by gas, spacer or mechanical adjuster.
Premix The method of engine lubrication on older two-stroke engines. Engine oil is mixed with the petrol in the fuel tank in a specific ratio. The fuel/oil mix is sometimes referred to as "petroil".
Primary drive Description of the drive from the crankshaft to the clutch. Usually by gear or chain.
PS Pfedestärke - a German interpretation of BHP.
PSI Pounds-force per square inch. Imperial measurement of tyre pressure and cylinder pressure measurement.
PTFE Polytetrafluoroethylene. A low friction substance.

Pulse secondary air injection system A process of promoting the burning of excess fuel present in the exhaust gases by routing fresh air into the exhaust ports.

Q

Quartz halogen bulb Tungsten filament surrounded by a halogen gas. Typically used for the headlight **(see illustration)**.

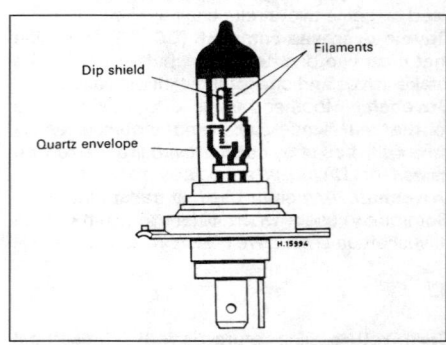

Quartz halogen headlight bulb construction

R

Rack-and-pinion A pinion gear on the end of a shaft that mates with a rack (think of a geared wheel opened up and laid flat). Sometimes used in clutch operating systems.
Radial play Up and down movement about a shaft.
Radial ply tyres Tyre plies run across the tyre (from bead to bead) and around the circumference of the tyre. Less resistant to tread distortion than other tyre types.
Radiator A liquid-to-air heat transfer device designed to reduce the temperature of the coolant in a liquid cooled engine.
Rake A feature of steering geometry - the angle of the steering head in relation to the vertical **(see illustration)**.

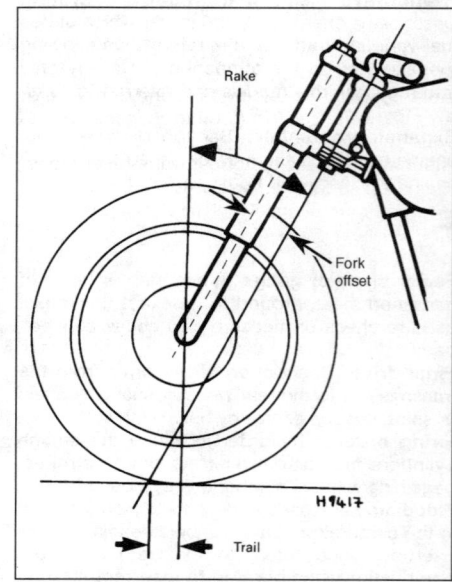

Steering geometry

Technical Terms Explained   REF•51

Rebore Providing a new working surface to the cylinder bore by boring out the old surface. Necessitates the use of oversize piston and rings.

Rebound damping A means of controlling the oscillation of a suspension unit spring after it has been compressed. Resists the spring's natural tendency to bounce back after being compressed.

Rectifier Device for converting the ac output of an alternator into dc for battery charging.

Reed valve An induction system commonly used on two-stroke engines.

Regulator Device for maintaining the charging voltage from the generator or alternator within a specified range.

Relay A electrical device used to switch heavy current on and off by using a low current auxiliary circuit.

Resistance Measured in ohms. An electrical component's ability to pass electrical current.

RON (Research Octane Number) A measure of a fuel's resistance to knock.

rpm revolutions per minute.

Runout The amount of wobble (in-and-out movement) of a wheel or shaft as it's rotated. The amount a shaft rotates 'out-of-true'. The out-of-round condition of a rotating part.

S

SAE (Society of Automotive Engineers) A standard for the viscosity of a fluid.

Sealant A liquid or paste used to prevent leakage at a joint. Sometimes used in conjunction with a gasket.

Service limit Term for the point where a component is no longer useable and must be renewed.

Shaft drive A method of transmitting drive from the transmission to the rear wheel.

Shell bearings Plain bearings consisting of two shell halves. Most often used as big-end and main bearings in a four-stroke engine. Often called bearing inserts.

Shim Thin spacer, commonly used to adjust the clearance or relative positions between two parts. For example, shims inserted into or under tappets or followers to control valve clearances. Clearance is adjusted by changing the thickness of the shim.

Short-circuit An electrical condition where current shorts to earth (ground) bypassing the circuit components.

Skimming Process to correct warpage or repair a damaged surface, eg on brake discs or drums.

Slide-hammer A special puller that screws into or hooks onto a component such as a shaft or bearing; a heavy sliding handle on the shaft bottoms against the end of the shaft to knock the component free.

Small-end bearing The bearing in the upper end of the connecting rod at its joint with the gudgeon pin.

Spalling Damage to camshaft lobes or bearing journals shown as pitting of the working surface.

Specific gravity (SG) The state of charge of the electrolyte in a lead-acid battery. A measure of the electrolyte's density compared with water.

Straight-cut gears Common type gear used on gearbox shafts and for oil pump and water pump drives.

Stanchion The inner sliding part of the front forks, held by the yokes. Often called a fork tube.

Stoichiometric ratio The optimum chemical air/fuel ratio for a petrol engine, said to be 14.7 parts of air to 1 part of fuel.

Sulphuric acid The liquid (electrolyte) used in a lead-acid battery. Poisonous and extremely corrosive.

Surface grinding (lapping) Process to correct a warped gasket face, commonly used on cylinder heads.

T

Tapered-roller bearing Tapered inner race of caged needle rollers and separate tapered outer race. Examples of taper roller bearings can be found on steering heads.

Tappet A cylindrical component which transmits motion from the cam to the valve stem, either directly or via a pushrod and rocker arm. Also called a cam follower.

TCS Traction Control System. An electronically-controlled system which senses wheel spin and reduces engine speed accordingly.

TDC Top Dead Centre denotes that the piston is at its highest point in the cylinder.

Thread-locking compound Solution applied to fastener threads to prevent slackening. Select type to suit application.

Thrust washer A washer positioned between two moving components on a shaft. For example, between gear pinions on gearshaft.

Timing chain See **Cam Chain**.

Timing light Stroboscopic lamp for carrying out ignition timing checks with the engine running.

Top-end A description of an engine's cylinder block, head and valve gear components.

Torque Turning or twisting force about a shaft.

Torque setting A prescribed tightness specified by the motorcycle manufacturer to ensure that the bolt or nut is secured correctly. Undertightening can result in the bolt or nut coming loose or a surface not being sealed. Overtightening can result in stripped threads, distortion or damage to the component being retained.

Torx key A six-point wrench.

Tracer A stripe of a second colour applied to a wire insulator to distinguish that wire from another one with the same colour insulator. For example, Br/W is often used to denote a brown insulator with a white tracer.

Trail A feature of steering geometry. Distance from the steering head axis to the tyre's central contact point.

Triple clamps The cast components which extend from the steering head and support the fork stanchions or tubes. Often called fork yokes.

Turbocharger A centrifugal device, driven by exhaust gases, that pressurises the intake air. Normally used to increase the power output from a given engine displacement.

TWI Abbreviation for Tyre Wear Indicator. Indicates the location of the tread depth indicator bars on tyres.

U

Universal joint or U-joint (UJ) A double-pivoted connection for transmitting power from a driving to a driven shaft through an angle. Typically found in shaft drive assemblies.

Unsprung weight Anything not supported by the bike's suspension (ie the wheel, tyres, brakes, final drive and bottom (moving) part of the suspension).

V

Vacuum gauges Clock-type gauges for measuring intake tract vacuum. Used for carburettor synchronisation on multi-cylinder engines.

Valve A device through which the flow of liquid, gas or vacuum may be stopped, started or regulated by a moveable part that opens, shuts or partially obstructs one or more ports or passageways. The intake and exhaust valves in the cylinder head are of the poppet type.

Valve clearance The clearance between the valve tip (the end of the valve stem) and the rocker arm or tappet/follower. The valve clearance is measured when the valve is closed. The correct clearance is important - if too small the valve won't close fully and will burn out, whereas if too large noisy operation will result.

Valve lift The amount a valve is lifted off its seat by the camshaft lobe.

Valve timing The exact setting for the opening and closing of the valves in relation to piston position.

Vernier caliper A precision measuring instrument that measures inside and outside dimensions. Not quite as accurate as a micrometer, but more convenient.

VIN Vehicle Identification Number. Term for the bike's engine and frame numbers.

Viscosity The thickness of a liquid or its resistance to flow.

Volt A unit for expressing electrical "pressure" in a circuit. Volts = current x ohms.

W

Water pump A mechanically-driven device for moving coolant around the engine.

Watt A unit for expressing electrical power. Watts = volts x current.

Wear limit see **Service limit**

Wet liner A liquid-cooled engine design where the pistons run in liners which are directly surrounded by coolant **(see illustration)**.

Wet liner arrangement

Wheelbase Distance from the centre of the front wheel to the centre of the rear wheel.

Wiring harness or loom Describes the electrical wires running the length of the motorcycle and enclosed in tape or plastic sheathing. Wiring coming off the main harness is usually referred to as a sub harness.

Woodruff key A key of semi-circular or square section used to locate a gear to a shaft. Often used to locate the alternator rotor on the crankshaft.

Wrist pin Another name for gudgeon or piston pin.

REF•52 Index

Note: *References throughout this index are in the form - "Chapter number" • "Page number"*

A

Abnormal noises – REF•40
About this manual – 0•8
Acknowledgements – 0•8
Air filter
    housing removal and installation – 4•6
    housing variable intake system information, testing, removal and installation – 4•7
    renewal – 1•14
Alternator check, removal and installation – 9•16
Asbestos – 0•9

B

Battery – 0•9
    charging – 9•3
    checks – REF•44, REF•45
    removal, installation, inspection and maintenance – 9•3
Bearings – REF•14
    fault finding – REF•16
    types – REF•16
Bleeding brake system – 7•16
Bodywork – 8•1 *et seq*
    fairing panels – 8•3
    mirrors – 8•2
    mudguard – 8•6
    seat – 8•1
    seat cowling – 8•2
Brake fluid
    change – 1•15
    level checks – 0•15
Brake light
    bulb renewal – 9•6
    switches – 9•7
Brake system check – 1•12
Brakes MOT test checks – REF•30
Brakes, wheels and tyres – 7•1 *et seq*
    calipers – 7•4, 7•10
    discs – 7•7, 7•11
    dual combined brake system (DCBS) components – 7•13
    hoses, pipes and unions – 1•19, 7•15
    master cylinder – 1•19, 7•8, 7•12
    pads – 1•7, 7•4, 7•9
    system bleeding – 7•16
    tyres – 7•24
    wheel bearings (and holder) – 7•20, (7•21)
    wheels – 7•18, 7•19, 7•21
Braking problems – REF•43
Buying spare parts – 0•12

C

Cables
    choke – 1•10, 4•19
    throttle – 1•10, 4•19
Calipers
    removal, overhaul and installation
        front – 7•4
        rear – 7•10
    seals renewal – 1•19
Camshaft drive gears removal, inspection and installation – 2•21
Camshafts and followers removal, inspection and installation 2•11
Catalytic converter – 4•22
Chains – REF•18
Charging system
    leakage and output test – 9•16
    testing general information and precautions – 9•16
Choke cable
    check – 1•10
    removal and installation – 4•19
Clutch
    fluid – 0•15, 1•15
    master cylinder removal, overhaul and installation – 2•29
    problems fault finding – REF•39
    removal, inspection and installation – 2•25
    slave cylinder removal, overhaul and installation – 2•30
    switch check and replacement – 9•13
Connecting rods bearings – 2•43
Contents – 0•2
Conversion factors – REF•26

Index REF•53

Coolant
    flushing and refilling – 1•18
    level check – 0•14
    temperature gauge check and replacement – 9•9
Cooling system – 3•1 *et seq*
    check – 1•11
    draining – 1•17
    fan – 3•2
    fan switch – 3•2
    hoses – 3•8
    pressure cap – 3•2
    radiators – 3•6
    reservoir – 3•2
    temperature gauge and sender – 3•4
    thermostat and housing – 3•5
    water pump – 3•7
Crankcase
    inspection and servicing – 2•41
    separation and reassembly – 2•38
Crankshaft and main bearings removal, inspection and installation – 2•48
Cylinder compression check – 1•18, REF•44
Cylinder head
    disassembly, inspection and reassembly – 2•17
    removal and installation – 2•15
Cylinder head covers removal and installation – 2•9

D

Daily (pre-ride) checks – 0•14 *et seq*
    brake fluid level – 0•15
    clutch fluid level – 0•15
    coolant level – 0•14
    engine/transmission oil level – 0•14
    final drive – 0•16
    legal – 0•16
    safety – 0•16
    steering – 0•16
    suspension – 0•16
    tyres – 0•16
Dimensions – 0•11
Diode check and replacement - 9•13
Discs
    inspection, removal and installation
        front – 7•7
        rear – 7•11
Drive chain
    removal, cleaning and installation – 6•18
    checks, adjustment and lubrication – 1•6, 1•11
Dual combined brake system (DCBS) components removal, overhaul and installation – 7•13
Dust protection – REF•4

E

Earth connections check – REF•47
Electrical problems fault finding – REF•38
Electrical system – 9•1 *et seq*
    alternator – 9•16
    battery – 9•3
    brake/tail light – 9•6, 9•7
    charging system – 9•16
    clutch switch – 9•13
    diode – 9•13
    fault finding – 9•2
    fuses – 9•4
    handlebar switches – 9•11
    headlight assembly – 9•6
    headlight bulbs – 9•5
    horn – 9•13
    ignition switch – 9•10
    instrument lights – 9•10
    instruments – 9•7, 9•8
    lighting system – 9•4
    MOT test checks – REF•27
    neutral switch – 9•12
    oil pressure switch – 9•10
    regulator/rectifier – 9•18
    sensors – 9•8
    sidelight bulbs – 9•5
    sidestand switch – 9•12
    starter motor – 9•14
    starter relay – 9•13
    tail light assembly – 9•6
    turn signal – 9•6, 9•7
    wiring diagrams – 9•20
Electricity – 0•9
Engine compression checking – REF•44
Engine fault finding – REF•36
Engine, clutch and transmission – 2•1 *et seq*
    camshafts and followers – 2•11
    camshaft drive gears – 2•21
    clutch – 2•25
    clutch master cylinder – 2•29
    clutch slave cylinder – 2•30
    clutch system bleeding – 2•31
    connecting rods – 2•44
    connecting rod bearings – 2•43
    crankcase – 2•38, 2•41
    crankshaft – 2•48
    cylinder heads – 2•15, 2•17
    cylinder head covers – 2•9
    disassembly and reassembly – 2•9
    gearchange mechanism – 2•36
    main bearings – 2•43, 2•48
    major engine repair – 2•6
    oil cooler and pipes – 2•35
    oil pan – 2•31
    oil pressure relief valve – 2•34
    oil pump – 2•32
    operations possible with the engine in the frame – 2•6
    operations requiring engine removal – 2•6
    pistons – 2•44
    piston rings – 2•47
    primary drive gear – 2•24
    removal and installation – 2•6
    running-in procedure – 2•57
    selector drum and forks – 2•42
    starter motor clutch – 2•22
    start-up after overhaul – 2•57
    transmission shafts – 2•50, 2•51
    valves/valve seats/valve guides – 2•17
Engine/transmission oil
    change – 1•9
    level check – 0•14
    pressure check – 1•19
Evaporative emission control (EVAP) system – 1•15, 4•23
Excessive exhaust smoke – REF•42
Exhaust system
    check – 1•11
    excessive smoke – REF•42
    MOT test checks – REF•28
    removal and installation – 4•20
Eye protection – REF•4

Index

F

Fairing panels removal and installation – 8•3
Fast idle wax unit – 4•19
Fasteners – REF•5
Fault finding – REF•35 *et seq*
    bearings – REF•14
    electrical system – 9•2
    fuel injection system – 4•9
Fault finding equipment – REF•44
Final drive
    check – 0•16
    MOT test checks – REF•28
Fire – 0•9, REF•4
Fluids recommended – 1•2, REF•23
Footrests removal and installation – 6•2
Forks
    disassembly, inspection and reassembly – 6•7
    oil change – 1•18
    removal and installation – 6•5
Frame, suspension and final drive – 6•1 *et seq*
    drive chain – 6•18
    footrests and brackets – 6•2
    forks – 6•5, 6•7
    frame – 6•2
    handlebars – 6•4
    rear shock absorber – 6•13
    rear sprocket coupling – 6•19
    rear suspension linkage – 6•12
    sprockets – 6•18
    stands – 6•4
    steering head bearings – 6•12
    steering stem – 6•11
    suspension – 6•14
    swingarm – 6•15, 6•16
Fuel and exhaust systems – 4•1 *et seq*
    air filter housing – 4•6, 4•7
    catalytic converter 4•22
    choke cable – 4•19
    evaporative emission control (EVAP) system
        components – 4•23
    exhaust system 4•20
    fuel filter – 4•6
    fuel gauge check – 9•9
    fuel injection system – 4•8, 4•9, 4•10
    fuel pump – 4•4
    fuel supply system – 4•4
    fuel tank – 4•2, 4•3
    pulse secondary air (PAIR) system – 4•21
    starter valves – 4•17
    throttle body assembly – 4•16
    throttle cables – 4•19
Fuel checks – 0•16
Fuel filter removal and installation – 4•6
Fuel injection system
    components check, removal and installation – 4•10
    fault diagnosis and checking – 4•9
    general information – 4•8
Fuel pump check, removal and installation – 4•4
Fuel supply system flow rate and pressure check – 4•4
Fuel system check - 1•11
Fuel tank
    cleaning and repair – 4•3
    raising, removal and installation – 4•2
Fumes – 0•9
Fuses check and removal – 9•4

G

Gaskets – REF•17
Gauges – REF•9
**Gearchange mechanism removal, inspection and
    installation** – 2•36
Gearchanging problems fault finding – REF•40

H

Hand protection – REF•4
Handlebar
    removal and installation – 6•4
    switches removal and installation – 9•11
Headlight
    aim check and adjustment – 1•12
    assembly removal and installation – 9•6
    bulb renewal – 9•5
    system check – 9•4
Horn check and replacement – 9•13
Hoses – 3•8, REF•19
HT coils check, removal and installation – 5•2

I

Identification numbers – 0•12
Idle speed check and adjustment – 1•7
Idle/reduction gears removal, inspection and installation – 2•22
Ignition switch check, removal and installation – 9•10
Ignition system – 5•1 *et seq*
    HT coils – 5•2
    timing – 5•4
Instruments
    bulbs renewal – 9•10
    check and renewal – 9•8
    cluster removal and installation – 9•7
Introduction – 0•4

L

**LCD (Liquid Crystal Display) Unit and sensors check and
    replacement** – 9•9
Legal and safety checks – 0•16
Lighting checks – 0•16, 9•4
Lubricants recommended – 1•2, REF•23

M

Main bearings general note – 2•43
Master cylinder
    removal, overhaul and installation
        front – 7•8
        rear – 7•12
Measuring tools and gauges – REF•9
Mirrors removal and installation – 8•2
MOT Test checks – REF•27
Mudguard removal and installation – 8•6

N

Neutral switch check, removal and installation – 9•12
Nuts and bolts tightness check – 1•13

Index

O

Oil
　　care – 1•10, REF•4
　　engine – 0•14, 1•2, 1•9, 1•19
　　forks   1•18, 6•1
Oil cooler removal and installation – 2•35
Oil pan removal and installation – 2•31
Oil pressure
　　relief valve – 2•34
　　switch check, removal and installation – 9•10
　　warning light comes on – REF•41
Oil pump removal, inspection and installation – 2•32
Oil seals – REF•16
Overheating fault finding – REF•39

P

Pads
　　renewal
　　　　front – 7•4
　　　　rear – 7•9
Performance data – 0•10
Piston rings installation – 2•47
Piston/connecting rod assemblies removal, inspection and installation – 2•44
Poor handling/stability – REF•42
Pressure cap check – 3•2
Primary drive gear removal, inspection and installation – 2•24
Pulse secondary air (PAIR) system components – 4•21

R

Radiator removal and installation – 3•6
Rear sprocket coupling inspection and overhaul – 6•19
Reference – REF•1 *et seq*
　　conversion factors – REF•26
　　fault finding – REF•35
　　fault finding equipment – REF•44
　　lubricants and fluids – REF•23
　　MOT test checks – REF•27
　　security – REF•20
　　storage – REF•32
　　technical terms explained – REF•48
　　tools and workshop tips – REF•2
Regulator/rectifier check and replacement – 9•18
Routine maintenance and servicing – 1•1 *et seq*
　　air filter – 1•14
　　brake fluid – 1•15
　　brake hoses – 1•19
　　brake master cylinder – 1•19
　　brake pads – 1•7
　　brake system – 1•12
　　caliper seals – 1•19
　　choke cable – 1•10
　　clutch fluid – 1•15
　　component location – 1•4
　　cooling system – 1•11, 1•17
　　cylinder compression – 1•18
　　drive chain – 1•6, 1•11
　　engine/transmission – 1•9
　　evaporative emission control (EVAP) system – 1•15
　　exhaust system – 1•11
　　front forks – 1•18
　　fuel system – 1•11
　　headlight aim – 1•12
　　idle speed – 1•7
　　non-scheduled maintenance – 1•18
　　nuts and bolts – 1•13
　　oil pressure – 1•19
　　schedule – 1•3
　　sidestand – 1•13
　　spark plugs – 1•8, 1•9, 1•17
　　stands, pivots and cables lubrication – 1•8
　　steering head bearings – 1•14, 1•18
　　swingarm and suspension linkage bearings – 1•18
　　throttle cables – 1•10
　　valve clearances – 1•15
　　wheel bearings – 1•18
　　wheels and tyres – 1•14
Running-in procedure – 2•57

S

Safety checks – 0•16
Safety first! – 0•9
Sealants – REF•17
Seat cowling removal and installation – 8•2
Seat removal and installation – 8•1
Security – REF•20
Selector drum and forks removal, inspection and installation – 2•42
Sensors check and replacement – 9•8
Shock absorber removal, inspection and installation – 6•13
Short-circuit tracing – REF•47
Sidecars MOT test checks – REF•31
Sidelight bulb renewal – 9•5
Sidestand
　　check – 1•13
　　switch check and replacement – 9•12
Signalling checks – 0•16
Spark plugs
　　gaps check and adjustment – 1•8
　　renewal – 1•9, 1•17
Speedometer and speed sensor check and replacement – 9•8
Sprockets check and renewal – 6•18
Stand
　　lubrication – 1•8
　　removal and installation – 6•4
Starter motor
　　clutch removal, inspection and installation – 2•22
　　disassembly, inspection and reassembly – 9•14
Starter relay check and replacement – 9•13
Starter valves – 4•17
Start-up after overhaul – 2•57
Steering
　　checks – 0•16
　　MOT test checks – REF•28
Steering head bearings
　　check and adjustment – 1•14
　　inspection and renewal – 6•12
　　re-greasing – 1•18
Steering stem removal and installation – 6•11
Storage – REF•32
Suspension
　　adjustments – 6•14
　　check – 0•16, 1•13
　　MOT test checks – REF•28
Swingarm
　　and suspension linkage bearings re-greasing – 1•18
　　inspection and bearing renewal – 6•16
　　removal and installation – 6•15

T

Tachometer check and replacement – 9•9
Tail light assembly removal and installation – 9•6
Technical terms explained – REF•48
Temperature gauge and sender check and replacement – 3•4
Thermostat and housing removal, check and installation – 3•5
Throttle bodies removal and installation – 4•16
Throttle cables
    check – 1•10
    removal and installation – 4•19
Tools – REF•2
Torque and leverage – REF•13
Transmission shafts
    disassembly, inspection and reassembly – 2•51
    removal and installation – 2•50
Turn signal
    bulbs renewal – 9•6
    circuit check – 9•6
Tyres
    general check – 0•16, 1•14
    general information and fitting – 7•24
    MOT test checks – REF•30
    pressures – 0•16
    tread depth – 0•16

V

Valve clearances check and adjustment – 1•15
Valves/valve seats/valves guides servicing – 2•17
Voltage checks – REF•46

W

Water pump check, removal and installation – 3•7
Weights – 0•11
Wheel bearings
    check – 1•18
    renewal – 7•20
Wheels
    alignment check – 7•18
    general check – 1•14
    inspection and repair – 7•18
    MOT test checks – REF•30
    removal and installation – 7•19, 7•21
Wiring diagrams – 9•20
Workshop equipment – REF•4

Notes

Notes